Birkhäuser

Pseudo-Differential Operators

Theory and Applications
Vol. 7

Managing Editor

M.W. Wong (York University, Canada)

Editorial Board

Luigi Rodino (Università di Torino, Italy)
Bert-Wolfgang Schulze (Universität Potsdam, Germany)
Johannes Sjöstrand (Université de Bourgogne, Dijon, France)
Sundaram Thangavelu (Indian Institute of Science at Bangalore, India)
Maciej Zworski (University of California at Berkeley, USA)

Pseudo-Differential Operators: Theory and Applications is a series of moderately priced graduate-level textbooks and monographs appealing to students and experts alike. Pseudo-differential operators are understood in a very broad sense and include such topics as harmonic analysis, PDE, geometry, mathematical physics, microlocal analysis, time-frequency analysis, imaging and computations. Modern trends and novel applications in mathematics, natural sciences, medicine, scientific computing, and engineering are highlighted.

Maurice A. de Gosson

Symplectic Methods
in Harmonic Analysis
and in Mathematical Physics

Birkhäuser

Maurice A. de Gosson
Faculty of Mathematics, NuHAG
University of Vienna
Nordbergstraße 15
1090 Vienna
Austria
maurice.degosson@gmail.com, maurice.de.gosson@univie.ac.at

2010 Mathematics Subject Classification: 35S, 81Q, 81S, 53D, 46B, 37J05, 70H05, 70H15

ISBN 978-3-7643-9991-7   e-ISBN 978-3-7643-9992-4
DOI 10.1007/978-3-7643-9992-4

Library of Congress Control Number: 2011934363

Cover design: deblik, Berlin

Springer Basel AG is part of Springer Science+Business Media

www.birkhauser-science.com

Contents

Part I   Symplectic Mechanics

1   Hamiltonian Mechanics in a Nutshell

2   The Symplectic Group

Part II Harmonic Analysis in Symplectic Spaces

Part III Pseudo-differential Operators and Function Spaces

Part IV Applications

Foreword

This book is partially based on a series of lectures I gave during the fall term 2009 at the NuHAG Institute at the University of Vienna. It therefore has a structure reminiscent of a "Lecture Notes" monograph. This means, among other things, that I have felt free to often repeat or recall some material, or definition, that has been given in a previous chapter – exactly as when one is confronted by a class, where it is appropriate to give a brief review of what has been said during last week's lecture. I hope that this will make the reading of this book easier and more enjoyable to a majority of students and colleagues.

To those of the readers, certainly numerous, who will find that this book is too long for what it is, and have already closed it, I say "good bye!". For the few who would have liked to read more about all or some topics I develop, and who deplore that I haven't expanded them more, I can only quote the great Austrian physicist Walter Thirring (I owe this citation to Karlheinz Gröchenig):

> ... There are three things that are easy to start but very difficult to finish. The first is a war. The second is a love affair. The third is a trill. To this may be added a fourth: a book.

Maurice A. de Gosson

*This work is dedicated with all my love to Charlyne,
for her invaluable support and help.*

Preface

Harmonic analysis is one of the most active and fastest growing parts of both pure and applied mathematics. It has gone far beyond its primary goal, which was to study the representation of functions or signals as superpositions of trigonometric functions (Fourier series). The interest in harmonic analysis has always been great because of the wealth of its applications, and it plays nowadays a central role in the study of signal theory and time-frequency analysis. Its interest in pure mathematics (especially in functional analysis) has been revived by the introduction of new functional spaces which are tools of choice for studying regularity properties of pseudo-differential operators and their applications to mathematical physics. Methods from symplectic geometry add power and scope to modern harmonic analysis; historically these methods were perhaps for the first time systematically used in Folland's seminal book [59].

The aim of the present book is to give a rigorous and modern treatment of various objects from harmonic analysis with a strong emphasis on the underlying symplectic structure (for instance symplectic and metaplectic covariance properties). More specifically we have in mind two audiences: the time-frequency community, and mathematical physicists interested in applications to quantum mechanics. The concepts and methods are presented in such a way that they should be easily accessible to students at the upper-undergraduate level (a certain familiarity with basic Fourier analysis and the elementary theory of distributions is assumed). Needless to say, this book can also be read with profit by more advanced readers, and can be used as a reference work by researchers in partial differential equations, harmonic analysis, and mathematical physics. (Several chapters are part of ongoing research and contain material that is usually not addressed in introductory texts. For instance Gromov's non-squeezing theorem from symplectic topology and its applications, or the theory of phase space pseudodifferential operators.)

Description of the book

This book is divided into parts and chapters, each devoted to a particular topic. They have been designed in such a way that the material of each chapter can be covered in a 90 minutes lecture (but this, of course, very much depends on the student's background). The parts can be read independently.

Part I: Symplectic Mechanics

- **Chapter 1** is intended to be a review of the main concepts from Hamiltonian mechanics; while it can be skipped by the reader wanting to advance rapidly in the mathematics of harmonic analysis on symplectic spaces, it is recommended as a reference for a better understanding of the reasons for which many concepts are introduced. For instance, the Hamiltonian approach leads to a very natural and "obvious" motivation for consideration of the Heisenberg–Weyl operators, and of the Weyl pseudo-differential calculus. Also, deformation quantization does not really make sense unless one understands the mechanical reasons which lie behind it. The main result of this first chapter is that Hamiltonian flows consist of symplectomorphisms (the physicist's canonical transformations). This is proven in detail using an elementary method, that of the "variational equation" (which is a misnomer, because there is per se nothing variational in that equation!). We also discuss other topics, such as Poisson brackets (which is helpful to understand the first steps of deformation quantization; Hamilton–Jacobi theory is also briefly discussed).

- In **Chapter 2** the basics of the theory of the symplectic group are developed in a self-contained way. Only an elementary knowledge of linear algebra is required for understanding of the topics of this chapter; the few parts where we invoke more sophisticated material such as differential forms can be skipped by the beginner. A particular emphasis is put on the machinery of free symplectic matrices and their generating functions, which are usually ignored in first courses. The consideration of this topic simplifies many calculations, and has the advantage of yielding the easiest approach to the theory of the metaplectic group. We also discuss classical topics, such as the identification of the unitary group with a subgroup of the symplectic group.

- In **Chapter 3** we refine our study of the symplectic group by introducing the notion of free symplectic matrix and its generating functions. Free symplectic matrices can be defined in several different ways. Their importance comes from the fact that they are in a sense the building blocks of the symplectic group: every symplectic matrix is the product (in infinitely many ways) of exactly two free symplectic matrices. This property in turn allows an easy construction of simple sets of generators for the symplectic group. Last – but certainly not least! – the notion of free symplectic matrix will be instrumental for our definition in Chapter 7 of the metaplectic representation.

- In **Chapter 4** we discuss the notion of symplectomorphism, which is a generalization to the non-linear case of the symplectic transformations introduced in the previous chapters. This leads us to define two very interesting groups $\mathrm{Symp}(2n, \mathbb{R})$ and $\mathrm{Ham}(2n, \mathbb{R})$, respectively the group of all symplectomorphisms, and that of all Hamiltonian symplectomorphisms. These groups, which are of great interest in current research in symplectic topology, are non-linear generalizations of the symplectic group $\mathrm{Sp}(2n, \mathbb{R})$. The group

Ham$(2n, \mathbb{R})$ will play an important role in our derivation of Schrödinger's equation for arbitrary Hamiltonian functions.

- In **Chapter 5** we introduce new and very powerful tools from symplectic geometry and topology: Gromov's symplectic non-squeezing theorem, and the associated notion of symplectic capacity. The importance of these concepts (which go back to the mid 1980s, and for which Gromov got the Abel Prize in 2009) in applications has probably not yet been fully realized in mathematical analysis, and even less in mathematical physics.

- In **Chapter 6** we address a topic which belongs to both classical and quantum mechanics, namely uncertainties principles, and we do this from a topological point of view. We begin by discussing uncertainty principles associated with a quasi-probability distribution from a quite general point of view (hence applicable both to the classical and quantum cases), and introduce the associated notion of covariance matrix. This enables us to reformulate the strong version of the uncertainty principle in terms of symplectic capacities. This approach to both classical and quantum uncertainties is new and due to the author. It seems to be promising because it allows us to analyze uncertainties which are more general than those usually considered in the literature, and has led to the definition of "quantum blobs", which are symplectically invariant subsets of phase space with minimum symplectic capacity one-half of Planck's constant h. We also prove a multi-dimensional Hardy uncertainty principle, which says that a function and its Fourier transform cannot be simultaneously dominated by too sharply peaked Gaussians.

Part II: Harmonic Analysis in Symplectic Spaces

- **Chapter 7** is devoted to a detailed study of the metaplectic group $\mathrm{Mp}(2n, \mathbb{R})$ as a unitary representation in $L^2(\mathbb{R}^n)$ of the two-fold covering of the symplectic group $\mathrm{Sp}(2n, \mathbb{R})$. The properties of $\mathrm{Sp}(2n, \mathbb{R})$, as exposed in Chapters 2 and 3, allow us to identify the generators of the metaplectic group as "quadratic Fourier transforms", generalizing the usual Fourier transforms. We construct with great care the projection (covering) mapping from the metaplectic group to the symplectic group, having in mind our future applications to the Wigner transform and the Schrödinger equation. In the forthcoming chapters we will use systematically the properties of the metaplectic group, in particular when establishing symplectic/metaplectic covariance formulas in Weyl calculus and the theory of the Wigner transform.

- In **Chapter 8** we study two companions, the Heisenberg–Weyl and Grossmann–Royer operators. These operators can in a sense be viewed as "quantized" versions of, respectively, translation and reflection operators and are symplectic Fourier transforms of each other. We also discuss the related notion of Heisenberg group and algebra which play such an important role in harmonic analysis in phase space; our approach starts with the canonical commutation relations of quantum mechanics. We also define and briefly discuss the affine variant of the metaplectic group, namely the inhomogeneous metaplectic group $\mathrm{AMp}(2n, \mathbb{R})$ which is an extension by the Heisenberg–Weyl operators of the metaplectic group $\mathrm{Mp}(2n, \mathbb{R})$.

- In **Chapter 9** we study in great detail various algebraic and functional properties of the cross-ambiguity and cross-Wigner functions, which are concisely defined using the Heisenberg–Weyl and Grossmann–Royer operators introduced in the previous chapter. We discuss the relations with the short-time Fourier transform used in signal and time-frequency analysis. We also prove a useful inversion formula for the cross-Wigner transform; this formula plays an important role in the theory of Feichtinger's modulation spaces which will be studied later in this book.

Part III: Pseudo-differential Operators and Function Spaces

- In **Chapter 10** we present the basics of Weyl calculus, in particular the definition of the Weyl correspondence which plays such an important role both in the theory of pseudodifferential operators and in modern quantum mechanics of which it is one of the pillars. The chapter begins with an introductory section where the need for "quantization" is briefly discussed. We prove various formulas (in particular formulas for the adjoint of a Weyl operator, and that for the twisted symbol of the composition of two operators).

- In **Chapter 11** we take a close look at the notion of coherent states (they are elementary Gaussian functions); the properties of the metaplectic group allow us to give very explicit formulas for their natural extension, the squeezed coherent states, which play a pivotal role both in harmonic analysis and in quantum mechanics (especially in the subdiscipline known as quantum optics). This leads us naturally to the consideration of anti-Wick operators (also called Toeplitz or Berezin operators), of which we give the main properties.

- In **Chapter 12** we review two venerable topics from functional analysis: the theory of Hilbert–Schmidt and the associated theory of trace class operators. This will allow us to give a precise meaning to the notion of mixed quantum state in Chapter 13. We discuss in some detail the delicate procedure of calculating the trace. In particular we state and prove a result making legitimate the integration of the kernel when the operator is a Weyl pseudodifferential operator.

- In **Chapter 13** we give a rigorous definition of the notion of mixed quantum state, and of the associated density operators (called density matrices in quantum mechanics). The relation between density operators and the Wigner transform is made clear and fully exploited. We discuss the very delicate notion of positivity for the density operator. This is done by introducing the Kastler–Loupias–Miracle-Sole conditions, which we relate to the uncertainty principle. We also apply Hardy's uncertainty principle in its multi-dimensional form to the characterization of sub-Gaussian mixed states; the results are stated concisely using the notion of symplectic capacity.

- **Chapter 14** is of a rather technical nature. We introduce Shubin's global symbol classes, and the associated pseudo-differential operators. Shubin classes are of a greater use in quantum-mechanics than the ordinary Hörmander classes because their definition takes into account global properties of polynomial decrease in phase space. We discuss the notion of asymptotic expansion of the symbols, and show that operators which are at first sight much more general can be reduced to the case of ordinary pseudo-differential operators. We also study the notion of τ-symbol of a pseudo-differential, and give formulas allowing one to switch between different values of the parameter τ.

Part IV: Applications

- In **Chapter 15** we study a great classic of quantum mechanics, in fact one of its pillars: Schrödinger's time-dependent equation. We begin by showing that this equation can be derived from the theory of the metaplectic group when associated to a quadratic Hamiltonian function. In the second part of the chapter we generalize our construction to arbitrary Hamiltonian functions by using Stone's theorem on strongly continuous one-parameter groups of unitary operators together with the characteristic property of symplectic covariance of Weyl pseudo-differential calculus.

- **Chapters 16 and 17** are an introduction to Feichtinger's theory of modulation spaces. The elements ψ of these spaces are functions (or distributions) on $\mathbb{R}^n$ characterized by the property that the cross-Wigner transform $W(\psi, \phi)$ belongs to some weighted Banach space of integrable functions on $\mathbb{R}^{2n}$ for every "window" ϕ. The simplest example is provided by the Feichtinger algebra $M^1(\mathbb{R}^n)$ which is the smallest Banach algebra containing the Schwartz functions and being invariant under the action of the inhomogeneous group (Chapter 16). Since Feichtinger's algebra is a Banach algebra it can be used with profit as a substitute for the Schwartz space; it allows in particular, together with its dual, to define a Gelfand triple. Modulation spaces play a crucial role in time-frequency analysis and in the theory of pseudodifferential operators. Their importance in quantum mechanics has only been recently realized, and is being very actively investigated.

- **Chapter 18** is an introduction to a new topic, which we have called *Bopp calculus*. Bopp operators are pseudodifferential operators of a certain type acting on phase space functions or distributions. They are associated in a natural way to the usual Weyl operators by "Bopp quantization rules", $x \longrightarrow x + \frac{1}{2}i\hbar\partial_p$, $p \longrightarrow p - \frac{1}{2}i\hbar\partial_x$. These rules are often used heuristically by physicists working in the area of deformation quantization; this chapter gives a rigorous justification of these manipulations. We note that the theory of Bopp operators certainly has many applications in pure mathematics and physics (Schrödinger equation in phase space).

- In **Chapter 19** we give a few applications of Bopp calculus. We begin by studying spectral properties of Bopp operators, which we relate to those of the corresponding standard Weyl operators. As an example we derive the energy levels and eigenfunctions of the magnetic operator (also called Landau operator). We thereafter show that Bopp pseudodifferential operators allow one to express deformation quantization in terms of a pseudodifferential theory; this has of course many technical and conceptual advantages since it allows us to easily prove deep results on "stargenvalues" and "stargenvectors". The book ends on a beginning: the application of Bopp operators to an emerging subfield of mathematics called "noncommutative quantum mechanics" (NCQM), which has its origins in the quest for quantum gravity.

Acknowledgement

This work has been financed by the Austrian Research Agency FWF (Projektnummer P20442-N13). I wish to express my thanks to Hans Feichtinger and Karlheinz Gröchenig for their help and support, and correcting my misunderstandings about modulation spaces. I am solely responsible for remaining errors. I also thank Franz Luef for many stimulating conversations about modulation spaces and related topics from time-frequency analysis; some parts of this book are based on joint work with him. Last – but not least! – many thanks to you, Charlyne, for having typed with competence and patience a great part of the manuscript and having given me the loving environment I needed to write this book.

Prologue

In this preliminary chapter we introduce some notation and recall basic facts from linear algebra and vector calculus.

Some notation

Let $\mathbb{K} = \mathbb{R}$ or $\mathbb{C}$.

- $M(m, \mathbb{K})$ is the algebra of all $m \times m$ matrices with entries in $\mathbb{K}$.
- $\mathrm{GL}(m, \mathbb{K})$ is the general linear group. It consists of all invertible matrices in $M(m, \mathbb{K})$.
- $\mathrm{SL}(m, \mathbb{K})$ is the special linear group: it is the subgroup of $\mathrm{GL}(m, \mathbb{K})$ consisting of all matrices with determinant equal to 1.
- $\mathrm{Sym}(m, \mathbb{K})$ is the vector space of all symmetric matrices in $M(m, \mathbb{K})$; it has dimension $\frac{1}{2}m(m+1)$; $\mathrm{Sym}_+(2n, \mathbb{R})$ is the subset of $\mathrm{Sym}(m, \mathbb{K})$ consisting of the positive definite symmetric matrices.

The elements of $\mathbb{R}^m$ should be viewed as column vectors

$$x = \begin{pmatrix} x_1 \\ \vdots \\ x_n \end{pmatrix}$$

when displayed; for typographic simplicity we will usually write $x = (x_1, \ldots, x_n)$ in the text. The Euclidean scalar product $\langle \cdot, \cdot \rangle$ and norm $|\cdot|$ on $\mathbb{R}^m$ are defined by

$$x \cdot y = x^T y = \sum_{j=1}^{m} x_j y_j.$$

The gradient operator in the variables $x_1, \ldots, x_n$ will be denoted by

$$\partial_x \quad \text{or} \quad \begin{pmatrix} \partial x_1 \\ \vdots \\ \partial x_m \end{pmatrix}.$$

Let f and g be differentiable functions $\mathbb{R}^m \longrightarrow \mathbb{R}^m$; in matrix form the chain rule is

$$\partial_x(g \circ f)(x) = (Df(x))^T \partial_x f(x) \tag{1}$$

where $Df(x)$ is the Jacobian matrix of f: if $f = (f_1, \ldots, f_m)$ is a differentiable mapping $\mathbb{R}^m \longrightarrow \mathbb{R}^m$ then

$$Df = \begin{pmatrix} \frac{\partial f_1}{\partial x_1} & \frac{\partial f_1}{\partial x_2} & \cdots & \frac{\partial f_1}{\partial x_m} \\ \frac{\partial f_2}{\partial x_1} & \frac{\partial f_2}{\partial x_2} & \cdots & \frac{\partial f_2}{\partial x_m} \\ \vdots & \vdots & \ddots & \vdots \\ \frac{\partial f_m}{\partial x_1} & \frac{\partial f_n}{\partial x_2} & \cdots & \frac{\partial f_m}{\partial x_m} \end{pmatrix}. \tag{2}$$

Let $y = f(x)$; we will indifferently use the notation $Df(x)$ for the Jacobian matrix at x. If f is invertible, the inverse function theorem says that

$$D(f^{-1})(y) = [Df(x)]^{-1}. \tag{3}$$

If $f : \mathbb{R}^m \longrightarrow \mathbb{R}$ is a twice continuously differentiable function, its Hessian calculated at a point x is the symmetric matrix of second derivatives

$$D^2 f(x) = \begin{pmatrix} \frac{\partial^2 f}{\partial x_1^2} & \frac{\partial^2 f}{\partial x_1 \partial x_2} & \cdots & \frac{\partial^2 f}{\partial x_1 \partial x_n} \\ \frac{\partial^2 f}{\partial x_2 \partial x_1} & \frac{\partial^2 f}{\partial x_2^2} & \cdots & \frac{\partial^2 f}{\partial x_2 \partial x_n} \\ \vdots & \vdots & \ddots & \vdots \\ \frac{\partial^2 f}{\partial x_n \partial x_1} & \frac{\partial^2 f}{\partial x_n \partial x_2} & \cdots & \frac{\partial^2 f}{\partial x_n^2} \end{pmatrix}. \tag{4}$$

Notice that the Jacobian and Hessian matrices are related by the formula

$$D_x(\partial_x f)(x) = D_x^2 f(x). \tag{5}$$

Also note the following useful formulae:

$$\langle A\partial_x, \partial_x \rangle \, e^{-\frac{1}{2}\langle Mx, x\rangle} = [\langle MAMx, x\rangle - \mathrm{Tr}(AM)] \, e^{-\frac{1}{2}\langle Mx, x\rangle}, \tag{6}$$

$$\langle Bx, \partial_x \rangle \, e^{-\frac{1}{2}\langle Mx, x\rangle} = \langle MBx, x\rangle \, e^{-\frac{1}{2}\langle Mx, x\rangle}, \tag{7}$$

where A, B, and M are symmetric matrices.

The space $\mathcal{S}(\mathbb{R}^n)$ and its dual $\mathcal{S}'(\mathbb{R}^n)$

Very useful classes of functions and distributions are the so-called Schwartz space $\mathcal{S}(\mathbb{R}^n)$ and its dual $\mathcal{S}'(\mathbb{R}^n)$, which is the space of tempered distributions. In our context they are better adapted than the space $C_o^\infty(\mathbb{R}^n)$ of infinitely differentiable functions with compact support (the latter is not invariant under Fourier transform).

Definition 1. The space $\mathcal{S}(\mathbb{R}^n)$ consists of all infinitely differentiable functions $f : \mathbb{R}^n \longrightarrow \mathbb{C}$ such that for every pair (α, β) of multi-indices there exists a constant $C_{\alpha\beta} \geq 0$ such that

$$|x^\alpha \partial_x^\beta f(x)| \leq C_{\alpha\beta} \ \text{ for } x \in \mathbb{R}^n.$$

This condition is equivalent to the existence of $C'_{\alpha\beta} \geq 0$ such that

$$|\partial_x^\beta (x^\alpha f)(x)| \leq C'_{\alpha\beta} \ \text{ for } x \in \mathbb{R}^n.$$

The proof of the equivalence of the two conditions above is left to the reader; it readily follows – after some tedious calculations – from Leibniz's rule for the derivatives of a product.

Clearly $C_0^\infty(\mathbb{R}^n) \subset \mathcal{S}(\mathbb{R}^n)$; the archetypical example of a function which belongs to $\mathcal{S}(\mathbb{R}^n)$ but not to $C_0^\infty(\mathbb{R}^n)$ is the Gaussian $f(x) = e^{-|x|^2}$; more generally the product of a Gaussian by a polynomial is in $\mathcal{S}(\mathbb{R}^n)$. Note that $\mathcal{S}(\mathbb{R}^n)$ actually is an algebra: the product of two elements of $\mathcal{S}(\mathbb{R}^n)$ is also in $\mathcal{S}(\mathbb{R}^n)$ (this readily follows from the chain rule). The formulae

$$\|f\|_{\alpha\beta}^{(1)} = \sup_{x \in \mathbb{R}^n} |x^\alpha \partial_x^\beta f(x)|,$$

$$\|f\|_{\alpha\beta}^{(2)} = \sup_{x \in \mathbb{R}^n} |\partial_x^\beta (x^\alpha f)(x)|$$

define equivalent families of semi-norms on $\mathcal{S}(\mathbb{R}^n)$; one shows that $\mathcal{S}(\mathbb{R}^n)$ becomes a Fréchet space for the topology thus defined.

The Fourier transform

Let $f : \mathbb{R}^n \longrightarrow \mathbb{C}$ be an absolutely integrable function

$$\|f\|_{L^1} = \int_{\mathbb{R}^n} |f(x)|dx < \infty;$$

we will write for short $f \in L^1(\mathbb{R}^n)$. By definition the Fourier transform $\mathcal{F}f = \widehat{f}$ is the function defined by

$$\widehat{f}(\xi) = \left(\tfrac{1}{2\pi}\right)^{n/2} \int_{\mathbb{R}^n} e^{-i\xi \cdot x} f(x)dx.$$

We will use in this book the following variant of the Fourier transform $\mathcal{F}$:

$$F\psi(x) = \left(\tfrac{1}{2\pi\hbar}\right)^{n/2} \int_{\mathbb{R}^n} e^{-ix \cdot x'} \psi(x')dx';$$

here $\hbar$ is a positive parameter, which one identifies in physics with Planck's constant divided by 2π: $\hbar = h/2\pi$ (the notation is due to the physicist Dirac). One proves (Riemann–Lebesgue lemma) that $\lim_{|\xi| \to \infty} \widehat{f}(\xi) = 0$.

One of the main properties of the Schwartz space (and of it dual) is that it is invariant by the Fourier transform:

$$\mathcal{F} : \mathcal{S}(\mathbb{R}^n) \longrightarrow \mathcal{S}(\mathbb{R}^n),$$
$$\mathcal{F} : \mathcal{S}'(\mathbb{R}^n) \longrightarrow \mathcal{S}'(\mathbb{R}^n).$$

This is in strong contrast with the case of $C_o^\infty(\mathbb{R}^n)$: the only compactly supported function (or distribution, for that matter) whose Fourier transform is also compactly supported is 0 (this is easily seen if one knows that the Fourier transform of a compactly supported function is analytic, and can thus never have compact support).

Proposition 2. *The Fourier transforms $\mathcal{F} : f \longmapsto \widehat{f}$ and $f \longmapsto Ff$ are invertible automorphisms of $\mathcal{S}(\mathbb{R}^n)$ which extends by duality into automorphisms of $\mathcal{S}'(\mathbb{R}^n)$ defined by*

$$\langle \widehat{f}, g \rangle = \langle f, \widehat{g} \rangle \ , \ \langle Ff, g \rangle = \langle f, Fg \rangle$$

for $f \in \mathcal{S}'(\mathbb{R}^n)$, $g \in \mathcal{S}(\mathbb{R}^n)$. The restriction of these automorphism to $L^2(\mathbb{R}^n)$ are unitary.

Proof. Let $f \in \mathcal{S}(\mathbb{R}^n)$; for $\alpha, \beta \in \mathbb{N}^n$ we have

$$\xi^\alpha \partial_\xi^\beta \widehat{f} = (-i)^{|\alpha|+|\beta|} \widehat{\partial_x^\alpha x^\beta f};$$

since $\partial_x^\alpha x^\beta f \in \mathcal{S}(\mathbb{R}^n)$ there exists a constant $C_{\alpha\beta} > 0$ such that $|\xi^\alpha \partial_\xi^\beta \widehat{f}(\xi)| \leq C_{\alpha\beta}$, hence $\widehat{f} \in \mathcal{S}(\mathbb{R}^n)$. That the Fourier transform is an invertible automorphism of $\mathcal{S}(\mathbb{R}^n)$ follows from the Fourier inversion formula. The two last statements easily follow from Plancherel's formula

$$\int_{\mathbb{R}^n} f(x)\widehat{g}(x)dx = \int_{\mathbb{R}^n} \widehat{f}(x)g(x)dx$$

and their proof is therefore left to the reader. $\square$

Part I

Symplectic Mechanics

Chapter 1

Hamiltonian Mechanics in a Nutshell

This chapter is an introduction to the basics of Hamiltonian mechanics, with an emphasis on its symplectic formulation. It thus motivates the symplectic techniques which will be developed in the forthcoming chapters. In fact, Hamiltonian mechanics is historically the main motivation for the study of the symplectic group in particular, and of symplectic geometry in general. For complements and an extended study the reader can consult with profit the treatises by Abraham–Marsden [2] and Arnol'd [3]; an elementary introduction at the undergraduate level is the classical book by Goldstein [63] and its re-editions. (This book is written for physicists, however, and the mathematics is not always rigorous.)

Historically, Hamiltonian mechanics goes back to the early work of Hamilton and Lagrange; its symplectic formulation (as exposed in this chapter) is relatively recent; see Arnol'd [3] and Abraham et al. [1] for detailed accounts.

1.1 Hamilton's equations

We will use the notation $x = (x_1, \ldots, x_n)$, $p = (p_1, \ldots, p_n)$ for elements of $\mathbb{R}^n$ and $z = (x, p)$ for elements of $\mathbb{R}^{2n}$ (the "phase space"). When using matrix notation, x, p, z will always be viewed as column vectors.

1.1.1 Definition of Hamiltonian systems

Let H ("the Hamiltonian") be a real-valued function in $C^\infty(\mathbb{R}^{2n})$; more generally we will consider "time-dependent Hamiltonians" $H \in C^\infty(\mathbb{R}^{2n} \times \mathbb{R})$, functions of z and t.

Definition 3. The system of $2n$ ordinary differential equations

$$\frac{dx_j}{dt} = \frac{\partial H}{\partial p_j}(x, p, t), \quad \frac{dp_j}{dt} = -\frac{\partial H}{\partial x_j}(x, p, t) \tag{1.1}$$

is called the *Hamilton equations* associated with H.

To simplify the discussion we will assume that for every $z_0 = (x_0, p_0)$ belonging to an open subset Ω of $\mathbb{R}^{2n}$, this system has a unique solution $t \longmapsto z(t) = (x(t), p(t))$ such that $z(0) = z_0$, defined for $-T \leq t \leq T$ where $T > 0$. (See Abraham–Marsden [2], Ch. 1, §2.1, for a general discussion of global existence and uniqueness, including the important notion of "flow box".)

A basic example is the following; we state it in the case $n = 1$:

$$H(x, p) = \frac{p^2}{2m} + U(x) \tag{1.2}$$

where m is a positive constant ("the mass") and U a smooth function ("the potential"). In this case Hamilton's equations are

$$\frac{dx}{dt} = \frac{p}{m} \ , \ \frac{dp}{dt} = -U'(x). \tag{1.3}$$

In physics one writes $v = p/m$ (it is the velocity) and dp/dt, so that these equations are just a restatement of Newton's second law, familiar from elementary physics; the quantity $p^2/2m$ is the "kinetic energy". (We have discussed in some detail the physical interpretation of Hamilton's equations in [65].)

This example motivates the following definition:

Definition 4. Let $t \longmapsto z(t)$ be a solution of Hamilton's equations. The number $E(t) = H(z(t))$ is called the energy along the solution curve through $z_0 = z(0)$ at time t. When H is time-independent, we have $H(z(t)) = H(z(0))$ for every t. More generally, any function which is constant along the curves $t \longmapsto z(t)$ is called a "constant of the motion".

That the energy E is a constant for time-independent Hamiltonians follows from the chain rule applied to $H(z(t))$, taking Hamilton's equations into consideration: setting $z = z(t)$ we have

$$\frac{d}{dt}H(z(t)) = \sum_{j=1}^{n} \frac{\partial H}{\partial x_j}(z)\frac{dx_j}{dt} + \frac{\partial H}{\partial p_j}(z)\frac{dp_j}{dt} = 0.$$

In the case of time-dependent Hamiltonians the same argument shows that

$$\frac{d}{dt}H(z(t), t) = \frac{\partial H}{\partial t}(z(t), t)$$

hence the energy $E(t) = H(z(t), t)$ is not a constant of the motion.

1.1.2   A simple existence and uniqueness result

Here is an existence result which is sufficient for many applications to physics. We assume that the Hamiltonian is time-independent and of the type

$$H(x, p) = \sum_{j=1}^{n} \frac{p_j^2}{2m_j} + U(x)$$

where $U \in C^{\infty}(\mathbb{R}^n)$.

Proposition 5. *If $U \geq a$ for some constant a, then every solution of Hamilton's equations*

$$\frac{dx_j}{dt} = \frac{p_j}{m_j} \quad , \quad \frac{dp_j}{dt} = -\frac{\partial U}{\partial x_j}(x)$$

$(1 \leq j \leq n)$ exists for all times (and is unique).

Proof. In view of the local existence theory for ordinary differential equations it suffices to show that the solutions $t \longmapsto z(t)$ remain in bounded sets for finite times. Since Hamilton's equations are insensitive to the addition of a constant to the Hamiltonian we may assume $a = 0$, and rescaling if necessary the momentum and position coordinates it is no restriction neither to assume $m_j = 1$ for $1 \leq j \leq n$. For notational simplicity we moreover assume $n = 1$. Let thus $t \to z(t) = (x(t), p(t))$ be a solution curve of the equations

$$\frac{dx}{dt} = p \quad , \quad \frac{dp}{dt} = -\frac{\partial U}{\partial x}(x)$$

and let $E = H(z(t))$ be the energy; since $H \geq U$ we have $E \geq U(x(t))$. In view of the triangle inequality

$$|x(t)| \leq |x(0)| + |x(t) - x(0)| \leq |x(0)| + \int_0^t \left| \tfrac{d}{dt} x(s) \right| ds;$$

since $\frac{d}{dt} x(s) = p(s)$ and

$$p(t) = \sqrt{2(E - U(x(t))} \leq \sqrt{2E} \tag{1.4}$$

we have:

$$|x(t)| \leq |x(0)| + \int_0^t |p(s)|\, ds \leq |x(0)| + \int_0^t \sqrt{2(E - U(x(s))}\, ds$$

so that

$$|x(t)| \leq |x(0)| + t\sqrt{2E}. \tag{1.5}$$

The inequalities (1.4) and (1.5) show that for t in any finite time-interval $[0, T]$ the functions $t \longmapsto x(t)$ and $t \longmapsto p(t) = x(t)$, and hence $t \longmapsto z(t)$, stay forever in a bounded set. $\qquad\square$

One can show (see [1], §4.1) that the conclusions of Proposition 5 still hold if one replaces the boundedness condition $U \geq a$ by the much weaker requirement

$$U(x) \geq a - b|x|^2 \quad \text{for} \quad b > 0$$

where a and b are some constants ($b > 0$). This condition cannot be very much relaxed; for instance one shows (ibid.) that already in the case $n = 1$ the solutions of the Hamilton equations for

$$H(x, p) = \frac{p^2}{2m} - \frac{\varepsilon^2}{8} x^{2+(4/\varepsilon)}$$

are not defined for all t if $\varepsilon > 0$.

1.2   Hamiltonian fields and flows

From now on we will use the following more compact notation, borrowed from mechanics: time derivatives (i.e., derivatives with respect to t) will be denoted by putting a dot over the letter standing for the function. For instance, $\dot{x}$ means dx/dt. Derivatives will in general be written as ∂_x, ∂_{x_j}, etc. instead of $\partial/\partial x$, $\partial/\partial x_j$. We will also freely use the notation ∂_x for the gradient $(\partial_{x_1}, \ldots, \partial_{x_n})$. Similarly, $\partial_z = (\partial_x, \partial_p)$ is the gradient in the $2n$ variables $z_1 = x_1, \ldots, z_n = x_n; z_{n+1} = p_1, \ldots, z_{2n} = p_n$.

The Hamilton equations (1.1) can be rewritten in compact form as

$$\dot{z} = J\partial_z H(z) \tag{1.6}$$

where J is the "standard symplectic matrix" defined by

$$J = \begin{pmatrix} 0 & I \\ -I & 0 \end{pmatrix}$$

where 0 and I are the $n \times n$ zero and identity matrices. That matrix will play an essential role in all of this book.

1.2.1   The Hamilton vector field

Assume first that H is a time-independent Hamiltonian function.

Definition 6. We call the vector field

$$X_H = J\partial_z H = (\partial_x H, -\partial_p H)$$

the "Hamilton vector field of H"; the operator $J\partial_z$ is called a "*symplectic gradient*".

It follows from the elementary theory of ordinary autonomous differential equations that the system (1.1) defines a flow (ϕ_t^H): by definition the function $t \longmapsto z(t) = \phi_t^H(z_0)$ is the solution of Hamilton's equations with $z(0) = z_0$ and we have

$$\phi_t^H \phi_{t'}^H = \phi_{t+t'}^H \ , \ \phi_0^H = I \tag{1.7}$$

when t, t' and $t + t'$ are in the interval $[-T, T]$. In particular each ϕ_t^H is a diffeomorphism such that $(\phi_t^H)^{-1} = \phi_{-t}^H$.

Definition 7. One says that (ϕ_t^H) is the flow generated by the Hamilton equations for H.

The Hamilton vector field is gradient-free:

$$\operatorname{div} X_H = \partial_x \left(\partial_p H \right) - \partial_p \left(\partial_x H \right) = 0$$

hence the flow (ϕ_t^H) is incompressible. This result is called "Liouville's theorem" in the physics literature. Incompressibility means that for every subset U of Ω and $t \in [-T, T]$ such that $\phi_t^H(U) \subset \Omega$ we have

$$\operatorname{Vol} \phi_t^H(U) = \operatorname{Vol} U.$$

Hamiltonian flows are thus volume-preserving. This property also follows from the fact that Hamiltonian flows consist of symplectomorphisms, as we will see in a moment, for we then have

$$(\phi_t^H)^* \operatorname{Vol} = \operatorname{Vol}$$

where Vol is the volume form (2.11).

When H is a time-dependent function, Hamilton's equations become a non-autonomous system of differential equations which we can write concisely as

$$\dot{z} = J\partial_z H(z, t). \tag{1.8}$$

One again writes $X_H = \partial_z H$, but X_H is not, strictly speaking, a vector field in the usual sense (because it depends on the parameter t). The "flow" (ϕ_t^H) generated by time-dependent Hamiltonian no longer has the group property: $\phi_t^H \phi_{t'}^H \neq \phi_{t+t'}^H$. It is often useful to replace the notion of flow as defined above by that of time-dependent flow $(\phi_{t,t'}^H)$: $\phi_{t,t'}^H$ is the function defined by the condition that $t \longmapsto z(t, t') = \phi_{t,t'}^H(z_0)$ is the solution of Hamilton's equations with $z(t') = z_0$. Obviously

$$\phi_{t,t'}^H = \phi_{t,0}^H \left(\phi_{t',0}^H\right)^{-1}$$

hence the group property (1.7) has then to be replaced by

$$\phi_{t,t'}^H \phi_{t',t''}^H = \phi_{t,t''}^H \;,\; \phi_{t,t}^H = I \tag{1.9}$$

for all admissible t, t', and t''. Notice that it follows in particular that $(\phi_{t,t'}^H)^{-1} = \phi_{t',t}^H$.

It is however always possible to reduce the study of a time-dependent Hamiltonian to the time-independent case. The price to pay is that we have to work in a phase space with dimension $2n + 2$ instead of $2n$. We define a new Hamiltonian function $\widetilde{H}$ by the formula

$$\widetilde{H}(x, p, t, E) = H(x, p, t) - E \tag{1.10}$$

where E is a new variable, viewed as conjugate to the time t which now has the status of a "position variable"; we could as well write the definition of $\widetilde{H}$ in the form

$$\widetilde{H}(x, p, x_{n+1}, p_{n+1}) = H(x, p, x_{n+1}) - p_{n+1}$$

but we will however stick to the notation (1.10). The function $\widetilde{H}$ is defined on the extended phase space $\mathbb{R}^{2n+2} \equiv \mathbb{R}^{2n} \times \mathbb{R}_E \times \mathbb{R}_t$. The associated Hamilton equations are, expressed in terms of the original Hamiltonian H:

$$\frac{dx}{dt'} = \partial_p H \quad , \quad \frac{dp}{dt'} = -\partial_x H,$$

$$\frac{dE}{dt'} = \frac{\partial H}{\partial t} \quad , \quad \frac{\partial t}{\partial t'} = 1,$$

where the parameter t' plays the role of a new "time"; since $\widetilde{H}$ does not explicitly contain that parameter, it is a "time-independent" Hamiltonian on the extended phase space. Notice that in view of the fourth equation above we may choose $t' = t$ so that the two first equations are just the Hamiltonian equations for H; as a bonus the third equation is just the familiar law for the variation of energy of a time-dependent Hamiltonian system:

$$\frac{dE}{dt} = \frac{d}{dt}H(x,p,t) = \frac{\partial}{\partial t}H(x,p,t)$$

(the second equality because of the chain rule and using the fact that x and p satisfy Hamilton's equations). We can now define the "extended Hamiltonian flow" $(\widetilde{\phi}_t^H)$ of H by the formula $\widetilde{\phi}_t^H = \phi_t^{\widetilde{H}}$. Notice that since $(\phi_t^{\widetilde{H}})$ is the flow determined by a time-independent Hamiltonian, $(\widetilde{\phi}_t^H)$ enjoys the one-parameter group property $\widetilde{\phi}_t^H \widetilde{\phi}_{t'}^H = \widetilde{\phi}_{t+t'}^H$ and $\widetilde{\phi}_0^H = I$ (the identity operator on the extended phase space $\mathbb{R}^{2n+2}$). Denote now by $(\phi_{t,t'}^H)$ the two-parameter family of canonical transformations of $\mathbb{R}^{2n}$ defined as follows: for fixed t' the function $z = \phi_{t,t'}^H(z')$ is the solution of Hamilton's equations for H taking the value z' at time t'. Thus:

$$\phi_{t,t'}^H = \phi_t^H \left(\phi_{t'}^H\right)^{-1}. \tag{1.11}$$

Clearly $\phi_{t,t}^H$ is the identity operator on $\mathbb{R}^{2n}$ and $\phi_{t,t'}^H \phi_{t',t''}^H = \phi_{t,t''}^H$, $(\phi_{t,t'}^H)^{-1} = \phi_{t',t}^H$. The two-parameter family $(\phi_{t,t'}^H)$ is sometimes called the "time-dependent flow"; it is related to the extended flow defined above by the simple formula

$$\widetilde{\phi}_t^H(z',t',E') = (\phi_{t,t'}^H(z'), t+t', E_{t,t'}) \tag{1.12}$$

with

$$E_{t,t'} = E' + H(\phi_{t,t'}^H(z'),t) - H(z',t'). \tag{1.13}$$

We refer to the paper [154] by Struckmeier for a discussion of some subtleties and difficulties related to the method just outlined. See Sardanashvily [139] for an up-to-date discussion of time-dependent Hamiltonian mechanics from the point of view of differential geometry.

1.2.2   The symplectic character of Hamiltonian flows

Recall that J (the "standard symplectic matrix") is defined by

$$J = \begin{pmatrix} 0 & I \\ -I & 0 \end{pmatrix}$$

where 0 and I are the $n \times n$ zero and identity matrices. Note that $\det J = 1$ and

$$J^2 = I \quad , \quad J^T = J^{-1} = -J$$

(the superscript ^{T} denotes transposition).

Definition 8. A real $2n \times 2n$ matrix is said to be symplectic if it satisfies the conditions

$$S^T J S = S J S^T = J. \tag{1.14}$$

The set of all symplectic matrices is denoted by $\mathrm{Sp}(2n, \mathbb{R})$.

A symplectic matrix is invertible because $\det(S^T J S) = \det J$ implies $\det(S)^2 = 1$ since $\det J = 1$. (We will actually see later that we must have $\det S = 1$ when S is a symplectic matrix; this property is in fact not quite obvious.) We will see in the next chapter that $\mathrm{Sp}(2n, \mathbb{R})$ is a group, in fact one of the classical Lie groups.

The main property of Hamiltonian flows – or, at least the one that distinguishes them from general flows – is that they consist of *symplectomorphisms* (also called "canonical transformations", especially in the physical literature). There are several ways to prove this; our approach makes use of the so-called *variational equation* satisfied by the Jacobian matrices of a Hamiltonian flow. We prove the result for time-independent Hamiltonians; the extension to the time-dependent case is straightforward (see Exercise 10 below).

Theorem 9. *Let* (ϕ_t^H) *be a Hamiltonian flow defined on* $\Omega \subset \mathbb{R}^{2n}$.

(i) *The Jacobian matrix* $S_t(z) = D\phi_t^H(z)$ *satisfies the "variational equation"*

$$\frac{d}{dt} S_t(z) = J D^2 H(\phi_t^H(z)) S_t(z) \quad , \quad S_t(z) = I. \tag{1.15}$$

(ii) *The Jacobian matrix* $D\phi_t^H(z)$ *is symplectic for every* $z \in \Omega$:

$$[D\phi_t^H(z)]^T J D\phi_t^H(z) = D\phi_t^H(z) J [D\phi_t^H(z)]^T = J.$$

Proof. (i) Taking Hamilton's equation into account the time-derivative of the Jacobian matrix $S_t(z)$ is

$$\frac{d}{dt} S_t(z) = \frac{d}{dt}(D\phi_t^H(z)) = D\left(\frac{d}{dt}\phi_t^H(z)\right),$$

that is

$$\frac{d}{dt} S_t(z) = D(X_H(\phi_t^H(z))).$$

Using the fact that $X_H = J\partial_z H$ together with the chain rule, we have

$$D(X_H(\phi_t^H(z))) = D(J\partial_z H)(\phi_t^H(z))$$
$$= JD(\partial_z H)(\phi_t^H(z))$$
$$= J(D^2 H)(\phi_t^H(z))D\phi_t^H(z),$$

hence $S_t(z)$ satisfies the variational equation (1.15). Statement (ii) follows from (i): set $S_t = S_t(z)$ and $A_t = (S_t(z))^T J S_t(z)$; using the product rule together with (1.15) we have

$$\frac{dA_t}{dt} = \frac{d(S_t)^T}{dt} J S_t + (S_t)^T J \frac{dS_t}{dt}$$
$$= (S_t)^T D^2 H(z) S_t - (S_t)^T D^2 H(z) S_t$$
$$= 0.$$

It follows that the matrix $S_t^T J S_t$ is constant in t, hence, in particular, $S_t^T J S_t = S_0^T J S_0 = J$ (because S_0 is the identity) so that $S_t \in \mathrm{Sp}(2n, \mathbb{R})$ for all $t \in \mathbb{R}$. $\square$

The following exercise is easy, but the result is useful: it shows that even for time-depending Hamiltonians the flow consists of symplectomorphisms.

Exercise 10. Reformulate (and prove) the conclusions of Theorem 9 in the case of a time-dependent flow determined by a time-dependent Hamiltonian.

1.2.3   Poisson brackets

There is another way of writing Hamilton's equations; it makes use of the notion of Poisson bracket. Let us introduce the following notation: for any pair of vectors (z, z') in $\mathbb{R}^{2n}$ we set

$$\sigma(z, z') = (z')^T J z = J z \cdot z'.$$

The scalar $\sigma(z, z')$ is called the *symplectic product* of z and z'; σ is the *standard symplectic form* on $\mathbb{R}^{2n}$. Observe that σ is a bilinear form on $\mathbb{R}^{2n}$ which is antisymmetric: $\sigma(z, z') = -\sigma(z', z)$. It satisfies in addition the following non-degeneracy condition: we can have $\sigma(z, z') = 0$ for all z' if and only if $z = 0$ (this because J is invertible).

Definition 11. Let $(f, g) \in C^\infty(\mathbb{R}^{2n}) \times C^\infty(\mathbb{R}^{2n})$. The Poisson bracket of f and g is the function

$$\{f, g\} = \sum_{j=1}^{n} \partial_{x_j} f \partial_{p_j} g - \partial_{x_j} g \partial_{p_j} f = \partial_x f \cdot \partial_p g - \partial_x g \cdot \partial_p f.$$

It is immediate to verify that Hamilton's equations can be rewritten, using Poisson brackets, as

$$\dot{x}_j = \{x_j, H\} \quad , \quad \dot{p}_j = \{p_j, H\}.$$

The following properties of the Poisson brackets are proven by straightforward calculations:

- Anticommutativity:

$$\{f, g\} = -\{g, f\};$$

- Linearity:

$$\{f, g + h\} = \{f, g\} + \{f, h\},$$
$$\{f + g, h\} = \{f, h\} + \{g, h\},$$
$$\{\lambda f, g\} = \{f, \lambda g\} = \lambda\{f, g\};$$

- Leibniz's law

$$\{f, gh\} = \{f, g\}h + g\{f, h\};$$

- Jacobi identity:

$$\{f, \{g, h\}\} + \{h, \{f, g\}\} + \{g, \{h, f\}\} = 0.$$

The relation between the symplectic product and Poisson brackets is instructive; it comes from the following property:

Proposition 12. *Let X_H and X_K be the Hamilton fields of H and K. The Poisson bracket of H and K is given by*

$$\{H, K\} = -\sigma(X_H, X_K). \tag{1.16}$$

Proof. It is obvious since $X_H = (\partial_p H, -\partial_x H)$, $X_K = (\partial_p K, -\partial_x K)$ so that

$$\sigma(X_H, X_K) = -\partial_x H \cdot \partial_p K + \partial_x K \cdot \partial_p H = -\{f, g\}. \qquad \square$$

Poisson brackets are useful in various circumstances; they are historically at the origin of quantization deformation (and of *prequantization*, which is an unphysical mathematical theory; see Wallach [158] for an introduction to this topic).

1.3 Additional topics

1.3.1 Hamilton–Jacobi theory

Here is one method that can be used (at least theoretically) to integrate Hamilton's equations; historically it is one of the first known resolution schemes. A complete rigorous treatment is to be found in, for instance, Abraham et al. [1].

Given an arbitrary Hamiltonian function H the associated Hamilton–Jacobi equation is the (usually non-linear) partial differential equation with unknown Φ:

$$\frac{\partial \Phi}{\partial t} + H(x, \partial_x \Phi, t) = 0. \tag{1.17}$$

The interest of this equation comes from the fact that the knowledge of a sufficiently general solution Φ yields the solutions of Hamilton's equations for H. At first sight it may seem strange that one replaces a system of ordinary differential equations by a non-linear partial differential equation, but this procedure is often the only available method!

Proposition 13. *Let $\Phi = \Phi(x,t,\alpha)$ be a solution of*

$$\frac{\partial \Phi}{\partial t} + H(x, \partial_x \Phi, t) = 0 \tag{1.18}$$

depending on n non-additive constants of integration $\alpha_1, \ldots, \alpha_n$, and such that

$$\det D^2_{x,\alpha} \Phi(x,t,\alpha) \neq 0. \tag{1.19}$$

Let $\beta_1, \ldots, \beta_n$ be constants; the functions $t \longmapsto x(t)$ and $t \longmapsto p(t)$ determined by the implicit equations

$$\partial_\alpha \Phi(x,t,\alpha) = \beta \quad , \quad p = \partial_x \Phi(x,t,\alpha) \tag{1.20}$$

are solutions of Hamilton's equations for H.

Proof. We assume $n = 1$ for notational simplicity; the proof extends to the general case without difficulty. Condition (1.19) implies, in view of the implicit function theorem, that the equation $\partial_\alpha \Phi(x,t,\alpha) = \beta$ has a unique solution $x(t)$ for each t; this defines a function $t \longmapsto x(t)$. Inserting $x(t)$ in the formula $p = \partial_x \Phi(x,t,\alpha)$ we also get a function $t \longmapsto p(t) = \partial_x \Phi(x(t),t,\alpha)$. Let us show that $t \longmapsto (x(t),p(t))$ is a solution of Hamilton's equations for H. Differentiating the equation (1.18) with respect to α yields, using the chain rule,

$$\frac{\partial^2 \Phi}{\partial \alpha \partial t} + \frac{\partial H}{\partial p} \frac{\partial^2 \Phi}{\partial \alpha \partial x} = 0; \tag{1.21}$$

differentiating the first equation (1.20) with respect to t yields

$$\frac{\partial^2 \Phi}{\partial x \partial \alpha} \dot{x} + \frac{\partial^2 \Phi}{\partial t \partial \alpha} = 0; \tag{1.22}$$

subtracting (1.22) from (1.21) we get

$$\frac{\partial^2 \Phi}{\partial x \partial \alpha} \left(\frac{\partial H}{\partial p} - \dot{x} \right) = 0,$$

hence we have proven that $\dot{x} = \partial_p H$ since $\partial^2 \Phi / \partial x \partial \alpha$ is assumed to be non-singular. To show that $\dot{p} = -\partial_x H$ we differentiate (1.18) with respect to x:

$$\frac{\partial^2 \Phi}{\partial x \partial t} + \frac{\partial H}{\partial x} + \frac{\partial H}{\partial p} \frac{\partial^2 \Phi}{\partial x^2} = 0 \tag{1.23}$$

and $p = \partial_x \Phi$ with respect to t:

$$\dot{p} = \frac{\partial^2 \Phi}{\partial t \partial x} + \frac{\partial^2 \Phi}{\partial x^2} \dot{x}. \tag{1.24}$$

Inserting the value of $\partial^2 \Phi / \partial x \partial t$ given by (1.24) in (1.23) yields

$$\frac{\partial H}{\partial x} + \frac{\partial^2 \Phi}{\partial x^2} \dot{x} - \frac{\partial H}{\partial p} \frac{\partial^2 \Phi}{\partial x^2} + \dot{p} = 0$$

hence $\dot{p} = -\partial_x H$ since $\dot{x} = \partial_p H$. $\square$

When the Hamiltonian is time-independent, the Hamilton–Jacobi equation is separable: inserting $\Phi = \Phi_0 - Et$ in (1.18) we get the 'reduced Hamilton–Jacobi equation':

$$H(x, \partial_x \Phi_0, t) = E \tag{1.25}$$

which is often easier to solve in practice; the energy E can be taken as a constant of integration.

Exercise 14.

(i) Let $H = \frac{1}{2m} p^2$ be the Hamiltonian of a particle with mass m moving freely along the x-axis. Use (1.25) to find a complete family of solutions of the time-dependent Hamilton–Jacobi equation for H.

(ii) Do the same with the harmonic oscillator Hamiltonian $H = \frac{1}{2m}(p^2 + m^2\omega^2 x^2)$.

1.3.2 The invariant volume form

In what follows H denotes a time-independent Hamiltonian function on $\mathbb{R}^{2n}$ and E a real number.

Definition 15. When non-empty the level set $\Sigma_E = \{z \in \mathbb{R}^{2n} : H(z) = E\}$ is called the energy shell for H corresponding to the energy level E. An energy shell is said to be regular if the gradient field $\partial_z H$ of the Hamiltonian is orthogonal to Σ_E at every point and moreover never vanishes on Σ_E.

When Σ_E is a regular energy shell, the formula

$$\mathcal{N}(z) = \frac{\partial_z H(z)}{|\partial_z H(z)|}$$

thus defines a unit normal field N to the energy shell Σ_E. We claim that the $(2n-1)$-form dV^{2n-1} defined by

$$dV_E^{2n-1}(X_1, \ldots, X_{2n-1}) = dV^{2n}(\mathcal{N}, X_1, \ldots, X_{2n-1}) \tag{1.26}$$

$(X_1, \ldots, X_{2n-1}$ tangent vector fields to Σ_E) is a volume form on Σ_E;

$$dV^{2n} = \frac{1}{n!} \sigma \wedge \cdots \wedge \sigma$$

(n factors) is the 'Liouville volume form' on $\mathbb{R}^{2n}$. This can be rewritten as:

$$dV^{2n} = dp_1 \wedge \cdots \wedge dp_n \wedge dx_1 \wedge \cdots \wedge dx_n.$$

All we have to do is to check that $dV_E^{2n-1}(z) \neq 0$ for each z. Choose linearly independent tangent vectors $X_1, \ldots, X_{2n-1}$ to Σ_E; since $\mathcal{N}(z)$ is orthogonal to each $X_j(z)$, the $2n$ vectors $N(z), X_1(z), \ldots, X_{2n-1}(z)$ are linearly independent. Since Vol^{2n} is a volume form on $\mathbb{R}^{2n}$ we must thus have

$$\mathrm{Vol}^{2n}(N(z), X_1(z), \ldots, X_{2n-1}(z)) \neq 0$$

at every z, which proves our claim.

In the notation and terminology of intrinsic differential geometry the volume form just constructed is the restriction to Σ_E of interior products of the Liouville form Vol^{2n} by the vector field N; it is thus the contraction (or interior product) of Vol^{2n} with N:

$$dV_E^{2n-1} = i_{\partial_z H/|\partial_z H|}\,\text{Vol}^{2n}\Big|_{\Sigma_E}.$$

Here is an elementary example. Let $H = \frac{1}{2}(p^2 + x^2)$ be the harmonic oscillator Hamiltonian function. The energy shells are the circles $S^1(\sqrt{2E})$. Formula (1.26) yields

$$dV_E^1(z)(X, P) = -\begin{vmatrix} \frac{x}{2E} & X \\ \frac{p}{2E} & P \end{vmatrix} = \frac{1}{2E}(pX - xP)$$

hence $dV_E^1 = (p\,dx - x\,dp)/2E$.

A drawback with the standard volume element dV_E^{2n-1} is the following: while the Liouville form Vol^{2n} is invariant under Hamiltonian flows (more generally, under the action of any symplectomorphism), this is not the case of dV_E^{2n-1}. We can however remedy this inconvenience by defining a volume element on Σ_E related in a simple way to dV_E^{2n-1} and which will be invariant under the flow (ϕ_t^H) determined by any Hamiltonian defining that energy shell. One shows that

Proposition 16. *Let Σ_E be a regular energy shell for the Hamiltonian H.*

(i) *The formula*

$$\sigma_E^{2n-1} = \frac{1}{|\partial_z H|}dV_E^{2n-1} \tag{1.27}$$

defines a volume form on Σ_E such that

$$\text{Vol}^{2n} = dH \wedge \sigma_E^{2n-1}. \tag{1.28}$$

(ii) *For every subset M of Σ_E we have*

$$\int_{M_t} \sigma_E^{2n-1} = \int_M \sigma_E^{2n-1} \tag{1.29}$$

where $M_t = \phi_t^H(M)$ is the image of M by the Hamiltonian flow (ϕ_t^H).

Problem 17. (Requires a good knowledge of intrinsic differential calculus.) Show that the form σ_E^{2n-1} is the only volume form ν_E^{2n-1} on Σ_E such that $\text{Vol}^{2n} = dH \wedge \nu_E^{2n-1}$; deduce from this the invariance property (1.29) of σ_E^{2n-1}. [Hint: use the fact that both Vol^{2n} and dH are invariant under the Hamiltonian flow (ϕ_t^H).]

The invariant volume form defines a measure μ_E of sets on the energy shell. If $\mathcal{U} \subset \Sigma_E$, then

$$\mu_E(\mathcal{U}) = \int_{\mathcal{U}} \sigma_E^{2n-1} \equiv \int_{\mathcal{U}} \frac{dV_E^{2n-1}}{|\partial_z H|} \tag{1.30}$$

when defined.

We are next going to prove a very interesting property relating measure of the energy shell to the volume of its interior. That result is sometimes called the *Cavalieri principle*. We begin by making the following remark: let H be a Hamiltonian and Σ_E a regular and compact energy shell. We claim that there exists $\varepsilon_0 > 0$ and a family of diffeomorphisms $(\varphi_\varepsilon)_{-\varepsilon_0 \leq \varepsilon \leq \varepsilon_0}$ of phase space such that φ_0 is the identity, and

$$z \in \Sigma_E \Longrightarrow H(\varphi_{\Delta E}(z)) = E + \Delta E. \tag{1.31}$$

Since Σ_E is a regular energy shell, the gradient $\partial_z H$ does not vanish on Σ_E; by continuity we thus have $\partial_z H(z) \neq 0$ in a whole neighborhood $\mathcal{U}$ of Σ_E. Since Σ_E is compact, we actually conclude the existence of $c > 0$ such that $|\partial_z H(z)| \geq c$ in that neighborhood. Set now $X = \partial_z H(z)/|\partial_z H(z)|^2$. This vector field is defined on $\mathcal{U}$; let $(\varphi_\varepsilon)_{-\varepsilon_0 \leq \varepsilon \leq \varepsilon_0}$ be its flow. Let $z \in \Sigma_E$; by the chain rule

$$\frac{d}{d\varepsilon} H(\varphi_\varepsilon(z)) = \partial_z H(\varphi_\varepsilon(z)) \frac{d\varphi_\varepsilon}{ds}(z) = \partial_z H(\varphi_\varepsilon(z)) \frac{\partial_z H(\varphi_\varepsilon(z))}{|\partial_z H(\varphi_\varepsilon(z))|^2}$$

which is equal to 1, hence (1.31) since $H(\varphi_0(z)) = H(z) = E$.

Proposition 18. *Let Σ_E be a regular and compact energy shell and $V(E) = \mathrm{Vol}(M_E)$ the volume of the set bounded by Σ_E. We have*

$$\frac{\partial V(E)}{\partial E} = \int_{\Sigma_E} \frac{dV_E^{2n-1}}{|\partial_z H|} = \sigma_E^{2n-1}(\Sigma_E). \tag{1.32}$$

Proof. For a point $z \in \Sigma_E$ let $z + \Delta z$ be the intersection of the normal through z with $\Sigma_{E+\Delta E}$; the length of the line segment $[z, z + \Delta z]$ is $|\Delta z|$. The difference $\Delta V = V(E + \Delta E) - V(E)$ is the volume of the phase space region bounded by Σ_E and $\Sigma_{E+\Delta E}$; we have

$$d(\Delta V) = |\Delta z| dV_E^{2n-1}. \tag{1.33}$$

With the notation introduced above, $\Delta z = \varphi_{\Delta E}(z) - z$, hence

$$\Delta z = \Delta E \left(\frac{d\varphi_\varepsilon}{d\varepsilon}(z) \right)_{\varepsilon=0} + O((\Delta E)^2)$$

$$= \Delta E \frac{\partial_z H(z)}{|\partial_z H(z)|^2} + O((\Delta E)^2)$$

so that we can rewrite (1.33) as

$$d(\Delta V) = \Delta E \left(\frac{1}{|\partial_z H(z)|} + O((\Delta E)^2) \right) dV_E^{2n-1}$$

hence, integrating over Σ_E and dividing by ΔE:

$$\frac{\Delta V}{\Delta E} = \int_{\Sigma_E} \left(\frac{1}{|\partial_z H(z)|} + O(\Delta E) \right) dV_E^{2n-1}$$

which yields (1.32) letting $\Delta E \to 0$. $\qquad\square$

1.3.3   The problem of "Quantization"

Quantum mechanics has its historical origins in the work of Bohr, Born, Heisenberg, Jordan, Pauli, von Neumann, Schrödinger, Weyl and Wigner in the mid 1920s. Its thrust is that physical phenomena are not continuous phenomena, but instead take place in very small but discrete increments – that is, quanta. Besides its great intrinsic interest as one of the pillars of modern Science, quantum mechanics has triggered interest in new mathematical concepts, one of the most important being the Weyl (also called the Weyl–Wigner–Moyal) formalism. In this chapter we study the basic definitions and properties of Weyl calculus from a modern point of view, where the notions of Heisenberg–Weyl operator and cross-Wigner transform play an essential role.

Already in the early years (1925–1926) of Quantum Mechanics physicists where confronted with the problem of ordering, which consisted of finding an unambiguous procedure for associating to a "classical observable" (in mathematics, we would speak about a *real symbol*) a self-adjoint operator. The oldest quantization procedure was actually suggested by Schrödinger who associated to the Hamiltonian function

$$H(x,p) = \frac{p^2}{2m} + U(x)$$

the partial differential operator

$$\widehat{H} = -\frac{\hbar^2}{2m}\frac{\partial^2}{\partial x^2} + U(x)$$

in the case $n = 1$. Schrödinger's empirical prescription thus consisted in the formal substitution $p \longrightarrow -i\hbar\partial/\partial x$ in the Hamiltonian function. So far, so good. But, asked physicists, what should one then do when confronted with more complicated cases? For instance, what should the operator corresponding to

$$H(x,p) = \frac{1}{2m}(p+x)^2 + U(x)$$

then be? The "obvious" guess,

$$\widehat{H} = -\frac{\hbar^2}{2m}(-i\hbar\partial_x + x)^2 + U(x),$$

is not obvious at all, because if we expand the square in the function H we have infinitely many possible choices for quantizing the product $2px$, because there are infinitely many ways to write that function. For instance, we can write

$$2px = \tau px + (1 - \tau)xp = 2xp$$

for every number τ. Physicists decided to make a King Solomon's Choice: they decided that the "right" choice was $\alpha = 1/2$. This corresponds to the "canonical quantization rules"

$$x \longrightarrow \widehat{X} \;\;,\;\; p \longrightarrow \widehat{P} \;\;,\;\; px \longrightarrow \frac{1}{2}(\widehat{X}\widehat{P} + \widehat{P}\widehat{X}) \tag{1.34}$$

where $\widehat{X}$ is the operator of multiplication by x and $\widehat{P} = -i\hbar\partial/\partial x$ (this prescription is actually the one corresponding to the first "obvious" guess above).

It turns out that the canonical quantization rules above are particularly interesting because they lead to a symplectically covariant theory. It is actually no more than a particular case of the Weyl quantization procedure. It is sometimes objected that the choice of Weyl quantization is in a sense ad hoc. One could as well define "τ-quantization" which corresponds to the more general choice

$$x \longrightarrow \widehat{x} \ , \ p \longrightarrow \widehat{p} \ , \ px \longrightarrow \tau\widehat{x}\widehat{p} + (1-\tau)\widehat{p}\widehat{x} \tag{1.35}$$

which is mathematically interesting by itself. However, there is one reason to prefer this choice, and to think it is the right choice, thus confirming Weyl's insight. It turns out that Weyl quantization not only leads to a symplectically covariant quantization procedure (and pseudodifferential operator calculus), but in addition it is the only possible choice if one insists on symplectic covariance (this fundamental fact will be proven later in this book).

Here is an exercise:

Problem 19. Consider the polynomial

$$(t_1 x_1 + \cdots + t_n x_n + \tau_1 D_{x_1} + \cdots + \tau_n D_{n_1})^N$$

in the variables $t, \tau \in \mathbb{R}^n$ with operator coefficients (x_j is viewed as the multiplication operator by x_j) and write it in the form

$$\sum_{|\alpha+\beta|=N} \frac{N!}{\alpha!\beta!} t^\alpha \tau^\beta A_{\alpha\beta}.$$

Then $A_{\alpha\beta}$ is the operator with the Weyl symbol $x^\alpha \xi^\beta$.

We will actually study the notion of τ-quantization in some detail in Chapter 14 using Shubin's theory of global pseudo-differential operators.

There is another way to see things. Quantization can be viewed as a "deformation" of Hamiltonian mechanics (a little bit in the same way as special relativity is seen as a deformation of Galilean relativity). Deformation quantization is one of the themes of the last part of this book.

Chapter 2

The Symplectic Group

This chapter is a review of the most basic concepts of the theory of the symplectic group, and of related concepts, such as symplectomorphisms or the machinery of generating functions.

We may well be witnessing the advent of a "symplectic revolution" in fundamental Science. In fact, since the late sixties there has been a burst of applications of symplectic techniques to mathematics and physics, and even to engineering or medical sciences (magnetic resonance imaging is a typical example). It seems on the other hand that it may be possible to recast a great deal of mathematics in symplectic terms: there is indeed a process of "symplectization of Science" as pointed out by Gotay and Isenberg [80].

Symplectic geometry differs profoundly from more traditional geometries (such as Euclidean geometry, or its refinement Riemannian geometry) because it appears somewhat counter-intuitive to the uninitiated. In symplectic geometry all vectors are "orthogonal" to themselves because the 'scalar product' is anti-symmetric. As a consequence, the notion of length in a symplectic space does not make sense; but instead the notion of area does. For instance, in the plane $\mathbb{R}^2$, the standard symplectic form is (up to the sign) the determinant function: if $z = (x, p)$, $z' = (x', p')$ are two vectors in $\mathbb{R}^2$, then $\det(z, z') = xp' - x'p$ represents the oriented area of the parallelogram built on the vectors z, z'. In higher dimensions the situation is similar: the symplectic product of two vectors is the sum of the algebraic areas of the parallelograms built on the projections of these vectors on the conjugate planes. Symplectic geometry is thus an 'areal' type of geometry; this quality is actually reflected in recent, deep, theorems which express the fact that this 'two-dimensionality' has quite dramatic consequences for the behavior of Hamiltonian flows, which are much more rigid than was thought before the mid-1980s, when Gromov [87] proved very deep results in symplectic topology. Gromov was eventually awarded (2009) the Abel prize (the equivalent of the Nobel prize for mathematics) for his discoveries.

2.1   Symplectic matrices

Recall that the "standard symplectic matrix" is $J = \begin{pmatrix} 0 & I \\ -I & 0 \end{pmatrix}$ where 0 and I are the $n \times n$ zero and identity matrices. we have $\det J = 1$ and $J^2 = I$, $J^T = J^{-1} = -J$.

2.1.1   Definition of the symplectic group

Definition 20. *The set of all symplectic matrices is denoted by* $\mathrm{Sp}(2n, \mathbb{R})$. *Thus* $S \in \mathrm{Sp}(2n, \mathbb{R})$ if and only if

$$S^T J S = S J S^T = J. \tag{2.1}$$

If S is symplectic then S^{-1} is also symplectic because

$$(S^{-1})^T J S^{-1} = -(S J S^{-1})^T = J$$

since $J^T = J^{-1} = -J$. The product of two symplectic matrices being obviously symplectic as well, symplectic matrices thus form a group; that group is denoted by $\mathrm{Sp}(2n, \mathbb{R})$ and is called the (real) *symplectic group*. The conditions (2.1) are actually redundant. In fact:

$$S \in \mathrm{Sp}(2n, \mathbb{R}) \iff S^T J S = J \iff S J S^T = J \tag{2.2}$$

as you are asked to prove in Exercise 21 below:

Exercise 21. Show that $S \in \mathrm{Sp}(2n, \mathbb{R})$ if and only $S^T \in \mathrm{Sp}(2n, \mathbb{R})$. [Hint: use the fact that $(S^{-1})^T J S^{-1} = J$].

The eigenvalues of a symplectic matrix are of a particular type:

Problem 22. (i) Show that the eigenvalues of a symplectic matrix occur in quadruples $(\lambda, \lambda^{-1}, \bar{\lambda}, \bar{\lambda}^{-1})$. [Hint: show that the characteristic polynomial P of a symplectic matrix is reflexive: $P(\lambda) = \lambda^{2n} P(\lambda^{-1})$.] (ii) Show that the determinant of a symplectic matrix is equal to 1. (iii) Show that the eigenvalues of a symplectic matrix S and those of its inverse S^{-1} are the same.

2.1.2   Symplectic block-matrices

It is often useful for practical purposes to use block-matrix notation and to write

$$S = \begin{pmatrix} A & B \\ C & D \end{pmatrix} \tag{2.3}$$

where the entries A, B, C, D are $n \times n$ matrices. Recalling that

$$J = \begin{pmatrix} 0 & I \\ -I & 0 \end{pmatrix}$$

one verifies by an explicit calculation, using the identities $S^T J S = J = S J S^T$, that this matrix is symplectic if and only the two following sets of equivalent conditions are satisfied:

$$A^T C, \ B^T D \quad \text{are symmetric, and} \quad A^T D - C^T B = I, \tag{2.4}$$

$$A B^T, \ C D^T \quad \text{are symmetric, and} \quad A D^T - B C^T = I. \tag{2.5}$$

Using the second set of conditions it follows that the inverse of a symplectic matrix S written in the form (2.3) is

$$S^{-1} = \begin{pmatrix} D^T & -B^T \\ -C^T & A^T \end{pmatrix}. \tag{2.6}$$

Notice that in the case $n = 1$ the formula above reduces to the familiar

$$S^{-1} = \begin{pmatrix} d & -b \\ -c & a \end{pmatrix}$$

which is true for every 2×2 matrix $S = \begin{pmatrix} a & b \\ c & d \end{pmatrix}$ such that $\det(ad - bc) = 1$.

Exercise 23. Verify in detail the formulas (2.4), (2.5), (2.6) above.

Exercise 24. Show, using the conditions (2.4), (2.5) that S is symplectic if and only if S^T is.

Exercise 25. Show that if $S = \begin{pmatrix} A & B \\ C & D \end{pmatrix}$ is symplectic, then $A A^T + B B^T$ is invertible. [Hint: calculate $(A + iB)(B^T + iA^T)$ and use the fact that $A B^T = B A^T$.]

2.1.3 The affine symplectic group

An interesting extension of $\mathrm{Sp}(2n, \mathbb{R})$ consists of the affine symplectic automorphisms. We denote by $\mathrm{T}(2n, \mathbb{R})$ the group of phase space translations: $T(z_0) \in \mathrm{T}(2n, \mathbb{R})$ is the mapping $z \longmapsto z + z_0$. Clearly $\mathrm{T}(2n, \mathbb{R})$ is isomorphic to $\mathbb{R}^n \oplus \mathbb{R}^n$ equipped with addition.

Definition 26. The affine (or inhomogeneous) symplectic group is the semi-direct product

$$\mathrm{ASp}(2n, \mathbb{R}) = \mathrm{Sp}(2n, \mathbb{R}) \ltimes \mathrm{T}(2n, \mathbb{R}).$$

Formally, the group law of the semi-direct product $\mathrm{ASp}(2n, \mathbb{R})$ is given by

$$(S, z)(S', z') = (S S', z + S z');$$

this is conveniently written in matrix form as

$$\begin{pmatrix} S & z \\ 0_{1 \times 2n} & 1 \end{pmatrix} \begin{pmatrix} S' & z' \\ 0_{1 \times 2n} & 1 \end{pmatrix} = \begin{pmatrix} S S' & S z' + z \\ 0_{1 \times 2n} & 1 \end{pmatrix}. \tag{2.7}$$

One immediately checks that $\mathrm{ASp}(2n,\mathbb{R})$ is identified with the set of all affine transformations F of $\mathbb{R}^n \oplus \mathbb{R}^n$ such that F can be factorized as a product $F = ST(z)$ for some $S \in \mathrm{Sp}(2n,\mathbb{R})$ and $z \in \mathbb{R}^n \oplus \mathbb{R}^n$. Since translations are symplectomorphisms in their own right, it follows that $\mathrm{ASp}(2n,\mathbb{R})$ is the group of all affine symplectomorphisms of the symplectic space $(\mathbb{R}^n \oplus \mathbb{R}^n, \sigma)$. We note the following useful relations:

$$ST(z) = T(Sz)S, \quad T(z)S = ST(S^{-1}z).$$

2.2 Symplectic forms

We have defined the symplectic group in terms of matrices. It turns out that $\mathrm{Sp}(2n,\mathbb{R})$ can be defined intrinsically in terms of a general algebraic notion, that of symplectic form:

2.2.1 The notion of symplectic form

We begin with a general definition:

Definition 27. A bilinear form on $\mathbb{R}^n \oplus \mathbb{R}^n$ (or, more generally, on any even-dimensional real vector space) is called a "symplectic form" if it is antisymmetric and non-degenerate. The special antisymmetric bilinear form σ on $\mathbb{R}^n \oplus \mathbb{R}^n$ defined by

$$\sigma(z,z') = p \cdot x' - p' \cdot x \tag{2.8}$$

for $z = (x,p)$, $z' = (x',p')$ is symplectic; it is called the "standard symplectic form on $\mathbb{R}^n \oplus \mathbb{R}^{n}$".

The antisymmetry condition means that we have

$$\sigma(z,z') = -\sigma(z',z)$$

for all z,z' in $\mathbb{R}^{2n}$. Notice that the antisymmetry implies in particular that all vectors z are isotropic, that is:

$$\sigma(z,z) = 0.$$

The non-degeneracy condition means that the condition $\sigma(z,z') = 0$ for all $z \in \mathbb{R}^{2n}$ is equivalent to $z = 0$.

Definition (2.8) of the standard symplectic form can be rewritten in a convenient way using the symplectic standard matrix

$$J = \begin{pmatrix} 0 & I \\ -I & 0 \end{pmatrix}$$

where 0 and I are the $n \times n$ zero and identity matrices. In fact

$$\sigma(z,z') = Jz \cdot z' = (z')^T J z. \tag{2.9}$$

Exercise 28. Show that the standard symplectic form is indeed non-degenerate.

Let s be a linear mapping $\mathbb{R}^n \oplus \mathbb{R}^n \longrightarrow \mathbb{R}^n \oplus \mathbb{R}^n$. The condition $\sigma(sz, sz') = \sigma(z, z')$ is equivalent to $S^T J S = J$ where S is the matrix of s in the canonical basis of $\mathbb{R}^n \oplus \mathbb{R}^n$ that is, to $S \in \mathrm{Sp}(2n, \mathbb{R})$. We can thus redefine the symplectic group by saying that it is the group of all linear automorphisms of $\mathbb{R}^n \oplus \mathbb{R}^n$ which preserve the standard symplectic form σ.

There are other more "exotic" symplectic forms which originate from physical problems (for instance from quantum gravity); here is one example that will be studied further when we discuss non-commutative quantum mechanics at this end of this book: set

$$\Omega = \begin{pmatrix} \hbar^{-1}\Theta & I \\ -I & \hbar^{-1}N \end{pmatrix}$$

where Θ and N are $n \times n$ real antisymmetric matrices, and I the $n \times n$ identity. One usually requires that Θ and N depend on $\hbar$ and that $\Theta = O(\hbar^2)$, $N = O(\hbar^2)$. From this viewpoint Ω can be viewed as perturbation of J: we have $\Omega = J + O(\hbar^2)$. One shows that if $\hbar$ is small enough then Ω is invertible. Since Ω is antisymmetric the formula

$$\omega(z, z') = z \cdot \Omega^{-1} z' = (\Omega^T)^{-1} z \cdot z'$$

defines a new symplectic form on $\mathbb{R}^n \oplus \mathbb{R}^n$ (see Dias and Prata [31]). Note that ω coincides with the standard symplectic form σ when $\Theta = N = 0$.

2.2.2 Differential formulation

There is another, slightly more abstract, way to define the standard symplectic form which has advantages if one has Hamiltonian mechanics on manifolds in mind. It consists in observing that we can view σ as an exterior two-form on $\mathbb{R}^n \oplus \mathbb{R}^n$, in fact:

$$\sigma = dp \wedge dx = \sum_{j=1}^{n} dp_j \wedge dx_j \tag{2.10}$$

where $dp_j \wedge dx_j$ is the wedge product of the coordinate one-forms dp_j and dx_j. This formula is a straightforward consequence of the relation

$$dp_j \wedge dx_j(x, p; x', p') = p_j x'_j - p'_j x_j.$$

With this identification the standard symplectic form is related to the Lebesgue volume form Vol on $\mathbb{R}^n \oplus \mathbb{R}^n$ by the formula

$$\mathrm{Vol} = (-1)^{n(n-1)/2} \frac{1}{n!} \underbrace{\sigma \wedge \sigma \wedge \cdots \wedge \sigma}_{n \text{ factors}}. \tag{2.11}$$

Using this approach one can express very concisely that a diffeomorphism f of $\mathbb{R}^n \oplus \mathbb{R}^n$ is a symplectomorphism:

$$f \in \mathrm{Symp}(2n, \mathbb{R}) \Longleftrightarrow f^*\sigma = \sigma$$

where $f^*\sigma$ is the pull-back of the two-form σ by the diffeomorphism f:

$$f^*\sigma(z_0)(z, z') = \sigma(f(z_0))Df(z_0)z, Df(z_0)z').$$

($Df(z_0)$ the Jacobian matrix at z_0.)

In particular one immediately sees that a symplectomorphism is volume-preserving since we then also have $f^* \,\mathrm{Vol} = \mathrm{Vol}$ in view of (2.11).

The language of differential form allows an elegant (and concise) reformulation of the previous definitions. For instance, part (i) of Theorem (9) can thus be re-expressed as

$$(\phi_t^H)^*\sigma = \sigma.$$

It tuns out that Hamilton's equations can be rewritten in a very neat and concise way using the notion of contraction of a differential form. They are in fact equivalent to the concise relation

$$\iota_{X_H}\sigma + d_z H = 0 \tag{2.12}$$

between the contraction of the symplectic form with the Hamilton field and the differential of the Hamiltonian; this is easily verified by writing this formula "in coordinates", in which case it becomes

$$\sigma(X_H(z,t), \cdot) + d_z H = 0. \tag{2.13}$$

Formula(2.12) is usually taken as the starting point of Hamiltonian mechanics on symplectic manifolds, which is a topic of great current interest.

It is quite easy to reconstruct a Hamiltonian function from its Hamilton vector field; in fact:

$$H(z,t) = H(0,t) - \int_0^1 \sigma(X_H(sz), z)ds. \tag{2.14}$$

This formula is an immediate consequence of the observation that we have, for fixed t,

$$\begin{aligned}
H(z,t) - H(0,t) &= \int_0^1 \frac{d}{ds}H(sz,t)ds \\
&= \int_0^1 \partial_z H(sz,t) \cdot z\, ds \\
&= -\int_0^1 \sigma(X_H(sz), z)ds.
\end{aligned}$$

Notice that formula (2.14) defines H up to the addition of a smooth function of t.

2.3 The unitary groups $U(n, \mathbb{C})$ and $U(2n, \mathbb{R})$

Let $U(n, \mathbb{C})$ denote the complex unitary group: $u \in U(n, \mathbb{C})$ if and only if $u \in \mathcal{M}(n, \mathbb{C})$ (the algebra of complex matrices of dimension n) and $u^*u = uu^* = I$ (the conditions $u^*u = I$ and $uu^* = I$ are actually equivalent).

2.3.1 A useful monomorphism

Writing the elements $Z \in \mathcal{M}(n, \mathbb{C})$ in the form $Z = A + iB$ where A and B are real matrices we define a mapping

$$\iota : \mathcal{M}(n, \mathbb{C}) \longrightarrow \mathcal{M}(2n, \mathbb{R})$$

by the formula:

$$\iota(A + iB) = \begin{pmatrix} A & -B \\ B & B \end{pmatrix}. \tag{2.15}$$

Lemma 29. *The mapping ι is an algebra monomorphism: ι is injective and $\iota(Z + Z') = \iota(Z) + \iota(Z')$, $\iota(\lambda Z) = \lambda \iota(Z)$ for $\lambda \in \mathbb{C}$, and $\iota(ZZ') = \iota(Z)\iota(Z')$.*

Proof. It is easy to verify that ι is an algebra homomorphism (we leave the direct calculations to the reader); that ι is injective immediately follows from its definition. $\qquad\square$

We will see below that ι is an isomorphism of the unitary group onto a certain subgroup of the symplectic group.

2.3.2 Symplectic rotations

Let us prove the main result of this section; it identifies $U(n, \mathbb{C})$ with a subgroup of $\mathrm{Sp}(2n, \mathbb{R})$:

Proposition 30. *The restriction of the mapping*

$$\iota : \mathcal{M}(n, \mathbb{C}) \longrightarrow \mathcal{M}(2n, \mathbb{R}) \tag{2.16}$$

defined above is an isomorphism of $U(n, \mathbb{C})$ onto a subgroup $U(2n, \mathbb{R})$ of $\mathrm{Sp}(2n, \mathbb{R})$.

Proof. It follows from conditions (2.4), (2.5) for the entries of a symplectic matrix that the block matrix

$$U = \begin{pmatrix} A & -B \\ B & A \end{pmatrix} \tag{2.17}$$

is in $U(2n, \mathbb{R})$ if and only if

$$AB^T = B^T A, \quad AA^T + BB^T = I, \tag{2.18}$$

or, equivalently

$$A^T B = BA^T, \quad A^T A + B^T B = I. \tag{2.19}$$

The equivalence of conditions (2.18) and (2.19) is proved by noting that $U \in U(2n, \mathbb{R})$ if and only if $U^T \in U(2n, \mathbb{R})$ which follows from the fact that the monomorphism (2.16) satisfies $\iota(u^*) = \iota(u)^T$ and that the unitary group is invariant under the operation of taking adjoints. $\qquad\square$

Exercise 31. Show that $u \in U(2n, \mathbb{R})$ if and only if $UJ = JU$ and that

$$U(2n, \mathbb{R}) = \text{Sp}(2n, \mathbb{R}) \cap O(2n, \mathbb{R}). \tag{2.20}$$

The identity above shows that $U(2n, \mathbb{R})$ (which is a copy of the unitary group) consists of *symplectic rotations*. It contains the group $O(n)$ of all symplectic matrices of the type

$$\begin{pmatrix} A & 0 \\ 0 & A \end{pmatrix} \quad \text{with } AA^T = A^T A = I.$$

It is immediately verified that $O(n)$ is the image in $U(2n, \mathbb{R})$ of the orthogonal group $O(n, \mathbb{R})$ by the monomorphism ι.

2.3.3 Diagonalization and polar decomposition

A positive-definite matrix can always be diagonalized using an orthogonal matrix. When this matrix is in addition symplectic we can use a symplectic rotation to perform this diagonalization:

Proposition 32. *Let* $S \in \text{Sp}(2n, \mathbb{R})$ *be positive definite (in particular* $S = S^T$*). There exists* $U \in U(2n, \mathbb{R})$ *such that* $S = U^T DU$ *where*

$$D = \text{diag}(\lambda_1, \ldots, \lambda_n; \lambda_1^{-1}, \ldots, \lambda_n^{-1})$$

where $\lambda_1, \ldots, \lambda_n$ *are the* n *smallest eigenvalues of* S.

Proof. The eigenvalues of a symplectic matrix occur in quadruples: if λ is an eigenvalue, then so are λ^{-1}, $\bar{\lambda}$, and $\bar{\lambda}^{-1}$ (Exercise 22). If $S > 0$ these eigenvalues occur in real pairs (λ, λ^{-1}) with $\lambda > 0$ and we can thus order them as follows:

$$\lambda_1 \leq \cdots \leq \lambda_n \leq \lambda_n^{-1} \leq \cdots \leq \lambda_1^{-1}.$$

Let now U be an orthogonal matrix such that $S = U^T DU$. We are going to show that $U \in U(2n, \mathbb{R})$. It suffices for this to show that we can write U in the form (2.17) with A and B satisfying (2.18). Let $e_1, \ldots, e_n$ be n orthonormal eigenvectors of U corresponding to the eigenvalues $\lambda_1, \ldots, \lambda_n$. Since $SJ = JS^{-1}$ (S is both symplectic and symmetric) we have, for $1 \leq k \leq n$,

$$SJe_k = JS^{-1}e_k = \frac{1}{\lambda_j} Je_k$$

hence $\pm Je_1,\ldots,\pm Je_n$ are the orthonormal eigenvectors of U corresponding to the remaining n eigenvalues $1/\lambda_1,\ldots,1/\lambda_n$. Write now the $2n\times n$ matrix $(e_1,\ldots,e_n)$ as

$$(e_1,\ldots,e_n) = \begin{pmatrix} A \\ B \end{pmatrix}$$

where A and B are $n \times n$ matrices; we have

$$(-Je_1,\ldots,-Je_n) = -J\begin{pmatrix} A \\ B \end{pmatrix} = \begin{pmatrix} -B \\ A \end{pmatrix}$$

hence U is indeed of the type

$$U = (e_1,\ldots,e_n; -Je_1,\ldots,-Je_n = \begin{pmatrix} A & -B \\ B & A \end{pmatrix}.$$

The conditions (2.18) are satisfied since $U^T U = I$. $\qquad\square$

The following consequence of the result above shows that one can take powers of symplectic matrices, and that these powers still are symplectic. In fact:

Corollary 33. *Let S be a positive definite symplectic matrix. Then:*

  (i) *For every $\alpha \in \mathbb{R}$ there exists a unique $R \in \mathrm{Sp}(2n,\mathbb{R})$, $R > 0$, $R = R^T$, such that $S = R^\alpha$. In particular $S^{1/2} \in \mathrm{Sp}(2n,\mathbb{R})$.*

  (ii) *Conversely, if $R \in \mathrm{Sp}(2n,\mathbb{R})$ is positive definite, then $R^\alpha \in \mathrm{Sp}(2n,\mathbb{R})$ for every $\alpha \in \mathbb{R}$.*

Proof of (i). Set $R = U^T D^{1/\alpha} U$; then $R^\alpha = U^T D U = S$.

Proof of (ii). It suffices to note that we have

$$R^\alpha = (U^T D U)^\alpha = U^T D^\alpha U \in \mathrm{Sp}(2n,\mathbb{R}). \qquad\square$$

This result allows us to prove a polar decomposition result for the symplectic group. We denote by $\mathrm{Sym}_+(2n,\mathbb{R})$ the set of all symmetric positive definite real $2n \times 2n$ matrices.

Proposition 34. *For every $S \in \mathrm{Sp}(2n,\mathbb{R})$ there exists a unique $U \in U(2n,\mathbb{R})$ and a unique $R \in \mathrm{Sp}(2n,\mathbb{R}) \cap \mathrm{Sym}_+(2n,\mathbb{R})$, such that $S = RU$ (resp. $S = UR$).*

Proof. The matrix $R = S^T S$ is symplectic and positive definite. Set $U = (S^T S)^{-1/2} S$; since $(S^T S)^{-1/2} \in \mathrm{Sp}(2n,\mathbb{R})$ in view of Corollary 33, we have $U \in \mathrm{Sp}(2n,\mathbb{R})$. On the other hand

$$UU^T = (S^T S)^{-1/2} S S^T (S^T S)^{-1/2} = I$$

so that we actually have

$$U \in \mathrm{Sp}(2n,\mathbb{R}) \cap O(2n,\mathbb{R}) = U(2n,\mathbb{R})$$

(cf. Exercise 31). That we can alternatively write $S = UR$ (with different choices of U and R) follows by applying the result above to S^T. The uniqueness statement follows from the generic uniqueness of polar decompositions. $\square$

We will see in Chapter 11, Subsection 11.3 that Proposition 34 can be refined by giving explicit formulas for the matrices R and U ("pre-Iwasawa factorization").

Exercise 35. Use the result above to prove that every $S \in \mathrm{Sp}(2n, \mathbb{R})$ has determinant 1.

One very important consequence of the results above is the connectedness of the symplectic group:

Corollary 36. *The symplectic group* $\mathrm{Sp}(2n, \mathbb{R})$ *is a connected Lie group.*

Proof. Let us set $\mathrm{Sp}_+(2n, \mathbb{R}) = \mathrm{Sp}(2n, \mathbb{R}) \cap \mathrm{Sym}_+(2n, \mathbb{R})$. In view of Proposition 34 above the mapping

$$f : \mathrm{Sp}(2n, \mathbb{R}) \longrightarrow \mathrm{Sp}_+(2n, \mathbb{R}) \times U(2n, \mathbb{R})$$

defined by $f(S) = RU$ is a bijection; both f and its inverse f^{-1} are continuous, hence f is a homeomorphism. Now $U(2n, \mathbb{R})$ is connected, and so is $\mathrm{Sp}_+(2n, \mathbb{R})$. It follows that $\mathrm{Sp}(2n, \mathbb{R})$ is also connected. $\square$

Exercise 37. Check that $\mathrm{Sp}_+(2n, \mathbb{R})$ is connected (use for instance Corollary 33).

2.4   Symplectic bases and Lagrangian planes

Symplectic bases in phase space are in a sense the analogues of orthonormal bases in Euclidean geometry.

2.4.1   Definition of a symplectic basis

Let δ_{ij} be the Kronecker index: $\delta_{ij} = 1$ if $i = j$ and $\delta_{ij} = 0$ if $i \neq j$.

Definition 38. A set $\mathcal{B}$ of vectors

$$\mathcal{B} = \{e_1, \ldots, e_n\} \cup \{f_1, \ldots, f_n\}$$

of $\mathbb{R}^n \oplus \mathbb{R}^n$ is called a *"symplectic basis"* of $(\mathbb{R}^n \oplus \mathbb{R}^n, \sigma)$ if we have

$$\sigma(e_i, e_j) = \sigma(f_i, f_j) = 0, \quad \sigma(f_i, e_j) = \delta_{ij} \quad \text{for} \ \ 1 \leq i, j \leq n. \tag{2.21}$$

Exercise 39. Check that a symplectic basis is a basis in the usual sense.

An obvious example of a symplectic basis is the following: choose

$$e_i = (c_i, 0), \ \ e_i = (0, c_i)$$

where (c_i) is the canonical basis of $\mathbb{R}^n$. (For instance, if $n = 1$, $e_1 = (1, 0)$ and $f_1 = (0, 1)$.) These vectors form the canonical symplectic basis

$$\mathcal{C} = \{e_1, \ldots, e_n\} \cup \{f_1, \ldots, f_n\}$$

of $(\mathbb{R}^n \oplus \mathbb{R}^n, \sigma)$.

A very useful result is the following; it is a symplectic variant of the Gram–Schmidt orthonormalization procedure in Euclidean geometry. It also shows that there are (infinitely many) non-trivial symplectic bases:

Proposition 40. *Let A and B be two (possibly empty) subsets of $\{1, \ldots, n\}$. For any two subsets $\mathcal{E} = \{e_i : i \in A\}$, $\mathcal{F} = \{f_j : j \in B\}$ of the symplectic space $(\mathbb{R}^n \oplus \mathbb{R}^n, \sigma)$ such that the elements of $\mathcal{E}$ and $\mathcal{F}$ satisfy the relations*

$$\omega(e_i, e_j) = \omega(f_i, f_j) = 0 \ , \ \omega(f_i, e_j) = \delta_{ij} \ for \ (i, j) \in A \times B, \tag{2.22}$$

there exists a symplectic basis $\mathcal{B}$ of $(\mathbb{R}^n \oplus \mathbb{R}^n, \sigma)$ containing these vectors.

For a proof we refer to de Gosson [67], §1.2.2.

Symplectic automorphisms take symplectic bases to symplectic bases: this is obvious from the definition. In fact, the symplectic group acts transitively on the set of all symplectic bases:

Exercise 41. Show that for any two symplectic bases $\mathcal{B}$ and $\mathcal{B}'$ there exists $S \in \mathrm{Sp}(2n, \mathbb{R})$ such that $\mathcal{B} = S(\mathcal{B}')$.

2.4.2 The Lagrangian Grassmannian

The group $\mathrm{Sp}(2n, \mathbb{R})$ not only acts on points of phase space $\mathbb{R}^n \oplus \mathbb{R}^n$ but also on subspaces of $\mathbb{R}^n \oplus \mathbb{R}^n$. Among these of particular interest are "Lagrangian planes":

Definition 42. A Lagrangian plane of the symplectic space $(\mathbb{R}^n \oplus \mathbb{R}^n, \sigma)$ is an n-dimensional linear subspace ℓ of $\mathbb{R}^n \oplus \mathbb{R}^n$ having the following property: if $(z, z') \in \ell \times \ell$ then $\sigma(z, z') = 0$. The set of all Lagrangian planes in $(\mathbb{R}^n \oplus \mathbb{R}^n, \sigma)$ is denoted by $\mathrm{Lag}(2n, \mathbb{R})$; it is called the Lagrangian Grassmannian of $(\mathbb{R}^n \oplus \mathbb{R}^n, \sigma)$.

Both "coordinate planes" $\ell_X = \mathbb{R}^n \times \{0\}$ and $\ell_P = \{0\} \times \mathbb{R}^n$ are Lagrangian, and so is the diagonal $\Delta = \{(x, x) : x \in \mathbb{R}^n\}$ of $\mathbb{R}^n \oplus \mathbb{R}^n$. If ℓ is a Lagrangian plane, so is $S\ell$ for every $S \in \mathrm{Sp}(2n, \mathbb{R})$: first ℓ and $S\ell$ have the same dimension n, and if $z_1 = Sz$ and $z_1' = Sz'$ are in $S\ell$ with z and z' in ℓ, then $\sigma(z_1, z_1') = \sigma(z, z') = 0$. In fact, we have the following much more precise result:

Proposition 43. *The group action*

$$\mathrm{Sp}(2n, \mathbb{R}) \times \mathrm{Lag}(2n, \mathbb{R}) \longrightarrow \mathrm{Lag}(2n, \mathbb{R})$$

defined by $(S, \ell) \longmapsto S\ell$ *is transitive. That is, for every pair* $(\ell, \ell') \in \mathrm{Lag}(2n, \mathbb{R}) \times \mathrm{Lag}(2n, \mathbb{R})$ *there exists* $S \in \mathrm{Sp}(2n, \mathbb{R})$ *such that* $\ell = S\ell'$.

Proof. Choose bases $\{e_1, \ldots, e_n\}$ and $\{e_1', \ldots, e_n'\}$ of ℓ and ℓ' respectively. Since ℓ and ℓ' are Lagrangian planes we have $\sigma(e_i, e_j) = \sigma(e_i', e_j') = 0$ so in view of Proposition 40 we can find vectors $f_1, \ldots, f_n$ and $f_1', \ldots, f_n'$ such that

$$\mathcal{B} = \{e_1, \ldots, e_n\} \cup \{f_1, \ldots, f_n\},$$
$$\mathcal{B}' = \{e_1', \ldots, e_n'\} \cup \{f_1', \ldots, f_n'\}$$

are symplectic bases of $(\mathbb{R}^n \oplus \mathbb{R}^n, \sigma)$. Defining $S \in \mathrm{Sp}(2n, \mathbb{R})$ by the condition $\mathcal{B} = S(\mathcal{B}')$ (see Exercise 41); we have $\ell = S\ell'$. $\qquad\square$

Exercise 44. Show that the result above is still true if one replaces $\mathrm{Sp}(2n, \mathbb{R})$ by the unitary group $U(2n, \mathbb{R})$.

Here you are supposed to prove the following refinement of Proposition 43:

Problem 45. Two Lagrangian planes ℓ and ℓ' are said to be transversal if $\ell \cap \ell' = 0$; equivalently $\ell \oplus \ell' = \mathbb{R}^n \oplus \mathbb{R}^n$. Prove that $\mathrm{Sp}(2n, \mathbb{R})$ acts transitively on the set of all transversal Lagrangian planes (hint: use Proposition 40). Does the property remain true if we replace $\mathrm{Sp}(2n, \mathbb{R})$ by $U(2n, \mathbb{R})$?

The Lagrangian Grassmannian has a natural topology which makes it into a compact and connected topological space.

Proposition 46. *The Lagrangian Grassmannian* $\mathrm{Lag}(2n, \mathbb{R})$ *is homeomorphic to the coset space* $U(2n, \mathbb{R})/O(n)$ *where* $O(n)$ *is the image of* $O(n, \mathbb{R})$ *by the restriction of the embedding* $U(n, C) \longrightarrow U(2n, \mathbb{R})$. *Hence* $\mathrm{Lag}(2n, \mathbb{R})$ *is both compact and connected.*

Proof. $U(2n, \mathbb{R})$ acts transitively on $\mathrm{Lag}(2n, \mathbb{R})$ (Exercise 44); the isotropy subgroup of $\ell_P = \{0\} \times \mathbb{R}^n$ is precisely $O(n)$. It follows that $\mathrm{Lag}(2n, \mathbb{R})$ is homeomorphic to $U(2n, \mathbb{R})/O(n)$. Since $U(2n, \mathbb{R})/O(n)$ is trivially homeomorphic to $U(n, \mathbb{C})/O(n, \mathbb{R})$, and the projection $U(n, \mathbb{C}) \longrightarrow U(n, \mathbb{C})/O(n, \mathbb{R})$ is continuous, $\mathrm{Lag}(2n, \mathbb{R})$ is compact and connected because $U(n, \mathbb{C})$ has these properties. $\qquad\square$

Chapter 3

Free Symplectic Matrices

Free symplectic matrices are in a sense the building blocks of the symplectic group. Not only do they form a system of generators of $\mathrm{Sp}(2n, \mathbb{R})$, but they can be described by so-called "generating functions", well known in Hamiltonian mechanics. Free symplectic matrices and their generating functions will play a crucial role in the definition of the metaplectic group in Chapter 7. A related interesting reading is the older paper of Burdet et al. [24].

3.1 Generating functions

Let us begin by giving a few equivalent definitions of the notion of free symplectic matrix.

3.1.1 Definition of a free symplectic matrix

Let us begin by giving a general definition.

Definition 47. Let $S \in \mathrm{Sp}(2n, \mathbb{R})$. We say that S is "free" if it satisfies any of the three following equivalent conditions:

(i) For a given pair $(p, p') \in \mathbb{R}^{2n}$ there exists a unique pair $(x, x') \in \mathbb{R}^n \times \mathbb{R}^n$ such that $(x, p) = S(x', p')$;

(ii) If $S = \begin{pmatrix} A & B \\ C & D \end{pmatrix}$ then $\det B \neq 0$;

(iii) Setting $(x, p) = S(x', p')$ we have

$$\det \left(\frac{\partial x}{\partial p'}(z_0) \right) \neq 0. \tag{3.1}$$

Exercise 48. Show that all three conditions (i), (ii), and (iii) in the definition above indeed are equivalent.

Here is a very useful geometric characterization of free symplectic matrices. Suppose that $S \in \mathrm{Sp}(2n, \mathbb{R})$ is free and set $(x, p) = S(x', p')$ as above. Identifying σ with the differential 2-form

$$dp \wedge dx = \sum_{j=1}^{n} dp_j \wedge dx_j,$$

we have $dp \wedge dx = dp' \wedge dx'$ and this is equivalent, by Poincaré's lemma, to the existence of a function $G \in C^\infty(\mathbb{R}^n \oplus \mathbb{R}^n)$ such that

$$pdx = p'dx' + dG(x', p').$$

The condition $\det(\partial x / \partial p') \neq 0$ implies, by the implicit function theorem, that we can locally solve the equation $x = x(x', p')$ in p', so that $p' = p'(x, x')$ and hence $G(x', p')$ is, for $(x', p') \in \mathcal{U}$, a function of x, x' only: $G(x', p') = G(x', p'(x, x'))$. Calling this function W:

$$W(x, x') = G(x', p'(x, x'))$$

we thus have

$$pdx = p'dx' + dW(x, x')$$
$$= p'dx' + \partial_x W(x, x')dx + \partial_{x'} W(x, x')dx'$$

which requires $p = \partial_x W(x, x')$ and $p' = -\partial_{x'} W(x, x')$ and f is hence free in $\mathcal{U}$. We will see in a moment that this function W, which is uniquely defined up to an additive constant, plays a very important role under the name of "generating function" of the free symplectic automorphism (or matrix) S.

Let us give another, purely geometric, definition of the notion of free symplectic matrix. This property will be used when we prove our main factorization result below.

Proposition 49. *A matrix $S \in \mathrm{Sp}(2n, \mathbb{R})$ is free if and only if we have $S\ell_P \cap \ell_P = \{0\}$ where $\ell_P = \{0\} \times \mathbb{R}^n$.*

Proof. The set $S\ell_P \cap \ell_P$ is described by the equations $(Bp, Dp) = (0, p)$. It reduces to $\{0\}$ if and only if the solution of these equations is $p = 0$, which is equivalent to $\det B \neq 0$. $\qquad\square$

The interest of this characterization is that it allows us to define a more general notion of free symplectic matrix: one says that $S \in \mathrm{Sp}(2n, \mathbb{R})$ is free with respect to a Lagrangian plane $\ell \in \mathrm{Lag}(2n, \mathbb{R})$ if we have $S\ell \cap \ell = \{0\}$; see de Gosson [67], §2.2.3 for a study of this notion.

3.1.2   The notion of generating function

Free symplectic matrices are "generated" by quadratic forms in (x, x'):

Proposition 50. (i) *Let $S \in \mathrm{Sp}(2n, \mathbb{R})$ be a free symplectic matrix. Then*

$$(x, p) = S(x', p') \iff \begin{cases} p = \partial_x W(x, x'), \\ p' = -\partial_{x'} W(x, x'), \end{cases} \tag{3.2}$$

where W is the quadratic form given by

$$W(x, x') = \tfrac{1}{2} DB^{-1} x^2 - B^{-1} x \cdot x' + \tfrac{1}{2} B^{-1} A x'^2 \tag{3.3}$$

and DB^{-1} and $B^{-1}A$ are symmetric matrices.
   (ii) *If conversely*

$$W(x, x') = \tfrac{1}{2} P x^2 - L x \cdot x' + \tfrac{1}{2} Q x'^2 \tag{3.4}$$

with $P = P^T$, $Q = Q^T$, and $\det L \neq 0$, then the matrix

$$S_W = \begin{pmatrix} L^{-1} Q & L^{-1} \\ P L^{-1} Q - L^T & P L^{-1} \end{pmatrix} \tag{3.5}$$

is a free symplectic matrix whose generating function in the sense above is (3.4).

Proof of (i). The matrices DB^{-1} and $B^{-1}A$ are symmetric in view of (2.4). We have

$$\partial_x W(x, x') = DB^{-1} x - (B^{-1})^T x',$$
$$\partial_{x'} W(x, x') = -B^{-1} x' + B^{-1} A x';$$

setting $p = \partial_x W(x, x')$ and $p' = -\partial_{x'} W(x, x')$ and solving in x and p we get $x = Ax' + Bp'$, $p = Cx' + Dp'$, that is $(x, p) = S(x', p')$.

Proof of (ii). To see this, it suffices to remark that we have $(x, p) = S(x', p')$ if and only if $p = Px - L^T x'$ and $p' = Lx - Qx'$, and to solve the equations $p = Px - L^T x'$ and $p' = Lx - Qx'$ in x, p. $\square$

Notation 51. If the free symplectic matrix S has generating function W we will write $S = S_W$.

Corollary 52. *Let $S_W \in \mathrm{Sp}(2n, \mathbb{R})$ be a free symplectic matrix. Then $(S_W)^{-1}$ is also free, and we have*

$$S_W^{-1} = S_{W^*} \quad with \quad W^*(x, x') = -W(x', x). \tag{3.6}$$

Proof. The inverse of S_W is the symplectic matrix

$$S_W^{-1} = \begin{pmatrix} D^T & -B^T \\ -C^T & A^T \end{pmatrix}$$

which is thus free since $\det(-B^T) = (-1)^n \det B$. In view of part (i) in Proposition 50 the inverse S_W^{-1} is generated by the function

$$\begin{aligned} W^*(x, x') &= -\tfrac{1}{2} A^T (B^T)^{-1} x^2 + (B^T)^{-1} x \cdot x' - \tfrac{1}{2}(B^T)^{-1} D^T x'^2 \\ &= -\tfrac{1}{2} B^{-1} A x^2 + B^{-1} x' \cdot x - \tfrac{1}{2} D B^{-1} x'^2 \\ &= -W(x', x) \end{aligned}$$

(recall that $A^T (B^T)^{-1} = B^{-1} A$ and $(B^T)^{-1} D^T = D B^{-1}$). $\square$

The statement in the following exercise implies that almost every symplectic matrix is free:

Exercise 53. Show that the set $\mathrm{Sp}_0(2n, \mathbb{R})$ of all free symplectic matrices has codimension 1 in $\mathrm{Sp}(2n, \mathbb{R})$. [Hint: there is a bijective correspondence between the set of all triples (P, L, Q) (P and Q symmetric, $\det L \neq 0$) and $\mathrm{Sp}_0(2n, \mathbb{R})$.]

The notion of generating function also makes sense for affine symplectic mappings:

Proposition 54. *Let* $F = T(z_0) S_W \in \mathrm{ASp}(2n, \mathbb{R})$.

(i) *A free generating function of* $f = T(z_0) S_W$ *is the function*

$$W_{z_0}(x, x') = W(x - x_0, x') + p_0 \cdot x \tag{3.7}$$

where $z_0 = (x_0, p_0)$.

(ii) *Conversely, if* W *is the generating function of* S_W *then any polynomial*

$$W_{z_0}(x, x') = W(x, x') + \alpha \cdot x + \alpha' \cdot x' \tag{3.8}$$

with $\alpha, \alpha' \in \mathbb{R}^n$ *is a generating function of an affine symplectic transformation* $T(z_0) S_W$ *with* $z_0 = (x_0, p_0) = (B\alpha, D\alpha + \beta)$.

Proof. Let W_{z_0} be defined by (3.7), and set $(x', p') = S(x'', p'')$, $(x, p) = T(z_0)(x', p')$. We have

$$\begin{aligned} pdx - p'dx' &= (pdx - p''dx'') + (p''dx'' - p'dx') \\ &= (pdx - (p - p_0)d(x - x_0) + dW(x'', x') \\ &= d(p_0 \cdot x + W(x - x_0, x')) \end{aligned}$$

which shows that W_{z_0} is a generating function. Finally, formula (3.8) is obtained by a direct computation, expanding the quadratic form $W(x - x_0, x')$ in its variables. $\square$

Corollary 55. *Let $f = [S_W, z_0]$ be a free affine symplectic transformation, and set $(x, p) = f(x', p')$. The function Φ_{z_0} defined by*

$$\Phi_{z_0}(x, x') = \tfrac{1}{2} p \cdot x - \tfrac{1}{2} p' \cdot x' + \tfrac{1}{2} \sigma(z, z_0) \tag{3.9}$$

is also a free generating function for f; in fact,

$$\Phi_{z_0}(x, x') = W_{z_0}(x, x') + \tfrac{1}{2} p_0 \cdot x_0. \tag{3.10}$$

Proof. Setting $(x'', p'') = S(x, p)$, the generating function W satisfies

$$W(x'', x') = \tfrac{1}{2} p'' \cdot x'' - \tfrac{1}{2} p' \cdot x'$$

in view of Euler's formula for homogeneous functions. Let Φ_{z_0} be defined by formula (3.9); in view of (3.7) we have

$$W_{z_0}(x, x') - \Phi_{z_0}(x, x') = \tfrac{1}{2} p_0 \cdot x - \tfrac{1}{2} p \cdot x_0 - \tfrac{1}{2} p_0 \cdot x_0$$

which is (3.10); this proves the corollary since all generating functions of a symplectic transformation are equal up to an additive constant. $\square$

3.1.3 Application to the Hamilton–Jacobi equation

The notion of generating function also makes sense for general symplectomorphisms; it has an interesting application to the Hamilton–Jacobi equation briefly discussed in the first chapter.

Definition 56. A symplectomorphism ϕ of $\mathbb{R}^n \oplus \mathbb{R}^n$ is said to be free in a neighborhood $\mathcal{U}$ of $z_0 \in \mathbb{R}^n \oplus \mathbb{R}^n$ when its Jacobian matrix $D\phi(z')$ is a free symplectic matrix for each $z' \in \mathcal{U}$, that is, if and only if $\det(\partial x / \partial p') \neq 0$.

Let H be a Hamiltonian function; we will use the notation H_{pp}, H_{xp}, and H_{xx} for the matrices of second derivatives of H in the corresponding variables; for instance

$$H_{xp} = \left(\frac{\partial^2 H}{\partial x_j \partial p_k} \right)_{1 \le j, k \le n} = H_{px}^T.$$

Let (ϕ_t^H) be the associated flow; we assume it is defined for every t in some interval $[-T, T]$, $T > 0$.

Proposition 57. *There exists $\varepsilon > 0$ such that the symplectomorphism ϕ_t^H is free at $z_0 \in \mathbb{R}^n \oplus \mathbb{R}^n$ for $0 < |t| \le \varepsilon$ if and only if $\det H_{pp}(z_0) \neq 0$.*

Proof. Let $t \longmapsto z(t) = (x(t), p(t))$ be the solution to Hamilton's equations $\dot{z} = J \partial_z H(z)$ with initial condition $z(t_0) = z_0$. A second-order Taylor expansion in t of the function $z(t)$ yields

$$z(t) = z_0 + t X_H(z_0) + O(t^2);$$

and hence, in particular

$$x(t) = x_0 + t\partial_p H(z_0) + O(t^2).$$

It follows that the Jacobian matrix of $x(t)$ with respect to the p variables is

$$\frac{\partial x(t)}{\partial p} = tH_{pp}(z_0) + O(t^2)$$

hence there exists $\varepsilon > 0$ such that $\partial x(t)/\partial p$ is invertible in the interval $[-\varepsilon, 0[\cap]0, \varepsilon]$ if and only if $H_{pp}(z_0)$ is invertible; this is equivalent to saying that ϕ_t^H is free at the point z_0. $\qquad\square$

Exercise 58. Justify the last sentence of the proof above!

The result above applies when the Hamiltonian H is of the "physical type"

$$H(z,t) = \sum_{j=1}^{n} \frac{1}{2m_j} p_j^2 + U(x)$$

since $H_{pp}(z_0)$ is in this case the diagonal matrix whose diagonal elements are the numbers $1/2m_j$, $1 \le j \le n$. In this case ϕ_t^H is free for small non-zero t near each z_0 where it is defined. More generally it also applies to all Hamiltonians of the type

$$H(z,t) = \frac{1}{2}A(x)p^2 + U(x)$$

where $A(x) > 0$ (i.e., positive-definite).

Here is an application of the result above; it shows that the generating function provides us with a way of solving explicitly the Hamilton–Jacobi equation:

Proposition 59. *Suppose again that* $\det H_{pp}(z) \neq 0$ *for all* z. *Then, the Hamilton–Jacobi equation*

$$\frac{\partial \Phi}{\partial t} + H\left(x, \partial_x \Phi, t\right) = 0 \ , \ \Phi(x,0) = \Phi_0(x) \qquad (3.11)$$

has a solution Φ *for* $0 < |t| < \varepsilon$ *given by*

$$\Phi(x,t) = \Phi_0(x') + W(x, x'; t) \qquad (3.12)$$

where x' *is defined by the condition*

$$(x, p) = \phi_t^H\left(x', \partial_x \Phi_0(x')\right) \qquad (3.13)$$

and W *is the generating function*

$$W(x, x'; t) = \int pdx - Hdt.$$

Proof. We assume that $n = 1$ for notational simplicity (the generalization to arbitrary dimension is straightforward replacing partial derivatives by gradients, etc.). Formula (3.13) uniquely defines x' for small values of t: writing $x = (x', \partial_x \Phi_0(x'), t)$ we have, by the chain rule,

$$\frac{dx}{dx'} = \frac{\partial x}{\partial x'} + \frac{\partial x}{\partial p'} \frac{\partial^2 \Phi_0}{\partial x'^2}.$$

The limit for $t \to 0$ of the Jacobian matrix $D_t^H(z')$ being the identity it follows that dx/dx' is different from zero in some interval $[-\alpha, \alpha]$, $\alpha > 0$, and hence the mapping $x' \longmapsto \phi_t^H(x', \partial_x \Phi_0(x'))$ is a local diffeomorphism for each fixed $t \in [-\alpha, \alpha]$. Obviously $\lim_{t \to 0} \Phi(x, t) = \Phi_0(x)$ since $x' \to x$ as $t \to 0$, so that the Cauchy condition is satisfied. To prove that Φ is a solution of Hamilton–Jacobi's equation one then notes that

$$\Phi(x + \Delta x, t + \Delta t) - \Phi(x, t) = \int_L p \, dx - H \, dt$$

where L is the line segment joining (x, p, t) to $(x + \Delta x, p + \Delta p, t + \Delta t)$; the values p and $p + \Delta p$ are determined by the relations $p = \partial_x W(x, x'; t)$ and

$$p + \Delta p = \partial_x W(x + \Delta x, x' + \Delta x'; t + \Delta t)$$

where we have set $\Delta x' = x'(x + \Delta x) - x'(x)$. Thus,

$$\Phi(x + \Delta x, t + \Delta t) - \Phi(x, t) = p\Delta x + \frac{1}{2}\Delta p \Delta x - \Delta t \int_0^1 H(x + s\Delta x, p + s\Delta p) ds$$

and hence

$$\frac{\Phi(x, t + \Delta t) - \Phi(x, t)}{\Delta t} = -\int_0^1 H(x, p + s\Delta p) ds.$$

Taking the limit $\Delta t \to 0$ and noting that $\Delta p \to 0$ we get

$$\frac{\partial \Phi}{\partial t}(x, t) = -H(x, p). \tag{3.14}$$

Similarly,

$$\Phi(x + \Delta x, t) - \Phi(x, t) = p\Delta x + \frac{1}{2}\Delta p \Delta x$$

and $\Delta p \to 0$ as $\Delta x \to 0$ so that

$$\frac{\partial \Phi}{\partial x}(x, t) = p. \tag{3.15}$$

Combining both relations (3.14) and (3.15) we see that Φ satisfies Hamilton–Jacobi's equation. $\qquad\square$

3.2　A factorization result

There are many factorization results for symplectic matrices. Here we will only be concerned with factorizations using free symplectic matrices (see de Gosson [67] for more results).

3.2.1　Statement and proof

In our context the main interest of the notion of free symplectic matrix comes from the following factorization result which says that every symplectic matrix can be written as the product of exactly two free symplectic matrices. Our proof makes use of the transitivity of the action of $\mathrm{Sp}(2n, \mathbb{R})$ on the Lagrangian Grassmannian $\mathrm{Lag}(2n, \mathbb{R})$.

Theorem 60. *For every* $S \in \mathrm{Sp}(2n, \mathbb{R})$ *there exist two free symplectic matrices* S_W *and* $S_{W'}$ *such that* $S = S_W S_{W'}$.

Proof. The symplectic group $\mathrm{Sp}(2n, \mathbb{R})$ acts transitively not only on the Lagrangian Grassmannian $\mathrm{Lag}(2n, \mathbb{R})$ but also on the subset of $\mathrm{Lag}(2n, \mathbb{R}) \times \mathrm{Lag}(2n, \mathbb{R})$ consisting of all pairs (ℓ, ℓ') such that $\ell \cap \ell' = \{0\}$ (see Problem 45). Let ℓ_P be the Lagrangian plane $\{0\} \times \mathbb{R}^n$ and, for given ℓ, choose ℓ' transversal to both ℓ_P and $S\ell$:

$$\ell' \cap \ell_P = \ell' \cap S\ell = \{0\}.$$

In view of the pair transitivity property there exists $S_1 \in \mathrm{Sp}(2n, \mathbb{R})$ such that $S_1(\ell_P, \ell') = (\ell', S\ell_P)$, that is $S_1\ell_P = \ell'$ and $S\ell_P = S_1\ell'$. Since $\mathrm{Sp}(2n, \mathbb{R})$ acts transitively on $\mathrm{Lag}(2n, \mathbb{R})$ we can find S_2' such that $\ell' = S_2'\ell_P$ and hence $S\ell_P = S_1 S_2'\ell_P$. It follows that there exists $S'' \in \mathrm{Sp}(2n, \mathbb{R})$ such that $S''\ell_P = \ell_P$ and $S = S_1 S_2' S''$. Set $S_2 = S_2' S''$; then $S = S_1 S_2$ and we have

$$S_1\ell_P \cap \ell_P = \ell' \cap \ell_P = 0,$$
$$S_2\ell_P \cap \ell_0 = S_2'\ell_P \cap \ell_P = \ell' \cap \ell_P = 0.$$

Hence S_1 and S_2 are free in view of Proposition 49; our claim follows since $S = S_1 S_2$. $\square$

The choice of S_W and $S_{W'}$ in the factorization $S = S_W S_{W'}$ is of course not unique; for instance the identity I can be written as $I = S_W S_{W*}$ for *every* generating function W!

Problem 61. Modify the proof of Theorem 60 to show that, more generally, for every $(S, \ell_0) \in \mathrm{Sp}(2n, \mathbb{R}) \times \mathrm{Lag}(2n, \mathbb{R})$ there exist two symplectic matrices S_1, S_2 such that $S = S_1 S_2$ and $S_1\ell_0 \cap \ell_0 = S_2\ell_0 \cap \ell_0 = 0$.

3.2.2 Application: generators of $\mathbf{Sp}(2n, \mathbb{R})$

If P and L are, respectively, a symmetric and an invertible $n \times n$ matrix, and L an invertible matrix, we set

$$V_P = \begin{pmatrix} I & 0 \\ -P & I \end{pmatrix} , \ U_P = \begin{pmatrix} -P & I \\ -I & 0 \end{pmatrix} , \ M_L = \begin{pmatrix} L^{-1} & 0 \\ 0 & L^T \end{pmatrix} . \tag{3.16}$$

Proposition 62. *Every free symplectic matrix* $S = M_L = \begin{pmatrix} A & B \\ C & D \end{pmatrix}$ *can be factored as*

$$S = V_{-DB^{-1}} M_{B^{-1}} U_{-B^{-1}A} \tag{3.17}$$

and

$$S = V_{-DB^{-1}} M_{B^{-1}} J V_{-B^{-1}A} . \tag{3.18}$$

Proof. We begin by noting that we can write

$$\begin{pmatrix} A & B \\ C & D \end{pmatrix} = \begin{pmatrix} I & 0 \\ DB^{-1} & I \end{pmatrix} \begin{pmatrix} B & 0 \\ 0 & DB^{-1}A - C \end{pmatrix} \begin{pmatrix} B^{-1}A & I \\ -I & 0 \end{pmatrix} \tag{3.19}$$

whether S is symplectic or not. If now S is symplectic, then the middle factor in the right-hand side of (3.19) also is symplectic, since the first and the third factors obviously are. Taking the condition $AD^T - BC^T = I$ in (2.5) into account, we have $DB^{-1}A - C = (B^T)^{-1}$ and hence

$$\begin{pmatrix} B & 0 \\ 0 & DB^{-1}A - C \end{pmatrix} = \begin{pmatrix} B & 0 \\ 0 & (B^T)^{-1} \end{pmatrix}$$

so that

$$S = \begin{pmatrix} I & 0 \\ DB^{-1} & I \end{pmatrix} \begin{pmatrix} B & 0 \\ 0 & (B^T)^{-1} \end{pmatrix} \begin{pmatrix} B^{-1}A & I \\ -I & 0 \end{pmatrix} . \tag{3.20}$$

The factorization (3.17) follows (both DB^{-1} and $B^{-1}A$ are symmetric, as a consequence of the relations $B^T D = D^T B$ and $B^T A = A^T B$ in (2.4)). Noting that

$$\begin{pmatrix} B^{-1}A & I \\ -I & 0 \end{pmatrix} = \begin{pmatrix} 0 & I \\ -I & 0 \end{pmatrix} \begin{pmatrix} I & 0 \\ B^{-1}A & I \end{pmatrix}$$

the factorization (3.18) follows as well. $\qquad\square$

Conversely, if a matrix S can be written in the form $V_{-P} M_L J V_{-Q}$, then it is a free symplectic matrix; in fact,

$$S = S_W = \begin{pmatrix} L^{-1}Q & L^{-1} \\ PL^{-1}Q - L^T & PL^{-1} \end{pmatrix} \tag{3.21}$$

as is checked by a straightforward calculation.

A consequence of these results is the following:

Corollary 63. *Each of the sets*

$$\{V_P, M_L, J : P = P^T, \ \det L \neq 0\}$$

and

$$\{U_P, M_L : P = P^T, \ \det L \neq 0\}$$

generates $\mathrm{Sp}(2n, \mathbb{R})$.

Proof. Every $S \in \mathrm{Sp}(2n, \mathbb{R})$ is the product of two free symplectic matrices. It now suffices to apply Proposition 62 above.   □

Chapter 4

The Group of Hamiltonian Symplectomorphisms

Symplectic diffeomorphisms, or symplectomorphisms as they are often called, are the "canonical transformations" which have been known and used by physicists for a long time. They generalize the linear (and affine) symplectic mappings we have been using so far. A basic reference for this chapter is Polterovich [133].

4.1 The group $\mathrm{Symp}(2n, \mathbb{R})$

The notion of symplectic matrix or automorphism can be generalized to the non-linear case, and leads to the notion of symplectomorphism.

4.1.1 Definition and examples

Recall that a diffeomorphism of $\mathbb{R}^n \oplus \mathbb{R}^n$ is an invertible mapping $\phi : \mathbb{R}^n \oplus \mathbb{R}^n \longrightarrow \mathbb{R}^n \oplus \mathbb{R}^n$ (or $\Omega \longrightarrow \Omega'$ where Ω and Ω' are open subsets of $\mathbb{R}^n \oplus \mathbb{R}^n$) such that both ϕ and its inverse ϕ^{-1} are infinitely differentiable.

Definition 64. Let Ω be an open subset of $\mathbb{R}^n \oplus \mathbb{R}^n$. Let $\mathbb{R}^n \oplus \mathbb{R}^n$ be equipped with the symplectic form σ. A diffeomorphism $\phi : \Omega \longrightarrow \phi(\Omega) \subset \mathbb{R}^n \oplus \mathbb{R}^n$ is called a symplectomorphism (or symplectic diffeomorphism) if its Jacobian matrix $D\phi(z)$ is symplectic at every point $z \in \Omega$: $D\phi(z) \in \mathrm{Sp}(2n, \mathbb{R})$, that is

$$D\phi(z)^T J D\phi(z) = D\phi(z) J D\phi(z)^T = J.$$

In differential notation a diffeomorphism ϕ is a symplectomorphism if and only if $\phi^* \sigma = \sigma$:

$$\phi^* \left(\sum_{j=1}^n dp_j \wedge dx_j \right) = \sum_{j=1}^n dp_j \wedge dx_j$$

where ϕ^* denotes the pull-back by ϕ.

A typical (but rather trivial) example is the following: let $f : \mathbb{R}^n \longrightarrow \mathbb{R}^n$ be a diffeomorphism; then the formula

$$\phi(z) = (f(x), Df(x)^{-1}p)$$

defines a symplectomorphism $\phi : \mathbb{R}^n \oplus \mathbb{R}^n \longrightarrow \mathbb{R}^n \oplus \mathbb{R}^n$.

Exercise 65. What is the symplectomoprphism ϕ above when f is linear?

Exercise 66. Show that the mapping $\phi : (r, \alpha) \longmapsto (x, p)$ where $x = \sqrt{2r}\cos\alpha$ and $p = \sqrt{2r}\sin\alpha$ is a symplectomorphism of some subset Ω of $\mathbb{R}^n \oplus \mathbb{R}^n$ onto its image. (The variables $x = \sqrt{2r}\cos\alpha$ and $p = \sqrt{2r}\sin\alpha$ are called "symplectic polar coordinates"; the reader is invited to verify that the usual change to polar variables $x = r\cos\alpha$, $p = r\sin\alpha$ is not a symplectomorphism.)

If ϕ and ψ are symplectomorphisms defined on $\mathbb{R}^n \oplus \mathbb{R}^n$ then $\phi\psi = \phi \circ \psi$ is also a symplectomorphism: in view of the chain rule the Jacobian matrix of $\phi\psi$ at a point z is namely

$$D(\phi\psi)(z) = D\phi(\psi(z))D\psi(z)$$

and is hence a product of symplectic matrices. Using the formula $D(\phi^{-1}) = (D\phi)^{-1}$ for the Jacobian of the inverse of a diffeomorphism, one sees also that the inverse of a symplectomorphism is also a symplectomorphism. Thus, $\text{Symp}(2n, \mathbb{R})$ is a group for the usual composition law $\phi\psi = \phi \circ \psi$.

Definition 67. The set of all symplectomorphisms defined on $\mathbb{R}^n \oplus \mathbb{R}^n$ and equipped with the natural composition law is denoted by $\text{Symp}(2n, \mathbb{R})$ and called the group of symplectomorphisms of the symplectic space $(\mathbb{R}^n \oplus \mathbb{R}^n, \sigma)$.

Clearly the following inclusions hold:

$$\text{Sp}(2n, \mathbb{R}) \subset \text{ISp}(2n, \mathbb{R}) \subset \text{Symp}(2n, \mathbb{R}).$$

The group $\text{Symp}(2n, \mathbb{R})$ is equipped with a topology by specifying the convergent sequences: let $(\phi_j)_{j \in \mathbb{N}}$ be a sequence of symplectomorphisms of $\mathbb{R}^n \oplus \mathbb{R}^n$; we will say that

$$\lim_{j \to \infty} \phi_j = \phi \ \text{in} \ \text{Symp}(2n, \mathbb{R})$$

if and only if for every compact set $\mathcal{K}$ in $\mathbb{R}^n \oplus \mathbb{R}^n$ the sequences $(\phi_{j|\mathcal{K}})$ and $(D(\phi_{j|\mathcal{K}}))$ converge uniformly towards $\phi_{|\mathcal{K}}$ and $D(\phi_{|\mathcal{K}})$, respectively.

Symplectomorphisms preserve phase-space volume: this is an immediate consequence of the fact that the Jacobian matrix of a symplectomorphism is symplectic and thus has determinant equal to 1. On the differential level this can be seen as follows: the volume form dz in $\mathbb{R}^n \oplus \mathbb{R}^n$ is proportional to $\sigma^{\wedge n} = \sigma \wedge \sigma \wedge \cdots \wedge \sigma$ (n factors) and

$$\phi^*\sigma^{\wedge n} = \phi^*\sigma \wedge \phi^*\sigma \wedge \cdots \wedge \phi^*\sigma = \sigma^{\wedge n}.$$

Of course, more generally, we have $\phi^*\sigma^{\wedge k} = \sigma^{\wedge k}$ for $1 \leq k \leq n$.

We will study in Section 4.2 the connected component of $\text{Symp}(2n, \mathbb{R})$; it is the group of *Hamiltonian symplectomorphisms*.

4.2 Hamiltonian symplectomorphisms

The notion of Hamiltonian symplectomorphism appears naturally when one studies Hamiltonian flows. Recall from Chapter 1 that the flow (ϕ_t^H) determined by Hamilton's equations

$$\frac{dx_j}{dt} = \frac{\partial H}{\partial p_j}(x,p,t) \; , \; \frac{dp_j}{dt} = -\frac{\partial H}{\partial x_j}(x,p,t)$$

consists of Hamiltonian symplectomorphisms since the Jacobian matrix $D\phi_t^H(z)$ is symplectic at every point z of $\mathbb{R}^n \oplus \mathbb{R}^n$ (Theorem 9(ii)).

4.2.1 Symplectic covariance of Hamiltonian flows

Hamilton's equations are covariant (i.e., they retain their form) under canonical transformations. Let us begin by proving the following general result about vector fields. We will write $\varphi\psi$ for $\varphi \circ \psi$ when φ and ψ are diffeomorphisms.

Lemma 68. *Let (φ_t^X) be the flow of some vector field X on $\mathbb{R}^m$. Let φ be a diffeomorphism $\mathbb{R}^m \longrightarrow \mathbb{R}^m$. The family (φ_t^Y) of diffeomorphisms defined by*

$$\varphi_t^Y = \varphi^{-1}\varphi_t^X\varphi \tag{4.1}$$

is the flow of the vector field $Y = (D\varphi)^{-1}(X \circ \varphi)$, that is:

$$Y(u) = D(\varphi^{-1})(\varphi(u))X(\varphi(u)) = [D\varphi(u)]^{-1}X(\varphi(u)). \tag{4.2}$$

Proof. Obviously φ_0^Y is the identity; in view of the chain rule

$$\frac{d}{dt}\varphi_t^Y(x) = D(\varphi^{-1})(\varphi_t^X(\varphi(x)))X(\varphi_t^X(\varphi(x))$$

$$= (D\varphi)^{-1}(\varphi_t^Y(x))X(\varphi(\varphi_t^Y(x)))$$

hence

$$\frac{d}{dt}\varphi_t^Y(x) = Y(\varphi_t^Y(x))$$

which we set out to prove. $\qquad\square$

Let us apply this lemma to the Hamiltonian case. We define the push-forward of a vector field X by a diffeomorphism ϕ by the formula

$$\phi^*X = (D\phi)^{-1}(X \circ \phi)$$

well known from elementary differential geometry.

Proposition 69. *Let ϕ be a symplectomorphism of $(\mathbb{R}^n \oplus \mathbb{R}^n, \sigma)$.*

(i) *We have*
$$X_{H \circ \phi}(z) = [D\phi(z)]^{-1}(X_H \circ \phi)(z), \tag{4.3}$$

that is
$$\phi^* X_H = X_{H \circ \phi}.$$

(ii) *The flows (ϕ_t^H) and $(\phi_t^{H \circ \phi})$ are conjugate by ϕ:*
$$\phi_t^{H \circ \phi} = \phi^{-1} \phi_t^H \phi. \tag{4.4}$$

Proof. Let us prove (i); part (ii) will follow in view of Lemma 68 above. Set $K = H \circ \phi$. By the chain rule

$$\partial_z K(z) = [D\phi(z)]^T (\partial_z H)(\phi(z))$$

hence the vector field $X_K = J \partial_z K$ is given by

$$X_K(z) = J[D\phi(z)]^T \partial_z H(\phi(z)).$$

Since $D\phi(z)$ is symplectic we have

$$J[D\phi(z)]^T = [D\phi(z)]^{-1} J$$

and hence
$$X_K(z) = [D\phi(z)]^{-1} J \partial_z H(\phi(z))$$

which is (4.3). $\square$

4.2.2   The group $\mathrm{Ham}(2n, \mathbb{R})$

Let us now define the notion of Hamiltonian symplectomorphism.

In what follows Hamilton functions are generically time-dependent.

Definition 70. We will say that a diffeomorphism ϕ of $(\mathbb{R}^n \oplus \mathbb{R}^n, \sigma)$ is a *Hamiltonian symplectomorphism* (or diffeomorphism) if there exists a real function $H \in C^\infty((\mathbb{R}^n \oplus \mathbb{R}^n) \times \mathbb{R})$ such that $\phi = \phi_1^H$ where (ϕ_t^H) is the flow generated by H. The set of all Hamiltonian symplectomorphisms is denoted by $\mathrm{Ham}(2n, \mathbb{R})$.

Choosing H constant it is clear that the identity is a Hamiltonian symplectomorphism.

The choice of a time-one map ϕ_1^H in the definition above is of course arbitrary, and can be replaced by any ϕ_a^H: if $\phi = \phi_a^H$ for some $a \neq 0$ then we also have $\phi = \phi_1^{H^a}$ where $H^a(z, t) = aH(z, at)$. In fact, setting $t^a = at$ we have

$$\frac{dz^a}{dt} = J \partial_z H^a(z^a, t) \iff \frac{dz^a}{dt^a} = J \partial_z H(z^a, t^a)$$

and hence $\phi_t^{H^a} = \phi_{at}^H$.

We are going to see in a moment that $\mathrm{Ham}(2n, \mathbb{R})$ is a connected and normal subgroup of $\mathrm{Symp}(2n, \mathbb{R})$ for the induced composition law, thus justifying the following definition:

Definition 71. The set $\mathrm{Ham}(2n, \mathbb{R})$ of all Hamiltonian symplectomorphisms equipped with the law $\phi\psi = \phi \circ \psi$ is called the group of Hamiltonian symplectomorphisms of the standard symplectic space $(\mathbb{R}^n \oplus \mathbb{R}^n, \sigma)$. The topology of $\mathrm{Ham}(2n, \mathbb{R})$ is the topology induced by that of the group $\mathrm{Symp}(2n, \mathbb{R})$ of symplectomorphisms of $(\mathbb{R}^n \oplus \mathbb{R}^n, \sigma)$.

We assume in what follows that all Hamiltonian flows are defined on $\mathbb{R}^n \oplus \mathbb{R}^n$ for all values of time. This may seem a strong condition, especially after our discussion in Chapter 1, Subsection 1.1.2: many interesting Hamiltonian functions do not generate flows that exist forever. This difficulty can be suppressed by the following trick: (see, e.g., Polterovich [133]). If ϕ_t^H is not defined for all values of t, we just replace H by the function $H\Theta$ where $\Theta \in C_0^\infty(\mathbb{R}^n \oplus \mathbb{R}^n)$ is a compactly supported infinitely differentiable function equal to 1 on some arbitrarily chosen subset Ω of phase space. Thus, $H(z,t)$ is equal to 0 for z lying outside $\mathrm{Supp}(\Theta) \subset \Omega$. The classical theory of differential systems tells us that the solutions of Hamilton's equations then exist for all times, and hence the flow $(\phi_t^{H\Theta})$ is defined for all t. The diffeomorphisms $\phi_t^{H\Theta}$ are the identity outside the support of Θ. Suppose in fact that the initial point z_0 lies outside $\mathrm{Supp}(\Theta)$, so that $H(z_0)$ is constant. The function $z(t) = z_0$ is a solution of Hamilton's equations because $\dot{z}_0 = 0$ and $\partial_p H(z_0, t) = \partial_x H(z_0, t) = 0$. Since we always assume uniqueness, $z(t) = z_0$ is *the* solution, and thus $\phi_t^H(z_0) = z_0$. Moreover, any solution curve starting at time $t = 0$ from a point z_0 inside $\mathrm{Supp}(\Theta)$ will stay forever inside $\mathrm{Supp}(\Theta)$ (otherwise the curve would stop at a point outside $\mathrm{Supp}(\Theta)$ in view of the previous argument, and could not leave the exterior of $\mathrm{Supp}(\Theta)$ even under time-reversal, which is a contradiction).

Let us first prove a preparatory result which is interesting in its own right.

Proposition 72. *Let (ϕ_t^H) and (ϕ_t^K) be Hamiltonian flows. Then:*

$$\phi_t^H \phi_t^K = \phi_t^{H\#K} \quad and \quad (\phi_t^H)^{-1} = \phi_t^{\bar{H}} \tag{4.5}$$

where $H\#K$ and $\bar{H}$ are the Hamiltonian functions defined by

$$H\#K(z,t) = H(z,t) + K((\phi_t^H)^{-1}(z), t).$$
$$\bar{H}(z,t) = -H(\phi_t^H(z), t).$$

Proof of the first identity (4.5). By the product and chain rule we have

$$\frac{d}{dt}(\phi_t^H \phi_t^K) = (\frac{d}{dt}\phi_t^H)\phi_t^K + (D\phi_t^H)\phi_t^K \frac{d}{dt}\phi_t^K \tag{4.6}$$
$$= X_H(\phi_t^H \phi_t^K) + (D\phi_t^H)\phi_t^K \circ X_K(\phi_t^K) \tag{4.7}$$

and it thus suffices to show that

$$(D\phi_t^H)\phi_t^K \circ X_K(\phi_t^K) = X_{K \circ (\phi_t^H)^{-1}}(\phi_t^K). \tag{4.8}$$

Writing

$$(D\phi_t^H)\phi_t^K \circ X_K(\phi_t^K) = (D\phi_t^H)((\phi_t^H)^{-1}\phi_t^H \phi_t^K) \circ X_K((\phi_t^H)^{-1}\phi_t^H \phi_t^K)$$

the equality (4.8) follows from the transformation formula (4.3) in Proposition 69.

Proof of the second identity (4.5). It is an easy consequence of the first, noting that $(\phi_t^H \phi_t^{\bar{H}})$ is the flow determined by the Hamiltonian

$$K(z,t) = H(z,t) + \bar{H}((\phi_t^H)^{-1}(z),t) = 0;$$

$\phi_t^H \phi_t^{\bar{H}}$ is thus the identity, so that $(\phi_t^H)^{-1} = \phi_t^{\bar{H}}$ as claimed. $\qquad\square$

Let us now show that $\mathrm{Ham}(2n,\mathbb{R})$ is a group, as claimed. In fact we will prove a little bit more:

Proposition 73. $\mathrm{Ham}(2n,\mathbb{R})$ *is a normal subgroup of the group* $\mathrm{Symp}(2n,\mathbb{R})$ *of all symplectomorphisms of the standard symplectic space* $(\mathbb{R}^n \oplus \mathbb{R}^n, \sigma)$.

Proof. Let us show that if $\phi, \psi \in \mathrm{Ham}(2n,\mathbb{R})$ then $\phi\psi^{-1} \in \mathrm{Ham}(2n,\mathbb{R})$. We have $\phi = \phi_1^H$ and $\psi = \phi_1^K$ for some Hamiltonians H and K. In view of the identities (4.5) we have

$$\phi\psi^{-1} = \phi_1^H(\phi_1^K)^{-1} = \phi_1^{H \# \bar{K}}$$

hence $\phi\psi^{-1} \in \mathrm{Ham}(2n,\mathbb{R})$. That $\mathrm{Ham}(2n,\mathbb{R})$ is a normal subgroup of $\mathrm{Symp}(2n,\mathbb{R})$ immediately follows from formula (4.4) in Proposition 69: if ψ is a symplectomorphism and $\phi \in \mathrm{Ham}(2n,\mathbb{R})$ then

$$\phi_1^{H \circ \psi} = \psi^{-1}\phi_1^H\psi \in \mathrm{Ham}(2n,\mathbb{R}) \tag{4.9}$$

which was to be proven. $\qquad\square$

We are now going to prove a deep and beautiful result due to Banyaga [5]. It essentially says that a path of time-one Hamiltonian symplectomorphisms passing through the identity at time zero is itself Hamiltonian; as a consequence $\mathrm{Ham}(2n,\mathbb{R})$ is a connected group.

Let $t \longmapsto \phi_t$ be a path in $\mathrm{Ham}(2n,\mathbb{R})$, defined for $0 \le t \le 1$ and starting at the identity: $\phi_0 = I$. We will call such a path a *one-parameter family of Hamiltonian symplectomorphisms*. Thus, each ϕ_t is equal to some symplectomorphism $\phi_1^{H_t}$.

Theorem 74. *Let* $t \longmapsto \phi_t$, $0 \le t \le 1$ *be a continuous curve in* $\mathrm{Ham}(2n,\mathbb{R})$ *such that* $\phi_0 = I$. *Then* (ϕ_t) *is the Hamiltonian flow determined by the Hamiltonian function*

$$H(z,t) = -\int_0^1 \sigma(X(uz,t),z)du \ \ with \ \ X = (\tfrac{d}{dt}\phi_t) \circ (\phi_t)^{-1}. \tag{4.10}$$

Proof. The starting point of the argument is the following: one begins by noting that if X_H is a Hamiltonian vector field, one can reconstruct H by the following method: first write

$$H(z,t) = H(z,0) + \int_0^1 \frac{d}{du} X_H(uz,t)\,du$$

$$= H(z,0) + \int_0^1 [\partial_z H(uz,t) \cdot z]\,du$$

(the second equality in view of the chain rule).

Next observe that since $\nabla_z H(uz,t) = -J^2\partial_z H(uz,t) = -JX_H(uz,t)$ we have

$$H(z,t) = H(z,0) - \int_0^1 \sigma(X_H(uz,t), z)\,du$$

where σ is the standard symplectic form. Let us now prove Banyaga's formula (4.10). By definition of X we have $\frac{d}{dt}\phi_t = X\phi_t$ so that all we have to do is to prove that X is a (time-dependent) Hamiltonian field. For this it suffices to show that the contraction $i_X\sigma$ of the symplectic form with X is an exact differential one-form, for then $i_X\sigma = -dH$ where H is given by (4.10). The ϕ_t being symplectomorphisms, they preserve the symplectic form σ and hence $\mathcal{L}_X\sigma = 0$. In view of Cartan's homotopy formula we have

$$\mathcal{L}_X\sigma = i_X d\sigma + d(i_X\sigma) = d(i_X\sigma) = 0$$

so that $i_X\sigma$ is closed; by Poincaré's lemma it is also exact. $\qquad\square$

Corollary 75. *The group* $\mathrm{Ham}(2n,\mathbb{R})$ *of Hamiltonian symplectomorphisms is a connected subgroup of the group* $\mathrm{Symp}(2n,\mathbb{R})$ *of symplectomorphisms of* $(\mathbb{R}^n \oplus \mathbb{R}^n, \sigma)$.

Proof. The connectedness of $\mathrm{Ham}(2n,\mathbb{R})$ follows from Theorem 74: let $t \longmapsto \phi_t$, $0 \le t \le 1$ be a continuous curve in $\mathrm{Ham}(2n,\mathbb{R})$ joining the identity to $\phi \in \mathrm{Ham}(2n,\mathbb{R})$; then (ϕ_t) is a Hamiltonian flow and hence $\phi = \phi_1 \in \mathrm{Ham}(2n,\mathbb{R})$. $\quad\square$

We encourage the reader to pay some attention to the following exercise.

Exercise 76. Show that $\mathrm{Sp}(2n,\mathbb{R}) \subset \mathrm{Ham}(2n,\mathbb{R})$ and that the Hamiltonian function of (ϕ_t^H) such that $\phi_t^H \in \mathrm{Sp}(2n,R)$ is of the type $H(z,t) = \frac{1}{2}M(t)z \cdot z$ for some real symmetric matrix $M(t)$ depending smoothly on t.

4.3 The symplectic Lie algebra

Let $\mathcal{M}(2n,\mathbb{R})$ be the algebra of all real $2n \times 2n$ matrices. The symplectic group $\mathrm{Sp}(2n,\mathbb{R})$ is a closed subgroup of the general linear group $GL(2n,\mathbb{R})$: we have $\mathrm{Sp}(2n,\mathbb{R}) = f^{-1}(0)$ where f is the continuous mapping $\mathcal{M}(2n,\mathbb{R}) \longrightarrow \mathcal{M}(2n,\mathbb{R})$

defined by $f(M) = S^T M S - J$ hence $\mathrm{Sp}(2n,\mathbb{R})$ is closed in $\mathcal{M}(2n,\mathbb{R})$; since $\mathrm{Sp}(2n,\mathbb{R}) \subset GL(2n,\mathbb{R})$ it is also closed in $GL(2n,\mathbb{R})$. It follows that $\mathrm{Sp}(2n,\mathbb{R})$ is a classical Lie group, and it thus makes sense to talk about its Lie algebra.

4.3.1 Matrix characterization of $\mathfrak{sp}(2n,\mathbb{R})$

The main result is the following:

Proposition 77. *The Lie algebra $\mathfrak{sp}(2n,\mathbb{R})$ of $\mathrm{Sp}(2n,\mathbb{R})$ consists of all $X \in \mathcal{M}(2n,\mathbb{R})$ such that*

$$XJ + JX^T = 0 \ (\textit{equivalently } X^T J + JX = 0). \tag{4.11}$$

Proof. Let (S_t) be a differentiable one-parameter subgroup of $\mathrm{Sp}(2n,\mathbb{R})$ and X a $2n \times 2n$ real matrix such that $S_t = \exp(tX)$. Since S_t is symplectic we have $S_t J (S_t)^T = J$, that is

$$\exp(tX) J \exp(tX^T) = J.$$

Differentiating both sides of this equality with respect to t and then setting $t = 0$ we get $XJ + JX^T = 0$, and applying the same argument to the transpose S_t^T we get $X^T J + JX = 0$ as well. Suppose conversely that X is such that $XJ + JX^T = 0$ and let us show that $X \in \mathfrak{sp}(2n,\mathbb{R})$. For this it suffices to prove that $S_t = \exp(tX)$ is in $\mathrm{Sp}(2n,\mathbb{R})$ for every t. The condition $X^T J + JX = 0$ is equivalent to $X^T = JXJ$ hence $S_t^T = \exp(tJXJ)$; since $J^2 = -I$ we have $(JXJ)^k = (-1)^{k+1} JX^k J$ and hence

$$\exp(tJXJ) = -\sum_{k=0}^{\infty} \frac{(-t)^k}{k!} (JXJ)^k = -Je^{-tX}J.$$

It follows that $S_t^T J S_t = (-Je^{-tX}J)Je^{tX} = J$ so that $S_t \in \mathrm{Sp}(2n,\mathbb{R})$ as claimed. $\qquad\square$

Note that if one writes $X \in \mathfrak{sp}(2n,\mathbb{R})$ in block matrix form then it has the form

$$X = \begin{pmatrix} U & V \\ W & -V^T \end{pmatrix}$$

where U, V, and W are $n \times n$ matrices such that

$$V = V^T \ \text{ and } \ W = W^T.$$

In particular $\mathfrak{sp}(2,\mathbb{R})$ consists of all 2×2 matrices with vanishing trace:

$$X \in \mathfrak{sp}(2,\mathbb{R}) \iff \mathrm{Tr}\, X = 0.$$

Exercise 78. Show that the dimension of $\mathrm{Sp}(2n,\mathbb{R})$ as a Lie group is $n(2n+1)$. [Hint: write $X \in \mathfrak{sp}(2n,\mathbb{R})$ in block-matrix form and then use the fact that a Lie group has the same dimension as its Lie algebra.]

Problem 79.

(i) Let $\Delta_{jk} = (\delta_{jk})_{1 \leq j,k \leq n}$ ($\delta_{jk} = 0$ if $j \neq k$, $\delta_{jk} = 1$). Show that the matrices

$$
X_{jk} = \begin{bmatrix} \Delta_{jk} & 0 \\ 0 & -\Delta_{jk} \end{bmatrix}, \; Y_{jk} = \frac{1}{2}\begin{bmatrix} 0 & \Delta_{jk} + \Delta_{kj} \\ 0 & 0 \end{bmatrix},
$$
$$
Z_{jk} = \frac{1}{2}\begin{bmatrix} 0 & 0 \\ \Delta_{jk} + \Delta_{kj} & 0 \end{bmatrix} \quad (1 \leq j \leq k \leq n)
$$

form a basis of $\mathfrak{sp}(n)$.

(ii) Show that every $Z \in \mathfrak{sp}(n)$ can be written in the form $[X, Y] = XY - YX$ with $X, Y \in \mathfrak{sp}(n)$.

4.3.2 The exponential mapping

One should be careful to note that the exponential mapping

$$
\exp : \mathfrak{sp}(2n, \mathbb{R}) \longrightarrow \mathrm{Sp}(2n, \mathbb{R})
$$

is neither surjective nor injective; for instance it is not hard to prove that if $S \in \mathrm{Sp}(2, \mathbb{R})$ can be written in the form $S = \exp X$ with $X \in \mathfrak{sp}(1)$ then we must have $\mathrm{Tr}\, S \geq -2$ (see de Gosson [67], p. 37). However, when conditions of positivity and symmetry are imposed, one has a much better situation. In fact, denoting by $\mathrm{Sym}(2n, \mathbb{R})$ the set of real symmetric $2n \times 2n$ matrices and by $\mathrm{Sym}_+(2n, \mathbb{R})$ the subset of $\mathrm{Sym}(2n, \mathbb{R})$ consisting of positive definite matrices, we have:

Proposition 80.

(i) *We have $S \in \mathrm{Sp}(2n, \mathbb{R}) \cap \mathrm{Sym}_+(2n, \mathbb{R})$ if and only if $S = \exp X$ with $X \in \mathfrak{sp}(n)$ and $X = X^T$.*

(ii) *The exponential mapping is a diffeomorphism*

$$
\exp : \mathfrak{sp}(2n, \mathbb{R}) \cap \mathrm{Sym}(2n, \mathbb{R}) \longrightarrow \mathrm{Sp}(2n, \mathbb{R}) \cap \mathrm{Sym}_+(2n, \mathbb{R}).
$$

Proof. If $X \in \mathfrak{sp}(2n, \mathbb{R})$ and $X = X^T$ then S is both symplectic and symmetric positive definite. Assume conversely that S is symplectic and symmetric positive definite. The exponential mapping is a diffeomorphism

$$
\exp : \mathrm{Sym}(2n, \mathbb{R}) \longrightarrow \mathrm{Sym}_+(2n, \mathbb{R})
$$

hence there exists a unique $X \in \mathrm{Sym}(2n, \mathbb{R})$ such that $S = \exp X$. Let us show that $X \in \mathfrak{sp}(2n, \mathbb{R})$. Since $S = S^T$ we have $SJS = J$ and hence $S = -JS^{-1}J$. Because $-J = J^{-1}$ it follows that

$$
\exp X = J^{-1}(\exp(-X))J = \exp(-J^{-1}XJ)
$$

and, $J^{-1}XJ$ being symmetric, we conclude that $X = J^{-1}XJ$ that is $JX = -XJ$, showing that $X \in \mathfrak{sp}(2n, \mathbb{R})$. $\qquad\square$

We now view the group $\mathrm{Ham}(2n, \mathbb{R})$ of all Hamiltonian symplectomorphisms of $(\mathbb{R}^n \oplus \mathbb{R}^n, \sigma)$ as a Lie subgroup of the group of all diffeomorphisms of $\mathbb{R}^n \oplus \mathbb{R}^n$.

Proposition 81. *The Hamiltonian Lie algebra* $\mathfrak{ham}(2n, \mathbb{R})$ *consists of all Hamiltonian vector fields* X_H *with Lie bracket* $[X_H, X_K]$ *such that*

$$[X_H, X_K] = X_{\{H,K\}} \tag{4.12}$$

where $\{H, K\}$ *is the Poisson bracket:*

$$\{H, K\} = \partial_x H \cdot \partial_p K - \partial_x K \cdot \partial_p H. \tag{4.13}$$

Proof. The Lie algebra $\mathfrak{ham}(2n, \mathbb{R})$ of $\mathrm{Ham}(2n, \mathbb{R})$ is just the tangent space to $\mathrm{Ham}(2n, \mathbb{R})$ at the identity, hence it is the algebra of all vector fields X on $\mathbb{R}^n \oplus \mathbb{R}^n$ such that

$$X(z) = \frac{d}{dt} \phi_t(z) \Big|_{t=0} .$$

In view of Theorem 74 there exists a Hamiltonian function $H \in C^\infty((\mathbb{R}^n \oplus \mathbb{R}^n) \times \mathbb{R})$ such that $\phi_t = \phi_t^H$ hence $X = X_H$. The proof of formula (4.13) readily follows from Proposition 12; we leave the details to the reader as an exercise. $\square$

Chapter 5

Symplectic Capacities

We are going to describe a deep topological principle, Gromov's symplectic non-squeezing theorem [87], alias the "principle of the symplectic camel". As we will see in the next chapter, the main tool allowing the application of Gromov's theorem to the study of classical and quantum uncertainties is the derived notion of *symplectic capacity*, which is a typically "classical" object.

5.1 Gromov's theorem and symplectic capacities

In addition to being volume-preserving, Hamiltonian flows have an unexpected additional property as soon as the number of degrees of freedom is superior to 1; this property is a consequence of the symplectic non-squeezing theorem which was proved in 1985 by M. Gromov [87].

5.1.1 Statement of Gromov's theorem

Gromov's non-squeezing theorem is very surprising and has many indirect consequences. Let us state it precisely. We denote by $Z_j(R)$ the cylinder in $\mathbb{R}^n \oplus \mathbb{R}^n$ defined by the condition: a point (x,p) is in $Z_j(R)$ if and only if $x_j^2 + p_j^2 \leq R^2$.

Theorem 82 (Gromov). *If there exists a symplectomorphism ϕ in $\mathbb{R}^n \oplus \mathbb{R}^n$ sending the ball $B(r)$ in some cylinder $Z_j(R)$, then we must have $r \leq R$.*

It is essential for the non-squeezing theorem to hold that the considered cylinder is based on an x_j, p_j plane (or, more generally, on a symplectic plane). For instance, if we replace the cylinder $Z_j(R)$ by the cylinder $Z_{12}(R) : x_1^2 + x_2^2 \leq R^2$ based on the x_1, x_2 plane, it is immediate to check that the linear symplectomorphism ϕ defined by $\phi(x,p) = (\lambda x, \lambda^{-1} p)$ sends $B(r)$ into $Z_{12}(R)$ as soon as $\lambda \leq r/R$. Also, one can always "squeeze" a large ball into a big cylinder using volume-preserving diffeomorphisms that are not canonical. Here is an example in the case $n = 2$ that is very easy to generalize to higher dimensions: define a linear mapping f by

$$f(x_1, x_2, p_1, p_2) = (\lambda x_1, \lambda^{-1} x_2, \lambda p_1, \lambda^{-1} p_2).$$

Clearly $\det f = 1$ and f is hence volume-preserving; f is however not symplectic if $\lambda \neq 1$. Choosing again $\lambda \leq r/R$, the mapping f sends $B(R)$ into $Z_1(r)$.

Gromov's theorem actually holds when $Z_j(R)$ is replaced by any cylinder with radius R based on a symplectic plane, i.e., a two-dimensional subspace $\mathcal{P}$ of $\mathbb{R}^n \oplus \mathbb{R}^n$ such that the restriction of σ to $\mathcal{P}$ is non-degenerate (equivalently, $\mathcal{P}$ has a basis $\{e, f\}$ such that $\sigma(e, f) \neq 0$). The planes $\mathcal{P}_j$ of coordinates x_j, p_j are of course symplectic, and given an arbitrary symplectic plane $\mathcal{P}$ it is easy to construct a linear symplectomorphism S_j such that $S_j(\mathcal{P}) = \mathcal{P}_j$. It follows that a symplectomorphism ϕ sends $B(r)$ in the cylinder $Z_j(R)$ if and only if $\phi \circ S_j$ sends $B(r)$ in the cylinder $Z_j(R)$ with the same radius based on $\mathcal{P}$.

Gromov's theorem obviously applies to Hamiltonian flows, since these consist of symplectomorphisms. Here is a dynamical description of the non-squeezing theorem. Assume that we are dealing with a subset $\Omega_t \subset \mathbb{R}^n \oplus \mathbb{R}^n$ at time t modelling a large number of points in phase space, subject to a Hamiltonian flow (ϕ_t^H). Suppose that $\Omega_0 = \Omega$ is a phase space ball $B(r) : |z - z_0| \leq r$ at time $t = 0$. The orthogonal projection of that ball on any plane of coordinates is a circle with area πr^2. As time evolves, Ω will distort and may take after a while a very different shape, while keeping constant volume (because Hamiltonian flows are volume preserving). In fact, since conservation of volume has nothing to do with conservation of shape, one might very well envisage that Ω becomes stretched in all directions by the Hamiltonian flow (ϕ_t^H), and eventually gets very thinly spread out over huge regions of phase space, so that the projections on any plane could *a priori* become arbitrarily small after some time t. (In fact, this possibility is perfectly consistent with a deep result of Katok [106] which says that, up to sets of arbitrarily small measure ε, any kind of phase-space spreading is a priori possible for a volume-preserving flow. However, Gromov's theorem implies that the areas of the projections of the set $\phi_t^H(B(r))$ on any plane of *conjugate coordinates* x_j, p_j (or, more generally, on any symplectic plane) will never decrease below its original value πr^2.

5.1.2 Proof of Gromov's theorem in the affine case

All known proofs (direct, or indirect) of Gromov's theorem are notoriously difficult, whatever method one uses (this might explain that it had not been discovered earlier, even in the more "physical" framework of Hamiltonian dynamics). We note that a related heuristic justification of Gromov's theorem is given by Hofer and Zehnder in [101] p. 34; their argument however relies on an assumption which is (if true) at least as difficult to prove as Gromov's theorem itself! We are going to be much more modest, and to give a proof (actually two) of Gromov's theorem for affine symplectomorphisms; a symplectomorphism is affine if it can be factorized as the product of a symplectic transformation (i.e., an element of $\mathrm{Sp}(2n, \mathbb{R})$) and a phase space translation. Both proofs are of an elementary nature (the second is shorter, but slightly more conceptual; also see [67] §3.7.2 for a variant of this proof).

Proposition 83. *If there exists an affine symplectomorphism ϕ in $\mathbb{R}^n \oplus \mathbb{R}^n$ sending a ball $B(r)$ inside the cylinder $Z_j(R)$, then we must have $r \leq R$. Equivalently, the intersection of $\phi(B(r))$ by an affine plane parallel to a plane of conjugate coordinates x_j, p_j passing through the center of $\phi(B(r))$ is an ellipse with area πr^2.*

First proof. It relies on the fact that the form $pdx = \sum_j p_j dx_j$ is a relative integral invariant of every symplectomorphism, that is: if ϕ is a symplectomorphism and γ a cycle (or loop) in $\mathbb{R}^n \oplus \mathbb{R}^n$ then

$$\oint_\gamma pdx = \oint_{\phi(\gamma)} pdx \tag{5.1}$$

(see for instance Arnol'd [3], §44, p. 239). It is of course no restriction to assume that the ball $B(r)$ is centered at the origin, and that ϕ is a symplectic transformation S. We claim that the ellipse $\Gamma_j = S(B(r)) \cap \mathcal{P}_j$, intersection of the ellipsoid $S(B(r))$ with any plane $\mathcal{P}_j$ of conjugate coordinates x_j, p_j has area πr^2; the proposition immediately follows from this property. Let γ_j be the curve bounding the ellipse Γ_j and orient it positively; the area enclosed by γ is then

$$\mathrm{Area}(\Gamma_j) = \oint_{\gamma_j} p_j dx_j = \oint_{\gamma_j} pdx \tag{5.2}$$

hence, using property (5.1),

$$\mathrm{Area}(\Gamma_j) = \oint_{S^{-1}(\gamma_j)} pdx = \pi r^2 \tag{5.3}$$

(because $S^{-1}(\gamma)$ is a big circle of $B(r)$); notice that the assumption that $\mathcal{P}_j$ is a plane of conjugate coordinates x_j, p_j is essential for the second equality (5.2) to hold, making use of formula (5.1) possible [more generally, the argument works when $\mathcal{P}_j$ is replaced by any symplectic plane].

Second proof. With the same notation as above we note that the set

$$S^{-1}[S(B(r)) \cap \mathcal{P}_j]$$

is a big circle of $B(r)$, and hence encloses a surface with area πr^2. Now, $\mathcal{P}_j$ is a symplectic space when equipped with the skew-linear form $\sigma_j = dp_j \wedge dx_j$ and the restriction of S to $\mathcal{P}_j$ is also canonical from $(\mathcal{P}_j, \sigma_j)$ to the symplectic plane $S(\mathcal{P}_j)$ equipped with the restriction of the symplectic form σ. Canonical transformations being volume- (here: area) preserving it follows that $S(B(r)) \cap \mathcal{P}_j$ also has area πr^2. $\qquad\square$

It would certainly be interesting to generalize the first proof to arbitrary symplectomorphisms, thus yielding a new proof of Gromov's theorem in the general case, in fact, a refinement of it! The difficulty comes from the following fact: the key to the proof in the linear case is the fact that we were able to derive the equality

$$\int_{\gamma_R} p_j dx_j = \pi R^2$$

by exploiting the fact that the inverse image of the x_j, p_j plane by S was a plane cutting $B(R)$ along a big circle, which thus encloses an area equal to πR^2. When one replaces the *linear* transformation S by a non-linear one, the inverse image of x_j, p_j plane will not generally be a plane, but rather a surface. It turns out that this surface is not quite arbitrary: it is a symplectic 2-dimensional manifold. If the following property holds:

> *The section of $B(r)$ by any symplectic surface containing the center of $B(r)$ has an area at least πr^2*

then we would have, by the same argument,

$$\int_{\gamma_R} p_j dx_j \geq \pi R^2,$$

hence we would have proved Gromov's theorem in the general case. We do not know any proof of this property; nor do we know whether it is true!

We urge the reader to notice that the assumption that we are cutting $S(B(r))$ with a plane of *conjugate coordinates* is essential, because it is this assumption that allowed us to identify the area of the section with action. Here is a counterexample which shows that the property does not hold for arbitrary sections of the ellipsoid $S(B(r))$. Taking $n = 2$ we define a symplectic matrix

$$S = \begin{pmatrix} \lambda_1 & 0 & 0 & 0 \\ 0 & \lambda_2 & 0 & 0 \\ 0 & 0 & 1/\lambda_1 & 0 \\ 0 & 0 & 0 & 1/\lambda_2 \end{pmatrix},$$

where $\lambda_1 > 0$, $\lambda_2 > 0$, and $\lambda_1 \neq \lambda_2$. The set $S(B(r))$ is defined by

$$\frac{1}{\lambda_1}x_1^2 + \frac{1}{\lambda_2}x_2^2 + \lambda_1 p_1^2 + \lambda_2 p_2^2 \leq r^2$$

and its section with the x_2, p_2 plane is the ellipse

$$\frac{1}{\lambda_1}x_1^2 + \lambda_1 p_1^2 \leq r^2$$

which has area πr^2 as predicted, but its section with the x_2, p_1 plane is the ellipse

$$\frac{1}{\lambda_1}x_1^2 + \lambda_2 p_2^2 \leq R^2$$

which has area $\pi(r^2 \sqrt{\lambda_1/\lambda_2})$ different from πr^2 since $\lambda_1 \neq \lambda_2$.

Why is Gromov's theorem sometimes called "the principle of the symplectic camel"? The reason is metaphoric: Gromov's non-squeezing theorem can be restated by saying that there is no way to deform a phase space ball using symplectomorphisms in such a way that we can make it pass through a circular hole in a plane of conjugate coordinates x_j, p_j if the area of that hole is smaller than that of the cross-section of that ball: the biblical camel is the ball $B(R)$ and the hole in the plane is the eye of the needle!

5.2 The notion of symplectic capacity

Gromov's non-squeezing theorem makes possible the definition of a very interesting and useful topological notion, that of symplectic capacity, which was defined by Ekeland and Hofer [41]. We refer to Hofer and Zehnder [101], Polterovich [133] for much more on the topic.

5.2.1 Definition and existence

The following definition is standard, and the most commonly accepted in the literature:

Definition 84. A normalized symplectic capacity" on $(\mathbb{R}^n \oplus \mathbb{R}^n, \sigma)$ is a function assigning to every subset Ω of $\mathbb{R}^n \oplus \mathbb{R}^n$ a number $c(\Omega) \geq 0$, or $+\infty$, and having the four properties (SC1)–(SC4) listed below:

(SC1) Symplectic invariance:

$$c(\phi(\Omega)) = c(\Omega) \quad \text{if} \quad \phi \in \mathrm{Symp}(2n, R); \tag{5.4}$$

(SC2) Monotonicity:

$$c(\Omega) \leq c(\Omega') \quad \text{if } \Omega \subset \Omega'; \tag{5.5}$$

(SC3) Conformality:

$$c(\lambda\Omega) = \lambda^2 c(\Omega) \quad \text{for } \lambda \in \mathbb{R}; \tag{5.6}$$

(SC4) Normalization:

$$c(B(R)) = \pi R^2 = c(Z_j(R)). \tag{5.7}$$

Exercise 85. (i) Show that if condition (5.7) holds for one index j then it holds for all. (ii) Show that the cylinder $Z_j(R) : x_j^2 + p_j^2 \leq R^2$ can be replaced by any cylinder $Z_j(R)$ with radius R based on a symplectic plane.

We will also often consider the weaker notion of linear symplectic capacity:

Definition 86. A linear symplectic capacity assigns to every subset Ω of $\mathbb{R}^n \oplus \mathbb{R}^n$ a number $c^{\mathrm{lin}}(\Omega) \geq 0$ or $+\infty$, and having the properties (SC2)–(SC4) above, property (SC1) being replaced by:

(**SC1Lin**) *A linear symplectic capacity c^{lin} is invariant under phase-space translations and under the action of $\mathrm{Sp}(2n,\mathbb{R})$.*

This definition can be restated by saying that a linear symplectic capacity is only invariant under the action of the affine (or: inhomogeneous) symplectic group $\mathrm{ASp}(2n,\mathbb{R})$:

$$c^{\mathrm{lin}}(\phi(\Omega)) = c(\Omega) \text{ for all } \phi \in \mathrm{Asp}(2n,\mathbb{R}). \tag{5.8}$$

(Recall that $\mathrm{Asp}(2n,\mathbb{R})$ consists of all products $ST(z)$ where $S \in \mathrm{Sp}(2n,\mathbb{R})$ and $T(z): z' \longmapsto z' + z$ is an arbitrary phase-space translation.)

Obviously symplectic capacities are unbounded (even if the symplectic capacity of an unbounded set can be bounded, cf. property (SC4). We have for instance

$$c(\mathbb{R}^n \oplus \mathbb{R}^n) = c^{\mathrm{lin}}(\mathbb{R}^n \oplus \mathbb{R}^n) = +\infty \tag{5.9}$$

as immediately follows from the double equality $c(B(R)) = c^{\mathrm{lin}}(B(R)) = \pi R^2$. However, if Ω is bounded then all its symplectic capacities are finite: there exists R such that a ball $B(R)$ contains Ω, and hence $c(\Omega) \leq c(B(R)) = \pi R^2$ in view of the monotonicity property (SC2). More generally, it follows from properties (SC2) and (SC4) that if $B(R) \subset \Omega \subset Z_j(R)$ then $c(\Omega) = \pi R^2$; this illustrates the fact that sets very different in shape and volume can have the same symplectic capacity.

Also note that a set Ω with non-empty interior cannot have symplectic capacity equal to zero: let Ω' be the interior of Ω, it is an open set, and if it is not empty it contains a (possibly) very small) ball $B(\varepsilon)$. Using again (SC2) we have $\pi \varepsilon^2 = c(B(\varepsilon)) \leq c(\Omega)$.

We will usually drop the qualification "normalized" in the definition above and just speak about "symplectic capacities"; one exception to this rule will be the Ekeland–Hofer capacities discussed in de Gosson and Luef [77].

The reader is urged to keep in mind that the notion of symplectic capacity is not directly related to that of volume. For instance, the function c_{Vol} defined by

$$c_{\mathrm{Vol}}(\Omega) = [\mathrm{Vol}(\Omega)]^{1/n}$$

obviously satisfies the properties (SC1)–(SC4) above *except* the second identity (SC4): we have $c_{\mathrm{Vol}}(Z_j(R)) \neq \pi R^2$ as soon as $n > 1$; now the second identity (SC4) is precisely the most characteristic and interesting property of a symplectic capacity, because it is related to Gromov's non-squeezing theorem.

There remains to prove the existence of symplectic capacities.

Proposition 87. *Let $\Omega \subset \mathbb{R}^n \oplus \mathbb{R}^n$ and set*

$$R_\sigma = \sup_{\phi \in \mathrm{Symp}(2n,\mathbb{R})} \{R: \ \phi(B(R)) \subset \Omega\}.$$

Set $c_{\min}(\Omega) = \pi R_\sigma^2$ if $R_\sigma < \infty$, $c_{\min}(\Omega) = \infty$ if $R_\sigma = \infty$. Then $c_{\min}$ is a symplectic capacity on the standard symplectic space $(\mathbb{R}^n \oplus \mathbb{R}^n, \sigma)$.

Proof. Let us prove the symplectic invariance property (SC1). Let $\phi \in \mathrm{Symp}(2n, \mathbb{R})$ and $\phi' \in \mathrm{Symp}(2n, \mathbb{R})$ be such that $g(B(R)) \subset \Omega$; then $(\phi \circ \phi')(B(R)) \subset \phi(\Omega)$ for every $\phi \in \mathrm{Symp}(2n, \mathbb{R})$ hence $c_{\min}(\phi(\Omega)) \geq c_{\min}(\Omega)$. To prove the opposite inequality we note that replacing Ω by $\phi^{-1}(\Omega)$ leads to $c_{\min}(\Omega)) \geq c_{\min}(\phi^{-1}(\Omega))$; since ϕ is arbitrary we have in fact $c_{\min}(\Omega)) \geq c_{\min}(\phi(\Omega))$ for every $\phi \in \mathrm{Symp}(2n, \mathbb{R})$. It follows that we have equality: $c_{\min}(\phi(\Omega)) = c_{\min}(\Omega)$. The monotonicity property (SC2) is of course trivially verified because a symplectomorphism sending $B(R)$ in Ω' also sends $B(R)$ in any set $\Omega' \supset \Omega$. Let us prove the conformality property (SC3). First note it trivially holds for $\lambda = 0$ so we may assume $\lambda \neq 0$. Let $\phi \in \mathrm{Symp}(2n, \mathbb{R})$ and define ϕ_λ by $\phi_\lambda(z) = \lambda\phi(\lambda^{-1}z)$; it is clear that $\phi_\lambda \in \mathrm{Symp}(2n, \mathbb{R})$. The condition $\phi(B(R)) \subset \Omega$ being equivalent to $\lambda^{-1}\phi_\lambda(\lambda B(R)) \subset \Omega$, that is to $\phi_\lambda(B(\lambda R)) \subset \lambda\Omega$, it follows from the definition of $c_{\min}$ that $c_{\min}(\lambda\Omega) = \pi(\lambda R_\sigma)^2 = \lambda^2 c_{\min}(\Omega)$. Let us finally prove that the normalization conditions (SC4) are satisfied by $c_{\min}$. The equality $c_{\min}(B(R)) = \pi R^2$ is obvious: every ball $B(r)$ with $R' \leq R$ is sent into $B(R)$ by the identity and if $R' \geq R$ there exists no $\phi \in \mathrm{Symp}(2n, \mathbb{R})$ such that $\phi(B(R')) \subset B(R)$ because symplectomorphisms are volume-preserving. There remains to show that $c_{\min}(Z_j(R)) = \pi R^2$. If $R' \leq R$ then the identity sends $B(R')$ in $Z_j(R)$ hence $c_{\min}(Z_j(R)) \leq \pi R^2$. Assume that $c_{\min}(Z_j(R)) > \pi R^2$; then there exists a ball $B(R')$ with $R' > R$ and a $\phi \in \mathrm{Symp}(2n, \mathbb{R})$ such that $\phi(B(R')) \subset Z_j(R)$ and this would violate Gromov's theorem. $\qquad \square$

We urge the reader to note that in the proof of the fact that $c_{\min}$ indeed is a symplectic capacity, we needed Gromov's theorem only at the very last step, when we wanted to prove that $c_{\min}(Z_j(R)) = \pi R^2$.

Definition 88. The symplectic capacity $c_{\min}$ is called the Gromov width (or symplectic width). The (possibly infinite) number R_σ such that $c_{\min}(\Omega) = \pi R_\sigma^2$ is called the symplectic radius of Ω.

The proof that the existence of symplectic capacities is actually *equivalent* to Gromov's theorem is proposed to the reader's sagacity in the problem below:

Problem 89. Show that the existence of a single symplectic capacity proves Gromov's non-squeezing theorem.

Gromov's non-squeezing theorem actually allows us to easily construct another symplectic capacity $c_{\max}$, distinct from $c_{\min}$ and there are thus infinitely many symplectic capacities because for every real λ in the closed interval $[0, 1]$ the formula

$$c_\lambda = \lambda c_{\max} + (1 - \lambda)c_{\min} \tag{5.10}$$

obviously defines a symplectic capacity, and we have $c_\lambda \neq c_{\lambda'}$ if $\lambda \neq \lambda'$. (More generally, we can always interpolate two arbitrary symplectic capacities to obtain new capacities). The symplectic capacity $c_{\max}$ is constructed as follows: suppose that no matter how large we choose r there exists no symplectomorphism sending Ω inside a cylinder $Z_j(r)$. We then set $c_{\max}(\Omega) = +\infty$. Suppose that, on the contrary, there are symplectomorphisms sending Ω inside some cylinder $Z_j(r)$ and let R be the infimum of all such r. Thus, by definition,

$$c_{\max}(\Omega) = \inf_{\phi}\{\pi r^2 : \phi(\Omega) \subset Z_j(r)\} = \pi R^2 \tag{5.11}$$

where ϕ again ranges over all the symplectomorphisms $\mathbb{R}^n \oplus \mathbb{R}^n \longrightarrow \mathbb{R}^n \oplus \mathbb{R}^n$. We leave it to the reader to verify, using again the non-squeezing theorem, that $c_{\max}$ indeed is a symplectic capacity. As the notation suggests, we have:

Proposition 90. *For every* $\Omega \subset \mathbb{R}^n \oplus \mathbb{R}^n$ *we have*

$$c_{\min}(\Omega) \leq c(\Omega) \leq c_{\max}(\Omega) \tag{5.12}$$

for every symplectic capacity c.

Proof. Suppose first $c_{\min}(\Omega) > c(\Omega)$ and set $c(\Omega) = \pi R^2$; thus $c_{\min}(\Omega) > \pi R^2$. It follows, by definition of $c_{\min}$, that there exists $\varepsilon > 0$ and a symplectomorphism ϕ such that $\phi(B(R+\varepsilon)) \subset \Omega$. But then, in view of the monotonicity property (SC2) of c we have $c(\phi(B(R+\varepsilon))) \leq c(\Omega)$, that is, in view of the symplectic invariance property (SC1), $c(B(R+\varepsilon)) \leq c(\Omega)$. We thus have, taking (SC4) into account, $c(B(R+\varepsilon)) = \pi(R+\varepsilon)^2 \leq c(\Omega)$, and this contradicts $c(\Omega) = \pi R^2$. The proof of the inequality $c(\Omega) \leq c_{\max}(\Omega)$ is similar; we leave the details to the reader. $\square$

Exercise 91. Prove the existence of a subset Ω of $\mathbb{R}^n \oplus \mathbb{R}^n$ such that $c_{\min}(\Omega) < c_{\max}(\Omega)$.

The existence of linear symplectic capacities is proven exactly in the same way as above. In fact, for $\Omega \subset \mathbb{R}^n \oplus \mathbb{R}^n$ set

$$c_{\min}^{\lin}(\Omega) = \sup_{\phi \in \ISp(2n,\mathbb{R})} \{\pi R^2 : \phi(B(R)) \subset \Omega\}, \tag{5.13}$$

$$c_{\max}^{\lin}(\Omega) = \inf_{\phi \in \ISp(2n,\mathbb{R})} \{\pi R^2 : \phi(\Omega) \subset Z_j(R)\}; \tag{5.14}$$

it is immediate to show that $c_{\min}^{\lin}$ and $c_{\max}^{\lin}$ are linear symplectic capacities, which can be interpreted as follows: for every $\Omega \subset \mathbb{R}^n \oplus \mathbb{R}^n$ the number $c_{\min}^{\lin}(\Omega)$ (which can be infinite) is the supremum of all the πR^2 of phase space balls $B(R)$ that can be sent in Ω using elements of $\ISp(2n,\mathbb{R})$; similarly $c_{\min}^{\lin}(\Omega)$ is the infimum of all πR^2 such that a cylinder $Z_j(R)$ can contain the deformation of Ω by elements of the inhomogeneous symplectic group $\ISp(2n,\mathbb{R})$ (the group generated by phase space translations and the elements of $\Sp(2n,\mathbb{R})$). We have moreover

$$c_{\min}^{\lin}(\Omega) \leq c^{\lin}(\Omega) \leq c_{\max}^{\lin}(\Omega) \tag{5.15}$$

for every $\Omega \subset \mathbb{R}^n \oplus \mathbb{R}^n$ and every linear symplectic capacity $c_{\lin}$; the proof is similar to that of the inequalities (5.12).

The homogeneity property (SC2) satisfied by every symplectic capacity (linear or not) together with the fact that $c(B(R)) = \pi R^2$ suggests that symplectic capacities have something to do with the notion of area. In fact, the following is true: the symplectic capacity $c_{\min}(\Omega)$ of a subset in the phase plane $\mathbb{R}^2$ is the area of Ω when the latter is connected (Siburg [148]; also see the proof in Hofer and Zehnder [101], §3.5, Theorem 4). Note that the result in general no longer holds when Ω is disconnected: suppose for instance that Ω is the union of two disjoint disks with radii R and R' such that $R' < R$. Then $c_{\min}(\Omega) = \pi R^2 < \text{Area}(\Omega)$. The symplectic capacity $c_{\max}(\Omega)$ is the area when Ω is simply connected. Summarizing, it follows from the inequalities (5.12) that:

Proposition 92. *Let c be a symplectic capacity on the phase plane $\mathbb{R}^2$. We have $c(\Omega) = \text{Area}(\Omega)$ when Ω is a connected and simply connected surface.*

The reader may easily convince himself that $c_{\min}(\Omega)$ is not the area when Ω is disconnected, and that $c_{\max}(\Omega)$ is in general not the area when Ω fails to be simply connected (a typical counterexample is the annulus $r \leq x^2 + p^2 \leq R^2$).

5.2.2   The symplectic capacity of an ellipsoid

As the title of this subsection suggests, the symplectic capacity of an ellipsoid is intrinsically attached to that ellipsoid, in the sense that it does not depend on the choice of symplectic capacity. To prove this we will need the following symplectic diagonalization theorem, which is very interesting by itself, and which we will use several times in this book. It was proven by Williamson [162] in 1936, and has been rediscovered many times since. It can be viewed as a partial result in the non-trivial topic of classification of quadratic forms. We are following the presentation in Folland's book [60] (Proposition 4.22); for an alternative proof using Lagrange multipliers see Hofer and Zehnder [101], §1.7.

Theorem 93 (Williamson). *Let M be a positive-definite symmetric real $2n \times 2n$ matrix.*

There exists $S \in \text{Sp}(2n, \mathbb{R})$ such that

$$S^T M S = \begin{pmatrix} \Lambda & 0 \\ 0 & \Lambda \end{pmatrix}, \quad \Lambda \text{ diagonal,} \tag{5.16}$$

the diagonal entries λ_j of Λ being defined by the condition

$$\pm i\lambda_j \text{ is an eigenvalue of } JM. \tag{5.17}$$

Proof. Let $\langle \cdot, \cdot \rangle_M$ be the scalar product on $\mathbb{C}^{2n}$ defined by $\langle z, z' \rangle_M = \langle Mz, z' \rangle$. Since both $\langle \cdot, \cdot \rangle_M$ and the symplectic form are non-degenerate we can find a unique invertible matrix K of order $2n$ such that

$$\langle z, Kz' \rangle_M = \sigma(z, z')$$

for all z, z'; that matrix satisfies $K^T M = J = -MK$. Since the σ is antisymmetric we must have $K = -K^M$ where $K^M = -M^{-1}K^T M$ is the transpose of K with respect to $\langle \cdot, \cdot \rangle_M$; it follows that the eigenvalues of $K = -M^{-1}J$ are of the type $\pm i\lambda_j$, $\lambda_j > 0$, and so are those of JM^{-1}. The corresponding complex eigenvectors occurring in conjugate pairs $e'_j \pm if'_j$, we thus obtain a $\langle \cdot, \cdot \rangle_M$-orthonormal basis $\{e'_i, f'_j\}_{1 \leq i,j \leq n}$ of $\mathbb{R}^n \oplus \mathbb{R}^n$ such that $Ke'_i = \lambda_i f'_i$ and $Kf'_j = -\lambda_j e'_j$. Notice that it follows from these relations that we have $K^2 e'_i = -\lambda_i^2 e'_i$ and $K^2 f'_j = -\lambda_j^2 f'_j$ and that the vectors of the basis $\{e'_i, f'_j\}_{1 \leq i,j \leq n}$ satisfy the relations

$$\sigma(e'_i, e'_j) = \langle e'_i, Ke'_j \rangle_M = \lambda_j \langle e'_i, f'_j \rangle_M = 0,$$
$$\sigma(f'_i, f'_j) = \langle f'_i, Kf'_j \rangle_M = -\lambda_j \langle f'_i, e'_j \rangle_M = 0,$$
$$\sigma(f'_i, e'_j) = \langle f'_i, Ke'_j \rangle_M = \lambda_i \langle f'_i, f'_j \rangle_M = -\lambda_i \delta_{ij}.$$

Setting $e_i = \lambda_i^{-1/2} e'_i$ and $f_j = \lambda_j^{-1/2} f'_j$, the basis $\{e_i, f_j\}_{1 \leq i,j \leq n}$ is symplectic. Let S be the element of $\mathrm{Sp}(2n, \mathbb{R})$ mapping the canonical symplectic basis to $\{e_i, f_j\}_{1 \leq i,j \leq n}$. The $\langle \cdot, \cdot \rangle_M$-orthogonality of $\{e_i, f_j\}_{1 \leq i,j \leq n}$ implies (5.16) with $\Lambda = \mathrm{diag}(\lambda_1, \ldots, \lambda_n)$. $\qquad\square$

Williamson's theorem allows us to calculate rather easily the symplectic capacity of an ellipsoid:

Proposition 94. *Let $W = \{z : Mz^2 \leq 1\}$ (M symmetric) be an ellipsoid in $\mathbb{R}^n \oplus \mathbb{R}^n$ and let c be an arbitrary linear symplectic capacity on $(\mathbb{R}^n \oplus \mathbb{R}^n, \sigma)$. Let $\lambda_{1,\sigma} \geq \lambda_{2,\sigma} \geq \cdots \geq \lambda_{n,\sigma}$ be the decreasing sequence of the moduli of the eigenvalues $\pm i\lambda$ of JM. We have*

$$c(W) = \frac{\pi}{\lambda_{1,\sigma}} = c^{\mathrm{lin}}(W) \tag{5.18}$$

where c^{lin} is any linear symplectic capacity.

Proof. Let us choose $S \in \mathrm{Sp}(2n, \mathbb{R})$ such that the matrix $S^T M S = \begin{pmatrix} \Lambda & 0 \\ 0 & \Lambda \end{pmatrix}$. The set $S^{-1}(W)$ is thus the ellipsoid described by $\Lambda x^2 + \Lambda p^2 \leq 1$, that is

$$\sum_{j=1}^{n} \lambda_{j,\sigma}(x_j^2 + p_j^2) \leq 1. \tag{5.19}$$

Since $c(S^{-1}(W)) = c(W)$ in view of the symplectic invariance (SC1) of symplectic capacities it is sufficient to assume that the ellipsoid W is represented by (5.19). In view of the obvious double inequality

$$\lambda_{1,\sigma}(x_1^2 + p_1^2) \leq \sum_{j=1}^{n} \lambda_{j,\sigma}(x_j^2 + p_j^2) \leq \lambda_{1,\sigma} \sum_{j=1}^{n}(x_j^2 + p_j^2) \tag{5.20}$$

we have $B(\lambda_{1,\sigma}^{-1/2}) \subset W \subset Z_1(\lambda_{1,\sigma}^{-1/2})$ hence, using the monotonicity property (SC2) of symplectic capacities,

$$c(B(\lambda_{1,\sigma}^{-1/2})) \subset c(W) \subset (Z(\lambda_{1,\sigma}^{-1/2})).$$

The equality $c(\mathcal{W}) = \pi/\lambda_{1,\sigma}$ follows in view of the normalization conditions (SC4) satisfied by any symplectic capacity; the formula $c^{\mathrm{lin}}(\mathcal{W}) = \pi/\lambda_{1,\sigma}$ follows as well since we have put the ellipsoid $\mathcal{W}$ in the form (5.19) using only a linear symplectomorphism. $\qquad\qquad\square$

5.3 Other symplectic capacities

In this section we briefly review two interesting symplectic capacities which cannot be directly deduced from Gromov's non-squeezing theorem, and which can thus be used to derive this theorem (cf. Problem 89). The first example, the Hofer–Zehnder capacity, is normalized (in the sense of (SC4)), while the second example provides us with a whole family of non-normalized symplectic capacities.

5.3.1 The Hofer–Zehnder capacity

We mentioned at the beginning of this chapter that the notion of symplectic capacity can be viewed as a generalization of the notion of action. This is most easily seen by using the Hofer–Zehnder capacity. In [101] (Chapter 3) Hofer and Zehnder construct a symplectic capacity c^{HZ} which measures sets in a dynamical way. It is defined as follows. Let Ω be an open set in $\mathbb{R}^n \oplus \mathbb{R}^n$ and consider the class $\mathcal{H}(\Omega)$ of all Hamiltonian functions $H \geq 0$ having the following three properties:

- *H vanishes outside Ω (and is hence bounded);*
- *The critical values of H are 0 and $\max H$;*
- *The flow (ϕ_t^H) has no constant periodic orbit with period $T \leq 1$.*

Then, by definition,

$$c^{\mathrm{HZ}}(\Omega) = \sup\{\max H : H \in \mathcal{H}(\Omega)\}. \qquad (5.21)$$

To prove that c^{HZ} satisfies the properties (SC1)–(SC4) is not straightforward; the proof uses several tricks from the theory of Hamiltonian systems. However, it does not rely on Gromov's theorem, and thus provides an alternative proof of this result (cf. Problem 89). The Hofer–Zehnder capacity has the property that whenever Ω is a compact convex set in phase space then

$$c^{\mathrm{HZ}}(\Omega) = \oint_{\gamma_{\mathrm{min}}} p\,dx \qquad (5.22)$$

where $p\,dx = p_1 dx_1 + \cdots + p_n dx_n$ and γ_{min} is the shortest (positively oriented) Hamiltonian periodic orbit carried by the boundary $\partial\Omega$ of Ω. (For (5.22) the condition that Ω be compact and convex is essential (see Hofer and Zehnder's very illustrative "Bordeaux bottle" example in [101], p. 99).

Notice that this formula generalizes the observation made earlier in this section that symplectic capacities agree with the usual notion of area in the case

$n = 1$ for connected and simply connected surfaces. In fact, let Ω be such a surface in the phase plane, and assume that the boundary $\gamma = \partial\Omega$ is smooth, and given the positive orientation. We then have, by Stoke's theorem,

$$\text{Area}(\Omega) = \tfrac{1}{2} \oint_\gamma (pdx - xdp) = \oint_\gamma pdx.$$

Formula (5.22) in particular implies the inequalities

$$c_{\min}(\Omega) \leq c^{\mathrm{HZ}}(\Omega) \leq \left| \oint_\gamma pdx \right| \tag{5.23}$$

for every periodic orbit γ on $\partial\Omega$.

5.3.2   The Ekeland–Hofer capacities

The normalized symplectic capacities we have been using so far do not generally allow us to distinguish between ellipsoids: formula (5.18) implies that if $\Omega_M : Mz^2 \leq 1$ and $\Omega_{M'} : M'z^2 \leq 1$ are such that M and M' have the same smallest symplectic eigenvalue, then $c(\Omega_M) = c(\Omega_{M'})$. This can however be achieved by introducing a slightly more general notion of symplectic capacity. To do this we slightly relax the normalization condition (SC4) for symplectic capacities and replace it with the weaker requirement:

(SC4bis) $c(B^{2n}(R)) > 0$ and $c(Z_j(R)) < \infty$.

In [41, 42] Ekeland and Hofer construct a sequence of generalized symplectic capacities c_k^{EH} having the following properties:

(EH1) *The sequence* $(c_k^{\mathrm{EH}})_{k \geq 1}$ *is increasing:*

$$c_1^{\mathrm{EH}}(\Omega) \leq c_2^{\mathrm{EH}}(\Omega) \leq \cdots \leq c_k^{\mathrm{EH}}(\Omega) \leq \cdots \tag{5.24}$$

  for all $\Omega \subset \mathbb{R}^n \oplus \mathbb{R}^n$;

(EH2) *If* Ω *is convex with boundary* $\partial\Omega$, *then*

$$c_k^{\mathrm{EH}}(\Omega) = c_k^{\mathrm{EH}}(\partial\Omega) \tag{5.25}$$

  (hence $c_k^{\mathrm{EH}}(\Omega)$ *is determined by the boundary* $\partial\Omega$*);*

(EH3) *If* Ω *is convex, then*

$$c_1^{\mathrm{EH}}(\Omega) = c^{\mathrm{HZ}}(\Omega) \text{ and } c_k^{\mathrm{EH}}(\Omega) = \oint_\gamma pdx \tag{5.26}$$

  where γ *is a periodic Hamiltonian orbit carried by* $\partial\Omega$.

The values of the capacities c_k^{EH} on balls and cylinders are given by the formulas

$$c_k^{\mathrm{EH}}(B^{2n}(r)) = \left[\frac{k+n-1}{n}\right] \pi r^2, \tag{5.27}$$

$$c_k^{\mathrm{EH}}(Z_j^{2n}(r)) = \pi r^2; \tag{5.28}$$

in the first formula $[x]$ is the integer part of $x \in \mathbb{R}$.

The Ekeland–Hofer capacities c_k^{EH} allow us to classify phase-space ellipsoids. In fact, it readily follows from their properties (EH1) and (EH3) that the non-decreasing sequence of numbers $c_k^{\mathrm{EH}}(\Omega_M)$ is determined as follows: if $\mathrm{Spec}_\sigma(M) = (\lambda_1^\sigma, \ldots, \lambda_n^\sigma)$ write the numbers $k\pi/\lambda_1^\sigma$ in increasing order *with repetition* if a number occurs several times; we thus obtain a sequence $c_1 \leq c_2 \leq \cdots$ and we have

$$c_k^{\mathrm{EH}}(\Omega_M) = c_k. \tag{5.29}$$

That the Ekeland–Hofer capacities c_k^{EH} can be used to distinguish between ellipsoids follows from:

Proposition 95. *An ellipsoid $\Omega_M : Mz^2 \leq 1$ is uniquely determined (up to a symplectic transformation) by the sequence of its Ekeland–Hofer capacities $c_k^{EH}(\Omega_M)$.*

Proof. Suppose that Ω_M and $\Omega_{M'}$ are two ellipsoids with $\lambda_j^\sigma = \lambda_j'^\sigma$ for $1 \leq j < k$ and $\lambda_k^\sigma > \lambda_k'^\sigma$. Then the multiplicity of λ_k^σ in the sequence of Ekeland–Hofer capacities is one higher for Ω_M than for $\Omega_{M'}$ hence not all of these capacities agree on Ω_M and $\Omega_{M'}$. $\qquad\square$

Chapter 6

Uncertainty Principles

Since the aim of the first part of this book is to discuss classical mechanics from the symplectic point of view, some readers might be a little surprised by the title of the present chapter because it has a certain quantum-mechanical connotation. In fact "uncertainty principles" are mostly studied within the realm of quantum mechanics, even if the study of uncertainties is also a part of classical statistical mechanics. However, as we have shown in our paper [72], the formalism of the uncertainty principle is actually not as "quantum" as it may seem, but appears in classical mechanics if one uses the notion of symplectic capacity. Even if everything we will do is "classical", in the sense that we do not invoke any quantum mechanical properties, we will however make use of the notation $\hbar$ as if we were "doing quantum mechanics"; the reader who feels uncomfortable with this irruption of a quantum-mechanical constant can view $\hbar$ as a parameter measuring some indeterminacy in classical measurement processes.

The textbook uncertainty principle of quantum mechanics is usually stated in the form $\Delta x_j \Delta p_j \geq \frac{1}{2}\hbar$ (the "Heisenberg inequalities"). It is unfortunate that even in many otherwise excellent mathematical texts it is only this weak form of the uncertainty principle that is studied. In fact, to limit ourselves to the Heisenberg inequalities has many disadvantages, the most obvious being that these inequalities are not preserved by linear transformations (except trivial ones). A better formulation consists in using the Robertson [137] and Schrödinger [142] inequalities

$$(\Delta X_\alpha)^2(\Delta P_\alpha)^2 \geq \Delta(X_\alpha, P_\alpha)^2 + \tfrac{1}{4}\hbar^2 \ , \ 1 \leq \alpha \leq n \tag{6.1}$$

which we will express in the form

$$c(\mathcal{W}_\Sigma) \geq \tfrac{1}{2}h \tag{6.2}$$

where $\mathcal{W}_\Sigma$ is a certain ellipsoid and c a symplectic capacity. We will be following rather closely the exposition in our review paper de Gosson and Luef [77].

6.1   The Robertson–Schrödinger inequalities

The Robertson–Schrödinger uncertainty principle is a strong version of the Heisenberg inequalities $\Delta x_j \Delta p_j \geq \frac{1}{2}\hbar$, to which it reduces when one neglects the contributions due to the covariances.

6.1.1   The covariance matrix

In what follows ρ is a real-valued function defined on $\mathbb{R}^n \oplus \mathbb{R}^n$ satisfying the normalization condition

$$\int_{\mathbb{R}^{2n}} \rho(z)dz = 1 \tag{6.3}$$

and such that

$$\int_{\mathbb{R}^{2n}} (1 + |z|^2)|\rho(z)|dz < \infty. \tag{6.4}$$

We do not assume that $\rho \geq 0$ so ρ is not in general a true probability density. (Having later applications to quantum mechanics in mind, ρ will typically be the Wigner transform of a mixed quantum state.) We will call such a function ρ a "quasi-distribution", and work with it exactly as we would with an ordinary probability density.

Exercise 96. Show that condition (6.4) implies that the Fourier transform $F\rho$ is twice continuously differentiable.

Let us introduce the following notation: we set $z_\alpha = x_\alpha$ if $1 \leq \alpha \leq n$ and $z_\alpha = p_{\alpha-n}$ if $n + 1 \leq \alpha \leq 2n$.

We define the covariances and variances associated with ρ in the usual way by the formulas

$$\Delta(Z_\alpha, Z_\beta) = \int_{\mathbb{R}^{2n}} (z_\alpha - \langle z_\alpha \rangle)(z_\beta - \langle z_\beta \rangle)\rho(z)dz \tag{6.5}$$

and

$$(\Delta Z_\alpha)^2 = \Delta(Z_\alpha, Z_\alpha) = \int_{\mathbb{R}^{2n}} (z_\alpha - \langle z_\alpha \rangle)^2 \rho(z)dz. \tag{6.6}$$

In the formulas above the moments $\langle z_\alpha^k \rangle$, $k = 1, 2$, are the averages with respect to ρ of the corresponding functions:

$$\langle z_\alpha^k \rangle = \int_{\mathbb{R}^{2n}} z_\alpha^k \rho(z)dz. \tag{6.7}$$

These moments are of course well defined in view of condition (6.4).

Let $Z_1, Z_2, \ldots, Z_{2n}$ be random variables on $\mathbb{R}^n \oplus \mathbb{R}^n$ whose values are the phase-space coordinates $z_1, z_2, \ldots, z_n$. Since the integral of ρ is equal to 1, formulae

(6.5) and 6.6) can be rewritten in the familiar form

$$\Delta(Z_\alpha, Z_\beta) = \langle z_\alpha z_\beta \rangle - \langle z_\alpha \rangle \langle z_\beta \rangle, \tag{6.8}$$

$$(\Delta Z_\alpha)^2 = \Delta(Z_\alpha, Z_\alpha) = \langle z_\alpha^2 \rangle - \langle z_\alpha \rangle^2. \tag{6.9}$$

The quantities (6.5), (6.6), and (6.7) are well defined in view of condition (6.4): the integrals above are all absolutely convergent in view of the trivial estimates

$$\left| \int_{\mathbb{R}^{2n}} z_\alpha \rho(z) dz \right| \leq \int_{\mathbb{R}^{2n}} (1 + |z|^2) |\rho(z)| dz < \infty,$$

$$\left| \int_{\mathbb{R}^{2n}} z_\alpha z_\beta \rho(z) dz \right| \leq \int_{\mathbb{R}^{2n}} (1 + |z|^2) |\rho(z)| dz < \infty.$$

Definition 97. We will call the symmetric $2n \times 2n$ matrix

$$\Sigma = [\Delta(Z_\alpha, Z_\beta)]_{1 \leq \alpha, \beta \leq 2n}$$

the covariance matrix associated with ρ. When $\det \Sigma \neq 0$ the inverse Σ^{-1} is called the precision (or information) matrix.

For instance, when $n = 1$, the covariance matrix is

$$\Sigma = \begin{pmatrix} \Delta X^2 & \Delta(X, P) \\ \Delta(P, X) & \Delta P^2 \end{pmatrix}$$

where the quantities ΔX^2 and $\Delta(X, P)$ are defined by

$$\Delta X^2 = \langle x^2 \rangle - \langle x \rangle^2 \, , \, \Delta P^2 = \langle p^2 \rangle - \langle p \rangle^2,$$
$$\Delta(X, P) = \langle xp \rangle - \langle x \rangle \langle p \rangle.$$

6.1.2 A strong version of the Robertson–Schrödinger uncertainty principle

The following result is essential. The first statement (i) was apparently first noted in Narcowich [128] (Lemma 2.3), and part (ii) goes back to Narcowich [126], Narcowich and O'Connell [129], and Yuen [167]. The third part (iii) is a way of expressing the symplectic covariance of the uncertainty principle.

Theorem 98. *Let Σ be a real symmetric $2n \times 2n$ matrix and $\hbar$ a real number. Then $\Sigma + \frac{i\hbar}{2} J$ is a Hermitian matrix. Suppose that there exists a real number $\hbar \neq 0$ such that $\Sigma + \frac{i\hbar}{2} J \geq 0$. Then:*

(i) *The matrix Σ must be positive definite and we have $\Sigma + \frac{i\hbar'}{2} J \geq 0$ for every $\hbar' \leq \hbar$;*

(ii) *The inequalities*

$$(\Delta X_\alpha)^2 (\Delta P_\alpha)^2 \geq \Delta(X_\alpha, P_\alpha)^2 + \tfrac{1}{4} \hbar^2 \tag{6.10}$$

hold for $1 \leq \alpha \leq n$;

(iii) *Let $S \in \mathrm{Sp}(2n, \mathbb{R})$ and define $(X_\alpha^S, P_\alpha^S) = S(X_\alpha, P_\alpha)$. Then*

$$(\Delta X_\alpha^S)^2 (\Delta P_\alpha^S)^2 \geq \Delta(X_\alpha^S, P_\alpha^S)^2 + \tfrac{1}{4}\hbar^2. \tag{6.11}$$

Proof. We begin by noting that the matrix $\Sigma + \frac{i\hbar}{2}J$ is Hermitian because Σ is real symmetric and $(iJ)^* = (-i)(-J) = iJ$.

Proof of (i). Let us begin by showing that Σ is non-negative. Suppose indeed that Σ has a negative eigenvalue λ, and let z_λ be a real eigenvector corresponding to λ (such an eigenvector exists because Σ is real and symmetric). Since $z_\lambda^T J z_\lambda = 0$ we have

$$z_\lambda^T \left(\Sigma + \frac{i\hbar}{2} J \right) z_\lambda = z_\lambda^T \Sigma z_\lambda = \lambda |z_\lambda|^2 < 0$$

which contradicts the assumption $\Sigma + \frac{i\hbar}{2}J \geq 0$. We next show that 0 cannot be an eigenvalue of Σ; this will prove the statement. Suppose indeed that 0 is an eigenvalue, and let z_0 be a real eigenvector. For $\varepsilon > 0$ set $z(\varepsilon) = (I + i\varepsilon J)z_0$. Using the relations $\Sigma z_0 = 0$, $z_0^T \Sigma = 0$, and $z_0^T J z_0 = \sigma(z_0, z_0) = 0$ we get, after a few calculations,

$$z(\varepsilon)^T \left(\Sigma + \frac{i\hbar}{2} J \right) z(\varepsilon) = \varepsilon \frac{1}{2} \hbar |z_0|^2 + \varepsilon^2 (J z_0)^T \Sigma (J z_0).$$

Choose now ε opposite in sign to $\frac{1}{2}\hbar$; then $\varepsilon \frac{1}{2}\hbar |z_0|^2 < 0$ and if $|\varepsilon|$ is small enough we have $z(\varepsilon)^T (\Sigma + \frac{i\hbar}{2}J)z(\varepsilon) < 0$, which contradicts the fact that $\Sigma + \frac{i\hbar}{2}J \geq 0$. To show that $\Sigma + \frac{i\hbar'}{2}J \geq 0$ for every $\hbar' \leq \hbar$ it suffices to set $\hbar' = r\hbar$ with $0 < r \leq 1$ and to note that

$$\Sigma + \frac{i\hbar'}{2} J = (1 - r)\Sigma + r \left(\Sigma + \frac{i\hbar}{2} J \right) \geq 0$$

because $(1 - r)\Sigma \geq 0$ and $\Sigma + \frac{i\hbar}{2}J \geq 0$.

Proof of (ii). The non-negativity of the Hermitian matrix $\Sigma + \frac{i\hbar}{2}J$ can be expressed in terms of the submatrices

$$\Sigma_\alpha = \begin{pmatrix} (\Delta X_\alpha)^2 & \Delta(X_\alpha, P_\alpha) + \frac{i\hbar}{2} \\ \Delta(P_\alpha, X_\alpha) - \frac{i\hbar}{2} & (\Delta P_\alpha)^2 \end{pmatrix}$$

which are non-negative provided that $\Sigma + \frac{i\hbar}{2}J$ is. Since

$$\mathrm{Tr}(\Sigma_\alpha) = (\Delta X_\alpha)^2 + (\Delta P_\alpha)^2 \geq 0$$

we have $\Sigma_\alpha \geq 0$ if and only if

$$\det \Sigma_\alpha = (\Delta X_\alpha)^2 (\Delta P_\alpha)^2 - \Delta(X_\alpha, P_\alpha)^2 - \tfrac{1}{4}\hbar^2 \geq 0$$

which is equivalent to the inequality (6.10).

Proof of (iii). Set $\Sigma^S = S^T \Sigma S$; it is the covariance matrix of the vector

$$(X_1^S, \ldots, X_n^S; P_1^S, \ldots, P_n^S) = S(X_1, \ldots, X_n; P_1, \ldots, P_n).$$

Since S is symplectic we have $S^T J S = J$ and hence

$$\Sigma^S + \frac{i\hbar}{2} J = S^T (\Sigma + \frac{i\hbar}{2} J) S \geq 0;$$

the inequalities (6.11) now follow from the inequalities (6.10). $\qquad\square$

At this point it is appropriate to notice that (except in the case $n = 1$) the condition $\Sigma + \frac{i\hbar}{2} J \geq 0$ is *not equivalent* to the uncertainty inequalities (6.10); it is in fact a *stronger* condition. It is actually easy to see why we have equivalence when $n = 1$: the covariance matrix is just

$$\Sigma = \begin{pmatrix} \Delta X^2 & \Delta(X, P) \\ \Delta(P, X) & \Delta P^2 \end{pmatrix}$$

and since

$$\mathrm{Tr}\left(\Sigma + \frac{i\hbar}{2} J\right) = \Delta X^2 + \Delta P^2 \geq 0$$

the condition $\Sigma + \frac{i\hbar}{2} J \geq 0$ is equivalent to $\det\left(\Sigma + \frac{i\hbar}{2} J\right) \geq 0$, that is to

$$\Delta X^2 \Delta P^2 - \left(\Delta(X, P)^2 + \frac{1}{4}\hbar^2\right) \geq 0$$

which is precisely (6.10) in the case $n = 1$. That this equivalence between $\Sigma + \frac{i\hbar}{2} J \geq 0$ and (6.10) is not true in higher dimensions is easily seen on the following counterexample. Take $n = 2$ and $\frac{1}{2}\hbar = 1$ and define a covariance matrix by

$$\Sigma = \begin{pmatrix} 1 & -1 & 0 & 0 \\ -1 & 1 & 0 & 0 \\ 0 & 0 & 1 & 0 \\ 0 & 0 & 0 & 1 \end{pmatrix}. \tag{6.12}$$

We thus have $(\Delta X_1)^2 = (\Delta X_2)^2 = 1$ and $(\Delta P_1)^2 = (\Delta P_2)^2 = 1$, and also $\Delta(X_1, P_1) = \Delta(X_2, P_2) = 0$ so that the inequalities (6.10) are trivially satisfied (they are in fact equalities). The matrix $\Sigma + iJ$ is nevertheless indefinite.

Exercise 99. Verify the indefiniteness of the 4×4 covariance matrix (6.12).

Let us introduce the following terminology:

Definition 100. Let Σ be a covariance matrix. The phase space ellipsoid

$$\mathcal{W}_\Sigma = \{z \in \mathbb{R}^{2n} : \tfrac{1}{2}\Sigma^{-1} z^2 \leq 1\}$$

is called the "Wigner ellipsoid" associated with Σ. The dual ellipsoid

$$\mathcal{W}_\Sigma^* = \{z \in \mathbb{R}^{2n} : \tfrac{1}{2}\Sigma z^2 \leq 1\}$$

of the Wigner ellipsoid is called the "precision (or information) ellipsoid" (cf. Definition 97).

We will see in a while that the strong uncertainty principle in the form $\Sigma + \frac{i\hbar}{2}J \geq 0$ can be expressed in terms of the notion of symplectic capacity. But we first have to introduce some material from symplectic topology.

6.1.3　Symplectic capacity and the strong uncertainty principle

We are now going to give an application of the notion of symplectic capacity to the strong uncertainty principle discussed in the beginning of this chapter. Another application will be given in Chapter 6.2 when we discuss Hardy's inequalities.

Let us return to the uncertainty principle in its strong Robertson–Schrödinger form $\Sigma + \frac{i\hbar}{2}J \geq 0$. The following geometric result is the key to our formulation in terms of symplectic capacities:

Proposition 101. *Let Σ be a positive-definite real $2n \times 2n$ matrix. The three following conditions are equivalent:*

(i) *The Hermitian matrix $\Sigma + \frac{i\hbar}{2}J$ is non-negative.*

(ii) *The symplectic capacity of the Wigner ellipsoid*

$$\mathcal{W}_\Sigma = \{z : \tfrac{1}{2}\Sigma^{-1}z^2 \leq 1\}$$

    is such that

$$c(\mathcal{W}_\Sigma) \geq \pi\hbar = \tfrac{1}{2}h. \tag{6.13}$$

(iii) *The symplectic capacity of the dual ellipsoid*

$$\mathcal{W}_\Sigma^* = \{z : \tfrac{1}{2}\Sigma z^2 \leq 1\}$$

    is such that

$$c(\mathcal{W}_\Sigma^*) \leq \frac{2\pi}{\hbar} = \frac{4\pi^2}{h}. \tag{6.14}$$

Proof. Setting $M = \frac{1}{2}\Sigma^{-1}$ the Wigner ellipsoid is the set of all $z \in \mathbb{R}^{2n}$ such that $Mz^2 \leq 1$ and the condition $\Sigma + \frac{i\hbar}{2}J \geq 0$ is equivalent to $\frac{1}{2}M^{-1} + \frac{i\hbar}{2}J \geq 0$; using a symplectic diagonalization of M this is equivalent to $\frac{1}{2}D^{-1} + \frac{i\hbar}{2}J \geq 0$ where

$$D = \begin{pmatrix} \Lambda^\sigma & 0 \\ 0 & \Lambda^\sigma \end{pmatrix} \quad , \quad \Lambda^\sigma = \operatorname{diag}(\lambda_1^\sigma, \dots, \lambda_n^\sigma).$$

It follows that the characteristic polynomial of $\frac{1}{2}M^{-1} + \frac{i\hbar}{2}J$ is the product $P(t) = P_1(t)\cdots P_n(t)$ where

$$P_\alpha(t) = t^2 - (\lambda_\alpha^\sigma)^{-1}t + \tfrac{1}{4}(\lambda_\alpha^\sigma)^{-2} - \tfrac{1}{4}\hbar^2.$$

The eigenvalues of the matrix $\frac{1}{2}M^{-1} + \frac{i\hbar}{2}J$ are thus the real numbers $\frac{1}{2}[(\lambda_j^\sigma)^{-1} \pm \hbar]$ hence that matrix is non-negative if and only if $\lambda_j^\sigma \leq \frac{1}{2}(\frac{1}{2}\hbar)^{-1}$ for every j, that is if and only if $\lambda_{\max}^\sigma \leq \frac{1}{2}(\frac{1}{2}\hbar)^{-1}$; this is equivalent to

$$c(\mathcal{W}_\Sigma) = \pi/\lambda_{\max}^\sigma \geq 2\pi \cdot \tfrac{1}{2}\hbar|$$

and to

$$c(\mathcal{W}_\Sigma^*) = 2\pi\lambda_{\min}^\sigma \leq \pi/\left(\tfrac{1}{2}\hbar\right)$$

in view of the discussion above; this proves the inequalities (6.13) and (6.14). $\quad\square$

When Σ is a covariance matrix, the condition $c(\mathcal{W}_\Sigma) \geq \frac{1}{2}h$ thus implies (but is not equivalent to) the Robertson–Schrödinger uncertainty inequalities (6.10), that is we have

$$(\Delta X_\alpha)^2(\Delta P_\alpha)^2 \geq \Delta(X_\alpha, P_\alpha)^2 + \tfrac{1}{4}\hbar^2$$

for $1 \leq \alpha \leq n$.

We have conjectured in [72] and [77] that the condition $c(\mathcal{W}_\Sigma) \geq \frac{1}{2}h$ might well be the "true" uncertainty principle to be used both in classical and quantum mechanics under certain conditions that we do not discuss here. A rather obvious advantage in using this symplectic formulation is that if $c(\mathcal{W}_\Sigma) \geq \frac{1}{2}h$ then we also have $c(f(\mathcal{W}_\Sigma)) \geq \frac{1}{2}h$ when f is an arbitrary symplectomorphism of $(\mathbb{R}^n \oplus \mathbb{R}^n, \sigma)$ (this because of the symplectic invariance property (SC1) of symplectic capacities), so that the uncertainty principle expressed in this form is de facto symplectically invariant (which is not true of the Schrödinger–Robertson inequalities which only retain their form under linear symplectic transformations).

6.2 Hardy's uncertainty principle

In this section we give another application of the notion of symplectic capacity. A folk metatheorem is that a function ψ and its Fourier transform $F\psi$ cannot be simultaneously sharply localized. An obvious manifestation of this "principle" is when ψ is of compact support: in this case the Fourier transform $F\psi$ can be extended into an entire function, and is hence never of compact support. A less trivial way to express this kind of trade-off between ψ and $F\psi$ was discovered in 1933 by G.H. Hardy [98]. Hardy showed, using the Phragmén–Lindelöf principle from complex analysis, that if a function $\psi \in L^2(\mathbb{R})$ and its Fourier transform

$$F\psi(p) = \frac{1}{\sqrt{2\pi\hbar}} \int_{-\infty}^{\infty} e^{-\frac{i}{\hbar}px}\psi(x)dx$$

satisfy, for $|x| + |p| \to \infty$, estimates of the type

$$\psi(x) = \mathcal{O}\left(e^{-\frac{a}{2\hbar}x^2}\right) \quad , \quad F\psi(p) = \mathcal{O}\left(e^{-\frac{b}{2\hbar}p^2}\right) \tag{6.15}$$

with $a, b > 0$, then the following holds true:

- *If $ab > 1$ then $\psi = 0$;*
- *If $ab = 1$ we have $\psi(x) = Ce^{-\frac{a}{2\hbar}x^2}$ for some complex constant C;*
- *If $ab < 1$ there exists a whole space of $\mathcal{S}$ solutions, containing the functions $\psi(x) = Q(x)e^{-\frac{a}{2\hbar}x^2}$ where Q is a polynomial (equivalently, $\mathcal{S}$ contains the* finite linear combinations of Hermite polynomials).

In this section we will generalize Hardy's uncertainty principle to an arbitrary number of dimensions. We will thereafter reformulate it in terms of the notion of symplectic capacity previously introduced.

6.2.1   Two useful lemmas

The following result, although being of an elementary nature, is very useful. We will see that it is a refined version of Williamson's diagonalization theorem [162] in the block-diagonal case.

We make the preliminary observation that if A and B are positive definite matrices then the eigenvalues of AB are real because AB has the same eigenvalues as the symmetric matrix $A^{1/2}BA^{1/2}$. The following result shows that A and B can be simultaneously diagonalized in a particular way:

Lemma 102. *Let A and B be two positive-definite $n \times n$ real matrices. There exists $L \in GL(n, \mathbb{R})$ such that*

$$L^T AL = L^{-1}B(L^T)^{-1} = \Lambda \tag{6.16}$$

where $\Lambda = \mathrm{diag}(\sqrt{\lambda_1}, \ldots, \sqrt{\lambda_n})$ is the diagonal matrix whose eigenvalues are the square roots of the eigenvalues $\lambda_1, \ldots, \lambda_n$ of AB.

Proof. We claim that there exists $R \in GL(n, \mathbb{R})$ such that

$$R^T AR = I \text{ and } R^{-1}B(R^T)^{-1} = D \tag{6.17}$$

where $D = \mathrm{diag}(\lambda_1, \ldots, \lambda_n)$. In fact, first choose $P \in GL(n, \mathbb{R})$ such that $P^T AP = I$ and set $B_1^{-1} = P^T B^{-1} P$. Since B_1^{-1} is symmetric, there exists $H \in O(n, \mathbb{R})$ such that $B_1^{-1} = H^T D^{-1}H$ where D^{-1} is diagonal. Set now $R = PH^T$; we have $R^T AR = I$ and also

$$R^{-1}B(R^T)^{-1} = HP^{-1}B(P^T)^{-1}H^T = HB_1H^T = D$$

hence the equalities (6.17). Let $\Lambda = \mathrm{diag}(\sqrt{\lambda_1}, \ldots, \sqrt{\lambda_n})$. Since

$$R^T AB(R^T)^{-1} = R^T AR(R^{-1}B(R^T)^{-1}) = D$$

the diagonal elements of D are indeed the eigenvalues of AB hence $D = \Lambda^2$. Setting $L = R\Lambda^{1/2}$ we have

$$L^T AL = \Lambda^{1/2}R^T AR\Lambda^{1/2} = \Lambda,$$

$$L^{-1}B(L^{-1})^T = \Lambda^{-1/2}R^{-1}B(R^T)^{-1}\Lambda^{-1/2} = \Lambda$$

hence our claim. $\square$

The result above is a precise statement of a classical theorem of Williamson [162] in the block-diagonal case. That theorem says that every positive-definite symmetric matrix can be diagonalized using symplectic matrices. More precisely: let M be a positive definite real $2n \times 2n$ matrix; the eigenvalues of JM are those of the antisymmetric matrix $M^{1/2}JM^{1/2}$ and are thus of the type $\pm i\lambda_j^\sigma$ with $\lambda_j^\sigma > 0$.

We have the following result, which relates Lemma 102 to Williamson's theorem:

Lemma 103. *Let $A, B > 0$. The symplectic spectrum $(\lambda_1^\sigma, \ldots, \lambda_n^\sigma)$ of $M = \begin{pmatrix} A & 0 \\ 0 & B \end{pmatrix}$ consists of the decreasing sequence $\sqrt{\lambda_1} \geq \cdots \geq \sqrt{\lambda_n}$ of square roots of the eigenvalues λ_j of AB.*

Proof. Let $(\lambda_1^\sigma, \ldots, \lambda_n^\sigma)$ be the symplectic spectrum of M. The λ_j^σ are the eigenvalues of

$$JM = \begin{pmatrix} 0 & B \\ -A & 0 \end{pmatrix};$$

they are thus the moduli of the zeroes of the polynomial

$$P(t) = \det(t^2 I + AB) = \det(t^2 I + D)$$

where $D = \operatorname{diag}(\lambda_1, \ldots, \lambda_n)$; these zeroes are the numbers $\pm i\sqrt{\lambda_j}$, $j = 1, \ldots, n$; the result follows. $\square$

Recall that when L is invertible the matrix

$$M_L = \begin{pmatrix} L^{-1} & 0 \\ 0 & L^T \end{pmatrix} \tag{6.18}$$

is in $\operatorname{Sp}(2n, \mathbb{R})$. Lemma 102 can be restated by saying that if (A, B) is a pair of symmetric positive definite matrices then there exists L such that

$$\begin{pmatrix} A & 0 \\ 0 & B \end{pmatrix} = M_{L^T} \begin{pmatrix} \Lambda & 0 \\ 0 & \Lambda \end{pmatrix} M_L. \tag{6.19}$$

This lemma is thus a precise version of Williamson's theorem for block-diagonal positive matrices: it is not at all obvious from the statement of Williamson's theorem that $\begin{pmatrix} A & 0 \\ 0 & B \end{pmatrix}$ can be diagonalized using only a block-diagonal symplectic matrix!

Lemma 102 allows us to give a simple proof of a multi-dimensional version of this theorem. The following elementary remark will be useful:

Lemma 104. *Let $n > 1$. For $1 \leq j \leq n$ let f_j be a function of $(x_1, \ldots, \widetilde{x}_j, \ldots, x_n) \in \mathbb{R}^{n-1}$ (the tilde $\tilde{\;}$ suppressing the term it covers), and g_j a function of $x_j \in \mathbb{R}$. If*

$$h = f_1 \otimes g_1 = \cdots = f_n \otimes g_n$$

then there exists a constant C such that $h = C(g_1 \otimes \cdots \otimes g_n)$.

Proof. Assume that $n = 2$; then

$$h(x_1, x_2) = f_1(x_2)g_1(x_1) = f_2(x_1)g_2(x_2).$$

If $g_1(x_1)g_2(x_2) \neq 0$ then

$$f_1(x_2)/g_2(x_2) = f_2(x_1)/g_1(x_1) = C$$

hence $f_1(x_2) = Cg_2(x_2)$ and $h(x_1, x_2) = Cg_1(x_1)g_2(x_2)$. If $g_1(x_1)g_2(x_2) = 0$ then $h(x_1, x_2) = 0$ hence $h(x_1, x_2) = Cg_1(x_1)g_2(x_2)$ in all cases. The general case follows by induction on the dimension n: suppose that

$$h = f_1 \otimes g_1 = \cdots = f_n \otimes g_n = f_{n+1} \otimes g_{n+1};$$

for fixed x_{n+1} the function $k = f_1 \otimes g_1 = \cdots = f_n \otimes g_n$ is given by

$$k(x, x_{n+1}) = C(x_{n+1})g_1(x_1) \cdots g_n(x_n).$$

Since we also have

$$k(x, x_{n+1}) = f_{n+1}(x_1, \ldots, x_n)g_{n+1}(x_{n+1})$$

it follows that $C(x_{n+1}) = C$. $\square$

6.2.2   Proof of the multi-dimensional Hardy uncertainty principle

We are going the use the lemmas above to prove the following extension of Hardy's uncertainty principle:

Theorem 105. *Let A and B be two real positive definite matrices and $\psi \in L^2(\mathbb{R}^n)$, $\psi \neq 0$. Assume that*

$$|\psi(x)| \leq C_A e^{-\frac{1}{2\hbar}Ax^2} \quad and \quad |F\psi(p)| \leq C_B e^{-\frac{1}{2\hbar}Bp^2} \tag{6.20}$$

for some constants $C_A, C_B > 0$. Then:

  (i) *The eigenvalues λ_j, $j = 1, \ldots, n$, of the matrix AB are all ≤ 1;*

  (ii) *If $\lambda_j = 1$ for all j, then $\psi(x) = Ce^{-\frac{1}{2\hbar}Ax^2}$ for some some complex constant C;*

  (iii) *If $\lambda_j < 1$ for some j then the space of functions satisfying (6.20) contains every $\psi(x) = Q(x)e^{-\frac{1}{2\hbar}Ax^2}$ where Q is a complex polynomial.*

Proof. It is of course no restriction to assume that $C_A = C_B = C$. Let L be as in Lemma 102 and order the eigenvalues of AB decreasingly: $\lambda_1 \geq \lambda_2 \geq \cdots \geq \lambda_n$. It suffices to show that $\lambda_1 \leq 1$. Setting $\psi_L(x) = \psi(Lx)$ we have

$$F\psi_L(p) = F\psi((L^T)^{-1}p);$$

in view of (6.16) in Lemma 102 condition (6.20) is equivalent to

$$|\psi_L(x)| \leq Ce^{-\frac{1}{2\hbar}\Lambda x^2} \quad and \quad |F\psi_L(p)| \leq Ce^{-\frac{1}{2\hbar}\Lambda p^2} \tag{6.21}$$

where $\Lambda = \mathrm{diag}(\lambda_1, \lambda_2, \ldots, \lambda_n)$. Setting $\psi_{L,1}(x_1) = \psi_L(x_1, 0, \ldots, 0)$ we have

$$|\psi_{L,1}(x_1)| \leq Ce^{-\frac{1}{2\hbar}\lambda_1 x_1^2}. \tag{6.22}$$

On the other hand, by the Fourier inversion formula,

$$\int F\psi_L(p)dp_2 \cdots dp_n = (2\pi\hbar)^{n/2} \int \cdots \int_{\mathbb{R}^n} e^{-\frac{i}{\hbar}p\cdot x}\psi_L(x)dxdp_2\cdots dp_n$$
$$= (2\pi\hbar)^{(n-1)/2}F\psi_{L,1}(p_1)$$

and hence we have the inequality

$$|F\psi_{L,1}(p_1)| \leq C_{L,1}e^{-\frac{1}{2\hbar}\lambda_1 p_1^2} \tag{6.23}$$

for some constant $C_{L,1} > 0$. Applying Hardy's uncertainty principle in one dimension to the inequalities (6.22) and (6.23) we must have $\lambda_1^2 \leq 1$ hence the assertion (i).

Proof of (ii). The condition $\lambda_j = 1$ for all j means that

$$|\psi_L(x)| \leq Ce^{-\frac{1}{2\hbar}x^2} \quad and \quad |F\psi_L(p)| \leq Ce^{-\frac{1}{2\hbar}p^2} \tag{6.24}$$

for some $C > 0$. Let us keep $x' = (x_2, \ldots, x_n)$ constant; the partial Fourier transform of ψ_L in the x_1 variable is $F_1\psi_L = (F')^{-1}F\psi_L$ where $(F')^{-1}$ is the inverse Fourier transform in the x' variables, hence there exists $C' > 0$ such that

$$|F_1\psi_L(x_1, x')| \leq \left(\tfrac{1}{2\pi\hbar}\right)^{\frac{n-1}{2}} \int |F\psi_L(p)|dp_2\cdots dp_n \leq C'e^{-\frac{1}{2\hbar}p_1^2}.$$

Since $|\psi_L(x)| \leq C(x')e^{-\frac{1}{2\hbar}x_1^2}$ with $C(x') \leq e^{-\frac{1}{2\hbar}x'^2}$ it follows from Hardy's theorem that we can write

$$\psi_L(x) = f_1(x')e^{-\frac{1}{2\hbar}x_1^2}$$

for some real C^∞ function f_1 on $\mathbb{R}^{n-1}$. Applying the same argument to the remaining variables $x_2, \ldots, x_n$ we conclude that there exist C^∞ functions f_j for $j = 2, \ldots, n$, such that

$$\psi_L(x) = f_j(x_1, \ldots, \widetilde{x}_j, \ldots, x_n)e^{-\frac{1}{2\hbar}x_1^2}. \tag{6.25}$$

In view of Lemma 104 above we have $\psi_L(x) = C_L e^{-\frac{1}{2\hbar}x^2}$ for some constant C_L; since $\Lambda = I = L^T A L$ we thus have $\psi(x) = C_L e^{-Ax^2/2\hbar}$ as claimed.

Proof of (iii). Assume that $\lambda_1 < 1$ for $j \in \mathcal{J}$, $\mathcal{J}$ a subset of $\{1, \ldots, n\}$. By the same argument as in the proof of part (ii) establishing formula (6.25), we infer, using Hardy's theorem in the case $ab < 1$, that

$$\psi_L(x) = f_j(x_1, \ldots, \widetilde{x}_j, \ldots, x_n)Q_j(x_j)e^{-\frac{1}{2\hbar}x_j^2}$$

where Q_j is a polynomial with degree 0 if $j \notin \mathcal{J}$. One concludes the proof using once again Lemma 104. $\qquad\square$

6.2.3 Geometric interpretation

Let us give a geometric interpretation of Theorem 105 in terms of the notion of symplectic capacity. We begin by making an obvious observation: Hardy's uncertainty principle can be restated by saying that if $\psi \neq 0$ then the conditions $\psi(x) = \mathcal{O}(e^{-\frac{1}{2\hbar}ax^2})$ *and* $F\psi(p) = \mathcal{O}(e^{-\frac{1}{2\hbar}bp^2})$ imply that the ellipse $\mathcal{W} : ax^2 + bp^2 \leq \hbar$ has area $\pi\hbar/\sqrt{ab} \geq \pi\hbar = \frac{1}{2}h$:

$$\mathrm{Area}(\mathcal{W}) \geq 1/2h.$$

More precisely:

> *If the area of the ellipse $\mathcal{W}$ is smaller than $\frac{1}{2}h$ then $\psi = 0$; if this area equals $\frac{1}{2}h$ then $\psi(x) = Ce^{-\frac{1}{2\hbar}ax^2}$ and if it is larger than $\frac{1}{2}h$ then the functions $\psi(x) = Q(x)e^{-\frac{1}{2\hbar}ax^2}$, Q a polynomial, belong to the set of functions satisfying $\mathrm{Area}(\mathcal{W}) > \frac{1}{2}h$.*

We can restate Hardy's theorem in a very simple geometric way in terms of the symplectic capacity (not the volume!) of an ellipsoid. Recall that all symplectic capacities agree on phase space ellipsoids in view of Proposition 94. Recall that $\pi\hbar = \frac{1}{2}h$.

Proposition 106. *Let $\psi \in L^2(\mathbb{R}^n)$, $\psi \neq 0$. Assume that there exist constants $C_A > 0$ and $C_B > 0$ such that*

$$|\psi(x)| \leq C_A e^{-\frac{1}{2\hbar}Ax^2} \quad \text{and} \quad |F\psi(p)| \leq C_B e^{-\frac{1}{2\hbar}Bp^2}. \tag{6.26}$$

Then the symplectic capacity of the ellipsoid

$$\mathcal{W} = \{(x,p) : Ax^2 + Bp^2 \leq \hbar\}$$

is such that $c(\mathcal{W}) \geq \frac{1}{2}h$.

Proof of (i). Setting $M = \begin{pmatrix} A & 0 \\ 0 & B \end{pmatrix}$ the equation of $\mathcal{W}$ is $Mz^2 \leq \hbar$. In view of formula (5.18) in Proposition 94 together with the conformality property of symplectic capacities we have $c(\mathcal{W}) = \pi\hbar/\lambda_1^{\sigma}$ where λ_1^{σ} is the smallest of all numbers λ such that $\pm i\lambda$ is an eigenvalue of JM. In view of Lemma 103 $\lambda_j^{\sigma} = \sqrt{\lambda_j}$ where the λ_j are the eigenvalues of AB, and by Theorem 105 we must have $\lambda_j \leq 1$; the result follows since $\pi\hbar = \frac{1}{2}h$. $\qquad\square$

The result above thus shows that Hardy's uncertainty principle has an important (and unexpected) geometrical meaning. It will be generalized later on in this book when we discuss the Wigner formalism.

Part II

Harmonic Analysis in Symplectic Spaces

Chapter 7

The Metaplectic Group

The metaplectic group is a unitary representation of the double cover of the symplectic group; it plays an essential role in Weyl pseudodifferential calculus, because it appears as a characteristic group of symmetries for Weyl operators. In fact – and this fact seems to be largely ignored in the literature – this property (called "symplectic covariance") actually is characteristic (in a sense that will be made precise) of Weyl calculus. Metaplectic operators of course have many other applications; they allow us, for instance, to give explicit solutions to the time-dependent Schrödinger equation with quadratic Hamiltonian, as will be shown later, but they are also used with profit in optics, engineering, and last but not least, in quantum mechanics.

7.1 The metaplectic representation

The idea behind the metaplectic representation of the symplectic group is that one can associate to every symplectic matrix a pair of unitary operators on $L^2(\mathbb{R}^n)$ differing by a sign. Technically this is achieved by constructing of a unitary representation of the (connected) double covering $\mathrm{Sp}_2(2n, \mathbb{R})$ of $\mathrm{Sp}(2n, \mathbb{R})$. This representation (which is not irreducible, see Exercise 138 in Chapter 8) is called the metaplectic group and is denoted by $\mathrm{Mp}(2n, \mathbb{R})$. Equivalently, the sequence

$$0 \longrightarrow \mathbb{Z}_2 \longrightarrow \mathrm{Mp}(2n, \mathbb{R}) \longrightarrow \mathrm{Sp}(2n, \mathbb{R}) \longrightarrow 0$$

is exact. In many texts the existence of the metaplectic representation is motivated by vague considerations about the uniqueness of the Schrödinger representation and the Heisenberg–Weyl operators $\widehat{T}(z)$ which will be studied in Chapter 8. Following this argument there must exist, for every $S \in \mathrm{Sp}(2n, \mathbb{R})$ a unitary operator $\widehat{S}$ such that $\widehat{S}\widehat{T}(z)\widehat{S}^{-1} = \widehat{T}(Sz)$. However this relation certainly does not characterize precisely $\widehat{S}$ since it is still true if we replace it by $c\widehat{S}$ with $|c| = 1$. At best one obtains in this way a projective representation of the symplectic group.

We are following closely de Gosson [67], Chapter 7.

7.1.1   A preliminary remark, and a caveat

In many texts the metaplectic group is presented as being the group of unitary operators generated by the following elementary unitary operators:

- The modified Fourier transform $\mu(J)$ defined by

$$\mu(J)\psi(x) = \left(\tfrac{1}{2\pi i\hbar}\right)^n \int_{\mathbb{R}^n} e^{-\frac{i}{\hbar}x\cdot x'}\,\psi(x')dx';$$

- The "chirps"

$$\mu(V_{-P})\psi(x) = e^{\frac{i}{2\hbar}Px\cdot x}\psi(x)$$

  where P is a real symmetric matrix;

- The rescaling operators

$$\mu(M_L)\psi(x) = \sqrt{\det L}\,\psi(Lx)$$

  where $\sqrt{\det L}$ is "some adequate" determination of the square root of $\det L$.

We emphasize that the notation $\mu(S)$ used above to denote metaplectic operators associated to a symplectic automorphism S is ambiguous, and should be avoided because it can lead to contradictions; unfortunately it is often found in the literature, even in the best treatises. We will not use it in this book.

7.1.2   Quadratic Fourier transforms

We have seen that the symplectic group $\mathrm{Sp}(2n,\mathbb{R})$ is generated by the free symplectic matrices

$$S = \begin{pmatrix} A & B \\ C & D \end{pmatrix} \in \mathrm{Sp}(2n,\mathbb{R})\ ,\quad \det B \neq 0.$$

To each such matrix we associated the generating function

$$W(x,x') = \tfrac{1}{2}DB^{-1}x^2 - B^{-1}x\cdot x' + \tfrac{1}{2}B^{-1}Ax'^2$$

and we showed that

$$(x,p) = S(x',p') \Longleftrightarrow p = \partial_x W(x,x')\ ,\ p' = -\partial_{x'}W(x,x').$$

Conversely, to every polynomial of the type

$$W(x,x') = \tfrac{1}{2}Px^2 - Lx\cdot x' + \tfrac{1}{2}Qx'^2 \tag{7.1}$$
$$\text{with}\ \ P = P^T\ ,\ Q = Q^T\ ,\ \text{and}\ \ \det L \neq 0$$

we can associate a free symplectic matrix, namely

$$S_W = \begin{pmatrix} L^{-1}Q & L^{-1} \\ PL^{-1}Q - L^T & PL^{-1} \end{pmatrix}. \tag{7.2}$$

We now associate an operator $\widehat{S}_{W,m}$ to every S_W by setting, for $\psi \in \mathcal{S}(\mathbb{R}^n)$,

$$\widehat{S}_{W,m}\psi(x) = \left(\tfrac{1}{2\pi i}\right)^{n/2} \Delta(W) \int_{\mathbb{R}^n} e^{iW(x,x')}\psi(x')dx'; \tag{7.3}$$

here $\arg i = \pi/2$ and the factor $\Delta(W)$ is defined by

$$\Delta(W) = i^m \sqrt{|\det L|}; \tag{7.4}$$

the integer m corresponds to a choice of $\arg \det L$:

$$m\pi \equiv \arg \det L \mod 2\pi. \tag{7.5}$$

Notice that we can rewrite Definition (7.3) in the form

$$\widehat{S}_{W,m}\psi(x) = \left(\tfrac{1}{2\pi}\right)^{n/2} \left(e^{-i\frac{\pi}{4}}\right)^{\mu} \Delta(W) \int_{\mathbb{R}^n} e^{iW(x,x')}\psi(x')dx' \tag{7.6}$$

where

$$\mu = 2m - n. \tag{7.7}$$

Definition 107.

(i) The operator $\widehat{S}_{W,m}$ is called a "quadratic Fourier transform" associated to the free symplectic matrix S_W.
(ii) The class modulo 4 of the integer m is called "Maslov index" of $\widehat{S}_{W,m}$. The quadratic Fourier transform corresponding to the choices $S_W = J$ and $m = 0$ is denoted by $\widehat{J}$.

We will not discuss the properties of the Maslov index in this book; it has been studied in a comprehensive way in de Gosson [67].

The generating function of J being simply $W(x,x') = -x \cdot x'$, it follows that

$$\widehat{J}\psi(x) = \left(\tfrac{1}{2\pi i}\right)^{n/2} \int_{\mathbb{R}^n} e^{-ix \cdot x'}\psi(x')dx' = i^{-n/2}F\psi(x) \tag{7.8}$$

for $\psi \in \mathcal{S}(\mathbb{R}^n)$; F is the usual unitary Fourier transform defined by

$$F\psi(x) = \left(\tfrac{1}{2\pi}\right)^{n/2} \int_{\mathbb{R}^n} e^{-ix \cdot x'}\psi(x')dx'.$$

It follows from the Fourier inversion formula

$$F^{-1}\psi(x) = \left(\tfrac{1}{2\pi}\right)^{n/2} \int_{\mathbb{R}^n} e^{ix \cdot x'}\psi(x')dx'$$

that the inverse $\widehat{J}^{-1}$ of $\widehat{J}$ is given by the formula

$$\widehat{J}^{-1}\psi(x) = \left(\tfrac{i}{2\pi}\right)^{n/2} \int_{\mathbb{R}^n} e^{ix \cdot x'}\psi(x')dx' = i^{n/2}F^{-1}\psi(x).$$

Note that the identity operator cannot be represented by an operator $\widehat{S}_{W,m}$ since it is not a free symplectic matrix.

Of course, if m is one choice of Maslov index, then $m+2$ is another equally good choice: to each function W formula (7.3) associates not one but *two* operators $\widehat{S}_{W,m}$ and $\widehat{S}_{W,m+2} = -\widehat{S}_{W,m}$ (this reflects the fact that the operators $\widehat{S}_{W,m}$ are elements of the two-fold covering group of $\mathrm{Sp}(2n,\mathbb{R})$).

Let us define operators $\widehat{V}_{-P}$ and $\widehat{M}_{L,m}$ by

$$\widehat{V}_{-P}\psi(x) = e^{\frac{i}{2}Px\cdot x}\psi(x) \quad , \quad \widehat{M}_{L,m}\psi(x) = i^m\sqrt{|\det L|}\psi(Lx). \qquad (7.9)$$

We have the following useful factorization result:

Proposition 108. *Let W be the quadratic form* (7.1).

(i) *We have the factorization*

$$\widehat{S}_{W,m} = \widehat{V}_{-P}\widehat{M}_{L,m}\widetilde{J}\widehat{V}_{-Q}; \qquad (7.10)$$

(ii) *The operators $\widehat{S}_{W,m}$ extend to unitary operators $L^2(\mathbb{R}^n) \longrightarrow L^2(\mathbb{R}^n)$ and the inverse of $\widehat{S}_{W,m}$ is*

$$\widehat{S}_{W,m}^{-1} = \widehat{S}_{W^*,m^*} \text{ with } W^*(x,x') = -W(x',x) \, , \, m^* = n - m. \qquad (7.11)$$

Proof. (i) By definition of $\widehat{J}$ we have

$$\widehat{J}\psi(x) = i^{-n/2}F\psi(x) = \left(\tfrac{1}{2\pi i}\right)^{n/2}\int_{\mathbb{R}^n} e^{-ix\cdot x'}\psi(x')dx';$$

the factorization (7.10) immediately follows noting that

$$\widehat{M}_{L,m}\widehat{J}\psi(x) = \left(\tfrac{1}{2\pi i}\right)^{n/2} i^m\sqrt{|\det L|}\int_{\mathbb{R}^n} e^{-iLx\cdot x'}\psi(x')dx'.$$

(ii) The operators $\widehat{V}_{-P}$ and $\widehat{M}_{L,m}$ are trivially unitary, and so is the modified Fourier transform $\widehat{J}$; (ii) We obviously have

$$(\widehat{V}_{-P})^{-1} = \widehat{V}_P \text{ and } (\widehat{M}_{L,m})^{-1} = \widehat{M}_{L^{-1},-m}$$

and $\widehat{J}^{-1}$ is given by

$$\widehat{J}^{-1}\psi(x) = \left(\tfrac{i}{2\pi\hbar}\right)^{n/2}\int_{\mathbb{R}^n} e^{\frac{i}{\hbar}x\cdot x'}\psi(x')dx'.$$

Writing

$$\widehat{S}_{W,m}^{-1} = \widehat{V}_Q\widehat{J}^{-1}\widehat{M}_{L^{-1},-m}\widehat{V}_P$$

and noting that

$$\widehat{J}^{-1}\widehat{M}_{L^{-1},-m}\psi(x) = \left(\tfrac{i}{2\pi}\right)^{n/2} i^{-m} \sqrt{|\det L^{-1}|} \int_{\mathbb{R}^n} e^{ix\cdot x'}\psi(L^{-1}x')dx'$$

$$= \left(\tfrac{1}{2\pi i}\right)^{n/2} i^{-m+n} \sqrt{|\det L|} \int_{\mathbb{R}^n} e^{iL^T x\cdot x'}\psi(x')dx'$$

$$= \widehat{M}_{-L^T,n-m}\widehat{J}\psi(x)$$

the inversion formulas (7.11) follow. $\qquad\square$

It follows from the proposition above that the operators $\widehat{S}_{W,m}$ form a subset of the group $\mathcal{U}(L^2(\mathbb{R}^n))$ of unitary operators acting on $L^2(\mathbb{R}^n)$, which is closed under the operation of inversion. They thus generate a subgroup of $\mathcal{U}(L^2(\mathbb{R}^n))$.

Definition 109. The subgroup of $\mathcal{U}(L^2(\mathbb{R}^n))$ generated by the quadratic Fourier transforms $\widehat{S}_{W,m}$ is called the "metaplectic group" and is denoted by $\mathrm{Mp}(2n,\mathbb{R})$. The elements of $\mathrm{Mp}(2n,\mathbb{R})$ are called "metaplectic operators".

Every $\widehat{S} \in \mathrm{Mp}(2n,\mathbb{R})$ is thus, by definition, a product $\widehat{S}_{W_1,m_1}\cdots\widehat{S}_{W_k,m_k}$ of metaplectic operators associated to free symplectic matrices. We will use the following result which considerably simplifies many arguments; it is the metaplectic analogue of Theorem 60 which says that every symplectic matrix can be written as the product of two free symplectic matrices:

Proposition 110. *Every $\widehat{S} \in \mathrm{Mp}(2n,\mathbb{R})$ can be written as a product of exactly two quadratic Fourier transforms: $\widehat{S} = \widehat{S}_{W,m}\widehat{S}_{W',m'}$. (Such a factorization is, however, never unique: for instance $I = \widehat{S}_{W,m}\widehat{S}_{W^*,m^*}$ for every generating function W.)*

Proof. See Leray [114], Ch. 1, or de Gosson [64]); the result follows from the existence of a natural projection $\mathrm{Mp}(2n,\mathbb{R}) \longrightarrow \mathrm{Sp}(2n,\mathbb{R})$ which will be established in the next section and the fact that every $S \in \mathrm{Sp}(2n,\mathbb{R})$ can be written as a product $S_W S_{W'}$. (See Exercise 111 below.) $\qquad\square$

Exercise 111. Prove Proposition 110 above.

Proposition 110 has the following immediate consequence:

Corollary 112. *The metaplectic group $\mathrm{Mp}(2n,\mathbb{R})$ is generated by the operators $\widehat{V}_{-P}$, $\widehat{M}_{L,m}$, and $\widehat{J}$.*

Proof. It follows from the definition above of $\mathrm{Mp}(2n,\mathbb{R})$ together with the fact that each $\widehat{S}_{W,m}$ is a product $\widehat{V}_{-P}\widehat{M}_{L,m}\widehat{J}\widehat{V}_{-Q}$ (formula (7.10)). $\qquad\square$

Exercise 113. Show that quadratic Fourier transform $\widehat{S}_{W,m}$ cannot be a local operator (a local operator on $\mathcal{S}'(\mathbb{R}^n)$ is an operator $\widehat{S}$ such that $\mathrm{Supp}(\widehat{S}\psi) \subset \mathrm{Supp}(\psi)$ for $\psi \in \mathcal{S}'(\mathbb{R}^n)$).

7.2 The projection π^{Mp}

It turns out that $\mathrm{Mp}(2n, \mathbb{R})$ is a double covering of the symplectic group $\mathrm{Sp}(2n, \mathbb{R})$ and hence a faithful representation of $\mathrm{Sp}(2n, \mathbb{R})$.

7.2.1 Precise statement

The main result of this subsection is the following, whose detailed proof is given in Section 7.3:

Theorem 114. *The mapping $\widehat{S}_{W,m} \longmapsto S_W$, which to the quadratic Fourier transform*

$$\widehat{S}_{W,m}\psi(x) = \left(\tfrac{1}{2\pi i}\right)^{n/2} \Delta(W) \int_{\mathbb{R}^n} e^{iW(x,x')}\psi(x')dx'$$

associates the free symplectic matrix with generating function W, extends into a surjective group homomorphism

$$\pi^{\mathrm{Mp}} : \mathrm{Mp}(2n, \mathbb{R}) \longrightarrow \mathrm{Sp}(2n, \mathbb{R});$$

that is

$$\pi^{\mathrm{Mp}}(\widehat{S}\widehat{S}') = \pi^{\mathrm{Mp}}(\widehat{S})\pi^{\mathrm{Mp}}(\widehat{S}')$$

and the kernel of π^{Mp} is

$$\ker(\pi^{\mathrm{Mp}}) = \{-I, +I\}.$$

Hence $\pi^{\mathrm{Mp}} : \mathrm{Mp}(2n, \mathbb{R}) \longrightarrow \mathrm{Sp}(2n, \mathbb{R})$ is a twofold covering of the symplectic group.

The last statement follows from the theory of covering groups: a covering group of a Lie group has discrete fiber, isomorphic to the kernel of the projection homomorphism.

Definition 115. We will call the homomorphism π^{Mp} the natural projection, or covering mapping of $\mathrm{Mp}(2n, \mathbb{R})$ onto $\mathrm{Sp}(2n, \mathbb{R})$.

Recalling that $\mathrm{Sp}(2n, \mathbb{R})$ is generated by the symplectic matrices J and

$$M_L = \begin{pmatrix} L^{-1} & 0 \\ 0 & L^T \end{pmatrix} \quad , \quad V_{-P} = \begin{pmatrix} I & 0 \\ P & I \end{pmatrix}$$

$(\det L \neq 0, P = P^T)$, the natural projection has in addition the following properties:

Proposition 116. We have:

$$\pi^{\mathrm{Mp}}(\widehat{J}) = J \quad , \quad \pi^{\mathrm{Mp}}(\widehat{M}_{L,m}) = M_L \quad , \quad \pi^{\mathrm{Mp}}(\widehat{V}_P) = V_P. \tag{7.12}$$

Proof. The formula $\pi^{\mathrm{Mp}}(\widehat{J}) = J$ is obvious since $\widehat{J}$ is a quadratic Fourier transform with $W(x, x') = -x \cdot x'$. Let us prove the second formula (7.12). We have $\widehat{M}_{L,m} = \widehat{J}^{-1}(\widehat{JM_{L,m}})$ and $\widehat{JM_{L,m}} = \widehat{S}_{W,m}$ with $W(x, x') = -(L^T)^{-1} x \cdot x'$. Since $\pi^{\mathrm{Mp}}(\widehat{J}^{-1}) = J^{-1} = -J$ it follows that

$$\pi^{\mathrm{Mp}}(\widehat{M}_{L,m}) = \pi^{\mathrm{Mp}}(\widehat{J}^{-1})\pi^{\mathrm{Mp}}(\widehat{JM_{L,m}})$$
$$= \begin{pmatrix} 0 & I \\ -I & 0 \end{pmatrix} \begin{pmatrix} 0 & L^T \\ L^{-1} & 0 \end{pmatrix}$$
$$= \begin{pmatrix} L^{-1} & 0 \\ 0 & L^T \end{pmatrix}$$

hence $\pi^{\mathrm{Mp}}(\widehat{M}_{L,m}) = M_L$. The formula $\pi^{\mathrm{Mp}}(\widehat{V}_P) = V_P$ is proven using a similar argument. $\qquad\square$

7.2.2 Dependence on $\hbar$

It is useful to have a parameter-dependent version of $\mathrm{Mp}(2n, \mathbb{R})$; in the applications to quantum mechanics that parameter is $\hbar$, Planck's constant h divided by 2π.

The main observation is that a covering group can be "realized" in many different ways. Instead of choosing π^{Mp} as a projection, we could as well have chosen any other mapping $\mathrm{Mp}(2n, \mathbb{R}) \longrightarrow \mathrm{Sp}(2n, \mathbb{R})$ obtained from π^{Mp} by composing it on the left with an inner automorphism of $\mathrm{Mp}(2n, \mathbb{R})$, or on the right with an inner automorphism of $\mathrm{Sp}(2n, \mathbb{R})$, or both. The point is here that the diagram

$$\begin{array}{ccc} \mathrm{Mp}(2n, \mathbb{R}) & \xrightarrow{\;F\;} & \mathrm{Mp}(2n, \mathbb{R}) \\ \pi^{\mathrm{Mp}} \downarrow & & \downarrow \pi^{\mathrm{Mp}'} \\ \mathrm{Sp}(2n, \mathbb{R}) & \xrightarrow[G]{} & \mathrm{Sp}(2n, \mathbb{R}) \end{array}$$

is commutative: $\pi'^{\mathrm{Mp}} \circ F = G \circ \pi^{\mathrm{Mp}}$, because for all such π'^{Mp} we will have $\mathrm{Ker}(\pi'^{\mathrm{Mp}}) = \{\pm I\}$ and

$$\pi'^{\mathrm{Mp}} : \mathrm{Mp}(2n, \mathbb{R}) \longrightarrow \mathrm{Sp}(2n, \mathbb{R})$$

will then also be a covering mapping. We find it particularly convenient to define a new projection by using the following inner automorphism of $\mathrm{Mp}(2n, \mathbb{R})$: for $\lambda > 0$ set $\widehat{M}_\lambda = \widehat{M}_{\lambda I, 0}$, that is

$$\widehat{M}_\lambda \psi(x) = \lambda^{n/2} \psi(\lambda x) \ , \ \ \psi \in L^2(\mathbb{R}^n)$$

and denote by M_λ the projection of $\widehat{M}_\lambda$ on $\mathrm{Sp}(2n, \mathbb{R})$:

$$M_\lambda(x, p) = (\lambda^{-1} x, \lambda p).$$

We have $\widehat{M}_\lambda \in \mathrm{Mp}(2n,\mathbb{R})$ and $M_\lambda \in \mathrm{Sp}(2n,\mathbb{R})$. For $\widehat{S} \in \mathrm{Mp}(2n,\mathbb{R})$ we define $\widehat{S}^\hbar \in \mathrm{Mp}(2n,\mathbb{R})$ by

$$\widehat{S}^\hbar = \widehat{M}_{1/\sqrt{\hbar}}\widehat{S}\widehat{M}_{\sqrt{\hbar}}. \tag{7.13}$$

The projection of $S^\hbar$ on $\mathrm{Sp}(2n,\mathbb{R})$ is then given by:

$$\pi^{\mathrm{Mp}}(\widehat{S}^\hbar) = S^\hbar = M_{1/\sqrt{\hbar}}SM_{\sqrt{\hbar}}.$$

We now define the new projection

$$\pi^{\mathrm{Mp}\,\hbar} : \mathrm{Mp}(2n,\mathbb{R}) \longrightarrow \mathrm{Sp}(2n,\mathbb{R})$$

by the formula

$$\pi^{\mathrm{Mp}\,\hbar}(\widehat{S}^\hbar) = M_{\sqrt{\hbar}}(\pi^{\mathrm{Mp}}(\widehat{S}^\hbar))M_{1/\sqrt{\hbar}}$$

which is of course equivalent to

$$\pi^{\mathrm{Mp}^\hbar}(\widehat{S}^\hbar) = \pi^{\mathrm{Mp}}(\widehat{S}).$$

Suppose for instance that $\widehat{S} = \widehat{S}_{W,m}$; it is easily checked using the fact that W is homogeneous of degree 2 in (x, x') that

$$\widehat{S}^\hbar_{W,m}\psi(x) = \left(\tfrac{1}{2\pi i \hbar}\right)^{n/2} \Delta(W) \int_{\mathbb{R}^n} e^{\frac{i}{\hbar}W(x,x')}\psi(x')\,dx'$$

or, equivalently,

$$\widehat{S}^\hbar_{W,m} = \hbar^{-n/2}\widehat{S}_{W/\hbar,m}.$$

Also,

$$(\widehat{S}^\hbar_{W,m})^{-1}\psi(x) = \left(\tfrac{i}{2\pi\hbar}\right)^{n/2} \Delta(W^*) \int_{\mathbb{R}^n} e^{\frac{i}{\hbar}W^*(x,x')}\psi(x')\,dx'.$$

The projection of $\widehat{S}^\hbar_{W,m}$ on $\mathrm{Sp}(2n,\mathbb{R})$ is the free matrix S_W:

$$\pi^{\mathrm{Mp}^\hbar}(\widehat{S}^\hbar_{W,m}) = S_W. \tag{7.14}$$

Exercise 117. Show that if $\hbar$ and $\hbar'$ are two positive numbers, then we have

$$\widehat{S}^\hbar_{W,m} = M_{\sqrt{\hbar'/\hbar}}\widehat{S}^{\hbar'}_{W,m}M_{\sqrt{\hbar/\hbar'}}.$$

(i.e., $\mathrm{Mp}(2n,\mathbb{R})$ and $\mathrm{Mp}(2n,\mathbb{R})$ are equivalent representations of the metaplectic group).

In what follows we will use the following convention, notation, and terminology:

Notation 118. The projection $\mathrm{Mp}(2n,\mathbb{R}) \longrightarrow \mathrm{Sp}(2n,\mathbb{R})$ will always be assumed to be the homomorphism

$$\pi^{\mathrm{Mp}^\hbar} : \mathrm{Mp}(2n,\mathbb{R}) \longrightarrow \mathrm{Sp}(2n,\mathbb{R})$$

and we will drop all the superscripts referring to $\hbar$: we will write π^{Mp} for $\pi^{\mathrm{Mp}^\hbar}$, $\widehat{S}_{W,m}$ for $\widehat{S}^\hbar_{W,m}$ and $\widehat{S}$ for $\widehat{S}^\hbar$.

7.3   Construction of π^{Mp}

We begin by giving a few definitions.

7.3.1   The group $\mathrm{Diff}^{(1)}(n)$

Let us denote the elements of the dual $(\mathbb{R}^n \oplus \mathbb{R}^n)^*$ of $\mathbb{R}^n \oplus \mathbb{R}^n$ by a, b, etc. Thus $a(z) = a(x,p)$ is the value of the linear form a at the point $z = (x,p)$.

To every a we associate a first-order linear partial differential operator A obtained by replacing formally p in $a(x,p)$ by D_x: $A = a(x, D_x)$, $D_x = -i\partial_x$; thus, if $a(x,p) = \alpha \cdot x + \beta \cdot p$ for $\alpha = (\alpha_1, \ldots, \alpha_n)$, $\beta = (\beta_1, \ldots, \beta_n)$ in $\mathbb{R}^n$, then

$$A = \alpha \cdot x + \beta \cdot D_x = \alpha \cdot x - i\beta \cdot \partial_x. \tag{7.15}$$

Obviously the sum of two operators of the type above is an operator of the same type, and so is the product of such an operator by a scalar. It follows that these operators form a $2n$-dimensional vector space, which we denote by $\mathrm{Diff}^{(1)}(n)$.

The vector spaces $\mathbb{R}^n \oplus \mathbb{R}^n$, $(\mathbb{R}^n \oplus \mathbb{R}^n)^*$ and $\mathrm{Diff}^{(1)}(n)$ are isomorphic since they all have the same dimension $2n$. The following result explicitly describes three canonical isomorphisms between these spaces:

Lemma 119.

(i) *The linear mappings*

$$\varphi_1 : \mathbb{R}^n \oplus \mathbb{R}^n \longrightarrow (\mathbb{R}^n \oplus \mathbb{R}^n)^* \quad , \quad \varphi_1 : z_0 \longmapsto a,$$
$$\varphi_2 : (\mathbb{R}^n \oplus \mathbb{R}^n)^* \longrightarrow \mathrm{Diff}^{(1)}(n) \quad , \quad \varphi_2 : a \longmapsto A,$$

where a is the unique linear form on $\mathbb{R}^n \oplus \mathbb{R}^n$ such that $a(z) = \sigma(z, z_0)$, are isomorphisms, hence so is their compose φ :

$$\varphi = \varphi_2 \circ \varphi_1 : \mathbb{R}^n \oplus \mathbb{R}^n \longrightarrow \mathrm{Diff}^{(1)}(n);$$

the latter associates to $z_0 = (x_0, p_0)$ the operator

$$A = \varphi(z_0) = p_0 \cdot x - x_0 \cdot D_x.$$

(ii) *Let $[A, B] = AB - BA$ be the commutator of $A, B \in \mathrm{Diff}^{(1)}(n)$; we have*

$$[\varphi(z_1), \varphi(z_2)] = -i\sigma(z_1, z_2) \tag{7.16}$$

for all $z_1, z_2 \in \mathbb{R}^n \oplus \mathbb{R}^n$.

Proof. (i) The vector spaces $\mathbb{R}^n \oplus \mathbb{R}^n$, $(\mathbb{R}^n \oplus \mathbb{R}^n)^*$, and $\mathrm{Diff}^{(1)}(n)$ having the same dimension, it suffices to show that $\ker(\varphi_1)$ and $\ker(\varphi_2)$ are zero. Now, $\varphi_1(z_0) = 0$ is equivalent to the condition $\sigma(z, z_0) = 0$ for all z, and hence to $z_0 = 0$ since a symplectic form is non-degenerate. If $\varphi_2(a) = 0$ then

$$A\psi = \varphi_2(a)\psi = 0 \quad \text{for all} \quad \psi \in \mathcal{S}(\mathbb{R}^n)$$

which implies $A = 0$ and thus $a = 0$.

(ii) Let $z_1 = (x_1, p_1)$, $z_2 = (x_2, p_2)$. We have

$$\varphi(z_1) = p_1 \cdot x - x_1 \cdot D_x \quad , \quad \varphi(z_2) = p_2 \cdot x - x_2 \cdot D_x$$

and hence

$$[\varphi(z_1), \varphi(z_2)] = i(x_1 \cdot p_2 - x_2 \cdot p_1)$$

which is precisely the commutation formula (7.16). $\square$

We are next going to show that the metaplectic group $\mathrm{Mp}(2n, \mathbb{R})$ acts by conjugation on $\mathrm{Diff}^{(1)}(n)$. This will allow us to explicitly construct a covering mapping $\mathrm{Mp}(2n, \mathbb{R}) \longrightarrow \mathrm{Sp}(2n, \mathbb{R})$.

Recall that the symplectic matrices

$$V_P = \begin{pmatrix} I & 0 \\ -P & I \end{pmatrix}, \ M_L = \begin{pmatrix} L^{-1} & 0 \\ 0 & L^T \end{pmatrix}, \ J = \begin{pmatrix} 0 & I \\ I & 0 \end{pmatrix}$$

generate the group $\mathrm{Sp}(2n, \mathbb{R})$.

Lemma 120. *For $z_0 = (x_0, p_0) \in \mathbb{R}^n \oplus \mathbb{R}^n$ define $A \in \mathrm{Diff}^{(1)}(n)$ by*

$$A = \varphi(z_0) = p_0 \cdot x - x_0 \cdot D_x.$$

(i) *Let $\{\widehat{J}, \widehat{M_{L,m}}, \widehat{V_P}\}$ be the set of generators of $\mathrm{Mp}(2n, \mathbb{R})$ defined in Corollary 112. We have:*

$$\widehat{J}A\widehat{J}^{-1} = -x_0 \cdot x - p_0 \cdot D_x = \varphi(Jz_0), \tag{7.17}$$

$$\widehat{M_{L,m}}A(\widehat{M_{L,m}})^{-1} = L^T p_0 \cdot x - L^{-1} x_0 \cdot D_x = \varphi(M_L z_0), \tag{7.18}$$

$$\widehat{V_P}A(\widehat{V_P})^{-1} = (p_0 + Px_0) \cdot x - x_0 \cdot D_x = \varphi(V_P z_0). \tag{7.19}$$

(ii) *If $A \in \mathrm{Diff}^{(1)}(n)$ and $\widehat{S} \in \mathrm{Mp}(2n, \mathbb{R})$, then $\widehat{S}A\widehat{S}^{-1} \in \mathrm{Diff}^{(1)}(n)$.*

(iii) *For every $\widehat{S} \in \mathrm{Mp}(2n, \mathbb{R})$ the mapping*

$$\Phi_{\widehat{S}} : \mathrm{Diff}^{(1)}(n) \longrightarrow \mathrm{Diff}^{(1)}(n) \quad , \quad A \longmapsto \widehat{S}A\widehat{S}^{-1}$$

is a vector space automorphism.

Proof of (i). Using the properties of the Fourier transform, it is immediate to verify that:

$$(x_0 \cdot D_x)\psi = \widehat{J}^{-1}(x_0 \cdot x)\widehat{J}\psi,$$

$$(p_0 \cdot x)\psi = -\widehat{J}^{-1}(p_0 \cdot D_x)\widehat{J}\psi$$

for $\psi \in \mathcal{S}(\mathbb{R}^n)$ hence (7.17). To prove (7.18) it suffices to remark that

$$\widehat{M_{L,m}}(p_0 \cdot x)(\widehat{M_{L,m}})^{-1}\psi(x) = (p_0 \cdot Lx)\psi(x)$$

and

$$\widehat{M}_{L,m}(x_0 \cdot D_x)(\widehat{M}_{L,m})^{-1}\psi(x) = x_0(L^{-1})^T D_x \psi(x).$$

Let us prove formula (7.19). Recalling that by definition

$$\widehat{V}_{-P}\psi(x) = e^{\frac{i}{2}Px \cdot x}\psi(x)$$

we have, since P is symmetric,

$$(x_0 \cdot D_x)\widehat{V}_{-P}\psi(x) = \widehat{V}_{-P}\left(Px_0 \cdot x)\psi(x) + (p_0 \cdot D_x)\psi(x)\right)$$

and hence

$$\widehat{V}_P A(\widehat{V}_{-P}\psi)(x) = ([p_0 + Px_0) \cdot x]\,\psi(x) - (x_0 \cdot D_x)\psi(x)$$

which is (7.19).

Proof of (ii). Property (ii) immediately follows since $\widehat{S}$ is a product of operators $\widehat{J}$, $\widehat{M}_{L,m}$, $\widehat{V}_P$. (iii) The mapping $\Phi_{\widehat{S}}$ is trivially a linear mapping $\mathrm{Diff}^{(1)}(n) \longrightarrow \mathrm{Diff}^{(1)}(n)$. If $B = \widehat{S}A\widehat{S}^{-1} \in \mathrm{Diff}^{(1)}(n)$, then we have also $A = \widehat{S}^{-1}B\widehat{S} \in \mathrm{Diff}^{(1)}(n)$ since $A = \widehat{S}^{-1}B(\widehat{S}^{-1})^{-1}$. It follows that $\Phi_{\widehat{S}}$ is surjective and hence bijective. $\qquad\square$

Since the operators $\widehat{J}$, $\widehat{M}_{L,m}$, $\widehat{V}_P$ generate $\mathrm{Mp}(2n,\mathbb{R})$ the lemma above shows that for every $\widehat{S} \in \mathrm{Mp}(2n,\mathbb{R})$ there exists a linear automorphism S of $\mathbb{R}^n \oplus \mathbb{R}^n$ such that $\Phi_{\widehat{S}}(A) = \widehat{a \circ S}$ that is

$$\Phi_{\widehat{S}}(\varphi(z_0)) = \varphi(Sz_0). \tag{7.20}$$

Let us show that the automorphism S preserves the symplectic form. For $z, z' \in \mathbb{R}^n \oplus \mathbb{R}^n$ we have, in view of the commutation formula (7.16),

$$\begin{aligned}
\sigma(Sz, Sz') &= i\left[\varphi(Sz), \varphi(Sz')\right] &&= i\left[\Phi_{\widehat{S}}\varphi(z), \Phi_{\widehat{S}}\varphi(z')\right] \\
&= i\left[\widehat{S}\varphi(z)\widehat{S}^{-1}, \widehat{S}\varphi(z')\widehat{S}^{-1}\right] = i\widehat{S}\left[\varphi(z), \varphi(z')\right]\widehat{S}^{-1} \\
&= \sigma(z, z')
\end{aligned}$$

hence $S \in \mathrm{Sp}(2n,\mathbb{R})$ as claimed.

7.3.2   Construction of the projection

We are now able to describe explicitly the natural projection of $\mathrm{Mp}(2n,\mathbb{R})$ onto $\mathrm{Sp}(2n,\mathbb{R})$.

Definition 121. The covering projection $\pi^{\mathrm{Mp}} : \mathrm{Mp}(2n,\mathbb{R}) \longrightarrow \mathrm{Sp}(2n,\mathbb{R})$ is the mapping π^{Mp} which to $\widehat{S} \in \mathrm{Mp}(2n,\mathbb{R})$ associates the element $\pi^{\mathrm{Mp}}(\widehat{S}) = S \in \mathrm{Sp}(2n,\mathbb{R})$ defined by (7.20), that is

$$S = \varphi^{-1}\Phi_{\widehat{S}}\varphi. \tag{7.21}$$

That the mapping π^{Mp} indeed is a covering mapping follows from:

Proposition 122.

(i) *The mapping π^{Mp} is a continuous group epimorphism of $\mathrm{Mp}(2n,\mathbb{R})$ onto $\mathrm{Sp}(2n,\mathbb{R})$ such that:*

$$\pi^{\mathrm{Mp}}(\widehat{J}) = J \ , \ \ \pi^{\mathrm{Mp}}(\widehat{M}_{L,m}) = M_L \ , \ \ \pi^{\mathrm{Mp}}(\widehat{V}_P) = V_P \tag{7.22}$$

and hence

$$\pi^{\mathrm{Mp}}(\widehat{S}_{W,m}) = S_W. \tag{7.23}$$

(ii) *We have $\ker(\pi^{\mathrm{Mp}}) = \{-I, +I\}$; hence $\pi^{\mathrm{Mp}} : \mathrm{Mp}(2n,\mathbb{R}) \longrightarrow \mathrm{Sp}(2n,\mathbb{R})$ is a two-fold covering map.*

Proof. (i) Let us first show that π^{Mp} is a group homomorphism. In view of the obvious identity $\Phi_{\widehat{S}}\Phi_{\widehat{S}'} = \Phi_{\widehat{S}\widehat{S}'}$ we have

$$\begin{aligned}
\pi^{\mathrm{Mp}}(\widehat{S}\widehat{S}') &= \varphi^{-1}\Phi_{\widehat{S}\widehat{S}'}\varphi \\
&= (\varphi^{-1}\Phi_{\widehat{S}}\varphi)(\varphi^{-1}\Phi_{\widehat{S}'}\varphi) \\
&= \pi^{\mathrm{Mp}}(\widehat{S})\pi^{\mathrm{Mp}}(\widehat{S}').
\end{aligned}$$

Let us next prove that π^{Mp} is surjective. We have seen in Corollary 63 of Proposition 62 that the matrices J, M_L, and V_P generate $\mathrm{Sp}(2n,\mathbb{R})$ when L and P range over, respectively, the invertible and symmetric real matrices of order n. It is thus sufficient to show that formulae (7.22) hold. Now, using (7.17), (7.18), and (7.19) we have

$$\varphi\Phi_{\widehat{J}}\varphi^{-1} = J \ , \ \varphi\Phi_{\widehat{M}_{L,m}}\varphi^{-1} = M_L \ , \ \varphi\Phi_{\widehat{V}_P}\varphi^{-1} = V_P$$

hence (7.12). Formula (7.23) follows since every quadratic Fourier transform $\widehat{S}_{W,m}$ can be factorized as

$$\widehat{S}_{W,m} = \widehat{V}_{-P}\widehat{M}_{L,m}\widehat{J}\widehat{V}_{-Q}$$

in view of Proposition 108 above. To establish the continuity of the mapping π^{Mp} we first remark that the isomorphism $\varphi : \mathbb{R}^n \oplus \mathbb{R}^n \longrightarrow \mathrm{Diff}^{(1)}(n)$ defined in Lemma 119 is trivially continuous, and so is its inverse. Since $\Phi_{\widehat{S}\widehat{S}'} = \Phi_{\widehat{S}}\Phi_{\widehat{S}'}$ it suffices to show that for every $A \in \mathrm{Diff}^{(1)}(n)$, $\psi_S(A)$ has A as limit when $\widehat{S} \to I$ in $\mathrm{Mp}(2n,\mathbb{R})$. Now, $\mathrm{Mp}(2n,\mathbb{R})$ is a group of continuous automorphisms of $\mathcal{S}(\mathbb{R}^n)$ hence, when $\widehat{S} \to I$ then $\widehat{S}^{-1}\psi \to \psi$ for every $\psi \in \mathcal{S}(\mathbb{R}^n)$, that is $A\widehat{S}^{-1}\psi \to A\psi$ and also $\widehat{S}A\widehat{S}^{-1}\psi \to \psi$. (ii) Suppose that $\varphi^{-1}\Phi_{\widehat{S}}\varphi = I$. Then $\widehat{S}A\widehat{S}^{-1} = A$ for every $A \in \mathrm{Diff}^{(1)}(n)$ and this is only possible if $\widehat{S}$ is multiplication by a constant c with $|c| = 1$ (see exercise below); thus $\ker(\pi^{\mathrm{Mp}}) \subset S^1$. In view of Lemma 110 we have $\widehat{S} = \widehat{S}_{W,m}\widehat{S}_{W',m'}$ for some choice of (W,m) and (W',m') hence the condition $\widehat{S} \in \ker(\pi^{\mathrm{Mp}})$ is equivalent to

$$\widehat{S}_{W',m'} = c(\widehat{S}_{W,m})^{-1} = c\widehat{S}_{W^*,m^*}$$

which is only possible if $c = \pm 1$ hence $\widehat{S} = \pm I$ as claimed. $\qquad\square$

Chapter 8

Heisenberg–Weyl and Grossmann–Royer Operators

The Heisenberg–Weyl operators (also sometimes called simply "Heisenberg operators") are in a sense the easiest way to access quantum mechanics, because their definition can be understood in terms of a simple Hamiltonian dynamics: they are the time-one evolution operator for the quantized displacement Hamiltonian. One can actually also define these operators in terms of the phase function of a Lagrangian manifold without invoking any quantization at all; we will not use this approach here and refer the interested reader to Chapter 5 in de Gosson [67]). Together with their cousins, the Grossmann–Royer operators, the Heisenberg–Weyl operators play a key role in the theory of Weyl pseudo-differential operators, and moreover allow us to simplify many statements and proofs. In particular they allow a neat definition of the cross-ambiguity and Wigner transforms as we will see in Chapter 9. We will also briefly discuss the notion of Weyl–Heisenberg frame, also called Gabor frame in time-frequency analysis.

8.1 Dynamical motivation, and definition

The Heisenberg–Weyl operators (also sometimes called Heisenberg operators) are the "quantized" variants of phase-space translations. The presentation we give is "dynamical": we start with a displacement Hamiltonian, which we then "quantize", as opposed with the usual approaches.

8.1.1 The displacement Hamiltonian

The phase space translation operators $T(z_0) : z \longmapsto z + z_0$ are symplectomorphisms (because the Jacobian matrix of a translation is the identity, and is hence symplectic). These translations are in fact even Hamiltonian symplectomorphisms: we have

$T(z_0) \in \mathrm{Ham}(2n, \mathbb{R})$. To see this define, for each z_0, a Hamiltonian function H_{z_0} by

$$H_{z_0}(z) = \sigma(z, z_0) = p \cdot x_0 - p_0 \cdot x.$$

The associated Hamilton equations are $\dot{x} = x_0$, $\dot{p} = p_0$ so that the flow is given by the formula

$$\phi_t^{H_{z_0}}(z') = z' + tz_0$$

and we have $T(z_0) = \phi_1^{H_{z_0}}$.

Definition 123. The function $H_{z_0} = \sigma(z, z_0)$ is called the displacement (or *translation*) Hamiltonian.

Let us look for a "quantized" version of the $T(z_0)$. For this we consider the Schrödinger equation

$$i\hbar \frac{\partial \psi}{\partial t} = \widehat{H}_{z_0} \psi \quad , \quad \psi(x, 0) = \psi_0(x) \tag{8.1}$$

where $\widehat{H}_{z_0}$ is the operator

$$\widehat{H}_{z_0} = \sigma(\hat{z}, z_0) = -i\hbar x_0 \cdot \partial_x - p_0 \cdot x \tag{8.2}$$

obtained from H_{z_0} by formally replacing p by $-i\hbar\partial_x$. The solution of (8.1) can be formally written as

$$\psi(x, t) = \widehat{T}(z_0, t)\psi_0(x) = e^{-\frac{it}{\hbar}\sigma(\hat{z}, z_0)}\psi_0(x). \tag{8.3}$$

Using for instance the method of characteristics, or a direct calculation, one sees that an explicit formula for this solution is given by

$$\widehat{T}(z_0, t)\psi_0(x) = e^{\frac{i}{\hbar}(tp_0 \cdot x - \frac{1}{2}t^2 p_0 \cdot x_0)}\psi_0(x - tx_0). \tag{8.4}$$

It is clear that $\widehat{T}(z_0, t)$ is a unitary operator on $L^2(\mathbb{R}^n)$: we have

$$\|\widehat{T}(z_0, t)\psi\|_{L^2} = \|\psi\|_{L^2} \tag{8.5}$$

for every $\psi \in L^2(\mathbb{R}^n)$.

The reader is invited to observe that we have here a case where the Schrödinger equation can be explicitly solved; it is actually a particular case of a more general situation; we will come back to this when we study the Schrödinger equation in Chapter 15.

8.1.2 The Heisenberg–Weyl operators

The considerations above lead us to the following definition:

Definition 124. The operator $\widehat{T}(z_0) = \widehat{T}(z_0, 1)$ is called the Heisenberg–Weyl (for short: HW) operator determined by z_0. Thus, explicitly,

$$\widehat{T}(z_0)\psi = e^{\frac{i}{\hbar}(p_0 \cdot x - \frac{1}{2} p_0 \cdot x_0)}\psi(x - x_0). \tag{8.6}$$

This formula can also be written

$$\widehat{T}(z_0)\psi = e^{-\frac{i}{\hbar}\sigma(\hat{z}, z_0)}\psi_0(x) = e^{\frac{i}{\hbar}\sigma(z_0, \hat{z})}\psi(x) \tag{8.7}$$

(cf. formula (8.3)).

Notice that it is clear from formula (8.4) that we have $\widehat{T}(z_0, t) = \widehat{T}(tz_0)$.

While ordinary translation operators obviously form an abelian group isomorphic to the additive group $\mathbb{R}^n \oplus \mathbb{R}^n$:

$$T(z)T(z') = T(z')T(z) = T(z + z'),$$

this is not true of the HW operators; in particular these operators do not commute. The following relations are considered by many mathematicians or physicists almost as "mythic", in the sense that they are supposed to contain the essence of quantum mechanics. This view is however questionable, because the HW operators (and thus their commutation relations) can be defined using only classical arguments (the Hamilton–Jacobi theory together with the notion of phase of a Lagrangian manifold: see de Gosson [66, 67]).

Proposition 125. *The Heisenberg–Weyl operators satisfy the relations*

$$\widehat{T}(z_0)\widehat{T}(z_1) = e^{\frac{i}{\hbar}\sigma(z_0, z_1)}\widehat{T}(z_1)\widehat{T}(z_0) \tag{8.8}$$

and

$$\widehat{T}(z_0 + z_1) = e^{-\frac{i}{2\hbar}\sigma(z_0, z_1)}\widehat{T}(z_0)\widehat{T}(z_1) \tag{8.9}$$

for all $z_0, z_1 \in \mathbb{R}^{2n}$.

Proof. Translations act on functions on $\mathbb{R}^n$ via the formula $T(z_0)\psi(x) = \psi(x - x_0)$ if $z_0 = (x_0, p_0)$. Let us prove formula (8.8). We have

$$\widehat{T}(z_0)\widehat{T}(z_1) = \widehat{T}(z_0)(e^{\frac{i}{\hbar}(p_1 \cdot x - \frac{1}{2} p_1 \cdot x_1)}T(z_1))$$

$$= e^{\frac{i}{\hbar}(p_0 \cdot x - \frac{1}{2} p_0 \cdot x_0)}e^{\frac{i}{\hbar}(p_1 \cdot (x - x_0) - \frac{1}{2} p_1 \cdot x_1)}T(z_0 + z_1)$$

and, similarly

$$\widehat{T}(z_1)\widehat{T}(z_0) = e^{\frac{i}{\hbar}(p_1 \cdot x - \frac{1}{2} p_1 \cdot x_1)}e^{\frac{i}{\hbar}(p_0 \cdot (x - x_1) - \frac{1}{2} p_0 \cdot x_0)}T(z_0 + z_1).$$

Defining the quantities

$$\Phi = p_0 \cdot x - \tfrac{1}{2}p_0 \cdot x_0 + p_1 \cdot (x - x_0) - \tfrac{1}{2}p_1 \cdot x_1,$$
$$\Phi' = p_1 \cdot x - \tfrac{1}{2}p_1 \cdot x_1 + p_0 \cdot (x - x_1) - \tfrac{1}{2}p_0 \cdot x_0,$$

we have

$$\widehat{T}(z_0)\widehat{T}(z_1) = e^{\frac{i}{\hbar}(\Phi-\Phi')}\widehat{T}(z_1)\widehat{T}(z_0)$$

and an immediate calculation yields

$$\Phi - \Phi' = p_0 \cdot x_1 - p_1 \cdot x_0 = \sigma(z_0, z_1)$$

which proves (8.8). Let us next prove formula (8.9). We have

$$\widehat{T}(z_0 + z_1) = e^{\frac{i}{\hbar}\Phi''}T(z_0 + z_1)$$

with

$$\Phi'' = (p_0 + p_1) \cdot x - \tfrac{1}{2}(p_0 + p_1) \cdot (x_0 + x_1).$$

On the other hand we have seen above that

$$\widehat{T}(z_0)\widehat{T}(z_1) = e^{\frac{i}{\hbar}\Phi}T(z_0 + z_1)$$

so that

$$\widehat{T}(z_0 + z_1) = e^{\frac{i}{\hbar}(\Phi''-\Phi)}\widehat{T}(z_0)\widehat{T}(z_1).$$

A straightforward algebraic calculation shows that

$$\Phi'' - \Phi = \tfrac{1}{2}p_1 \cdot x_0 - \tfrac{1}{2}p_0 \cdot x_1 = -\tfrac{1}{2}\sigma(z_0, z_1)$$

hence formula (8.9). $\square$

Exercise 126. Prove formally the formulas (8.8) and (8.9) using the differential expression (8.7) of the Heisenberg–Weyl operators.

We note that the HW operators act on functions defined on "configuration space" $\mathbb{R}^n$ while the translations $T(z_0)$ act on phase space $\mathbb{R}^n \oplus \mathbb{R}^n$. It is not difficult to remedy at this dissymmetry: it suffices to define, for a function $\Psi \in \mathcal{S}(\mathbb{R}^n \oplus \mathbb{R}^n)$,

$$\widehat{T}(z_0)\Psi(z) = e^{\frac{i}{\hbar}(p_0 \cdot x - \frac{1}{2}p_0 \cdot x_0)}\Psi(z - z_0); \tag{8.10}$$

this definition of course extends to $\Psi \in \mathcal{S}'(\mathbb{R}^n \oplus \mathbb{R}^n)$ and one immediately verifies, using the argument in the proof of the proposition above, that this redefinition of the Heisenberg–Weyl operators again satisfy the relations (8.8) and (8.9). We will use a variant of this idea in Chapter 18 when we study phase-space pseudo-differential operators ("Bopp calculus").

Exercise 127. Verify that the operators (8.10) satisfy the same relations (8.8) and (8.9) as the HW operators.

8.1.3 The symplectic covariance property of the HW operators

The phase space translation operators $T(z_0)$ satisfy the intertwining formula $ST(z_0)S^{-1} = T(Sz_0)$ for every $S \in \mathrm{Sp}(2n, \mathbb{R})$. It is therefore perhaps not so surprising that we have a similar formula for the HW operators. In fact:

Theorem 128. *Let* $\widehat{S} \in \mathrm{Mp}(2n, \mathbb{R})$ *and* $S = \pi^{\mathrm{Mp}}(\widehat{S})$. *We have*

$$\widehat{S}\widehat{T}(z_0)\widehat{S}^{-1} = \widehat{T}(Sz_0) \tag{8.11}$$

for every $z_0 \in \mathbb{R}^{2n}$.

Proof. To prove formula (8.11) it is sufficient to assume that $\widehat{S}$ is a quadratic Fourier transform $\widehat{S}_{W,m}$ since every $\widehat{S} \in \mathrm{Mp}(2n, \mathbb{R})$ is a product of two such operators. Suppose indeed we have shown that

$$\widehat{T}(S_W z_0) = \widehat{S}_{W,m}\widehat{T}(z_0)\widehat{S}_{W,m}^{-1}; \tag{8.12}$$

writing an arbitrary element S of $\mathrm{Mp}(2n, \mathbb{R})$ as a product $S_{W,m}S_{W',m'}$ we have

$$\begin{aligned}
\widehat{T}(Sz_0) &= \widehat{S}_{W,m}(\widehat{S}_{W',m'}\widehat{T}(z_0)\widehat{S}_{W',m'}^{-1})\widehat{S}_{W,m}^{-1} \\
&= \widehat{S}_{W,m}\widehat{T}(S_{W'}z_0)\widehat{S}_{W,m}^{-1} \\
&= \widehat{T}(S_W S_{W'}z_0) \\
&= \widehat{S}_{W,m}\widehat{S}_{W',m'}\widehat{T}(z_0)(\widehat{S}_{W,m}\widehat{S}_{W',m'})^{-1} \\
&= \widehat{S}\widehat{T}(z_0)\widehat{S}^{-1}.
\end{aligned}$$

Let us set out to prove (8.12); equivalently:

$$\widehat{T}(z_0)\widehat{S}_{W,m} = \widehat{S}_{W,m}\widehat{T}(S_W^{-1}z_0). \tag{8.13}$$

For $\psi \in \mathcal{S}(\mathbb{R}^n)$ set

$$g(x) = \widehat{T}(z_0)\widehat{S}_{W,m}\psi(x).$$

By definition of $\widehat{S}_{W,m}$ and $\widehat{T}(z_0)$ we have

$$g(x) = \left(\tfrac{1}{2\pi i\hbar}\right)^{n/2} \Delta(W)e^{-\frac{1}{2\hbar}p_0 \cdot x_0} \int_{\mathbb{R}^n} e^{\frac{i}{\hbar}(W(x-x_0,x')+p_0 \cdot x)}\psi(x')dx'.$$

In view of formula (3.7) in Proposition 54, the function

$$W_0(x,x') = W(x - x_0, x') + p_0 \cdot x \tag{8.14}$$

is a generating function of the free affine symplectomorphism $T(z_0)S_W$, hence we have just shown that

$$\widehat{T}(z_0)\widehat{S}_{W,m} = e^{\frac{i}{2\hbar}p_0 \cdot x_0}\widehat{S}_{W_0,m} \tag{8.15}$$

where $\widehat{S}_{W_0,m}$ is one of the metaplectic operators associated to W_0. Let us now set

$$h(x) = \widehat{S}_{W,m}\widehat{T}(S_W^{-1}z_0)\psi(x) \text{ and } (x_0', p_0') = \widehat{S}_{W,m}^{-1}(x_0, p_0);$$

we have

$$h(x) = \left(\tfrac{1}{2\pi i\hbar}\right)^{n/2}\Delta(W)\int_{\mathbb{R}^n} e^{\frac{i}{\hbar}W(x,x')}e^{-\frac{i}{2\hbar}p_0'\cdot x_0'}e^{\frac{i}{\hbar}p_0'\cdot x'}\psi(x'-x_0')\,dx'$$

that is, performing the change of variables $x' \longmapsto x' + x_0'$:

$$h(x) = \left(\tfrac{1}{2\pi i\hbar}\right)^{n/2}\Delta(W)\int_{\mathbb{R}^n} e^{\frac{i}{\hbar}W(x,x'+x_0')}e^{\frac{i}{2\hbar}p_0'\cdot x_0'}e^{\frac{i}{\hbar}p_0'\cdot x'}\psi(x')\,dx'.$$

We will thus have $h(x) = g(x)$ as claimed, if we show that

$$W(x, x' + x_0') + \tfrac{1}{2}p_0'\cdot x_0' + p_0'\cdot x' = W_0(x, x') - \tfrac{1}{2}p_0\cdot x_0$$

that is

$$W(x, x' + x_0') + \tfrac{1}{2}p_0'\cdot x_0' + p_0'\cdot x' = W(x - x_0, x') + p_0\cdot x - \tfrac{1}{2}p_0\cdot x_0.$$

Replacing x by $x + x_0$ this amounts to proving that

$$W(x + x_0, x' + x_0') + \tfrac{1}{2}p_0'\cdot x_0' + p_0'\cdot x' = W(x, x') + \tfrac{1}{2}p_0\cdot x_0 + p_0\cdot x.$$

But this equality immediately follows from Proposition 54 and its Corollary 55.
$\square$

Exercise 129. Give an alternative proof of formula (8.11) using the generators $\widehat{V}_{-P}$, $\widehat{M}_{L,m}$, $\widehat{J}$ of $\mathrm{Mp}(2n, \mathbb{R})$ defined by (7.9).

The inhomogeneous metaplectic group $\mathrm{AMp}(2n, \mathbb{R})$ is an extension by the HW operators of the metaplectic group $\mathrm{Mp}(2n, \mathbb{R})$. It is the analogue at the "quantized" level of the inhomogeneous symplectic group $\mathrm{ASp}(2n, \mathbb{R})$ (see Definition 26). Its construction requires the use of the Heisenberg–Weyl operators.

8.2 The Heisenberg group

The Heisenberg group is a venerable topic, closely related to the Heisenberg–Weyl operators. It has played an important role in the development of quantum mechanics following ideas of Heisenberg and Weyl. We shortly discuss it for the sake of completeness even if we will not really use it in this book. Excellent references for the Heisenberg group are the books by Schempp [140] and Stein [153] (Chapter 12); very readable presentations are also given in Folland [59] and Gröchenig [82].

8.2.1 The canonical commutation relations

Consider the textbook "quantum operators" $\widehat{X}_j$, $\widehat{P}_j$ on $\mathcal{S}(\mathbb{R}^n)$ defined, for $\psi \in \mathcal{S}(\mathbb{R}^n)$ by

$$\widehat{X}_j\psi = x_j\psi \quad , \quad \widehat{P}_j\psi = -i\hbar\frac{\partial\psi}{\partial x_j}.$$

These operators satisfy the commutation relations

$$[\widehat{X}_i, \widehat{X}_j] = [\widehat{P}_i, \widehat{P}_j] = 0 \quad , \quad [\widehat{X}_i, \widehat{P}_j] = i\hbar\delta_{ij}I \tag{8.16}$$

thus justifying the following definition:

Definition 130. A *"Heisenberg algebra"* is a Lie algebra $\mathfrak{h}_n$ with a basis

$$\{\widehat{X}_1, \ldots, \widehat{X}_n; \widehat{P}_1, \ldots, \widehat{P}_n; \widehat{T}\}$$

whose elements satisfy the so-called canonical commutation relations (for short: CCR)

$$\begin{aligned}
[\widehat{X}_i, \widehat{X}_j] &= 0 , & [\widehat{P}_i, \widehat{P}_j] &= 0 , \\
[\widehat{X}_i, \widehat{P}_j] &= \delta_{ij}\widehat{T}, & [\widehat{X}_i, \widehat{T}] &= 0 , \\
[\widehat{P}_i, \widehat{T}] &= 0
\end{aligned} \tag{8.17}$$

for $1 \le i, j \le n$; $\hbar$ is a constant identified in Physics with Planck's constant h divided by 2π.

In the realization (8.16) of the CCR one usually chooses for $\widehat{T}$ multiplication of functions by the imaginary number $i\hbar$: $\widehat{T}\psi = i\hbar\psi$. (The operator $\widehat{T}$ can be viewed as "setting the quantum scale".)

Writing $\widehat{U}$ and $\widehat{U}'$ in the basis $\{\widehat{X}_1, \ldots, \widehat{X}_n; \widehat{P}_1, \ldots, \widehat{P}_n; \widehat{T}\}$ we have

$$\widehat{U} = \sum_{i=1}^{n} x_i\widehat{X}_i + p_i\widehat{P}_i + t\widehat{T} \quad , \quad \widehat{U}' = \sum_{i=1}^{n} x_i'\widehat{X}_i + p_i'\widehat{P}_i + t'\widehat{T}.$$

Setting $z = (x, p)$, $z' = (x', p')$ the CCR are then immediately seen to be equivalent to the relation

$$[\widehat{U}, \widehat{U}'] = \sigma(z, z')\widehat{T}. \tag{8.18}$$

Formula (8.18) quite explicitly shows that the CCR are intimately related to the choice of the symplectic structure on $\mathbb{R}^n \oplus \mathbb{R}^n$; this observation will be fully exploited in Chapters 18 and 19 when we motivate and discuss phase space pseudo-differential operators.

Exercise 131. Verify that the Heisenberg algebra $\mathfrak{h}_n$ really is a Lie algebra.

8.2.2 Heisenberg group and Schrödinger representation

Let us now describe the simply connected Lie group $\mathbb{H}_n$ corresponding to the Lie algebra $\mathfrak{h}_n$. We begin by recalling that the exponential mapping $\exp : \mathfrak{g} \longrightarrow G$ from a Lie algebra $\mathfrak{g}$ to a Lie group G does not in satisfy the relation $\exp X \exp Y = \exp(X + Y)$ if $XY \neq YX$. Now, there is a formula, called the Baker–Campbell–Hausdorff formula, that says that under some conditions (which we assume to hold), there exists a $C(X, Y) \in \mathfrak{g}$ such that

$$e^X e^Y = e^{C(X,Y)}$$

where $C(X, Y)$ has a series expansion of the type

$$C(X, Y) = X + Y + \frac{1}{2}[X, Y] + \sum_{j=1}^{\infty} C_j(X, Y)$$

where the $C_j(X, Y)$ are linear combinations of commutators of higher order. Since all Lie brackets in $\widehat{U}, \widehat{U}'$ of length superior to 2 vanish, hence the Baker–Campbell–Hausdorff formula reduces (if $\widehat{U}$ and $\widehat{U}'$ are sufficiently close to zero) to the simple formula

$$\exp(\widehat{U}) \exp(\widehat{U}') = \exp(\widehat{U} + \widehat{U}' + \tfrac{1}{2}[\widehat{U}, \widehat{U}'])$$

that is, in view of formula (8.18),

$$\exp(\widehat{U}) \exp(\widehat{U}') = \exp(\widehat{U} + \widehat{U}' + \tfrac{1}{2}\sigma(z, z')\widehat{T}). \tag{8.19}$$

The exponential being a diffeomorphism of a neighborhood $\mathcal{U}$ of zero in $\mathfrak{h}_n$ onto a neighborhood of the identity in $\mathbb{H}_n$ we can identify $\widehat{U}, \widehat{U}'$, for small z, z', t, t', with the exponentials $\exp(\widehat{U})$, $\exp(\widehat{U}')$, and $\exp(\widehat{U}) \exp(\widehat{U}')$ with the element $(z, t)\maltese(z', t)$ of $\mathbb{R}^{2n+1} = \mathbb{R}^{2n} \times \mathbb{R}$ defined by

$$(z, t)\maltese(z', t) = (z + z', t + t' + \tfrac{1}{2}\sigma(z, z')). \tag{8.20}$$

It turns out that this formula defines a (non-commutative) group law:

Definition 132. The set $\mathbb{R}^{2n} \times \mathbb{R}$ equipped with the group law

$$(z, t)\maltese(z', t) = (z + z', t + t' + \tfrac{1}{2}\sigma(z, z'))$$

is called the $(2n + 1)$-dimensional Heisenberg group $\mathbb{H}_n$. [In physics it is often called the Weyl group.]

Obviously $(0, 0) \in \mathbb{R}^{2n} \times \mathbb{R}$ is a unit for the composition law $\maltese$, and each (z, t) is invertible with inverse $(-z, -t)$ (the latter property immediately follows from the fact that $\sigma(z, z) = 0$).

Exercise 133. Verify that the law $\maltese$ is associative, so it really defines a group structure on $\mathbb{R}^{2n+1}$.

At this point we remark that some authors use the following variant of Definition 132:

Definition 134. The reduced Heisenberg group $\mathbb{H}_n^{\mathrm{red}}$ is the set $\mathbb{R}^{2n} \times S^1$ equipped with the law

$$(z, u) \lozenge (z', u') = (zz', uu' e^{\frac{i}{2\hbar} \sigma(z, z')}). \tag{8.21}$$

It turns out that $\mathbb{H}_n$ is the universal covering group of this "exponentiated" version $\mathbb{H}_n^{\mathrm{red}}$ of $\mathbb{H}_n$. Define in fact a projection $\pi : \mathbb{H}_n \longrightarrow \mathbb{H}_n^{\mathrm{red}}$ by the formula $\pi(z, t) = (z, e^{it})$. We have

$$\pi\left[(z, t)(z', t')\right] = \pi(z + z', t + t' + \tfrac{i}{2}\sigma(z, z'))$$
$$= (z + z', e^{it} e^{it'} e^{\frac{i}{2}\sigma(z, z')})$$
$$= \pi(z, t)\pi(z', t')$$

so that π is a group homomorphism; it is in addition trivially surjective (because every $u \in S^1$ is of the type e^{it} for some $t \in \mathbb{R}$). Now the kernel $\ker \pi = \pi^{-1}\{(0, 1)\}$ is defined by $z = 0$ and $e^{it} = 1$, that is $t \in 2\pi\mathbb{Z}$; it is thus a discrete subgroup of $\mathbb{H}_n$ so that $\mathbb{H}_n$ is indeed a covering group of $\mathbb{H}_n^{\mathrm{red}}$; since $\mathbb{H}_n$ is simply connected (it is just $\mathbb{R}^{2n} \times \mathbb{R}$ as a set) it is thus the universal covering of $\mathbb{H}_n^{\mathrm{red}}$.

Since the Lie algebra $\mathfrak{g}$ of the universal covering group of a Lie group G is isomorphic to that of G itself, we see that the Lie algebra of $\mathbb{H}_n^{\mathrm{red}}$ is just $\mathfrak{h}_n$.

There is a useful identification of $\mathbb{H}_n$ with a subgroup $\mathbb{H}_n^{\mathrm{pol}}$ of $\mathrm{GL}(2n + 2, \mathbb{R})$. That group (the "polarized Heisenberg group") consists of all $(2n + 2) \times (2n + 2)$ upper-triangular matrices of the type

$$M(z, t) = \begin{pmatrix} 1 & p_1 & \cdots & p_n & t \\ 0 & 1 & \cdots & 0 & x_1 \\ \vdots & \vdots & \vdots & \vdots & \vdots \\ 0 & 0 & \cdots & 1 & x_n \\ 0 & 0 & \cdots & 0 & 1 \end{pmatrix}$$

(the entries of the principal diagonal are each equal to 1); we find it convenient to write these matrices for short as

$$M(z, t) = \begin{pmatrix} 1 & p^T & t \\ 0 & 1 & x \\ 0 & 0 & 1 \end{pmatrix}.$$

One easily checks that the determinant of $M(z, t)$ is 1:

$$\det M(z, t) = 1$$

and that its inverse is given by the formula

$$M(z, t)^{-1} = \begin{pmatrix} 1 & -p^T & -t + p \cdot x \\ 0 & 1 & -x \\ 0 & 0 & 1 \end{pmatrix};$$

we have moreover

$$M(z,t)M(z',t') = M(z+z', t+t'+p\cdot x'). \tag{8.22}$$

Exercise 135. Show that the mapping $\phi : \mathbb{H}_n^{\mathrm{pol}} \longrightarrow \mathbb{H}_n$ defined by

$$\phi(M(z,t)) = \left(z, t - \tfrac{1}{2}p\cdot x\right) \tag{8.23}$$

is a group isomorphism.

Exercise 136. Show that the Lie algebra $\mathfrak{h}_n^{\mathrm{pol}}$ of $\mathbb{H}_n^{\mathrm{pol}}$ consists of all matrices

$$X^{\mathrm{pol}}(z,t) = \begin{pmatrix} 0 & p^T & t - \frac{1}{2}p\cdot x \\ 0 & 0 & x \\ 0 & 0 & 0 \end{pmatrix}.$$

The polarized version of the Heisenberg group is useful in many applications (for instance medical imaging); see for instance Schempp [140, 141].

8.2.3 The Stone–von Neumann theorem

Recall that a unitary representation of a topological group G is a pair $(\mathcal{H}, \pi)$ where $\mathcal{H}$ is a Hilbert space and π is a strongly continuous homomorphism of G into the group $U(\mathcal{H})$ of all unitary operators on $\mathcal{H}$. "Strong continuity" refers to the following property:

If $\lim_{j\to\infty} g_j = g$ in G then $\lim_{j\to\infty} \pi(g_j)\psi = \pi(g)\psi$ for every $\psi \in \mathcal{H}$.

Given a representation $(\mathcal{H}, \pi)$ of G the functions $g \longmapsto (\psi|\pi(g)\phi)_{\mathcal{H}}$ are called the "representation coefficients" of $(\mathcal{H}, \pi)$.

Let us introduce some terminology:

Definition 137.

(i) Two representations $(\mathcal{H}_1, \pi_1)$ and $(\mathcal{H}_2, \pi_2)$ are equivalent if there exists a unitary operator $U : \mathcal{H}_1 \longrightarrow \mathcal{H}_2$ such that $U\pi_1(g)U^{-1} = \pi_2(g)$ for all $g \in G$.

(ii) A representation $(\mathcal{H}, \pi)$ is said to be irreducible if $\{0\}$ and $\mathcal{H}$ are the only closed subspaces of $\mathcal{H}$ invariant under all the operators $\pi(g)$, $g \in G$.

It is easy to see that the metaplectic representation is not irreducible.

Exercise 138. Let $L^2_{\mathrm{even}}(\mathbb{R}^n)$ (resp. $L^2_{\mathrm{odd}}(\mathbb{R}^n)$) be the vector subspaces of $L^2(\mathbb{R}^n)$ consisting of all even (resp. odd) functions. Let $\widehat{S} \in \mathrm{Mp}(2n, \mathbb{R})$. Show that

$$L^2(\mathbb{R}^n) = L^2_{\mathrm{even}}(\mathbb{R}^n) \oplus L^2_{\mathrm{odd}}(\mathbb{R}^n) \quad \text{and} \quad \widehat{S}(L^2_{\mathrm{even}}(\mathbb{R}^n)) \subset L^2_{\mathrm{even}}(\mathbb{R}^n)$$

and

$$\widehat{S}(L^2_{\mathrm{odd}}(\mathbb{R}^n)) \subset L^2_{\mathrm{odd}}(\mathbb{R}^n).$$

[Hint: use Proposition 110.]

The Heisenberg–Weyl operators are related in the following way to the Heisenberg group:

Proposition 139. *Let $\mathcal{U}(L^2(\mathbb{R}^n))$ be the group of all unitary operators on $L^2(\mathbb{R}^n)$. The mapping $\rho : \mathbb{H}_n \longrightarrow \mathcal{U}(L^2(\mathbb{R}^n))$ defined by*

$$\rho(z,t) = e^{\frac{i}{\hbar}t}\widehat{T}(z) \tag{8.24}$$

is a unitary representation of the Heisenberg group $\mathbb{H}_n$.

Proof. The operators $\rho(z,t)$ are obviously unitary on $L^2(\mathbb{R}^n)$. Let us show that ρ is a group homomorphism, that is

$$\rho\left[(z,t)\maltese(z',t')\right] = \rho(z,t)\rho(z',t').$$

By definition of the multiplication on $\mathbb{H}_n$ and formula (8.9) we have

$$\rho\left[(z,t)\maltese(z',t')\right] = \rho\left[(z+z',t+t'+\tfrac{1}{2}\sigma(z,z'))\right] = e^{\frac{i}{\hbar}(t+t'+\frac{1}{2}\sigma(z,z'))}\widehat{T}(z+z')$$
$$= e^{\frac{i}{\hbar}(t+t')}\widehat{T}(z)\widehat{T}(z') \qquad = \rho(z,t)\rho(z',t'). \qquad \square$$

This result leads us to the following definition:

Definition 140. The representation $(L^2(\mathbb{R}^n),\rho)$ is called the Schrödinger representation of the Heisenberg group.

One can prove (see for instance [67, 59, 82, 158]) that the Schrödinger representation is irreducible, that is:

The only closed subspaces of $L^2(\mathbb{R}^n)$ which are invariant under every operator $\rho(z,t)$ (or, equivalently, $\widehat{T}(z)$) are $\{0\}$ or $L^2(\mathbb{R}^n)$ itself.

In view of a result of Schur (Schur's Lemma) this condition is equivalent to:

If A is a bounded operator on $L^2(\mathbb{R}^n)$ commuting with ρ (that is $A\rho = \rho A$) then $A = \lambda I$ for some $\lambda \in \mathbb{C}$.

There is a deep result of Stone and von Neumann about the uniqueness of the Schrödinger representation. It is in fact so well known that it has acquired the status of a "folk theorem" which has led to the following usual (mis-)quotation:

The Schrödinger representation is the only *irreducible representation of $\mathbb{H}_n$ up to trivial transformations such as rescalings.*

Here is one statement of the theorem of Stone and von Neumann:

Theorem 141. *Every irreducible representation $(L^2(\mathbb{R}^n),\pi)$ of the Heisenberg group $\mathbb{H}_n$ is equivalent to one of the following:*
   (i) $\pi(z,t) = e^{\frac{i}{\hbar}z_0 \cdot z}$ *where $z_0 \in \mathbb{R}^{2n}$;*
   (ii) $\pi(z,t) = \widehat{T}_\lambda(z,t)$ *where $\widehat{T}(z,t) = \widehat{T}(\lambda z,t)$.*

A more complete statement is the following (Folland, [59], §5):

Theorem 142. *Let $(\mathcal{H}, \pi)$ be a unitary representation of the Heisenberg group $\mathbb{H}_n$ such that $\pi(0,t) = e^{\frac{i}{\hbar}t}I$ for some $\hbar \neq 0$. Then $\mathcal{H} = \bigoplus \mathcal{H}_j$ where the $\mathcal{H}_j$ are pairwise orthogonal subspaces of $\mathcal{H}$ each invariant under π and such that the restrictions $\pi_j = \pi_{|\mathcal{H}_j}$ are unitarily equivalent to $(L^2(\mathbb{R}^n), \rho)$ for every j. In particular, if $(\mathcal{H}, \pi)$ is irreducible then π is equivalent to $(L^2(\mathbb{R}^n), \rho)$.*

We will see later on, when we discuss pseudo-differential operators on phase space, that it is perfectly possible to construct non-trivial representations of the Heisenberg group which are distinct from the Schrödinger representation provided that one replaces $L^2(\mathbb{R}^n)$ by other Hilbert spaces.

8.3   The Grossmann–Royer operators

We introduce in this section the Grossmann–Royer operators which are a kind of reflection operators. Their definition goes back to the work of Grossmann [88] and Royer [138]. These operators are not universally known, and this is indeed very unfortunate since their use allows one to considerably simplify many proofs. In addition they allow an alternative definition of Weyl operators in terms of the symbol, as we will see in Chapter 10

8.3.1   The symplectic Fourier transform

We begin by introducing the notion of symplectic Fourier transform, which is a "twisted" form of the usual Fourier transform on $\mathbb{R}^{2n}$. We will often use it when dealing later on with Weyl calculus.

Definition 143. The symplectic Fourier transform F_σ is defined, for $a \in \mathcal{S}(\mathbb{R}^n \oplus \mathbb{R}^n)$, by

$$F_\sigma a(z) = \left(\tfrac{1}{2\pi\hbar}\right)^n \int_{\mathbb{R}^{2n}} e^{-\frac{i}{\hbar}\sigma(z,z')} a(z')dz'. \tag{8.25}$$

We will often use the shorthand notation $a_\sigma = F_\sigma a$.

The following propositions give the main properties of the symplectic Fourier transform. Recall that the standard ($\hbar$-dependent) Fourier transform on $\mathcal{S}(\mathbb{R}^n \oplus \mathbb{R}^n)$ is given by

$$Fa(z) = \left(\tfrac{1}{2\pi\hbar}\right)^n \int_{\mathbb{R}^{2n}} e^{-\frac{i}{\hbar}z \cdot z'} a(z')dz'. \tag{8.26}$$

Proposition 144.

  (i) *The Fourier transforms F_σ and F are related by the formula*

$$F_\sigma a(z) = Fa(Jz) = F(a \circ J)(z). \tag{8.27}$$

  In particular F_σ is a linear automorphism

$$F_\sigma : \mathcal{S}(\mathbb{R}^n \oplus \mathbb{R}^n) \longrightarrow \mathcal{S}(\mathbb{R}^n \oplus \mathbb{R}^n)$$

which extends by duality into an automorphism

$$F_\sigma : \mathcal{S}'(\mathbb{R}^n \oplus \mathbb{R}^n) \longrightarrow \mathcal{S}'(\mathbb{R}^n \oplus \mathbb{R}^n).$$

(ii) *The symplectic Fourier transform is involutive and unitary:*

$$F_\sigma \circ F_\sigma = I \quad , \quad \|F_\sigma a\|_{L^2(\mathbb{R}^{2n})} = \|a\|_{L^2(\mathbb{R}^{2n})}. \tag{8.28}$$

Proof of (i). Writing $\sigma(z, z') = Jz \cdot z'$ where J is the standard symplectic matrix, we have,

$$F_\sigma a(z) = \left(\tfrac{1}{2\pi\hbar}\right)^n \int_{\mathbb{R}^{2n}} e^{-\frac{i}{\hbar} Jz \cdot z'} a(z') dz' = Fa(Jz)$$

hence the first equality (8.27). Setting $z' = Jz''$ in the integral the second formula formula (8.27) follows as well since we have $Jz \cdot Jz'' = z \cdot z''$.

Proof of (ii). The equality $F_\sigma \circ F_\sigma = I$ follows from the usual Fourier inversion formula written in the form $F(Fa)(z) = a(-z)$:

$$F_\sigma(F_\sigma a)(z) = F(Fa)(-z) = a(z).$$

It follows that the symplectic Fourier transform is both involutive and unitary. $\square$

In particular the symplectic Fourier transform F_σ is its own inverse: $F_\sigma^{-1} = F_\sigma$ on $\mathcal{S}'(\mathbb{R}^n \oplus \mathbb{R}^n)$. Thus:

$$F_\sigma a(z) = \left(\tfrac{1}{2\pi\hbar}\right)^n \int_{\mathbb{R}^{2n}} e^{-\frac{i}{\hbar}\sigma(z,z')} a(z') dz',$$

$$a(z) = \left(\tfrac{1}{2\pi\hbar}\right)^n \int_{\mathbb{R}^{2n}} e^{-\frac{i}{\hbar}\sigma(z,z')} F_\sigma a(z') dz'.$$

More generally, the symplectic Fourier transform behaves well under the action of the symplectic group:

Proposition 145. *For $a \in \mathcal{S}'(\mathbb{R}^n \oplus \mathbb{R}^n)$ and $S \in \mathrm{Sp}(2n, \mathbb{R})$ we have*

$$F_\sigma a(Sz) = F_\sigma(a \circ S)(z). \tag{8.29}$$

Proof. It suffices to assume that $a \in \mathcal{S}(\mathbb{R}^n \oplus \mathbb{R}^n)$, in which case

$$F_\sigma a(Sz) = \left(\tfrac{1}{2\pi\hbar}\right)^n \int_{\mathbb{R}^{2n}} e^{-\frac{i}{\hbar}\sigma(Sz,z')} a(z') dz';$$

since we have

$$\sigma(Sz, z') = \sigma(S^{-1}Sz, S^{-1}z') = \sigma(z, S^{-1}z')$$

because S^{-1} is symplectic, it follows, setting $z'' = S^{-1}z'$, that

$$F_\sigma a(Sz) = \left(\tfrac{1}{2\pi\hbar}\right)^n \int_{\mathbb{R}^{2n}} e^{-\frac{i}{\hbar}\sigma(z,S^{-1}z')} a(z') dz'$$

$$= \left(\tfrac{1}{2\pi\hbar}\right)^n \int_{\mathbb{R}^{2n}} e^{-\frac{i}{\hbar}\sigma(z,z'')} a(Sz'') dz'$$

which proves (8.29). $\square$

F_σ also satisfies the following variants of the Plancherel formula:

Proposition 146. *The symplectic Fourier transform satisfies the Plancherel formula*

$$(F_\sigma a|b)_{L^2(\mathbb{R}^{2n})} = (a|F_\sigma b)_{L^2(\mathbb{R}^{2n})} \tag{8.30}$$

or, equivalently, using the distributional brackets $\langle\langle \cdot, \cdot \rangle\rangle$ on $\mathcal{S}(\mathbb{R}^n \oplus \mathbb{R}^n)$:

$$\int_{\mathbb{R}^{2n}} a(z)F_\sigma b(z)dz = \int_{\mathbb{R}^{2n}} F_\sigma a(-z)b(z)dz, \tag{8.31}$$

$$\langle\langle a, b_\sigma \rangle\rangle = \langle\langle a_\sigma^\vee, b \rangle\rangle = \langle\langle a_\sigma, b^\vee \rangle\rangle \tag{8.32}$$

where $a_\sigma^\vee(z) = a_\sigma(-z)$.

Proof. It is a straightforward consequence of the fact that F_σ is a unitary involution:

$$(F_\sigma a|b)_{L^2(\mathbb{R}^{2n})} = (F_\sigma^2 a|F_\sigma b)_{L^2} = (a|F_\sigma b)_{L^2}. \qquad \square$$

8.3.2 Definition of the Grossmann–Royer operators

The simplest – or perhaps the most convincing! – way of introducing the Grossmann–Royer operators is to express them as the conjugate of a reflection operator by a Heisenberg–Weyl operator:

Definition 147. The Grossmann–Royer operator $\widehat{T}_{\mathrm{GR}}(z_0)$ is the operator

$$\widehat{T}_{\mathrm{GR}}(z_0) : \mathcal{S}(\mathbb{R}^n) \longrightarrow \mathcal{S}(\mathbb{R}^n)$$

defined by the formulae

$$\widehat{T}_{\mathrm{GR}}(0)\psi(x) = \psi(-x) \tag{8.33}$$

and

$$\widehat{T}_{\mathrm{GR}}(z_0) = \widehat{T}(z_0)\widehat{T}_{\mathrm{GR}}(0)\widehat{T}(z_0)^{-1}. \tag{8.34}$$

The following properties are straightforward (but useful!) consequences of the definition:

Proposition 148. *The Grossmann–Royer operators are linear and unitary involutions of $\mathcal{S}(\mathbb{R}^n)$ (and hence of $\mathcal{S}'(\mathbb{R}^n)$), and the action of $\widehat{T}_{GR}(z_0)$, $z_0 = (x_0, p_0)$, is explicitly given by the formula*

$$\widehat{T}_{GR}(z_0)\psi(x) = e^{\frac{2i}{\hbar}p_0 \cdot (x-x_0)}\psi(2x_0 - x) \tag{8.35}$$

for any function (or distribution) $\psi : \mathbb{R}^n \longrightarrow \mathbb{C}$.

Proof. The linearity of $\widehat{T}_{\mathrm{GR}}(z_0)$ is obvious. That $\widehat{T}_{\mathrm{GR}}(z_0)$ is an involution, that is

$$\widehat{T}_{\mathrm{GR}}(z_0)\widehat{T}_{\mathrm{GR}}(z_0) = I,$$

follows from the sequence of equalities

$$\widehat{T}_{\mathrm{GR}}(z_0)\widehat{T}_{\mathrm{GR}}(z_0) = \widehat{T}(z_0)\widehat{T}_{\mathrm{GR}}(0)\widehat{T}(z_0)^{-1}\widehat{T}(z_0)\widehat{T}_{\mathrm{GR}}(0)\widehat{T}(z_0)^{-1}$$

$$= \widehat{T}(z_0)\widehat{T}_{\mathrm{GR}}(0)\widehat{T}_{\mathrm{GR}}(0)\widehat{T}(z_0)^{-1} = \widehat{T}(z_0)\widehat{T}(z_0)^{-1}.$$

Setting $x' = 2x_0 - x$, we have

$$\|\widehat{T}_{\mathrm{GR}}(z_0)\psi\|_{L^2}^2 = \int_{\mathbb{R}^n} |\psi(2x_0 - x)|^2 dx = \|\psi\|_{L^2}^2$$

hence $\widehat{T}_{\mathrm{GR}}(z_0)$ is also unitary. Formula (8.35) follows from Definition (8.34) by a straightforward calculation which is left to the reader as a pleasant exercise. $\square$

The following result shows that the operators $\widehat{T}_{\mathrm{GR}}(z_0)$ and $\widehat{T}(z_0)$ are intimately related by the symplectic Fourier transform:

Proposition 149. *Let* $\psi \in \mathcal{S}'(\mathbb{R}^n)$. *We have*

$$\widehat{T}_{GR}(z_0)\psi(x) = 2^{-n} F_\sigma[\widehat{T}(\cdot)\psi(x)](-z_0) \tag{8.36}$$

where F_σ is the symplectic Fourier transform.

Proof. Since $\widehat{T}_{\mathrm{GR}}(z_0)$ and F_σ are continuous automorphisms of $\mathcal{S}'(\mathbb{R}^n)$ it is sufficient to assume that $\psi \in \mathcal{S}(\mathbb{R}^n)$. Formula (8.36) follows from (8.35): using the explicit expressions of $\sigma(z_0, z')$ and $\widehat{T}(z')\psi(x)$ the right-hand side of (8.36) is

$$A = \left(\tfrac{1}{4\pi\hbar}\right)^n \int_{\mathbb{R}^{2n}} e^{\frac{i}{\hbar}\sigma(z_0,z')}\widehat{T}(z')\psi(x)dz'$$

$$= \left(\tfrac{1}{4\pi\hbar}\right)^n \int_{\mathbb{R}^{2n}} e^{\frac{i}{\hbar}(p_0\cdot x' - p'\cdot x_0 + p'\cdot x - \frac{1}{2}p'\cdot x')}\psi(x - x')dz'$$

$$= \left(\tfrac{1}{4\pi\hbar}\right)^n \int_{\mathbb{R}^n} \left(\int_{\mathbb{R}^n} e^{\frac{i}{\hbar}p'\cdot(x - x_0 - \frac{1}{2}x')}dp'\right) e^{\frac{i}{\hbar}p_0\cdot x'}\psi(x)dx'.$$

Now, in the distributional sense,

$$\int_{\mathbb{R}^2} e^{\frac{i}{\hbar}p'\cdot(x - x_0 - \frac{1}{2}x')}dp' = (2\pi\hbar)^n \delta(x - x_0 - \tfrac{1}{2}x')$$

and hence, setting $y = \tfrac{1}{2}x'$:

$$A = 2^{-n} \int_{\mathbb{R}^n} \delta(x - x_0 - \tfrac{1}{2}x')e^{\frac{i}{\hbar}p_0\cdot x'}\psi(x)dx'$$

$$= \int_{\mathbb{R}^n} \delta(y + x_0 - x)e^{\frac{2i}{\hbar}p_0\cdot y}\psi(x)dy$$

$$= e^{\frac{2i}{\hbar}p_0\cdot(x - x_0)}\psi(-x + 2x_0)$$

which proves (8.35). $\square$

We have seen that the Grossmann–Royer operators are involutions; more generally we have the following result which is a generalization of the fact that the product of two reflections is a translation:

Proposition 150. *The Grossmann–Royer operators satisfy the product formula*

$$\widehat{T}_{GR}(z_0)\widehat{T}_{GR}(z_1) = e^{-\frac{2i}{\hbar}\sigma(z_0,z_1)}\widehat{T}(2(z_0 - z_1)) \tag{8.37}$$

for all $z_0, z_1 \in \mathbb{R}^{2n}$.

Proof. We have

$$\widehat{T}_{\mathrm{GR}}(z_0)\widehat{T}_{\mathrm{GR}}(z_1)\psi(x) = \widehat{T}_{\mathrm{GR}}(z_0)\left[e^{\frac{2i}{\hbar}p_1\cdot(2x_0-x-x_1)}\psi(2x_1 - x)\right]$$
$$= e^{\frac{2i}{\hbar}p_0\cdot(x-x_0)}e^{\frac{2i}{\hbar}p_1\cdot(x-x_1)}\psi(2x_1 - (2x_0 - x))$$
$$= e^{\frac{i}{\hbar}\Phi}\psi(x - 2(x_0 - x_1))$$

with

$$\Phi = 2[(p_0 - p_1)x - p_0x_0 - p_1x_1 + 2p_1x_0].$$

On the other hand

$$\widehat{T}(2(z_0 - z_1))\psi(x) = e^{\frac{i}{\hbar}\Phi'}\psi(x - 2(x_0 - x_1))$$

with

$$\Phi' = 2((p_0 - p_1)x - (p_0 - p_1)(x_0 - x_1)).$$

We have $\Phi - \Phi' = -2\sigma(z_0, z_1)$ hence the result. $\qquad\square$

8.3.3   Symplectic covariance

We have previously seen that the Heisenberg–Weyl operators $\widehat{T}(z_0)$ satisfy the intertwining formula

$$\widehat{S}\widehat{T}(z_0)\widehat{S}^{-1} = \widehat{T}(Sz_0).$$

We are going to see that a similar relation holds for the Grossmann–Royer operators.

Proposition 151. *Let $\widehat{S} \in \mathrm{Mp}(2n, \mathbb{R})$ and $S = \pi^{\mathrm{Mp}}(S)$. We have*

$$\widehat{S}\widehat{T}_{GR}(z_0)\widehat{S}^{-1} = \widehat{T}_{GR}(Sz_0) \tag{8.38}$$

for every $z_0 \in \mathbb{R}^{2n}$.

Proof. To prove formula (8.38) recall (Proposition 149) that we have

$$\widehat{T}_{\mathrm{GR}}(z_0) = \widehat{T}(z_0)\widehat{T}_{\mathrm{GR}}(0)\widehat{T}(z_0)^{-1}.$$

It follows that

$$\widehat{T}_{\mathrm{GR}}(Sz_0) = \widehat{S}\widehat{T}(z_0)(\widehat{S}_{\mathrm{GR}}^{-1}\widehat{T}(0)\widehat{S})\widehat{T}(z_0)^{-1}\widehat{S}^{-1}.$$

It thus suffices to show that $\widehat{S}_{\mathrm{GR}}^{-1}\widehat{T}(0)\widehat{S} = \widehat{T}_{\mathrm{GR}}(0)$, and as above it is no restriction to assume that $\widehat{S} = \widehat{S}_{W,m}$. For $\psi \in \mathcal{S}(\mathbb{R}^n)$ we have, since $\widehat{T}_{\mathrm{GR}}(0)$ is just a reflection operator,

$$\begin{aligned}
\widehat{T}_{\mathrm{GR}}(0)\widehat{S}_{W,m}\psi(x) &= \left(\tfrac{1}{2\pi i\hbar}\right)^{n/2}\Delta(W)\int_{\mathbb{R}^n} e^{\frac{i}{\hbar}W(-x,x')}\psi(x')\,dx' \\
&= \left(\tfrac{1}{2\pi i\hbar}\right)^{n/2}\Delta(W)\int_{\mathbb{R}^n} e^{\frac{i}{\hbar}W(-x,-x'')}\psi(-x'')\,dx' \\
&= \widehat{S}_{W,m}\widehat{T}_{\mathrm{GR}}(0)\psi(x),
\end{aligned}$$

the last equality because $W(-x,-x'') = W(x,x'')$ since W is a quadratic form. For the same reason we have $W(-x,x') = W(x,-x')$ and hence

$$\begin{aligned}
\widehat{T}_{\mathrm{GR}}(0)\widehat{S}_{W,m}\psi(x) &= \left(\tfrac{1}{2\pi i\hbar}\right)^{n/2}\Delta(W)\int_{\mathbb{R}^n} e^{\frac{i}{\hbar}W(x,-x'')}\psi(x'')\,dx' \\
&= \left(\tfrac{1}{2\pi i\hbar}\right)^{n/2}\Delta(W)\int_{\mathbb{R}^n} e^{\frac{i}{\hbar}W(x,x'')}\psi(-x'')\,dx' \\
&= \left(\tfrac{1}{2\pi i\hbar}\right)^{n/2}\Delta(W)\int_{\mathbb{R}^n} e^{\frac{i}{\hbar}W(x,x'')}\widehat{T}_{\mathrm{GR}}(0)\psi(x'')\,dx',
\end{aligned}$$

that is

$$\widehat{T}_{\mathrm{GR}}(0)\widehat{S}_{W,m}\psi(x) = \widehat{S}_{W,m}\widehat{T}_{\mathrm{GR}}(0)\psi$$

which proves our claim. $\qquad\square$

Exercise 152. Show that for every $\widehat{S} \in \mathrm{Mp}(2n,\mathbb{R})$ there exists $\widehat{S}' \in \mathrm{Mp}(2n,\mathbb{R})$ such that

$$\widehat{S}\widehat{T}_{\mathrm{GR}}(z_0) = \widehat{T}(Sz_0)\widehat{S}'.$$

[Hint: Consider first the case $\widehat{S} = \widehat{S}_{W,m}$.]

8.4 Weyl–Heisenberg frames

This section aims at being a modest introduction to frame theory, especially to Weyl–Heisenberg frames (which are also called Gabor frames in signal theory and time-frequency analysis[1]). Our main sources are Christensen [26] and Gröchenig [82]. For a detailed study of the topic of varying the lattice of Gabor frames see Feichtinger and Kaiblinger [53].

---

[1] Since Weyl–Heisenberg frames will be defined using Heisenberg–Weyl operators, it would be better to call them "Heisenberg–Weyl frames". We are however complying here with the commonly accepted terminology.

8.4.1 The notion of frame

Frames are generalizations of the notion of basis in a Hilbert space.

Definition 153. A frame in a Hilbert space $\mathcal{H}$ is a sequence $(\psi_j)_j = (\psi_j)_{j\in\mathbb{N}}$ in $\mathcal{H}$ for which there exist $a, b > 0$ such that

$$a\|\psi\|_{\mathcal{H}}^2 \le \sum_j |(\psi|\psi_j)_{\mathcal{H}}|^2 \le b\|\psi\|_{\mathcal{H}}^2 \tag{8.39}$$

for all $\psi \in \mathcal{H}$. The numbers a and b are called the lower and upper frame bounds, respectively. If $a = b$ then $(\psi_j)_j$ is called a tight frame; if $a = b = 1$ it is called a normalized tight frame.

An orthonormal basis of $\mathcal{H}$ is a normalized tight frame since we always have the identity

$$\sum_j |(\psi|\psi_j)_{\mathcal{H}}|^2 = \|\psi\|_{\mathcal{H}}^2.$$

Notice that in the definition above one does not require that the vectors ψ_j are linearly independent, even less that they form an orthonormal set. A tight frame can of course always be normalized, replacing each ψ_j by $a^{-1/2}\psi_j$.

Fundamental tools in the study of frames are the following operators:

- The *frame operator*: it is the operator $\widehat{\mathcal{F}}$ on $\mathcal{H}$ defined by

$$\widehat{\mathcal{F}}\psi = \sum_j (\psi|\psi_j)_{\mathcal{H}}\psi_j \tag{8.40}$$

  for $\psi \in \mathcal{H}$;

- The *synthesis* (or *reconstruction*) operator: it is the operator $\widehat{\mathcal{R}} : \ell^2(\mathbb{N}) \longrightarrow \mathcal{H}$ defined by

$$\widehat{\mathcal{R}}[(c_j)_j] = \sum_j c_j\psi_j;$$

- The *coefficient* (or *analysis*) operator: it is the operator $\widehat{\mathcal{C}} : \mathcal{H} \longrightarrow \ell^2(\mathbb{N})$ defined by

$$\widehat{\mathcal{C}}\psi = ((\psi|\psi_j)_{\mathcal{H}})_j .$$

It is immediate to check that $\widehat{\mathcal{C}} = \widehat{\mathcal{R}}^*$ and that $\widehat{\mathcal{F}} = \widehat{\mathcal{C}}\widehat{\mathcal{C}}^*$. In particular $\widehat{\mathcal{F}}$ is thus a positive and self-adjoint operator. Notice that when $(\psi_j)_j$ is an orthonormal basis the frame operator is the identity. One shows – but we will not do it here – that the series in the right-hand side of formula (8.40) is unconditionally convergent, that is, $\sum_j (\psi|\psi_{\epsilon(j)})_{\mathcal{H}}\psi_{\sigma(j)} < \infty$ for every permutation ϵ of $\mathbb{N}$, in which case the limit is the same regardless of ϵ. For a detailed study of the notion of unconditional convergence see [82, §5.3].

Proposition 154. *The frame operator $\widehat{\mathcal{F}}$ is an invertible positive self-adjoint bounded operator on the complex Hilbert space $\mathcal{H}$ such that*

$$a\|\psi\|_{\mathcal{H}}^2 \le (\widehat{\mathcal{F}}\psi|\psi)_{\mathcal{H}} \le b\|\psi\|_{\mathcal{H}}^2 \tag{8.41}$$

for every $\psi \in \mathcal{H}$. Equivalently

$$aI \le \widehat{\mathcal{F}} \le bI \tag{8.42}$$

which means $\widehat{\mathcal{F}} - aI \ge 0$ and $bI - \widehat{\mathcal{F}} \ge 0$.

Proof. As noticed above, the self-adjointness and positivity statements follow from the formula $\widehat{\mathcal{F}} = \mathcal{C}\mathcal{C}^*$. It can also be seen directly by noting that we have by definition of $\widehat{\mathcal{F}}$,

$$(\widehat{\mathcal{F}}\psi|\psi)_{\mathcal{H}} = \sum_j (\psi|\psi_j)_{\mathcal{H}}(\psi_j|\psi)_{\mathcal{H}} = \sum_j |(\psi|\psi_j)_{\mathcal{H}}|^2 . \tag{8.43}$$

The double inequality (8.41) immediately follows from (8.39). In view of (8.41) we have, if $\|\psi\|_{\mathcal{H}} = 1$,

$$0 < 1 - b^{-1}(\widehat{\mathcal{F}}\psi|\psi)_{\mathcal{H}} \le \frac{b-a}{b} < 1$$

hence the operator norm of $I - b^{-1}\widehat{\mathcal{F}}\psi$ satisfies

$$\|I - b^{-1}\widehat{\mathcal{F}}\psi\| = \sup_{\|\psi\|_{\mathcal{H}}=1} |(I - b^{-1}\widehat{\mathcal{F}}\psi|\psi)_{\mathcal{H}} < 1.$$

It follows that $b^{-1}\widehat{\mathcal{F}} = I - (I - b^{-1}\widehat{\mathcal{F}})$ is invertible, hence $\widehat{\mathcal{F}}$ is also invertible. $\quad\square$

Using the frame operator we can construct a new frame, the so-called dual frame $(\widehat{\mathcal{F}}^{-1}\psi_j)_j$:

Proposition 155. *Let $(\psi_j)_j$ be a frame with frame operator $\widehat{\mathcal{F}}$. Then $(\widehat{\mathcal{F}}^{-1}\psi_j)_j$ is also a frame with frame bounds b^{-1} and a^{-1} and frame operator $\widehat{\mathcal{F}}^{-1}$.*

Proof. Let us show that

$$b^{-1}\|\psi\|_{\mathcal{H}}^2 \le \sum_j |(\psi|\widehat{\mathcal{F}}^{-1}\psi_j)_{\mathcal{H}}|^2 \le a^{-1}\|\psi\|_{\mathcal{H}}^2 ; \tag{8.44}$$

this will prove that $(\widehat{\mathcal{F}}^{-1}\psi_j)_{j\in J}$ is a frame with bounds b^{-1} and a^{-1}. We first observe that since $(\widehat{\mathcal{F}}^{-1})^* = \widehat{\mathcal{F}}^{-1}$ we have

$$\sum_j |(\psi|\widehat{\mathcal{F}}^{-1}\psi_j)_{\mathcal{H}}|^2 = \sum_j |(\widehat{\mathcal{F}}^{-1}\psi|\psi_j)_{\mathcal{H}}|^2$$

and hence, replacing ψ with $\widehat{\mathcal{F}}^{-1}\psi$ in (8.43),

$$\sum_j |(\psi|\widehat{\mathcal{F}}^{-1}\psi_j)_{\mathcal{H}}|^2 = (\psi|\widehat{\mathcal{F}}^{-1}\psi)_{\mathcal{H}} = (\widehat{\mathcal{F}}^{-1}\psi|\psi)_{\mathcal{H}}.$$

Because $\widehat{\mathcal{F}}$ and the identity commute with $\widehat{\mathcal{F}}^{-1}$ we can multiply both sides of (8.42) which yields $\widehat{\mathcal{F}}^{-1} \leq a^{-1}I$ and $\widehat{\mathcal{F}}^{-1} \geq b^{-1}I$ so that $b^{-1}I \leq \widehat{\mathcal{F}}^{-1} \leq a^{-1}I$ and this is equivalent to (8.45).

$$b^{-1}\|\psi\|_{\mathcal{H}}^2 \leq (\widehat{\mathcal{F}}^{-1}\psi|\psi)_{\mathcal{H}} \leq a^{-1}\|\psi\|_{\mathcal{H}}^2. \tag{8.45}$$

Let us show that $(\widehat{\mathcal{F}}^{-1}\psi|\psi)_{\mathcal{H}} = \sum_j |(\psi|\widehat{\mathcal{F}}^{-1}\psi_j)_{\mathcal{H}}|^2$ for every $\psi \in \mathcal{H}$, the identity (8.44) will follow. Replacing ψ with $\widehat{\mathcal{F}}\psi$ this equality is equivalent to $(\psi|\widehat{\mathcal{F}}\psi)_{\mathcal{H}} = \sum_j |(\widehat{\mathcal{F}}\psi|\psi_j)_{\mathcal{H}}|^2$, which is just (8.43). There remains to prove that $\widehat{\mathcal{F}}^{-1}$ is the frame operator for $(\widehat{\mathcal{F}}^{-1}\psi_j)_{j \in J}$. Let us denote by $\widehat{\mathcal{F}}'$ this frame operator; by definition,

$$\begin{aligned}
\widehat{\mathcal{F}}'\psi &= \sum_j (\psi|\widehat{\mathcal{F}}^{-1}\psi_j)_{\mathcal{H}}\widehat{\mathcal{F}}^{-1}\psi_j \\
&= \widehat{\mathcal{F}}^{-1}\sum_j (\widehat{\mathcal{F}}^{-1}\psi|\psi_j)_{\mathcal{H}}\psi_j \\
&= \widehat{\mathcal{F}}^{-1}\widehat{\mathcal{F}}(\widehat{\mathcal{F}}^{-1}\psi)
\end{aligned}$$

hence $\widehat{\mathcal{F}}' = \widehat{\mathcal{F}}^{-1}$. $\square$

Problem 156. Show that if the bounds a and b are optimal for the frame $(\psi_j)_j$ then b^{-1} and a^{-1} are optimal bounds for the dual frame $(\widehat{\mathcal{F}}^{-1}\psi_j)_j$.

The following result shows why frames are important in harmonic analysis:

Proposition 157. *Let $(\psi_j)_j$ be a frame with frame operator $\widehat{\mathcal{F}}$. We have*

$$\psi = \sum_j (\psi|\widehat{\mathcal{F}}^{-1}\psi_j)_{\mathcal{H}}\psi_j = \sum_j (\psi|\psi_j)_{\mathcal{H}}\widehat{\mathcal{F}}^{-1}\psi_j, \tag{8.46}$$

both series being unconditionally convergent for every $\psi \in \mathcal{H}$.

Proof. We have $\psi = \widehat{\mathcal{F}}[\widehat{\mathcal{F}}^{-1}\psi]$ hence, by Definition (8.40) of $\widehat{\mathcal{F}}$,

$$\psi = \sum_j (\widehat{\mathcal{F}}^{-1}\psi|\psi_j)_{\mathcal{H}}\psi_j = \sum_j (\psi|\widehat{\mathcal{F}}^{-1}\psi_j)_{\mathcal{H}}\psi$$

which proves the first equality (8.46). The proof of the unconditional convergence of the series is omitted (see Christensen [26]). $\square$

This result shows that given a frame $(\psi_j)_j$ we can always write an arbitrary function $\psi \in \mathcal{H}$ as an unconditionally convergent series

$$\psi = \sum_j c_j \psi_j \;\; , \;\; (c_j)_j \in \ell^2(\mathbb{N})$$

with coefficients $c_j = (\widehat{\mathcal{F}}^{-1}\psi|\psi_j)_{\mathcal{H}}$. However, except in the case where $(\psi_j)_j$ is a basis, these coefficients are not unique; it turns out that the choice $c_j = (\widehat{\mathcal{F}}^{-1}\psi|\psi_j)_{\mathcal{H}}$ is in a sense optimal: If $(\psi_j)_j$ is a frame with frame operator $\widehat{\mathcal{F}}$ then $\psi = \sum_j c_j \psi_j$ with $(c_j)_j \in \ell^2(\mathbb{N})$ implies that

$$\sum_j |c_j|^2 \geq \sum_j |(\widehat{\mathcal{F}}^{-1}\psi|\psi_j)_{\mathcal{H}}|^2$$

(see [82], Proposition 5.1.4, for a proof of this inequality). Let us give the following definition:

Definition 158. Let $(\psi_j)_j$ be a frame with frame operator $\widehat{\mathcal{F}}$. If each $\psi \in \mathcal{H}$ has a unique decomposition $\psi = \sum_j c_j \psi_j$ (with coefficients thus given by $c_j = (\widehat{\mathcal{F}}^{-1}\psi|\psi_j)_{\mathcal{H}}$) one says that $(\psi_j)_j$ is a *Riesz basis* of $\mathcal{H}$.

Of course the frame operator of a Riesz basis is the identity. Moreover:

Proposition 159. *A frame $(\psi_j)_j$ of $\mathcal{H}$ is a Riesz basis if and only if there exists an orthonormal basis $(\phi_j)_j$ of $\mathcal{H}$ and an invertible bounded operator T on $\mathcal{H}$ such that $\psi_j = T\phi_j$ for each $j \in \mathbb{N}$.*

For a proof of this result we refer to [82], Proposition 5.1.5, where alternative necessary and sufficient conditions are also given.

8.4.2 Weyl–Heisenberg frames

Weyl–Heisenberg frames are traditionally defined in time-frequency analysis using the shift (= translation) and modulation operators T_x and E_ω given by

$$T_{x_0}\psi(x) = \psi(x - x_0) \;\; , \;\; E_{\omega_0}\psi(x) = e^{2\pi i \omega_0 \cdot x}\psi(x). \tag{8.47}$$

Exercise 160. Show that these operators satisfy the commutation relation $T_x E_\omega = e^{-2\pi i \omega \cdot x} E_\omega T_x$.

These operators are related to the Heisenberg–Weyl operators by the simple (but important) formula

$$\widehat{T}(z) = e^{\pi i p \cdot x} T_x E_p = E_{p/2} T_x E_{p/2} \tag{8.48}$$

when $\hbar = 1/2\pi$.

Consider Weyl–Heisenberg frames in terms of *lattices*. A lattice in phase space $\mathbb{R}^{2n}$ is a cocompact discrete subgroup of $\mathbb{R}^{2n}$; more precisely:

Definition 161. A lattice in $\mathbb{R}^{2n}$ is a discrete subgroup $\Lambda = M(\mathbb{Z}^{2n})$ where $M \in GL(2n, \mathbb{R})$. The number $\mathrm{Vol}(\Lambda) = \det M$ is called the volume of the lattice Λ. The lattice $\Lambda^o = (M^{-1})^*(\mathbb{Z}^{2n})$ is called the dual lattice of Λ. When $M = \alpha S$ with $S \in \mathrm{Sp}(2n, \mathbb{R})$ then Λ is called a symplectic lattice.

It is convenient in practice to choose $M = \begin{pmatrix} A & 0_{n\times n} \\ 0_{n\times n} & B \end{pmatrix}$ with $A, B \in GL(n, \mathbb{R})$; we then have $\Lambda = A\mathbb{Z}^n \times B\mathbb{Z}^n$. In many applications one even makes the simpler choice $M = \begin{pmatrix} \alpha I_{n\times n} & 0_{n\times n} \\ 0_{n\times n} & \beta I_{n\times n} \end{pmatrix}$ in which case the lattice is just $\Lambda = \alpha\mathbb{Z}^n \times \beta\mathbb{Z}^n$.

Definition 162. Let $\phi \in L^2(\mathbb{R}^n)$, $\phi \neq 0$ ("window") and a lattice Λ. The set

$$\mathcal{G}(\phi, \Lambda) = \{\widehat{T}(z)\phi : z \in \Lambda\}$$

is called a Weyl–Heisenberg (or Gabor) system. If $\mathcal{G}(\phi, \Lambda)$ is a frame, it is called a Weyl–Heisenberg (or Gabor) frame. The corresponding frame operator is denoted by $\mathcal{F}_{\phi,\Lambda}$; it is given for $\psi \in L^2(\mathbb{R}^n)$ by

$$\mathcal{F}_{\phi,\Lambda}\psi = \sum_{z\in\Lambda} (\psi|\widehat{T}(z)\phi)_{L^2(\mathbb{R}^n)} \widehat{T}(z)\phi \tag{8.49}$$

and the right-hand side of this equality is called a Gabor expansion.

In Chapter 9 we will study in detail the cross-ambiguity function of a pair (ψ, ϕ) of elements of $L^2(\mathbb{R}^n)$; it is the function on $L^2(\mathbb{R}^{2n})$ defined by

$$A(\psi, \phi)(z) = \left(\tfrac{1}{2\pi\hbar}\right)^n (\psi|\widehat{T}(z)\phi)_{L^2(\mathbb{R}^n)}. \tag{8.50}$$

Given a Weyl–Heisenberg frame $\mathcal{G}(\phi, \Lambda)$ it is easy to characterize the frame operator in terms of the cross-ambiguity function: we have

$$\mathcal{F}_{\phi,\Lambda}\psi = (2\pi\hbar)^n \sum_{z\in\Lambda} A(\psi, \phi)(z)\widehat{T}(z)\phi. \tag{8.51}$$

In fact by Definition (8.40) of the frame operator we have

$$\mathcal{F}_{\phi,\Lambda}\psi = \sum_{z\in\Lambda} (\psi|\widehat{T}(z)\phi)_{L^2(\mathbb{R}^n)} \widehat{T}(z)\phi,$$

hence (8.51). The cross-ambiguity function has the following symplectic covariance property: for every $\widehat{S} \in \mathrm{Mp}(2n, \mathbb{R})$ with projection $S = \pi^{\mathrm{Mp}}(\widehat{S})$ on $\mathrm{Sp}(2n, \mathbb{R})$ we have

$$A(\widehat{S}\psi, \widehat{S}\phi)(z) = A(\psi, \phi)(S^{-1}z); \tag{8.52}$$

this important property follows at once from formula (8.11) which we find convenient to rewrite here in the form

$$\widehat{T}(S^{-1}z_0) = \widehat{S}^{-1}\widehat{T}(z_0)\widehat{S} \tag{8.53}$$

for further use.

Proposition 163. *Let* $\mathcal{G}(\phi,\Lambda)$ *be a Weyl–Heisenberg frame and* $\widehat{S} \in \mathrm{Mp}(2n,\mathbb{R})$ *with projection* $S = \pi^{\mathrm{Mp}}(\widehat{S})$. *Then* $\mathcal{G}(\widehat{S}\phi, S\Lambda)$ *is also a Weyl–Heisenberg frame and the frame operators of* $\mathcal{G}(\phi,\Lambda)$ *and* $\mathcal{G}(\widehat{S}\phi, S\Lambda)$ *are related by the metaplectic conjugation formula*

$$\mathcal{F}_{\widehat{S}\phi, S\Lambda} = \widehat{S}\mathcal{F}_{\phi,\Lambda}\widehat{S}^{-1}. \tag{8.54}$$

Proof. In view of formulas (8.51) and (8.52) we have

$$\mathcal{F}_{\phi,\Lambda}\widehat{S}^{-1}\psi = \sum_{z \in \Lambda} A(\psi, \widehat{S}\phi)(Sz)\widehat{T}(z)\phi = \sum_{z \in S\Lambda} A(\psi, \widehat{S}\phi)(z)\widehat{T}(S^{-1}z)\phi$$

and hence, using (8.53),

$$\mathcal{F}_{\phi,\Lambda}\widehat{S}^{-1}\psi = \sum_{z \in S\Lambda} A(\psi, \widehat{S}\phi)(z)\widehat{S}^{-1}\widehat{T}(z_0)\widehat{S}\phi$$

$$= \widehat{S}^{-1}\left(\sum_{z \in S\Lambda} A(\psi, \widehat{S}\phi)(z)\widehat{T}(z_0)\widehat{S}\phi\right)$$

$$= \widehat{S}^{-1}\mathcal{F}_{\widehat{S}\phi, S\Lambda}$$

proving formula (8.54). $\square$

8.4.3 A useful "dictionary"

In the time-frequency literature Weyl–Heisenberg frames are defined using the shift and modulation operators (8.47) and the Gabor expansion defined above is then replaced with

$$\psi = \sum_{(x,\omega) \in \Lambda} (\psi | T_x E_\omega \phi) T_x E_\omega \psi. \tag{8.55}$$

It is therefore useful to have a "dictionary" allowing one to pass from one definition to the other. Recalling that in the case $\hbar = 1/2\pi$ the operators $\widehat{T}(z)$ and $T_x E_\omega$ are related by the formula $\widehat{T}(z) = e^{\pi i p \cdot x} T_x E_\omega$ $(\omega = p)$ and hence

$$\mathcal{F}_{\phi,\Lambda}\psi = \sum_{z \in \Lambda}(\psi|\widehat{T}(z)\phi)_{L^2}\widehat{T}(z)\phi = \sum_{z \in \Lambda}(\psi|T_x E_p \phi)_{L^2}T_x E_\omega \phi$$

because we have $(\psi|\widehat{T}(z)\phi)_{L^2} = e^{-\pi i p \cdot x}(\psi|T_x E_\omega \phi)_{L^2}$. Thus, if $\hbar = 1/2\pi$ both Definitions (8.49) and (8.55) coincide. How about the general case? We observe that the reduction to the former case can be easily made by setting $p = 2\pi\hbar\omega$: we then have $\widehat{T}(x,p) = e^{\pi i \omega \cdot x} T_x E_\omega$ so that (8.49) and (8.55) again are the same. Thus:

$$\widehat{T}(x, 2\pi\hbar p) = e^{\pi i \omega \cdot x} T_x E_p.$$

It follows that we have the equality

$$\sum_{z\in\Lambda}(\psi|T_x E_p\phi)_{L^2}T_x E_p\phi = \sum_{z\in\Lambda}(\psi|\widehat{T}(x,2\pi\hbar p)\phi)_{L^2}\widehat{T}(x,2\pi\hbar p)\phi. \qquad (8.56)$$

Defining the new lattice $\Lambda^\hbar = \begin{pmatrix} I & 0 \\ 0 & 2\pi\hbar I \end{pmatrix}\Lambda$ we get the important formula

$$\sum_{z\in\Lambda}(\psi|T_x E_p\phi)_{L^2}T_x E_p\phi = \sum_{z\in\Lambda^\hbar}(\psi|\widehat{T}(z)\phi)_{L^2}\widehat{T}(z)\phi \qquad (8.57)$$

valid for arbitrary $\hbar$.

It is useful to rescale the window ϕ by using some dilation, to make it more adequate for a quantum-mechanical use. For example, it is customary in time-frequency analysis to work with the Gaussian window $\psi_0(x) = 2^{n/4}e^{-\pi x^2}$ while one prefers to use the "coherent state" $\psi_0^\hbar(x) = (\pi\hbar)^{-n/4}e^{-x^2/2\hbar}$ in quantum mechanics; both functions are normalized in $L^2(\mathbb{R}^n)$ and we have $\psi_0^\hbar = \psi_0$ when $\hbar = 1/2\pi$. Introducing the unitary scaling operator $\widehat{M}_\lambda$ defined for $\lambda > 0$ by

$$\widehat{M}_\lambda\psi(x) = \lambda^{n/2}\psi(\lambda x)$$

we then have $\widehat{M}_{\sqrt{2\pi\hbar}}\psi_0 = \psi_0^\hbar$ and $\widehat{M}_{1/\sqrt{2\pi\hbar}}\psi_0^\hbar = \psi_0$. We notice that the operator $\widehat{M}_\lambda$ belongs to the metaplectic group $\mathrm{Mp}(2n,\mathbb{R})$; in fact $\widehat{M}_\lambda = \widehat{M}_{L,0}$ with $L = \lambda I$, so that $\widehat{M}_\lambda$ has projection

$$M_\lambda = \begin{pmatrix} \lambda I & 0 \\ 0 & \lambda^{-1}I \end{pmatrix} \in \mathrm{Sp}(2n,\mathbb{R}). \qquad (8.58)$$

A simple calculation shows that we have

$$\widehat{T}(x,2\pi\hbar p)\widehat{M}_{\sqrt{2\pi\hbar}} = e^{\pi i p\cdot x}T_x E_p$$

and hence

$$\sum_{z\in\Lambda}(\psi|\widehat{T}(z)\phi)_{L^2}\widehat{T}(z)\phi = \sum_{z\in\Lambda}(\psi|T_x E_p\phi)_{L^2}T_x E_\omega\phi.$$

Proposition 164. *For $\psi,\phi \in L^2(\mathbb{R}^n)$ set $\psi^\hbar = \widehat{M}_{\sqrt{2\pi\hbar}}\psi$ and $\phi^\hbar = \widehat{M}_{\sqrt{2\pi\hbar}}\phi$. Let $\mathcal{G}(\phi,\Lambda)$ be a Gabor system with frame operator $\mathcal{F}_{\phi,\Lambda}$. Then $\mathcal{G}(\phi^\hbar,\sqrt{2\pi\hbar}\Lambda)$ is a Gabor system with frame operator $\mathcal{F}_{\phi,\sqrt{2\pi\hbar}\Lambda}$ such that*

$$\mathcal{F}_{\phi,\sqrt{2\pi\hbar}\Lambda}\psi^\hbar = \sum_{z\in\sqrt{2\pi\hbar}\Lambda}(\psi^\hbar|\widehat{T}(z)\phi^\hbar)_{L^2}\widehat{T}(z)\phi^\hbar. \qquad (8.59)$$

Proof. We have

$$\mathcal{F}_{\phi,\Lambda}\psi = \sum_{z\in\Lambda^{\hbar}} (\psi|\widehat{T}(z)\phi)_{L^2}\widehat{T}(z)\phi$$

$$= \sum_{z\in\Lambda^{\hbar}} (\psi|\widehat{T}(z)\widehat{M}_{\sqrt{2\pi\hbar}}\phi^{\hbar})_{L^2}\widehat{T}(z)\widehat{M}_{\sqrt{2\pi\hbar}}\phi^{\hbar}$$

that is

$$\mathcal{F}_{\phi,\Lambda}\psi = \sum_{z\in\Lambda^{\hbar}} (\psi|\widehat{M}_{\sqrt{2\pi\hbar}}\widehat{T}(M^{-1}_{\sqrt{2\pi\hbar}}z))_{L^2}\widehat{M}_{\sqrt{2\pi\hbar}}\widehat{T}(M^{-1}_{\sqrt{2\pi\hbar}}z)\phi^{\hbar}$$

since $\widehat{T}(z)\widehat{M}_{\sqrt{2\pi\hbar}} = \widehat{M}_{\sqrt{2\pi\hbar}}\widehat{T}(M^{-1}_{\sqrt{2\pi\hbar}}z)$ in view of formula (8.53). Since the adjoint of $\widehat{M}_{\sqrt{2\pi\hbar}}$ is $\widehat{M}^{-1}_{\sqrt{2\pi\hbar}}$ the formula above is equivalent to

$$\widehat{M}^{-1}_{\sqrt{2\pi\hbar}}\mathcal{F}_{\phi,\Lambda}\psi = \sum_{z\in\Lambda^{\hbar}} (\widehat{M}^{-1}_{\sqrt{2\pi\hbar}}\psi|\widehat{T}(M^{-1}_{\sqrt{2\pi\hbar}}z))_{L^2}\widehat{T}(M^{-1}_{\sqrt{2\pi\hbar}}z)\phi^{\hbar}$$

$$= \sum_{z\in M_{\sqrt{2\pi\hbar}}\Lambda^{\hbar}} (\psi^{\hbar}|\widehat{T}(z)\phi^{\hbar})_{L^2}\widehat{T}(z)\phi^{\hbar}.$$

Observing that by definition of $\Lambda^{\hbar}$ and formula (8.58)

$$M^{-1}_{\sqrt{2\pi\hbar}}\Lambda^{\hbar} = \begin{pmatrix} \sqrt{2\pi\hbar}I & 0 \\ 0 & (\sqrt{2\pi\hbar})^{-1}I \end{pmatrix}\begin{pmatrix} I & 0 \\ 0 & 2\pi\hbar I \end{pmatrix}\Lambda = \sqrt{2\pi\hbar}\Lambda$$

we thus have

$$\widehat{M}^{-1}_{\sqrt{2\pi\hbar}}\mathcal{F}_{\phi,\Lambda}\psi = \sum_{z\in\sqrt{2\pi\hbar}\Lambda} (\psi^{\hbar}|\widehat{T}(z)\phi^{\hbar})_{L^2}\widehat{T}(z)\phi^{\hbar}.$$

Taking into account the metaplectic conjugation formula (8.54) we have

$$\widehat{S}\mathcal{F}_{\widehat{S}\phi,S\Lambda} = \widehat{S}\mathcal{F}_{\phi,\Lambda}$$

and

$$\widehat{M}^{-1}_{\sqrt{2\pi\hbar}}\mathcal{F}_{\phi,\Lambda}\psi = \mathcal{F}_{\phi^{\hbar},\sqrt{2\pi\hbar}\Lambda}\widehat{M}^{-1}_{\sqrt{2\pi\hbar}}\psi = \mathcal{F}_{\phi^{\hbar},\sqrt{2\pi\hbar}\Lambda}\psi^{\hbar},$$

hence formula (8.59). $\qquad\square$

The result above allows us to restate the following *necessary* condition for a Weyl–Heisenberg system to be a frame in terms of $\hbar$:

Proposition 165. *Let $\Lambda_{\alpha\beta} = \alpha\mathbb{Z}^n \times \beta\mathbb{Z}^n$ be a lattice in $\mathbb{R}^{2n}$, and $\mathcal{G}(\phi,\Lambda_{\alpha\beta})$ the corresponding Gabor system.*

(i) *If $\mathcal{G}(\phi,\Lambda_{\alpha\beta})$ is a frame for $L^2(\mathbb{R}^n)$ then we have $\alpha\beta \le 2\pi\hbar$;*

(ii) *$\mathcal{G}(\phi,\Lambda_{\alpha\beta})$ is a Riesz basis for $L^2(\mathbb{R}^n)$ if and only if it is a frame and $\alpha\beta = 2\pi\hbar$;*

(iii) *$\mathcal{G}(\phi,\Lambda_{\alpha\beta})$ is an orthonormal basis for $L^2(\mathbb{R}^n)$ if and only if it is a tight frame and $\|\phi\|_{L^2} = 1$ and $\alpha\beta = 2\pi\hbar$.*

For a proof of this result (in the case $2\pi\hbar = 1$) see [82], §7.5. The study and even the statement of *sufficient* conditions for $\mathcal{G}(\phi, \Lambda_{\alpha\beta})$ to be a frame is rather complicated; see Theorem 6.5.1 in [82] which proves a condition due to Walnut [159].

A very interesting situation occurs when one chooses a Gaussian window because Gaussians play a somewhat privileged role in both time-frequency analysis and quantum mechanics. Unfortunately at the time we are writing one has a simple necessary and sufficient condition only in the case $n = 1$:

Proposition 166. *Let* $\psi_0^\hbar(x) = (\pi\hbar)^{-1/4} e^{-x^2/2\hbar}$ *with* $x \in \mathbb{R}$ *and* $\Lambda_{\alpha\beta} = \alpha\mathbb{Z} \times \beta\mathbb{Z}$. *The Gabor system* $\mathcal{G}(\psi_0^\hbar, \Lambda_{\alpha\beta})$ *is a frame for* $L^2(\mathbb{R}^2)$ *if and only if* $\alpha\beta < 2\pi\hbar$.

The proof of this result uses methods from complex analysis (see Lyubarski [120] and Seip and Wallstén [146]).

Chapter 9

Cross-ambiguity and Wigner Functions

The Heisenberg–Weyl and Grossmann–Royer operators allow us to define in a particular simple way two classical objects from symplectic harmonic analysis, namely the cross-ambiguity and Wigner functions, which are symplectic Fourier transforms of each other. Wigner introduced the eponymic distribution in 1932 as a substitute for a phase space probability density, but he did that in an ad hoc way, a kind of "lucky guess" one could say. It has since then been realized that the Wigner distribution (and its companion, the cross-ambiguity function) have a very natural meaning in Weyl calculus, and that they can be simply defined in terms of the Grossmann–Royer and Heisenberg–Weyl operators of last chapter.

9.1 The cross-ambiguity function

The cross-ambiguity function is a venerable object much used in harmonic analysis, and having many applications in signal theory, time-frequency analysis, and engineering (for instance radar theory). We have already briefly encountered it when discussing Weyl–Heisenberg frames.

9.1.1 Definition of $A(\psi, \phi)$

Recall that the Heisenberg–Weyl operator $\widehat{T}(z_0)$ is defined by

$$\widehat{T}(z_0)\psi = e^{\frac{i}{\hbar}(p_0 \cdot x - \frac{1}{2}p_0 \cdot x_0)}\psi_0(x - x_0)$$

where $\psi \in \mathcal{S}(\mathbb{R}^n)$ (or $\psi \in \mathcal{S}'(\mathbb{R}^n)$).

Definition 167. Let ψ and ϕ be in $\mathcal{S}(\mathbb{R}^n)$. The function $(\psi, \phi) \longrightarrow A(\psi, \phi)$ defined by

$$A(\psi, \phi)(z) = \left(\tfrac{1}{2\pi\hbar}\right)^n (\psi|\widehat{T}(z)\phi)_{L^2(\mathbb{R}^n)} \tag{9.1}$$

is called the cross-ambiguity function (or transform). The function $A\psi = A(\psi, \psi)$ given by

$$A\psi(z) = \left(\tfrac{1}{2\pi\hbar}\right)^n (\psi|\widehat{T}(z)\psi)_{L^2(\mathbb{R}^n)} \tag{9.2}$$

is called the (auto-)ambiguity function or sometimes, the Wigner–Fourier transform of ψ.

Notice that formula (9.1) can be rewritten in the form

$$A(\psi, \phi)(z) = \left(\tfrac{1}{2\pi\hbar}\right)^n (\widehat{T}(z)\psi^\vee|\phi^\vee)_{L^2(\mathbb{R}^n)} \tag{9.3}$$

where $\psi^\vee(x) = \psi(-x)$; this formula is somewhat useful and shows that

$$\overline{A(\psi, \phi)} = A(\psi^\vee, \phi^\vee). \tag{9.4}$$

When $\hbar = 1/2\pi$ and p is viewed as a frequency the function $A\psi$ is also called "radar ambiguity function" or "Woodward ambiguity function".

Having in mind our applications to the case where ψ might be a tempered distribution, we notice that Definition (9.1) might as well be written

$$A(\psi, \phi)(-z) = \left(\tfrac{1}{2\pi\hbar}\right)^n \langle \widehat{T}(z)\psi, \overline{\phi} \rangle \tag{9.5}$$

where $\langle \widehat{T}(z)\psi, \overline{\phi} \rangle$ is the distributional pairing of $\widehat{T}(z)\psi$ and $\overline{\phi}$. This formula defines unambiguously $A(\psi, \phi)$ when $\psi \in \mathcal{S}'(\mathbb{R}^n)$ and $\phi \in \mathcal{S}(\mathbb{R}^n)$.

Exercise 168. Check the "following polarization identity" satisfied by the cross-ambiguity function:

$$\mathrm{Re}\, A(\psi, \phi) = \frac{1}{4}\left[A(\psi + \phi) - A(\psi - \phi)\right]. \tag{9.6}$$

Exercise 169. In Chapter 8 we discussed the Stone–von Neumann theorem from the point of representation theory. Show that the cross-ambiguity function is related to the notion of "representation coefficient" for the Schrödinger representation.

The cross-ambiguity function of $(\psi, \phi) \in \mathcal{S}(\mathbb{R}^n) \times \mathcal{S}(\mathbb{R}^n)$ is explicitly given by the formula

$$A(\psi, \phi)(z) = \left(\tfrac{1}{2\pi\hbar}\right)^n \int_{\mathbb{R}^n} e^{-\frac{i}{\hbar}p\cdot x'}\, \psi(x' + \tfrac{1}{2}x)\overline{\phi(x' - \tfrac{1}{2}x)}dx'. \tag{9.7}$$

In fact, by definition of $\widehat{T}(z)$ we have, setting $z = (x, p)$,

$$A(\psi, \phi)(z) = \left(\tfrac{1}{2\pi\hbar}\right)^n \int_{\mathbb{R}^n} e^{\frac{i}{\hbar}(p\cdot x'' - \frac{1}{2}p\cdot x)}\, \psi^\vee(x'' - x)\overline{\phi^\vee(x'')}dx'' \tag{9.8}$$

which is precisely (9.8) performing the change of variables $x'' = -x' + \tfrac{1}{2}x$. The ambiguity function is thus given by

$$A\psi(z) = \left(\tfrac{1}{2\pi\hbar}\right)^n \int_{\mathbb{R}^n} e^{-\frac{i}{\hbar}p\cdot x'}\, \psi(x' + \tfrac{1}{2}x)\overline{\psi(x' - \tfrac{1}{2}x)}dx'. \tag{9.9}$$

9.1.2 Elementary properties of the cross-ambiguity function

Here are some elementary continuity properties of the cross-ambiguity function:

Proposition 170. *The cross-ambiguity function has the following properties:*

(i) *It is a continuous mapping* $\mathcal{S}(\mathbb{R}^n) \times \mathcal{S}(\mathbb{R}^n) \longrightarrow \mathcal{S}(\mathbb{R}^n \oplus \mathbb{R}^n)$;

(ii) *That mapping extends into a continuous mapping*

$$A : L^2(\mathbb{R}^n) \times L^2(\mathbb{R}^n) \longrightarrow C^0(\mathbb{R}^n \oplus \mathbb{R}^n) \cap L^\infty(\mathbb{R}^n \oplus \mathbb{R}^n)$$

such that

$$\|A(\psi, \phi)\|_\infty \leq \|\psi\|_{L^2} \|\phi\|_{L^2}. \tag{9.10}$$

Proof of (i). In view of formula (9.8) and the fact that multiplication by the exponential $e^{-\frac{i}{2\hbar}p \cdot x}$ is a mapping $\mathcal{S}(\mathbb{R}^n \oplus \mathbb{R}^n) \longrightarrow \mathcal{S}(\mathbb{R}^n \oplus \mathbb{R}^n)$ it suffices to show that the function F defined by

$$F(z) = \int_{\mathbb{R}^n} e^{\frac{i}{\hbar}p \cdot x'} \psi(x' - x)\overline{\phi(x')} dx'$$

is in $\mathcal{S}(\mathbb{R}^n \oplus \mathbb{R}^n)$ if ψ and ϕ are in $\mathcal{S}(\mathbb{R}^n)$. Now, F is (up to a constant factor) the partial Fourier transform in x' of the mapping $(x, x') \longmapsto f(x, x') = \psi(x' - x)\overline{\phi(x')}$; since $f \in \mathcal{S}(\mathbb{R}^n \times \mathbb{R}^n)$ the claim follows.

Proof of (ii). Using Definition (9.3) of $A(\psi, \phi)$ we have

$$|A(\psi, \phi)(z_0)| \leq \left(\tfrac{1}{2\pi\hbar}\right)^n \int_{\mathbb{R}^n} |\widehat{T}(z_0)\psi(-x)||\phi(-x)| dx$$

$$\leq \int_{\mathbb{R}^n} |\psi(-x - x_0)||\phi(-x)| dx$$

$$= \int_{\mathbb{R}^n} |\psi(x - x_0)||\phi(x)| dx,$$

hence, using the Cauchy–Schwarz inequality,

$$|A(\psi, \phi)(z_0)|^2 \leq \left(\int_{\mathbb{R}^n} |\psi(x - x_0)|^2 dx\right)^2 \left(\int_{\mathbb{R}^n} |\phi(x)|^2 dx\right)^2$$

that is

$$|A(\psi, \phi)(z_0)| \leq \|\psi\|_{L^2} \|\phi\|_{L^2}.$$

Taking the supremum with respect to z_0 we get the inequality (9.10). The fact that $A(\psi, \phi)$ is continuous follows since $\mathcal{S}(\mathbb{R}^n)$ is dense in $L^2(\mathbb{R}^n)$. $\qquad\square$

9.2   The cross-Wigner transform

The cross-Wigner function (or transform) is closely related to the cross-ambiguity function using a symplectic Fourier transform. It is an object of choice for studying quantum mechanics in phase space. The definition we give is a first illustration of the usefulness of Grossmann–Royer operators.

9.2.1   Definition and first properties of $W(\psi, \phi)$

Replacing the Heisenberg–Weyl operator by the Grossmann–Royer operator

$$\widehat{T}_{\mathrm{GR}}(z_0)\psi(x) = e^{\frac{2i}{\hbar}p_0 \cdot (x - x_0)}\psi(2x_0 - x)$$

in Definition 167 yields the cross-Wigner transform:

Definition 171. Let ψ and ϕ be as in $\mathcal{S}(\mathbb{R}^n)$. The function $(\psi, \phi) \longrightarrow W(\psi, \phi)$ defined by

$$W(\psi, \phi)(z) = \left(\tfrac{1}{\pi\hbar}\right)^n \left(\widehat{T}_{\mathrm{GR}}(z)\psi|\phi\right)_{L^2(\mathbb{R}^n)} \tag{9.11}$$

is called the cross-Wigner transform, or function (it is sometimes also called the Wigner–Moyal distribution). The function $W\psi = W(\psi, \psi)$ is called the "*Wigner transform*" of ψ (or "Wigner–Blokhintsev transform").

As in the case for the cross-ambiguity function, we might rewrite this formula in terms of distribution brackets as

$$W(\psi, \phi)(z) = \left(\tfrac{1}{\pi\hbar}\right)^n \langle \widehat{T}_{\mathrm{GR}}(z)\psi, \overline{\phi}\rangle. \tag{9.12}$$

As for the cross-ambiguity function, we have an analytic expression for the cross-Wigner function.

The formulas below are very often taken as a *definition* in the literature.

Proposition 172. *The cross-Wigner transform is given by the explicit formula*

$$W(\psi, \phi)(z) = \left(\tfrac{1}{2\pi\hbar}\right)^n \int_{\mathbb{R}^n} e^{-\frac{i}{\hbar}p \cdot y}\psi(x + \tfrac{1}{2}y)\overline{\phi(x - \tfrac{1}{2}y)}dy, \tag{9.13}$$

hence the Wigner transform is given by

$$W\psi(z) = \left(\tfrac{1}{2\pi\hbar}\right)^n \int_{\mathbb{R}^n} e^{-\frac{i}{\hbar}p \cdot y}\psi(x + \tfrac{1}{2}y)\overline{\psi(x - \tfrac{1}{2}y)}dy. \tag{9.14}$$

Proof. We have

$$(\widehat{T}_{\mathrm{GR}}(z_0)\psi|\phi)_{L^2} = \int_{\mathbb{R}^n} e^{\frac{2i}{\hbar}p_0 \cdot (x - x_0)}\psi(2x_0 - x)\overline{\phi(x)}dx.$$

Setting $y = 2(x_0 - x)$ this is

$$(\widehat{T}_{\mathrm{GR}}(z_0)\psi|\phi)_{L^2} = 2^{-n} \int_{\mathbb{R}^n} e^{-\frac{i}{\hbar}p_0 \cdot y}\psi(x_0 + \tfrac{1}{2}y)\overline{\phi(x_0 - \tfrac{1}{2}y)}dx$$
$$= (\pi\hbar)^n W(\psi, \phi)(z_0)$$

proving (9.14) in view of (9.11). $\square$

Clearly $(\psi, \phi) \longmapsto W(\psi, \phi)$ is a sesquilinear mapping (as is the cross-ambiguity function); it follows that we have the polarization identity

$$\operatorname{Re} W(\psi, \phi) = \frac{1}{4}\left[W(\psi + \phi) - W(\psi - \phi)\right] \tag{9.15}$$

(cf. formula (9.6) for the cross-ambiguity function).

The following property is obvious:

$$W(\psi, \phi) = \overline{W(\phi, \psi)}; \tag{9.16}$$

and hence, in particular

$W\psi$ is always a real function.

The fact that $W\psi$ is real has far-reaching consequences in quantum mechanics; this property allows $W\psi$ to be viewed as the substitute for a probability density; for instance, we will see later that $W\psi$ has the "correct marginals" in the sense that

$$\int_{\mathbb{R}^n} W(\psi, \phi)(z)dp = \psi(x)\overline{\phi(x)}, \tag{9.17}$$

$$\int_{\mathbb{R}^n} W(\psi, \phi)(z)dx = F\psi(p)\overline{F\phi(p)}. \tag{9.18}$$

However, the Wigner function is not in general positive. In fact, a classical result of Hudson [103] (also see Janssen [104]) tells us that $W\psi$ is non-negative if and only if ψ is a Gaussian, in fact:

$$W\psi \geq 0 \Leftrightarrow \psi(x) = Ce^{M(x-x_0)^2}$$

where M is a complex matrix with negative real eigenvalues and $x_0 \in \mathbb{R}^n$.

Exercise 173. Assume that $\psi \in \mathcal{S}(\mathbb{R}^n)$ is an odd function: $\psi(-x) = -\psi(x)$. Show that $W\psi$ takes negative values. [Hint: calculate $W\psi(0)$.]

We also mention the following tensor-product properties: if $x = (x', x'')$ with $x' \in \mathbb{R}^k$, $x'' \in \mathbb{R}^{n-k}$ and $\psi' \in \mathcal{S}(\mathbb{R}^k)$, $\psi'' \in \mathcal{S}(\mathbb{R}^{n-k})$, then

$$A(\psi' \otimes \psi'') = A'\psi' \otimes A''\psi'' \tag{9.19}$$

and

$$W(\psi' \otimes \psi'') = W'\psi' \otimes W''\psi'' \tag{9.20}$$

where A and A' (resp. W' and W'') are the cross-ambiguity (resp. Wigner) transforms on $\mathcal{S}(\mathbb{R}^k)$ and $\mathcal{S}(\mathbb{R}^{n-k})$, in that order. More generally, we have

$$A(\psi' \otimes \psi'', \phi' \otimes \phi'') = A'(\psi', \phi') \otimes A''(\psi'', \phi'') \tag{9.21}$$

and

$$W(\psi' \otimes \psi'', \phi' \otimes \phi'') = W'(\psi', \phi') \otimes W''(\psi'', \phi''). \tag{9.22}$$

These formulas are useful for the study of so-called partial traces in quantum mechanics.

9.2.2   Translations of Wigner transforms

The following result describes the behavior of the (cross) Wigner transform under translations and Heisenberg–Weyl operators. Recall that $T(z_0)$ is the translation operator $z \longmapsto z+z_0$; it acts on functions or distributions F on $\mathbb{R}^{2n}$ by $T(z_0)F(z) = F(z - z_0)$.

Proposition 174.

(i) *For every $\psi \in L^2(\mathbb{R}^n)$ and $z_0 \in \mathbb{R}^{2n}$ we have*

$$W(\widehat{T}(z_0)\psi, \widehat{T}(z_0)\phi)(z) = T(z_0)W(\psi, \phi)(z). \tag{9.23}$$

In particular

$$W(\widehat{T}(z_0)\psi)) = T(z_0)W\psi. \tag{9.24}$$

(ii) *More generally, if $\psi, \phi \in L^2(\mathbb{R}^n)$, then*

$$W(\widehat{T}(z_0)\psi, \widehat{T}(z_1)\phi)(z) = e^{-\frac{i}{\hbar}[\sigma(z,z_0-z_1)+\frac{1}{2}\sigma(z_0,z_1)]} W(\psi, \phi)(z - \langle z \rangle) \tag{9.25}$$

where $\langle z \rangle = \frac{1}{2}(z_0 + z_1)$.

Proof. The statements in (i) follow from (ii). Let us prove formula (9.25). We will use the notation $\langle x \rangle = \frac{1}{2}(x_0 + x_1)$ and $\langle p \rangle = \frac{1}{2}(p_0 + p_1)$. By definition of the Weyl–Heisenberg operators we have

$$\widehat{T}(z_0)\psi(x + \tfrac{1}{2}y) = e^{\frac{i}{\hbar}[p_0 \cdot (x+\frac{1}{2}y) - \frac{1}{2}p_0 \cdot x_0]} \psi(x - x_0 + \tfrac{1}{2}y),$$

$$\widehat{T}(z_1)\phi(x - \tfrac{1}{2}y) = e^{\frac{i}{\hbar}[p_1 \cdot (x-\frac{1}{2}y) - \frac{1}{2}p_1 \cdot x_1]} \phi(x - x_1 - \tfrac{1}{2}y)$$

and hence

$$\widehat{T}(z_0)\psi(x + \tfrac{1}{2}y)\overline{\widehat{T}(z_1)\phi(x - \tfrac{1}{2}y)}$$
$$= e^{\frac{i}{\hbar}\delta(z_0,z_1)} e^{\frac{i}{\hbar}\langle p \rangle \cdot y} \psi(x - x_0 + \tfrac{1}{2}y)\overline{\phi(x - x_1 - \tfrac{1}{2}y)}$$

with

$$\delta(z_0, z_1) = (p_0 - p_1) \cdot x - \tfrac{1}{2}(p_0 \cdot x_0 - p_1 \cdot x_1).$$

It follows that we have

$$W(\widehat{T}(z_0)\psi, \widehat{T}(z_1)\phi)(z) = \left(\tfrac{1}{2\pi\hbar}\right)^n e^{\frac{i}{\hbar}\delta(z_0,z_1)}$$
$$\times \int_{\mathbb{R}^n} e^{-\frac{i}{\hbar}(p-\langle p\rangle)\cdot y}\psi(x - x_0 + \tfrac{1}{2}y)\overline{\phi(x - x_0 - \tfrac{1}{2}y)}dy.$$

Performing the change of variables $y' = x_1 - x_0 + y$ in the integral this equality becomes

$$W(\widehat{T}(z_0)\psi, \widehat{T}(z_1)\phi)(z) = \left(\tfrac{1}{2\pi\hbar}\right)^n e^{\frac{i}{\hbar}\Delta}$$
$$\times \int_{\mathbb{R}^n} e^{-\frac{i}{\hbar}(p-\langle p\rangle)\cdot y}\psi(x - \langle x\rangle + \tfrac{1}{2}y)\overline{\phi(x - \langle x\rangle - \tfrac{1}{2}y)}dy$$

where the phase Δ is given by

$$\Delta = (p_0 - p_1)\cdot x - (x_0 - x_1)\cdot p + \tfrac{1}{2}(p_1\cdot x_0 - p_0\cdot x_1)$$
$$= -\sigma(z, z_0 - z_1) - \tfrac{1}{2}\sigma(z_0, z_1)$$

hence formula (9.25). $\qquad\qquad\qquad\qquad\qquad\qquad\qquad\qquad\qquad\qquad\square$

The following particular case of formula (9.25) will be important in our study of a certain pseudo-differential calculus in Chapter 18:

$$W(\widehat{T}(z_0)\psi, \phi)(z) = e^{-\frac{i}{\hbar}\sigma(z,z_0)}W(\psi,\phi)(z - \tfrac{1}{2}z_0).$$

9.3 Relations between $A(\psi,\phi)$, $W(\psi,\phi)$, and the STFT

The cross-ambiguity and cross-Wigner transform are related by a symplectic Fourier transform; in addition, both can be expressed in terms of the short-time Fourier transform used in signal theory and time-frequency analysis.

9.3.1 Two simple formulas

The definitions (and explicit expressions) of the cross-ambiguity and Wigner functions are formally very similar. In this section we analyze these similarities in detail, which leads us to some rather surprising results.

As claimed in the beginning of the chapter, the cross-ambiguity and Wigner functions are symplectic transforms of each other. Let us prove this.

Proposition 175. *Let* $(\psi,\phi) \in \mathcal{S}'(\mathbb{R}^n) \times \mathcal{S}'(\mathbb{R}^n)$.

(i) *We have*

$$A(\psi,\phi) = F_\sigma W(\psi,\phi) \tag{9.26}$$

where F_σ *is the symplectic Fourier transform; in particular* $A\psi = F_\sigma W\psi$.

(ii) *We have*

$$A(\psi,\phi)(z) = 2^{-n}W(\psi,\phi^{\vee})(\tfrac{1}{2}z) \qquad (9.27)$$

where $\phi^{\vee}(x) = \phi(-x)$.

Proof of (i). It is sufficient to assume that ψ and ϕ are in $\mathcal{S}(\mathbb{R}^n)$. Set $A = (2\pi\hbar)^{2n}F_\sigma W(\psi,\phi)$; by definition of F_σ and $W(\psi,\phi)$ we have

$$A(z) = \iiint_{\mathbb{R}^{3n}} e^{-\frac{i}{\hbar}\left[\sigma(z,z')+p'\cdot y\right]}\psi(x'+\tfrac{1}{2}y)\overline{\phi(x'-\tfrac{1}{2}y)}dp'dx'dy$$

$$= \iiint_{\mathbb{R}^{3n}} e^{-\frac{i}{\hbar}p'\cdot(y-x)}e^{-\frac{i}{\hbar}p\cdot x'}\psi(x'+\tfrac{1}{2}y)\overline{\phi(x'-\tfrac{1}{2}y)}dp'dx'dy.$$

In view of the "Fourier inversion formula"

$$\int_{\mathbb{R}^n} e^{-\frac{i}{\hbar}p'\cdot(y-x)}dp' = (2\pi\hbar)^n\,\delta(x-y)$$

we can rewrite the expression A as

$$A = (2\pi\hbar)^n \iint_{\mathbb{R}^{2n}} \delta(x-y)e^{-\frac{i}{\hbar}p\cdot x'}\psi(x'+\tfrac{1}{2}y)\overline{\phi(x'-\tfrac{1}{2}y)}dx'dy$$

$$= (2\pi\hbar)^n \int_{\mathbb{R}^n} e^{-\frac{i}{\hbar}p\cdot x'}\psi(x'+\tfrac{1}{2}x)\overline{\phi(x'-\tfrac{1}{2}x)}dx'$$

hence $A = A(\psi,\phi)$. [The calculation of the integral in p' is formal and should be viewed in the distributional sense; the reader willing to attain full rigor might want to redo the calculation using a distributional pairing $\langle F_\sigma W(\psi,\phi),\Phi\rangle$ with $\Phi \in \mathcal{S}(\mathbb{R}^n \oplus \mathbb{R}^n)$.]

Proof of (ii). We have, by definition of the cross-Wigner transform,

$$W(\psi,\phi)(\tfrac{1}{2}z) = \left(\tfrac{1}{2\pi\hbar}\right)^n \int_{\mathbb{R}^n} e^{-\frac{i}{2\hbar}p\cdot y}\psi(\tfrac{1}{2}x+\tfrac{1}{2}y)\overline{\phi(\tfrac{1}{2}x-\tfrac{1}{2}y)}dy$$

that is, setting $x' = \tfrac{1}{2}y$,

$$W(\psi,\phi^{\vee})(\tfrac{1}{2}z) = \left(\tfrac{1}{\pi\hbar}\right)^n \int_{\mathbb{R}^n} e^{-\frac{i}{\hbar}p\cdot x'}\psi(\tfrac{1}{2}x+x')\overline{\phi(\tfrac{1}{2}x-x')}dy'$$

$$= \left(\tfrac{1}{\pi\hbar}\right)^n \int_{\mathbb{R}^n} e^{-\frac{i}{\hbar}p\cdot y'}\psi(y'+\tfrac{1}{2}x)\overline{\phi^{\vee}(x'-\tfrac{1}{2}x)}dy'$$

hence (9.27) in view of formula (9.7). $\square$

We claim that:

Proposition 176. *If* $(\psi,\psi') \in \mathcal{S}'(\mathbb{R}^n) \times \mathcal{S}'(\mathbb{R}^n)$ *then*

$$W\psi = W\psi' \iff \psi = e^{i\alpha}\psi' \;,\quad \alpha \in \mathbb{R}. \qquad (9.28)$$

Proof. It is sufficient to assume that ψ and ϕ are in $\mathcal{S}'(\mathbb{R}^n)$. That both ψ and $e^{i\alpha}\psi$ ($\alpha \in \mathbb{R}$) have the same Wigner transform immediately follows from the definition of $W\psi$. Suppose conversely that $W\psi = W\psi'$ and, for fixed x, set

$$f(y) = \psi(x + \tfrac{1}{2}y)\overline{\psi}(x - \tfrac{1}{2}y),$$
$$f'(y) = \psi'(x + \tfrac{1}{2}y)\overline{\psi'}(x - \tfrac{1}{2}y).$$

The equality $W\psi = W\psi'$ is then equivalent to the equality of the Fourier transforms of f and f' and hence $f = f'$ that is

$$\psi(x + \tfrac{1}{2}y)\overline{\psi}(x - \tfrac{1}{2}y) = \psi'(x + \tfrac{1}{2}y)\overline{\psi'}(x - \tfrac{1}{2}y)$$

for all x, y; taking $y = 0$ we get $|\psi|^2 = |\psi'|^2$ which proves (9.28). $\qquad\square$

There remains the question of the invertibility of the Wigner transform. We will deal with that question in a moment.

We will see later (formula 10.26) that the cross-Wigner transform enjoys the following very nice symplectic covariance property: if $\widehat{S}$ is a metaplectic operator with projections, the symplectic matrix S is then

$$W(\widehat{S}\psi, \widehat{S}\phi)(z) = W(\psi, \phi)(S^{-1}z). \tag{9.29}$$

9.3.2 The short-time Fourier transform

A mathematical object closely related to the Wigner function is the short-time Fourier transform used in signal theory and time-frequency analysis:

Definition 177. Let $\phi \in \mathcal{S}(\mathbb{R}^n)$. The short-time Fourier transform (STFT) (or windowed Fourier transform, or Gabor transform) with window $\phi \in \mathcal{S}(\mathbb{R}^n)$ is the mapping $V_\phi : \mathcal{S}(\mathbb{R}^n) \longrightarrow \mathcal{S}(\mathbb{R}^n \oplus \mathbb{R}^n)$ defined by

$$V_\phi\psi(z) = \int_{\mathbb{R}^n} e^{-2\pi i p \cdot x'}\psi(x')\overline{\phi(x' - x)}dx'. \tag{9.30}$$

We note the following rescaling result, whose (trivial) proof is left to the reader as an exercise:

Lemma 178. *For real* $\lambda \neq 0$ *set* $\psi_\lambda(x) = \psi(\lambda x)$. *We have*

$$V_{\phi_\lambda}\psi_\lambda(x, p) = \lambda^{-n}V_\phi\psi(\lambda x, \lambda^{-1}p). \tag{9.31}$$

Taking $\lambda = \sqrt{2\pi\hbar}$ it is easy to see that the STFT and the cross-Wigner transform are related by the formulae

$$W(\psi, \phi)(z) = \left(\tfrac{2}{\pi\hbar}\right)^{n/2} e^{\frac{2i}{\hbar}p \cdot x} V_{\phi^\vee_{\sqrt{2\pi\hbar}}}\psi_{\sqrt{2\pi\hbar}}\left(z\sqrt{\tfrac{2}{\pi\hbar}}\right) \tag{9.32}$$

where $\phi^\vee(x) = \phi(-x)$; equivalently:

$$V_\phi\psi(z) = \left(\tfrac{2}{\pi\hbar}\right)^{-n/2} e^{-i\pi p \cdot x} W(\psi_{1/\sqrt{2\pi\hbar}}, \phi^\vee_{1/\sqrt{2\pi\hbar}})\left(z\sqrt{\tfrac{\pi\hbar}{2}}\right). \tag{9.33}$$

Exercise 179. Verify in detail formulae (9.32) and (9.33) using the definitions.

Exercise 180. Prove the Moyal identity

$$\langle V_\phi \psi, V_{\phi'} \psi' \rangle = \langle \psi, \psi' \rangle \overline{\langle \phi, \phi' \rangle} \tag{9.34}$$

for the short-time Fourier transform. (This formula is called by some authors the "orthogonality relation for the STFT". It is familiar in representation theory, and apparently goes back to the 1940s; it is thus posterior to the Wigner transform.)

In particular, if one takes $\hbar = 1/2\pi$ (which is the standard choice in time-frequency analysis) one gets

$$W(\psi, \phi)(z) = 2^n e^{4\pi i p \cdot x} V_{\phi^\vee} \psi(2z) \tag{9.35}$$

and

$$V_\phi \psi(z) = 2^{-n} e^{-i\pi p \cdot x} W(\psi, \phi^\vee)(\tfrac{1}{2} z). \tag{9.36}$$

These formulas will be used in Chapters 16 and 17 when we study the theory of modulation spaces (they are usually defined in terms of the STFT).

9.3.3 The Cohen class

The lack of positivity of the Wigner distribution $W\psi$ which makes its interpretation as a true probability density problematic has led to search for alternative distributions $Q\psi$. One of the most famous examples is Husimi's distribution, which is the convolution of the Wigner transform with a Gaussian function. More generally, we will say following Gröchenig [82], §4.5, that $Q\psi$ belongs to the Cohen class if it is of the type $Q\psi = W\psi * \theta$ for some distribution $\theta \in \mathcal{S}'(\mathbb{R}^n \oplus \mathbb{R}^n)$. The following result gives sufficient conditions for a distribution to belong to Cohen's class:

Proposition 181. *Let* $Q : \mathcal{S}(\mathbb{R}^n) \times \mathcal{S}(\mathbb{R}^n) \longrightarrow \mathcal{S}(\mathbb{R}^n \oplus \mathbb{R}^n)$ *be a sesquilinear form and set* $Q\psi = Q(\psi, \psi)$. *If* Q *is such that*

$$Q\psi(z - z_0) = Q(\widehat{T}(z_0)\psi)(z), \tag{9.37}$$

$$|Q(\psi, \phi)(0, 0)| \le \|\psi\|_{L^2(\mathbb{R}^n)} \|\phi\|_{L^2(\mathbb{R}^n)} \tag{9.38}$$

for all ψ, ϕ *in* $L^2(\mathbb{R}^n)$, *then there exists* $\theta \in \mathcal{S}'(\mathbb{R}^n \oplus \mathbb{R}^n)$ *such that* $Q\psi = W\psi * \theta$ *for all* $\psi \in \mathcal{S}(\mathbb{R}^n)$.

Proof. (Cf. [82], Theorem 4.5.1.) The condition (9.38) ensures us that $(\psi, \phi) \longmapsto Q(\psi, \phi)(0, 0)$ is a bounded sesquilinear form. Hence, by Riesz's representation theorem there exists a bounded operator $\widehat{A}$ on $L^2(\mathbb{R}^n)$ such that $Q(\psi, \phi)(0, 0) = \langle \widehat{A}\psi, \phi \rangle$. In view of the covariance formula (9.37) we have

$$Q\psi(z_0) = Q(\widehat{T}(-z_0)\psi)(0)$$
$$= \langle \widehat{A}\widehat{T}(-z_0)\psi, \widehat{T}(-z_0)\psi \rangle.$$

In view of Schwartz's kernel theorem there exists a distribution $K \in \mathcal{S}'(\mathbb{R}^n \times \mathbb{R}^n)$ such that $\langle \widehat{A}\psi, \phi \rangle = \langle\langle K, \psi \otimes \overline{\phi} \rangle\rangle$ for all $\psi, \phi \in \mathcal{S}(\mathbb{R}^n)$ ($\langle\langle \cdot, \cdot \rangle\rangle$ the distributional bracket on $\mathcal{S}(\mathbb{R}^{2n})$) so that we can write

$$Q\psi(z_0) = \langle\langle K, \widehat{T}(-z_0)\psi \otimes \overline{\widehat{T}(-z_0)\psi} \rangle\rangle$$
$$= \iint_{\mathbb{R}^n \times \mathbb{R}^n} K(x,y)\widehat{T}(-z_0)\psi(x)\overline{\widehat{T}(-z_0)\psi(y)}dxdy.$$

By definition of the Weyl–Heisenberg operators we have

$$\widehat{T}(-z_0)\psi(x) = e^{\frac{i}{\hbar}(-p_0\cdot x - \frac{1}{2}p_0\cdot x_0)}\psi(x+x_0)$$
$$\overline{\widehat{T}(-z_0)\psi(y)} = e^{-\frac{i}{\hbar}(-p_0\cdot y - \frac{1}{2}p_0\cdot x_0)}\overline{\psi(y+x_0)}$$

and hence

$$Q\psi(z_0) = \iint_{\mathbb{R}^n \times \mathbb{R}^n} K(x,y)e^{-\frac{i}{\hbar}p_0\cdot(x-y)}\psi(x+x_0)\overline{\psi(y+x_0)}dxdy. \tag{9.39}$$

On the other hand, for every $\theta \in \mathcal{S}'(\mathbb{R}^n \oplus \mathbb{R}^n)$ we have

$$(W\psi * \theta)(z_0) = \int_{\mathbb{R}^n} W\psi(z_0 - z)\theta(z)dz$$

(the integral being interpreted in the distributional sense) hence, in view of the definition of the Wigner transform,

$$(W\psi * \theta)(z_0) = \left(\tfrac{1}{2\pi\hbar}\right)^n \iiint_{\mathbb{R}^n \times \mathbb{R}^n \times \mathbb{R}^n} e^{-\frac{i}{\hbar}(p_0-p)\cdot y'}$$
$$\times \psi(x_0 - x' + \tfrac{1}{2}y')\overline{\psi(x_0 - x' - \tfrac{1}{2}y')}\theta(x',p')dpdx'dy'$$

that is, calculating the integral in the p variables,

$$(W\psi * \theta)(z_0) = \left(\tfrac{1}{2\pi\hbar}\right)^{n/2} \iint_{\mathbb{R}^n \times \mathbb{R}^n} F_2^{-1}\theta(x',y')e^{-\frac{i}{\hbar}p_0\cdot y'}$$
$$\times \psi(x_0 - x' + \tfrac{1}{2}y')\overline{\psi(x_0 - x' - \tfrac{1}{2}y')}\theta(x',p')dx'dy'$$

where $F_2^{-1}\theta$ is the partial inverse Fourier transform of θ in the second set of variables. Making the change of variables $x' = -\tfrac{1}{2}(x+y)$, $y' = x - y$ we have $dx'dy' = dxdy$ and the equality above becomes

$$(W\psi * \theta)(z_0) = \left(\tfrac{1}{2\pi\hbar}\right)^{n/2} \iint_{\mathbb{R}^n \times \mathbb{R}^n} F_2^{-1}\theta(x, x-y)e^{-\frac{i}{\hbar}p_0\cdot(x-y)}$$
$$\times \psi(x+x_0)\overline{\psi(y+x_0)}dxdy. \tag{9.40}$$

Comparing formulas (9.39) and (9.40) we see that $Q\psi = W\psi * \theta$ where θ is determined by the equality

$$K(x,y) = \left(\tfrac{1}{2\pi\hbar}\right)^{n/2} F_2^{-1}\theta(x, x-y)$$

that is

$$\theta(x,p) = (2\pi\hbar)^{n/2} \int_{\mathbb{R}^n} e^{-\frac{i}{\hbar}p\cdot y} K(x, x-y) dy. \qquad \square$$

9.4 The Moyal identity

Moyal's identity is a fundamental formula which will allow us later in this book to construct wavepacket transforms; it has many uses in quantum mechanics and time-frequency analysis.

9.4.1 Statement and proof

The *Moyal identity*, is valid for both the cross-Wigner and ambiguity functions. It shows, in particular, that up to a constant factor the mapping $\psi \longmapsto W(\psi, \phi)$ is, for each fixed $\phi \in L^2(\mathbb{R}^n)$, a partial isometry of $L^2(\mathbb{R}^n)$ onto a closed subspace of $L^2(\mathbb{R}^{2n})$. This fact will be used in Chapter 18, where these mappings will be used to intertwine ordinary Weyl calculus with "Bopp calculus". Because of its importance in both practical and theoretical considerations we dignify the result as a theorem:

Theorem 182. *The cross-Wigner and cross-ambiguity functions satisfies the "Moyal identity"*

$$(W(\psi,\phi)|W(\psi',\phi'))_{L^2(\mathbb{R}^{2n})} = \left(\tfrac{1}{2\pi\hbar}\right)^n (\psi|\psi')_{L^2(\mathbb{R}^n)}\overline{(\phi|\phi')_{L^2(\mathbb{R}^n)}} \qquad (9.41)$$

and

$$(A(\psi,\phi)|A(\psi',\phi'))_{L^2(\mathbb{R}^{2n})} = \left(\tfrac{1}{2\pi\hbar}\right)^n (\psi|\psi')_{L^2(\mathbb{R}^n)}\overline{(\phi|\phi')_{L^2(\mathbb{R}^n)}} \qquad (9.42)$$

for all $(\psi,\phi) \in L^2(\mathbb{R}^n) \times L^2(\mathbb{R}^n)$. Equivalently:

$$\langle W(\psi,\phi), W(\psi',\phi')\rangle = \left(\tfrac{1}{2\pi\hbar}\right)^n \langle \psi,\overline{\psi'}\rangle\langle \phi,\overline{\phi'}\rangle. \qquad (9.43)$$

In particular

$$\|W\psi\|_{L^2(\mathbb{R}^{2n})} = \left(\tfrac{1}{2\pi\hbar}\right)^{n/2} \|\psi\|_{L^2(\mathbb{R}^n)}. \qquad (9.44)$$

Proof. It suffices to prove formula (9.41) since formula (9.42) immediately follows using the fact that the symplectic Fourier transform is unitary (recall that the cross-Wigner and ambiguity functions are symplectic Fourier transforms of each other). Also, the equivalence of formulas (9.41) and (9.43) is obvious. Let us prove (9.41). The scalar product

$$A = (2\pi\hbar)^{2n}(W(\psi,\phi)|W(\psi',\phi'))_{L^2(\mathbb{R}^{2n})}$$

is given by the expression

$$\int_{\mathbb{R}^{4n}} e^{-\frac{i}{\hbar}p\cdot(y-y')}\psi(x+\tfrac{1}{2}y)\overline{\psi'(x+\tfrac{1}{2}y')}\,\overline{\phi(x-\tfrac{1}{2}y)}\phi'(x-\tfrac{1}{2}y')\,dy\,dy'\,dx\,dp.$$

The integral in p (interpreted as a distributional bracket) yields

$$\int_{\mathbb{R}^n} e^{-\frac{i}{\hbar}p\cdot(y-y')}dp = (2\pi\hbar)^n \delta(y-y')$$

and hence

$$A = (2\pi\hbar)^n \int_{\mathbb{R}^{3n}} \psi(x+\tfrac{1}{2}y)\overline{\psi'(x-\tfrac{1}{2}y)}\overline{\phi(x+\tfrac{1}{2}y)}\phi'(x-\tfrac{1}{2}y)\,dy'\,dx.$$

Setting $u = x + \tfrac{1}{2}y$ and $v = x - \tfrac{1}{2}y$ we have $du\,dv = dx\,dy$ hence

$$A = (2\pi\hbar)^n \left(\int_{\mathbb{R}^n} \psi(u)\overline{\psi'(u)}du \right) \left(\int_{\mathbb{R}^n} \overline{\phi(v)}\phi'(v)dv \right)$$

which is (9.41).							$\square$

Using Moyal's identity we can prove:

Proposition 183. *The cross-Wigner and ambiguity transforms* $(\psi,\phi) \longrightarrow W(\psi,\phi)$ *and* $(\psi,\phi) \longrightarrow A(\psi,\phi)$ *extends into bilinear mappings*

$$W : \mathcal{S}'(\mathbb{R}^n) \times \mathcal{S}'(\mathbb{R}^n) \longrightarrow \mathcal{S}'(\mathbb{R}^n), \tag{9.45}$$

$$A : \mathcal{S}'(\mathbb{R}^n) \times \mathcal{S}'(\mathbb{R}^n) \longrightarrow \mathcal{S}'(\mathbb{R}^n), \tag{9.46}$$

and we have

$$W : L^2(\mathbb{R}^n) \times L^2(\mathbb{R}^n) \longrightarrow L^2(\mathbb{R}^n \oplus \mathbb{R}^n) \cap C^0(\mathbb{R}^n \oplus \mathbb{R}^n), \tag{9.47}$$

$$A : L^2(\mathbb{R}^n) \times L^2(\mathbb{R}^n) \longrightarrow L^2(\mathbb{R}^n \oplus \mathbb{R}^n) \cap C^0(\mathbb{R}^n \oplus \mathbb{R}^n). \tag{9.48}$$

We have the following inequalities:

$$\|W(\psi,\phi)\|_\infty \le \left(\tfrac{1}{\pi\hbar}\right)^n \|\psi\|_{L^2(\mathbb{R}^n)}\|\phi\|_{L^2(\mathbb{R}^n)}, \tag{9.49}$$

$$\|A(\psi,\phi)\|_\infty \le \left(\tfrac{1}{\pi\hbar}\right)^n \|\psi\|_{L^2(\mathbb{R}^n)}\|\phi\|_{L^2(\mathbb{R}^n)}. \tag{9.50}$$

Proof. Let us show that $W(\psi,\phi)$ and $A(\psi,\phi)$ are square integrable if ψ and ϕ are. The property for the Wigner transform W follows from Moyal's identity (9.42). That it is also true for the ambiguity function A follows from the relation $A = F_\sigma W$ and the fact that the symplectic Fourier transform is an automorphism of

$L^2(\mathbb{R}^n \oplus \mathbb{R}^n)$. Let us show the continuity property. We have, using (9.41), the integral definition of $W(\psi, \phi)$, and the Cauchy–Schwarz inequality,

$$\|W(\psi, \phi)\|_\infty = \sup_{z \in \mathbb{R}^{2n}} |W(\psi, \phi)(z)|$$

$$\leq \left(\tfrac{1}{2\pi\hbar}\right)^n \int_{\mathbb{R}^n} |\psi(x + \tfrac{1}{2}y)||\phi(x - \tfrac{1}{2}y)|dy$$

$$\leq \left(\tfrac{1}{2\pi\hbar}\right)^n \left(\int_{\mathbb{R}^n} |\psi(x + \tfrac{1}{2}y)|^2 dy\right)^{1/2} \left(\int_{\mathbb{R}^n} |\phi(x - \tfrac{1}{2}y)|^2 dy\right)^{1/2},$$

hence the inequality (9.49) since we have

$$\left(\int_{\mathbb{R}^n} |\psi(x + \tfrac{1}{2}y)|^2 dy\right)^{1/2} = 2^{n/2}\|\psi\|_{L^2(\mathbb{R}^n)},$$

$$\left(\int_{\mathbb{R}^n} |\phi(x - \tfrac{1}{2}y)|^2 dy\right)^{1/2} = 2^{n/2}\|\phi\|_{L^2(\mathbb{R}^n)}.$$

The inequality (9.50) is proven by a similar argument. $\qquad\square$

Let us mention, without proof (see [82], Theorem 3.3.2) that the cross-Wigner transform satisfies the "Lieb inequality"

$$\int_{\mathbb{R}^{2n}} |W(\psi, \phi)|^q(z)dz \leq C_q \left[\|\psi\|_{L^q(\mathbb{R}^n)}\|\psi\|_{L^q(\mathbb{R}^n)}\right]^q$$

for $2 \leq q < \infty$. Here $L^q(\mathbb{R}^n)$ is the space of all complex functions on $\mathbb{R}^n$ such that

$$\|\psi\|_{L^q(\mathbb{R}^n)} = \left(\int_{\mathbb{R}^n} |\psi(x)|^q dx\right)^{1/q} < \infty.$$

9.4.2   An inversion formula

Moyal's formula also allows us to prove the following important inversion formula for the Wigner function; it is important because it also yields a formula for the adjoint mapping of $\psi \longmapsto W(\psi, \phi)$ and because it will allow us later in this book to prove a fundamental property of modulation spaces.

Proposition 184. *Let $(\phi, \gamma) \in L^2(\mathbb{R}^n) \times L^2(\mathbb{R}^n)$ be such that $(\gamma|\phi)_{L^2(\mathbb{R}^n)} \neq 0$. For every $\psi \in \mathcal{S}'(\mathbb{R}^n)$ we have*

$$\psi(x) = \frac{2^n}{(\gamma|\phi)_{L^2(\mathbb{R}^n)}} \int_{\mathbb{R}^{2n}} W(\psi, \phi)(z_0)\widehat{T}_{GR}(z_0)\gamma(x)dz_0. \tag{9.51}$$

Proof. Let us denote by $\chi(x)$ the right-hand side of (9.51):

$$\chi(x) = \frac{2^n}{(\gamma|\phi)_{L^2(\mathbb{R}^n)}} \int_{\mathbb{R}^{2n}} W(\psi, \phi)(z_0)\widehat{T}_{\mathrm{GR}}(z_0)\gamma(x)dz_0.$$

This function is well defined since $W(\psi, \phi) \in L^2(\mathbb{R}^n \oplus \mathbb{R}^n)$ in view of Moyal's identity (9.41). For any $\alpha \in \mathcal{S}(\mathbb{R}^n)$ we have

$$(\chi|\alpha) = \frac{2^n}{(\gamma|\phi)_{L^2}} \int_{\mathbb{R}^{2n}} W(\psi, \phi)(z)(\widehat{T}_{\mathrm{GR}}(z)\gamma|\alpha)_{L^2} dz.$$

Recalling (formula (9.11)) that by definition

$$W(\psi, \phi)(z) = \left(\tfrac{1}{\pi\hbar}\right)^n (\widehat{T}_{\mathrm{GR}}(z)\psi|\phi)_{L^2},$$

we thus have the sequence of equalities

$$\begin{aligned}
(\chi|\alpha) &= \frac{(2\pi\hbar)^n}{(\gamma|\phi)_{L^2}} \int_{\mathbb{R}^{2n}} W(\psi, \phi)(z)W(\gamma, \alpha)dz \\
&= \frac{(2\pi\hbar)^n}{(\gamma|\phi)_{L^2}} \int_{\mathbb{R}^{2n}} W(\psi, \phi)(z)\overline{W(\alpha, \gamma)}dz \\
&= \frac{(2\pi\hbar)^n}{(\gamma|\phi)_{L^2}} (W(\psi, \phi)|W(\alpha, \gamma))_{L^2}
\end{aligned}$$

(the second equality because $W(\gamma, \alpha) = \overline{W(\alpha, \gamma)}$). Applying Moyal's identity (9.41) to $(W(\psi, \phi)|W(\alpha, \gamma))$ we get

$$(\chi|\alpha)_{L^2} = \left(\frac{1}{2\pi\hbar}\right)^n \frac{(2\pi\hbar)^n}{(\gamma|\phi)} (\psi|\alpha)_{L^2}\overline{(\phi|\gamma)_{L^2}} = (\psi|\alpha)_{L^2}.$$

Since this identity holds for all $\alpha \in \mathcal{S}(\mathbb{R}^n)$ we have $\chi = \psi$ almost everywhere, which proves formula (9.51). $\qquad\square$

An interesting consequence of the result above is the following; it will be useful in our study of modulation spaces:

Corollary 185. *Let $\phi \in \mathcal{S}(\mathbb{R}^n)$, $\phi \neq 0$. The following properties are equivalent:*

(i) $\psi \in \mathcal{S}(\mathbb{R}^n)$;

(ii) $W(\psi, \phi) \in \mathcal{S}(\mathbb{R}^n \oplus \mathbb{R}^n)$;

(iii) *For every $N \geq 0$ there exists $C_N \geq 0$ such that*

$$|W(\psi, \phi)| \leq C_N(1 + |z|)^{-N}.$$

Proof. We already know that (i)$\Longrightarrow$(ii). That (ii)$\Longrightarrow$(iii) is obvious in view of the definition of $\mathcal{S}(\mathbb{R}^n \oplus \mathbb{R}^n)$. There remains to prove that (iii)$\Longrightarrow$(i). It is easily verified that the function χ defined by

$$\chi(x) = \frac{2^n}{\|\phi\|_{L^2}} \int_{\mathbb{R}^{2n}} W(\psi, \phi)(z)\widehat{T}_{\mathrm{GR}}(z)\phi(x)dz$$

is in $\mathcal{S}(\mathbb{R}^n)$; but then $\chi = \psi$ in view of the proof of Proposition 184 since (iii) implies in particular that $W(\psi, \phi) \in L^2(\mathbb{R}^n \oplus \mathbb{R}^n)$. Thus $\psi \in \mathcal{S}(\mathbb{R}^n)$ and we are done. $\qquad\square$

Exercise 186. Check that we indeed have $\chi \in \mathcal{S}(\mathbb{R}^n)$ by using one of the definitions of the space $\mathcal{S}(\mathbb{R}^n)$.

One also proves the following extension of Proposition 184:

Corollary 187. *Assume that* $W(\psi, \phi) = O(|z|^m)$ *for* $|z| \to \infty$ *($m \in \mathbb{R}$). The inversion formula (9.51) still holds under the assumption* $\psi \in \mathcal{S}'(\mathbb{R}^n)$.

Proof. The result is obtained by refining the proof above, cf. Gröchenig [82], Corollary 11.2.7. $\qquad\qquad\square$

Here is an interesting application of the Moyal identity to orthonormal bases, which does not seem to be universally known, and which we will use several times in this book (in particular to establish a trace formula for a product of two Weyl operators, and to study the spectral properties of the "Bopp operators" that will be introduced in Chapter 18). This result shows that to each orthonormal basis of $L^2(\mathbb{R}^n)$ we can associate an orthonormal basis of $L^2(\mathbb{R}^n \oplus \mathbb{R}^n)$ using the cross-Wigner transform. Since we will prove this result in a more general setting in Chapter 18 (Theorem 442) we just outline the argument here.

Proposition 188. *Let* $(\psi_j)_j$ *be an orthonormal basis of* $L^2(\mathbb{R}^n)$. *The vectors* $\Phi_{j,k} = (2\pi\hbar)^{n/2} W(\psi_j, \psi_k)$ *form an orthonormal basis of* $L^2(\mathbb{R}^n \oplus \mathbb{R}^n)$.

Proof. We have, using Moyal's identity (9.41),

$$
\begin{aligned}
(\Phi_{j,k} | \Phi_{j',k'})_{L^2} &= (2\pi\hbar)^n (W(\psi_j, \psi_k) | W(\psi_{j'}, \psi_{k'}))_{L^2} \\
&= (\psi_j | \psi_{j'})_{L^2} \overline{(\psi_k | \psi_{k'})_{L^2}}
\end{aligned}
$$

hence the $\Phi_{j,k}$ form an orthonormal system. There remains to show that if $\Psi \in L^2(\mathbb{R}^n \oplus \mathbb{R}^n)$ is orthogonal to the family $(\Phi_{j,k})_{j,k}$ then $\Psi = 0$. This is done using the properties of the adjoint of the linear mapping $\psi_j \longmapsto W(\psi_j, \psi_k)$ and formula (9.51) above. $\qquad\qquad\square$

9.5 Continuity and growth properties

In this section we study some global properties of the cross-Wigner and ambiguity transforms.

9.5.1 Continuity of $A(\psi, \phi)$ and $W(\psi, \phi)$

We begin by stating two formulas, the proof of which is left to the reader (they are obtained quite easily by using Leibniz's formula for the product of two functions).

Let $\psi \in \mathcal{S}(\mathbb{R}^n)$ (or $\mathcal{S}'(\mathbb{R}^n)$) and $\alpha, \beta \in \mathbb{N}^n$ two multi-indices. We define $\partial_x^\alpha = \partial_{x_1}^{\alpha_1} \cdots \partial_{x_n}^{\alpha_n}$ and $x^\beta = x^{\beta_1} \cdots x^{\beta_n}$. We have:

$$\partial_x^\alpha x^\beta \widehat{T}(z)\psi = \sum_{\gamma \leq \alpha, \delta \leq \beta} c_{\alpha\beta\gamma\delta} x^\delta p^\gamma \widehat{T}(z)(\partial_x^{\alpha-\gamma} x^{\beta-\delta}\psi), \tag{9.52}$$

$$\partial_x^\alpha x^\beta \widehat{T}_{\mathrm{GR}}(z)\psi = \sum_{\gamma \leq \alpha, \delta \leq \beta} d_{\alpha\beta\gamma\delta} x^\delta p^\gamma \widehat{T}_{\mathrm{GR}}(z)(\partial_x^{\alpha-\gamma} x^{\beta-\delta}\psi) \tag{9.53}$$

where the $c_{\alpha\beta\gamma\delta}$ and $d_{\alpha\beta\gamma\delta}$ are complex constants and $\gamma \leq \alpha$ means $\gamma_j \leq \alpha_j$ for $j = 1, 2, \ldots, n$.

We will need the following result which is genuinely interesting in its own right:

Proposition 189. *The maps $z \longmapsto \widehat{T}(z)$ and $z \longmapsto \widehat{T}_{GR}(z)$ are strongly continuous on $\mathcal{S}(\mathbb{R}^n)$ and weakly $*$-continuous on $\mathcal{S}'(\mathbb{R}^n)$.*

Proof. Let us prove the two claims for the map $z \longmapsto \widehat{T}(z)$; the proof for $z \longmapsto \widehat{T}_{\mathrm{GR}}(z)$ is identical, replacing the coefficient $c_{\alpha\beta\gamma\delta}$ by $d_{\alpha\beta\gamma\delta}$. Let $\psi \in \mathcal{S}(\mathbb{R}^n)$. Since $\widehat{T}(z)$ acts linearly on functions we have to show that, for all multi-indices α and β, we have

$$\lim_{|z| \to 0} \|\partial_x^\alpha x^\beta (\widehat{T}(z)\psi - \psi)\|_\infty = 0$$

where $\|\psi\|_\infty = \sup |\psi|$. In view of formula (9.52) we have

$$\|\partial_x^\alpha x^\beta (\widehat{T}(z)\psi - \psi)\|_\infty \leq \|(\widehat{T}(z)(\partial_x^\alpha x^\beta \psi) - (\partial_x^\alpha x^\beta \psi)\|_\infty$$
$$+ \sum_{\substack{0 < \gamma \leq \alpha \\ 0 < \delta \leq \beta}} c_{\alpha\beta\gamma\delta} |x^\delta p^\gamma| \, \|\widehat{T}(z)(\partial_x^{\alpha-\gamma} x^{\beta-\delta}\psi)\|_\infty.$$

It is clear that

$$\lim_{|z| \to 0} |x^\delta p^\gamma| \, \|\widehat{T}(z)(\partial_x^{\alpha-\gamma} x^{\beta-\delta}\psi)\|_\infty = 0$$

since $\gamma \neq 0$, $\delta \neq 0$. There remains to show that

$$\lim_{|z| \to 0} \|(\widehat{T}(z)(\partial_x^\alpha x^\beta \psi) - (\partial_x^\alpha x^\beta \psi)\|_\infty = 0.$$

This is clear if ψ is compactly supported, i.e., $\psi \in C_0^\infty(\mathbb{R}^n)$. The convergence in the general case follows from the density of $C_0^\infty(\mathbb{R}^n)$ in $\mathcal{S}(\mathbb{R}^n)$. There remains to prove the statement about $\mathcal{S}'(\mathbb{R}^n)$. Assume that $\psi \in \mathcal{S}'(\mathbb{R}^n)$ and $\phi \in \mathcal{S}(\mathbb{R}^n)$; we have

$$\lim_{|z| \to 0} \langle \widehat{T}(z)\psi, \phi \rangle = \lim_{|z| \to 0} \langle \psi, \widehat{T}(-z)\phi \rangle = \langle \psi, \phi \rangle$$

and hence $\widehat{T}(z)$ is weakly $*$-continuous on $\mathcal{S}'(\mathbb{R}^n)$ as claimed. $\qquad \square$

Let us now prove the main result of this subsection.

Theorem 190. *Let $\psi \in \mathcal{S}'(\mathbb{R}^n)$ and $\phi \in \mathcal{S}(\mathbb{R}^n)$. Then:*

(i) *Both $A(\psi, \phi)$ and $W(\psi, \phi)$ are continuous functions:*

$$A(\psi, \phi) \in C^0(\mathbb{R}^n \oplus \mathbb{R}^n) \quad and \quad W(\psi, \phi) \in C^0(\mathbb{R}^n \oplus \mathbb{R}^n); \tag{9.54}$$

(ii) *There exist constants $C, C' > 0$ and μ, μ' such that*

$$|A(\psi, \phi)(z)| \le C(1 + |z|)^{\mu}, \tag{9.55}$$

$$|W(\psi, \phi)(z)| \le C'(1 + |z|)^{\mu'} \tag{9.56}$$

for all $z \in \mathbb{R}^{2n}$.

Proof of (i). The continuity statements for $A(\psi, \phi)$ and $W(\psi, \phi)$ are immediate consequences of Proposition 189.

Proof of (ii). In view of formula (9.27) relating $A(\psi, \phi)$ and $W(\psi, \phi)$ it is sufficient to prove the estimate for the cross-ambiguity function. Recalling that

$$A(\psi, \phi)(z) = \left(\tfrac{1}{2\pi\hbar}\right)^n \langle \widehat{T}(-z)\psi, \overline{\phi} \rangle$$

(formula (9.5)), we have, since $\psi \in \mathcal{S}'(\mathbb{R}^n)$ and using inequality (9.52),

$$|A(\psi, \phi)(z)| \le C \sum_{\substack{|\alpha| \le M \\ |\beta| \le N}} \|\partial_x^\alpha x^\beta \widehat{T}(z)\overline{\phi}\|_\infty$$

$$\le C' \sum_{\substack{|\alpha| \le M \\ |\beta| \le N}} \sum_{\substack{\gamma \le \alpha \\ \delta \le \beta}} x^\delta p^\gamma \|\partial_x^{\alpha-\gamma} x^{\beta-\delta} \overline{\phi}\|_\infty.$$

Since $\phi \in \mathcal{S}(\mathbb{R}^n)$ we have $\|\partial_x^{\alpha-\gamma} x^{\beta-\delta} \overline{\phi}\|_\infty < \infty$ so that the right-hand side of the last inequality is a polynomial in $z = (x, p)$; the estimate $|A(\psi, \phi)(z)| \le C(1+|z|)^{\mu}$ follows. $\qquad\qquad \square$

9.5.2 Decay properties of $A(\psi, \phi)$ and $W(\psi, \phi)$

We are going to see that it suffices with a decay property of $A(\psi, \phi)$ and $W(\psi, \phi)$ which does not invoke derivatives to prove that ψ and ϕ are both in $\mathcal{S}(\mathbb{R}^n)$. This result will have several pleasant consequences, which will be fully exploited when we study modulation spaces. We begin with a preparatory result.

Lemma 191. *Let $\phi \in \mathcal{S}(\mathbb{R}^n)$ and let Ψ be a function defined on $\mathbb{R}^{2n}$ and such that for every $N \ge 0$ there exists $C_N > 0$ such that*

$$|\Psi(z)| \le C_N(1 + |z|)^{-N}. \tag{9.57}$$

Then, the function θ defined by

$$\theta(x) = \int_{\mathbb{R}^{2n}} \Psi(z_0)\widehat{T}_{GR}(z_0)\phi(x)dz_0 \tag{9.58}$$

is in $\mathcal{S}(\mathbb{R}^n)$.

Proof. The integral in the right-hand side of (9.58) is obviously absolutely convergent in view of the assumption (9.57). Things do not change for

$$\partial_x^\alpha x^\beta \theta(x) = \int_{\mathbb{R}^{2n}} \Psi(z_0)(\partial_x^\alpha x^\beta \widehat{T}_{\mathrm{GR}}(z_0))\phi(x)dz_0$$

as is easily checked using the estimate (9.53) (we leave the task of writing explicit inequalities to the reader as a technical exercise); it follows that $\|\partial_x^\alpha x^\beta \theta(x)\|_\infty < \infty$ so that $\theta \in \mathcal{S}(\mathbb{R}^n)$ as claimed. $\qquad\square$

Let us now prove:

Proposition 192. *Let $\phi \in \mathcal{S}(\mathbb{R}^n)$ and $\psi \in \mathcal{S}'(\mathbb{R}^n)$. Then the following four properties are equivalent:*

(i) *$\psi \in \mathcal{S}(\mathbb{R}^n)$;*
(ii) *$W(\psi,\phi) \in \mathcal{S}(\mathbb{R}^n \oplus \mathbb{R}^n)$;*
(iii) *$A(\psi,\phi) \in \mathcal{S}(\mathbb{R}^n \oplus \mathbb{R}^n)$;*
(iv) *For every $N \geq 0$ there exist $C_N, C_N' > 0$ such that*

$$|W(\psi,\phi)(z)| \leq C_N(1 + |z|)^{-N}, \tag{9.59}$$

$$|A(\psi,\phi)(z)| \leq C_N'(1 + |z|)^{-N}. \tag{9.60}$$

Proof. That (i)$\Longrightarrow$(ii) was established in Proposition 170. We have (ii)$\Longrightarrow$(iii) because $A(\psi,\phi)$ and $W(\psi,\phi)$ are (symplectic) Fourier transforms of each other (alternatively use formula (9.27). That (ii) or (iii)$\Longrightarrow$(iv) is obvious by definition of $\mathcal{S}(\mathbb{R}^n \oplus \mathbb{R}^n)$. Let us show that (iv)$\Longrightarrow$(i). Choosing $\Psi = W(\psi,\phi)$ in Lemma 191 above we have $\psi \in \mathcal{S}(\mathbb{R}^n)$ in view of the inversion formula (9.51). $\qquad\square$

An interesting consequence of this result is:

Corollary 193. *For every $\phi \in \mathcal{S}(\mathbb{R}^n)$, $\phi \neq 0$, the seminorms $\psi \longmapsto \|\psi\|_\phi$ where*

$$\|\psi\|_{s,\phi} = \sup_{z \in \mathbb{R}^{2n}} [(1 + |z|)^s W(\psi,\phi)(z)]$$

define the topology of $\mathcal{S}(\mathbb{R}^n)$.

Proof. That the $\|\cdot\|_{s,\phi}$ are seminorms on $\mathcal{S}(\mathbb{R}^n)$ is obvious in view of the equivalence (i)$\Longleftrightarrow$(iv) in the proposition above. Let us show that the seminorms $\|\cdot\|_{s,\phi}$ are equivalent to the usual seminorms

$$\|\psi\|_{\alpha\beta} = \sup_{x \in \mathbb{R}^n} |\partial_x^\beta x^\alpha \psi(x)| = \|\partial_x^\beta x^\alpha \psi\|$$

on $\mathcal{S}(\mathbb{R}^n)$. By the inversion formula (9.51) we have

$$|\partial_x^\beta x^\alpha \psi(x)| \leq \frac{2^n}{\|\phi\|_{L^2}^2} \int_{\mathbb{R}^{2n}} |W(\psi,\phi)(z_0)||\partial_x^\beta x^\alpha \widehat{T}_{\mathrm{GR}}(z_0)\phi(x)|dz$$

and hence, using the estimate (9.53), there exists a polynomial function P such that

$$\|\partial_x^\beta x^\alpha \psi\|_\infty \leq C \int_{\mathbb{R}^{2n}} |W(\psi,\phi)(z_0)|P(z_0)|dz;$$

note that the integral on the right side is absolutely convergent in view of the estimates (9.59). Setting $m = \deg P(z)$ we have

$$\|\partial_x^\beta x^\alpha \psi\|_\infty \leq C' \int_{\mathbb{R}^{2n}} |W(\psi,\phi)(z)(1+|z|)^m dz$$

$$\leq C' \int_{\mathbb{R}^{2n}} |W(\psi,\phi)(z)(1+|z|)^{m+2n+1} dz$$

$$\times \int_{\mathbb{R}^{2n}} (1+|z|)^{-2n-1} dz$$

that is

$$\|\partial_x^\beta x^\alpha \psi(x)\|_\infty \leq C'' \|\psi\|_{m+2n+1,\phi}.$$

But this estimate implies that the identity operator $\psi \longmapsto \psi$ from $\mathcal{S}(\mathbb{R}^n)$ equipped with the topology defined by the seminorms $\|\psi\|_{s,\phi}$ onto $\mathcal{S}(\mathbb{R}^n)$ equipped with the usual topology is continuous. The open mapping theorem for the Fréchet space $\mathcal{S}(\mathbb{R}^n)$ then implies that the identity operator is an isomorphism, and we are thus done. $\square$

Chapter 10

The Weyl Correspondence

This chapter introduces and discusses the basics of "Weyl correspondence", or "Weyl quantization" as it is also called. It is today the preferred quantization procedure in physics, for reasons that will be discussed (one of the main features of the Weyl correspondence is that it is the only pseudo-differential quantization procedure which is symplectically covariant). It is in a sense the natural generalization of the Schrödinger correspondence rule $xp \longrightarrow \frac{1}{2}(\widehat{px} + \widehat{xp})$ from the early days of quantum mechanics as discussed in Section 1.3.3 of the first chapter.

10.1 The Weyl correspondence

The material developed in the previous two chapters gives us all the elements we need to define the Weyl correspondence $a \overset{\text{Weyl}}{\longleftrightarrow} \widehat{A}$ in a simple way, without having to invoke pseudo-differential calculus (our approach, which is common in quantum mechanics and time-frequency analysis) has the advantage of immediately allowing the use of quite general symbols. It also immediately makes explicit the relation between the Weyl correspondence and related objects such as the Wigner transform.

We begin by defining the notion of Weyl operator in terms of the Heisenberg–Weyl and Grossmann–Royer operators. We will thereafter give equivalent definitions using the cross-ambiguity and Wigner functions.

10.1.1 First definitions and properties

Recall the definitions of the Heisenberg–Weyl and Grossmann–Royer operators:

$$\widehat{T}(z_0)\psi = e^{\frac{i}{\hbar}(p_0 \cdot x - \frac{1}{2}p_0 \cdot x_0)}\psi(x - x_0),$$

$$\widehat{T}_{\mathrm{GR}}(z_0)\psi(x) = e^{\frac{2i}{\hbar}p_0 \cdot (x - x_0)}\psi(2x_0 - x).$$

Definition 194. Let $a \in \mathcal{S}(\mathbb{R}^n \oplus \mathbb{R}^n)$ and $a_\sigma = F_\sigma a$ be the symplectic Fourier transform of a. The Weyl operator with symbol a is the linear operator $\widehat{A} : \mathcal{S}(\mathbb{R}^n) \longrightarrow \mathcal{S}(\mathbb{R}^n)$ defined by

$$\widehat{A}\psi(x) = \left(\tfrac{1}{2\pi\hbar}\right)^n \int_{\mathbb{R}^{2n}} a_\sigma(z_0)\widehat{T}(z_0)\psi(x)dz_0; \tag{10.1}$$

equivalently,

$$\widehat{A}\psi(x) = \left(\tfrac{1}{\pi\hbar}\right)^n \int_{\mathbb{R}^{2n}} a(z_0)\widehat{T}_{\mathrm{GR}}(z_0)\psi(x)dz_0. \tag{10.2}$$

We will write $\widehat{A} \overset{\mathrm{Weyl}}{\longleftrightarrow} a$ or $a \overset{\mathrm{Weyl}}{\longleftrightarrow} \widehat{A}$ ("Weyl correspondence"). The function a_σ is called the "twisted" (or covariant) symbol of $\widehat{A}$.

The fact that $a \in \mathcal{S}(\mathbb{R}^n \oplus \mathbb{R}^n)$ ensures us that $\widehat{A} : \mathcal{S}(\mathbb{R}^n) \longrightarrow \mathcal{S}(\mathbb{R}^n)$ is clear. In fact:

Proposition 195. *If* $a \in \mathcal{S}(\mathbb{R}^n \oplus \mathbb{R}^n)$ *then* $\widehat{A} \overset{\mathrm{Weyl}}{\longleftrightarrow} a$ *is a continuous operator* $\mathcal{S}(\mathbb{R}^n) \longrightarrow \mathcal{S}(\mathbb{R}^n)$ *and hence extends into a continuous operator* $\mathcal{S}'(\mathbb{R}^n) \longrightarrow \mathcal{S}'(\mathbb{R}^n)$.

Proof. We only sketch the proof and leave the details to the reader as an exercise. If $a \in \mathcal{S}(\mathbb{R}^n \oplus \mathbb{R}^n)$ and $\psi \in \mathcal{S}(\mathbb{R}^n)$ then the function $z_0 \longmapsto a(z_0)(x^\alpha \partial_x^\alpha)\widehat{T}_{\mathrm{GR}}(z_0)\psi$ belongs to $\mathcal{S}(\mathbb{R}^n \oplus \mathbb{R}^n)$ for all multi-indices $\alpha \in \mathbb{N}^n$, hence $|(x^\alpha \partial_x^\alpha)\widehat{A}\psi(x)| < \infty$ for all $\alpha \in \mathbb{N}^n$; the property follows. $\qquad\square$

We can rewrite these definition in slightly more compact form as

$$\widehat{A} = \left(\tfrac{1}{2\pi\hbar}\right)^n \int_{\mathbb{R}^{2n}} a_\sigma(z_0)\widehat{T}(z_0)dz_0$$

and

$$\widehat{A} = \left(\tfrac{1}{\pi\hbar}\right)^n \int_{\mathbb{R}^{2n}} a(z_0)\widehat{T}_{\mathrm{GR}}(z_0)dz_0$$

where the integrals are interpreted as Bochner integrals (that is, operator-valued integrals).

Exercise 196. Prove the equivalence of Definitions (10.1) and (10.2).

These formulas already make quite "obvious" the fact that we will be able to extend the definition of the Weyl correspondence to much larger classes of symbols than the Schwartz functions. This can be seen by rewriting the definition of $\widehat{A}$ in terms of the distributional brackets $\langle\langle \cdot, \cdot \rangle\rangle$ on $\mathbb{R}^{2n}$:

$$\widehat{A}\psi = \left(\tfrac{1}{2\pi\hbar}\right)^n \langle\langle a_\sigma(\cdot), \widehat{T}(\cdot)\psi \rangle\rangle = \left(\tfrac{1}{\pi\hbar}\right)^n \langle\langle a(\cdot), \widehat{T}_{\mathrm{GR}}(\cdot)\psi \rangle\rangle. \tag{10.3}$$

The notation $\widehat{A} \overset{\mathrm{Weyl}}{\longleftrightarrow} a$ or $a \overset{\mathrm{Weyl}}{\longleftrightarrow} \widehat{A}$ introduced in the definition above suggests that the Weyl correspondence is one-to-one. In the following Proposition we show that this indeed the case:

Proposition 197. *The Weyl correspondence* $a \overset{\text{Weyl}}{\longleftrightarrow} \widehat{A}$ *is linear and one-to-one:*

(i) *If* $a \overset{\text{Weyl}}{\longleftrightarrow} \widehat{A}$ *and* $a' \overset{\text{Weyl}}{\longleftrightarrow} \widehat{A}$ *then* $a = a'$.

(ii) *In particular* $1 \overset{\text{Weyl}}{\longleftrightarrow} I$ *where* I *is the identity operator on* $\mathcal{S}'(\mathbb{R}^n)$.

Proof of (i). The linearity of the correspondence $a \overset{\text{Weyl}}{\longleftrightarrow} \widehat{A}$ is obvious; to show that it is one-to-one it thus suffices to show that if $\widehat{A}\psi = 0$ for all $\psi \in \mathcal{S}(\mathbb{R}^n)$ then $a = 0$. But $\widehat{A}\psi = 0$ is equivalent, by the second formula (10.3), to $\langle a(\cdot), \widehat{T}_{\mathrm{GR}}(\cdot)\psi \rangle = 0$, that is to $a = 0$ since $\psi \in \mathcal{S}(\mathbb{R}^n)$ is arbitrary.

Proof of (ii). To show that if $a = 1$ then $\widehat{A}$ is the identity it suffices to note that by the second formula (10.3) we have

$$\widehat{A}\psi(x) = \left(\tfrac{1}{\pi\hbar}\right)^n \langle\langle 1, \widehat{T}_{\mathrm{GR}}(\cdot)\psi(x)\rangle\rangle$$

$$= \int_{\mathbb{R}^n} \left(\int_{\mathbb{R}^n} e^{\frac{2i}{\hbar} p_0 \cdot (x - x_0)} dp_0 \right) \psi(2x_0 - x) dx_0$$

$$= \int_{\mathbb{R}^n} \delta(2(x - x_0))\psi(2x_0 - x) dx_0 = \psi(x)$$

so that $\widehat{A}\psi = \psi$ for all $\psi \in \mathcal{S}(\mathbb{R}^n)$. By continuity, using the density of $\mathcal{S}(\mathbb{R}^n)$, we also have $\widehat{A}\psi = \psi$ for all $\psi \in \mathcal{S}'(\mathbb{R}^n)$ hence $\widehat{A} = I$ on $\mathcal{S}'(\mathbb{R}^n)$. $\square$

Property (ii) in the proposition above is a particular case of the following interesting result which determines explicitly the Weyl symbol of the Heisenberg–Weyl operator:

Proposition 198. *The operator with Weyl symbol* $z \mapsto e^{-\frac{i}{\hbar}\sigma(z, z_0)}$ *is the Heisenberg–Weyl operator* $\widehat{T}(z_0)$.

Proof. Let us write $a_{z_0}(z) = e^{-\frac{i}{\hbar}\sigma(z, z_0)}$. Let $\widehat{A}_{z_0}$ be the operator with Weyl symbol a_{z_0}. We have, by the second formula (10.3),

$$\widehat{A}_{z_0}\psi(x) = \left(\tfrac{1}{\pi\hbar}\right)^n \iint_{\mathbb{R}^n \times \mathbb{R}^n} e^{-\frac{i}{\hbar}\sigma(z, z')} e^{\frac{2i}{\hbar} p_0 \cdot (x - x_0)} \psi(2x_0 - x) dp_0 dx_0$$

$$= \left(\tfrac{1}{\pi\hbar}\right)^n \iint_{\mathbb{R}^n \times \mathbb{R}^n} e^{\frac{i}{\hbar} p' \cdot x_0} e^{\frac{i}{\hbar} p_0 \cdot (x' + 2x_0 - 2x)} \psi(2x_0 - x) dp_0 dx_0$$

$$= \left(\tfrac{1}{\pi\hbar}\right)^n \int_{\mathbb{R}^n} e^{\frac{i}{\hbar} p' \cdot x_0} \left[\int_{\mathbb{R}^n} e^{\frac{i}{\hbar} p_0 \cdot (x' + 2x_0 - 2x)} dp_0 \right] \psi(2x_0 - x) dx_0$$

$$= 2^n \int_{\mathbb{R}^n} e^{\frac{i}{\hbar} p' \cdot x_0} \delta(x' + 2x_0 - 2x)\psi(2x_0 - x) dx_0.$$

Setting $y = 2x_0$ we get

$$\widehat{A}_{z_0}\psi(x) = \int_{\mathbb{R}^n} e^{\frac{i}{\hbar} p' \cdot x_0} \delta(y - (2x - x'))\psi(2x_0 - x) dx_0 = e^{\frac{i}{\hbar}(p' \cdot x - \frac{1}{2} p' \cdot x')} \psi(x - x')$$

which concludes the proof. $\square$

Taking into account the differential expression (8.7) of the Heisenberg–Weyl operator we thus have

$$e^{-\frac{i}{\hbar}\sigma(\hat{z},z_0)} \overset{\text{Weyl}}{\longleftrightarrow} e^{-\frac{i}{\hbar}\sigma(z,z_0)}. \tag{10.4}$$

As a consequence of Schwartz's kernel theorem, every continuous operator $\mathcal{S}(\mathbb{R}^n \oplus \mathbb{R}^n) \longrightarrow \mathcal{S}'(\mathbb{R}^n \oplus \mathbb{R}^n)$ is a Weyl operator for a suitable symbol (usually a tempered distribution); we will prove this in a moment, but let us for the moment show that the notion of Weyl operator makes sense even when the symbol is not a function, but a distribution. A first constraint comes from the fact that the symplectic Fourier transform $a_\sigma = F_\sigma a$ must be defined. This is however a minor limitation because F_σ is an isomorphism $\mathcal{S}'(\mathbb{R}^n \oplus \mathbb{R}^n) \longrightarrow \mathcal{S}'(\mathbb{R}^n \oplus \mathbb{R}^n)$ and we are thus authorized to use any tempered distributions as a symbol. Since we have, for $\psi, \phi \in \mathcal{S}(\mathbb{R}^n)$,

$$\langle \widehat{A}\psi, \phi \rangle = \iint_{\mathbb{R}^{2n} \times \mathbb{R}} a_\sigma(z_0) \widehat{T}(z_0) \psi(x) \phi(x) dz_0 dx, \tag{10.5}$$

$\widehat{A}$ is defined in the weak sense by:

$$\langle \widehat{A}\psi, \phi \rangle = \langle a_\sigma, \Phi \rangle \quad , \quad \Phi(z_0) = \langle \widehat{T}(z_0)\psi, \phi \rangle. \tag{10.6}$$

Proposition 199. *If* $a \in \mathcal{S}'(\mathbb{R}^n \oplus \mathbb{R}^n)$ *then* $\widehat{A} : \mathcal{S}(\mathbb{R}^n) \longrightarrow \mathcal{S}'(\mathbb{R}^n)$.

Proof. The condition $a \in \mathcal{S}'(\mathbb{R}^n \oplus \mathbb{R}^n)$ is equivalent to $a_\sigma \in \mathcal{S}'(\mathbb{R}^n \oplus \mathbb{R}^n)$ hence it is sufficient to show that Φ defined in (10.6) is in $\mathcal{S}(\mathbb{R}^n \oplus \mathbb{R}^n)$. But this follows from Lemma 170. $\square$

Let us now return to formula (10.8) which, as we said, can be taken as an alternative definition of the Weyl correspondence. Choosing $\phi = \psi$ this formula implies, since $W\psi = W(\psi, \psi)$:

$$(\widehat{A}\psi|\psi)_{L^2(\mathbb{R}^n)} = \int_{\mathbb{R}^{2n}} a(z) W\psi(z) dz.$$

More about the properties of the Weyl correspondence in a moment; we are first going to rewrite the definition of $\widehat{A}$ in two equivalent ways using the cross-Wigner and ambiguity transforms.

10.1.2 Definition using the Wigner transform

The following result can be taken as an alternative definition of the Weyl operator with symbol a; it also makes obvious the fact that the Weyl correspondence extends to symbols $a \in \mathcal{S}'(\mathbb{R}^n \oplus \mathbb{R}^n)$.

Proposition 200. *Let $(\psi, \phi) \in \mathcal{S}(\mathbb{R}^n) \times \mathcal{S}(\mathbb{R}^n)$ and assume that $\widehat{A} \overset{\text{Weyl}}{\longleftrightarrow} a$ is a mapping $\mathcal{S}(\mathbb{R}^n) \longrightarrow L^2(\mathbb{R}^n)$. We have both*

$$(\widehat{A}\psi|\phi)_{L^2(\mathbb{R}^n)} = \int_{\mathbb{R}^{2n}} a_\sigma(z) A(\psi, \phi)(-z) dz \tag{10.7}$$

and

$$(\widehat{A}\psi|\phi)_{L^2(\mathbb{R}^n)} = \int_{\mathbb{R}^{2n}} a(z) W(\psi, \phi)(z) dz. \tag{10.8}$$

Proof. It is sufficient to assume that $a \in \mathcal{S}(\mathbb{R}^n \oplus \mathbb{R}^n)$. We have

$$(\widehat{A}\psi|\phi)_{L^2} = \left(\tfrac{1}{2\pi\hbar}\right)^n \int_{\mathbb{R}^n} \left(\int_{\mathbb{R}^{2n}} a_\sigma(z_0) \int_{\mathbb{R}^n} \widehat{T}(z_0)\psi(x) dz_0 \right) \overline{\phi(x)} dx$$

and hence, by Fubini's theorem,

$$(\widehat{A}\psi|\phi)_{L^2} = \left(\tfrac{1}{2\pi\hbar}\right)^n \int_{\mathbb{R}^{2n}} a_\sigma(z_0) \left(\int_{\mathbb{R}^n} \widehat{T}(z_0)\psi(x)\overline{\phi(x)} dx \right) dz_0$$

that is, in view of Definition (9.1) of the cross-Wigner transform,

$$(\widehat{A}\psi|\phi)_{L^2} = \int_{\mathbb{R}^{2n}} a_\sigma(z_0) A(\psi, \phi)(-z_0) dz_0$$

which is formula (10.7). In view of formula (9.26) we have

$$A(\psi, \phi)(-z_0) = F_\sigma W(\psi, \phi)(-z_0)$$

and hence, by Plancherel's formula (8.31) for the symplectic Fourier transform

$$(\widehat{A}\psi|\phi)_{L^2} = \int_{\mathbb{R}^{2n}} F_\sigma a(z_0) A(\psi, \phi)(-z_0) dz_0$$
$$= \int_{\mathbb{R}^{2n}} a(z_0) F_\sigma A(\psi, \phi)(-z_0) dz_0$$

which is formula (10.8) since $F_\sigma A(\psi, \phi) = W(\psi, \phi)$ (see formula (9.26)). $\square$

The formulas above can be easily extended to the distributional case using Proposition 199. In fact:

Corollary 201. *Let $(\psi, \phi) \in \mathcal{S}(\mathbb{R}^n) \times \mathcal{S}(\mathbb{R}^n)$ and assume that $\widehat{A} \overset{\text{Weyl}}{\longleftrightarrow} a$ with $a \in \mathcal{S}'(\mathbb{R}^n \oplus \mathbb{R}^n)$. Then*

$$\langle \widehat{A}\psi, \overline{\phi} \rangle = \langle\langle a_\sigma, A(\psi, \phi)^\vee \rangle\rangle = \langle\langle a, W(\psi, \phi) \rangle\rangle \tag{10.9}$$

where $A(\psi, \phi)^\vee(z) = A(\psi, \phi)(-z)$ and $\langle\langle \cdot, \cdot \rangle\rangle$ is the distributional bracket for distributions on $\mathbb{R}^n \oplus \mathbb{R}^n$.

Proof. It is an immediate consequence of formulas (10.7), (10.8) using Proposition 199. $\square$

10.1.3   Probabilistic interpretation

As we already have discussed, the Wigner function is not in general positive. In fact, as mentioned earlier, a classical result of Hudson [103] (also see Janssen [104]) tells us that $W\psi$ is non-negative if and only if ψ is a Gaussian. An illustration of this generic non-positivity of the Wigner transform is provided by the following easy exercise:

Exercise 202. Assume that $\psi \in \mathcal{S}(\mathbb{R}^n)$ is an odd function: $\psi(-x) = -\psi(x)$. Show that $W\psi$ takes negative values. [Hint: calculate $W\psi(0)$.]

However, there are several good reasons for which it might be adequate to view $W\psi$ as a good substitute for a true probability density. Here is a first motivation for viewing the Wigner function as a "quasi probability" density:

Proposition 203. *Assume that* $\psi, \phi \in L^1(\mathbb{R}^n) \cap L^2(\mathbb{R}^n)$. *We have*

$$\int_{\mathbb{R}^n} W(\psi, \phi)(z)dp = \psi(x)\overline{\phi(x)}, \tag{10.10a}$$

$$\int_{\mathbb{R}^n} W(\psi, \phi)(z)dx = F\psi(p)\overline{F\phi(p)} \tag{10.10b}$$

hence, in particular

$$\int_{\mathbb{R}^n} W\psi(z)dp = |\psi(x)|^2 \; , \; \int_{\mathbb{R}^n} W\psi(z)dx = |F\psi(p)|^2. \tag{10.11}$$

Proof. Let us prove the first formula (10.10). Noting that in view of the "inverse Fourier transform formula", which we write sloppily as

$$\int_{\mathbb{R}^n} e^{-\frac{i}{\hbar}p\cdot y}dp = (2\pi\hbar)^n \delta(y)$$

we have

$$\int_{\mathbb{R}^n} W(\psi, \phi)(z)dp = \int_{\mathbb{R}^n} \delta(y)\psi(x + \tfrac{1}{2}y)\overline{\phi(x - \tfrac{1}{2}y)}dy$$

$$= \int_{\mathbb{R}^n} \delta(y)\psi(x)\overline{\phi(x)}dy$$

$$= \psi(x)\overline{\phi(x)}$$

as claimed. Let us prove the second formula (10.10). Setting $x' = x + \tfrac{1}{2}y$ and $x'' = x - \tfrac{1}{2}y$ in the right-hand of the equality

$$\int_{\mathbb{R}^n} W\psi(z)dx = \left(\tfrac{1}{2\pi\hbar}\right)^n \iint_{\mathbb{R}^n \times \mathbb{R}^n} e^{-\frac{i}{\hbar}p\cdot y}\psi(x + \tfrac{1}{2}y)\overline{\phi(x - \tfrac{1}{2}y)}dydx$$

we get

$$\int_{\mathbb{R}^n} W\psi(z)dx = \left(\tfrac{1}{2\pi\hbar}\right)^n \iint_{\mathbb{R}^n\times\mathbb{R}^n} e^{-\frac{i}{\hbar}p\cdot x'}\psi(x')\overline{e^{-\frac{i}{\hbar}p\cdot x''}\phi(x'')}dx'dx''$$

$$= \left(\tfrac{1}{2\pi\hbar}\right)^n \int_{\mathbb{R}^n} e^{-\frac{i}{\hbar}p\cdot x'}\psi(x')dx'\overline{\int_{\mathbb{R}^n} e^{-\frac{i}{\hbar}p\cdot x''}\phi(x'')dx''} = F\psi(p)\overline{F\phi(p)}.$$

$$\square$$

It immediately follows from any of the two formulae (10.11) above that

$$\int_{\mathbb{R}^{2n}} W\psi(z)dz = \|\psi\|^2_{L^2(\mathbb{R}^n)} = \|F\psi\|^2_{L^2(\mathbb{R}^n)}. \tag{10.12}$$

If the function ψ is normalized: $\|\psi\|^2_{L^2} = 1$, then so is $W\psi(z)$:

$$\int_{\mathbb{R}^{2n}} W\psi(z)dz = 1 \quad \text{if} \quad \|\psi\|^2_{L^2(\mathbb{R}^n)} = 1.$$

If in addition $W\psi \geq 0$ it would thus be a probability density; but again, this is only the case when ψ is a Gaussian function.

These two formulas are just particular cases of a (generalized) Radon transforms of the Wigner distribution $W\psi$, corresponding to integration along the particular Lagrangian planes $\ell_P = 0 \times \mathbb{R}^n$ and $\ell_X = 0 \times \mathbb{R}^n$, respectively. More generally one could define that transform as the function defined by

$$R_\ell(u) = \int_\ell W\psi(z)d\mu_\ell(z)$$

where $d\mu_\ell(z)$ is the Euclidean measure on the Lagrangian plane ℓ. The Radon transform was defined by the Austrian mathematician J. Radon [134] in 1917 (see the English translation Radon and Parks [135]). For a mathematically rigorous treatment of its various extensions (which belong to the area of integral geometry) see Helgason's book [99].

Having the probabilistic interpretation of the Wigner transform $W\psi$ in mind formulas (10.7) and (10.9) can be interpreted as giving the average value of the symbol a when $\|\psi\|^2_{L^2} = 1$. This point of view is totally consistent with the one adopted in quantum mechanics, from which the following definition comes:

Definition 204. Let $\psi \neq 0$ be viewed as a "quantum state", and let a be a real symbol, viewed as a "classical observable". Then

$$\langle \widehat{A} \rangle_\psi = \frac{(\widehat{A}\psi|\psi)_{L^2(\mathbb{R}^n)}}{(\psi|\psi)_{L^2(\mathbb{R}^n)}} \tag{10.13}$$

is called the average value of the corresponding "quantum observable" in the quantum state ψ.

Formula (10.13) can be viewed as saying that the expectation value $\langle \widehat{A} \rangle_\psi$ of the Weyl operator $\widehat{A}$ in the state ψ is the average of the symbol a with respect to the quasi-probability distribution $W\psi$.

10.1.4   The kernel of a Weyl operator

The following result will be studied further from a slightly different point of view in Section 10.4 when we study the Weyl correspondence from the point of view of pseudo-differential operators. Recall that the kernel of an operator $A : \mathcal{S}(\mathbb{R}^n) \longrightarrow \mathcal{S}'(\mathbb{R}^n)$ is a distribution $K \in \mathcal{S}'(\mathbb{R}^n \times \mathbb{R}^n)$ such that

$$\langle A\psi, \phi \rangle = \langle K, \phi \otimes \psi \rangle$$

for all $\phi, \psi \in \mathcal{S}(\mathbb{R}^n)$. We may also write

$$A\psi(x) = \int_{\mathbb{R}^n} K(x,y)\psi(y)dy$$

where the integral is to be understood in the distributional sense. Every linear operator $A : \mathcal{S}(\mathbb{R}^n) \longrightarrow \mathcal{S}'(\mathbb{R}^n)$ has a kernel $K \in \mathcal{S}'(\mathbb{R}^n \times \mathbb{R}^n)$ provided it is continuous from $\mathcal{S}(\mathbb{R}^n)$ to $\mathcal{S}'(\mathbb{R}^n)$ (this is Schwartz's kernel theorem; for a refinement see [82]).

Proposition 205. (i) Let $\widehat{A} \overset{\text{Weyl}}{\longleftrightarrow} a$. The kernel of $\widehat{A}$ and its symbol a are related by the following formulas:

$$K_{\widehat{A}}(x,y) = \left(\tfrac{1}{2\pi\hbar}\right)^n \int_{\mathbb{R}^n} e^{\frac{i}{\hbar}p\cdot(x-y)}a(\tfrac{1}{2}(x+y),p)dp, \tag{10.14}$$

$$a(x,p) = \int_{\mathbb{R}^n} e^{-\frac{i}{\hbar}p\cdot y}K_{\widehat{A}}(x+\tfrac{1}{2}y, x-\tfrac{1}{2}y)dy. \tag{10.15}$$

Proof of (i). Let us express $\widehat{A}$ in terms of its symbol and the Grossmann–Royer operators (formula (10.2)):

$$\widehat{A}\psi(x) = \left(\tfrac{1}{\pi\hbar}\right)^n \int_{\mathbb{R}^{2n}} a(z_0)\widehat{T}_{\text{GR}}(z_0)\psi(x)dz$$

$$= \left(\tfrac{1}{\pi\hbar}\right)^n \int_{\mathbb{R}^{2n}} a(z_0)e^{\frac{2i}{\hbar}p_0\cdot(x-x_0)}\psi(2x-x_0)dz.$$

Setting $y = 2x - x_0$ and $p = p_0$ we get

$$\widehat{A}\psi(x) = \left(\tfrac{1}{2\pi\hbar}\right)^n \iint_{\mathbb{R}^n \times \mathbb{R}^n} e^{\frac{i}{\hbar}p\cdot(x-y)}a(\tfrac{1}{2}(x+y),p)\psi(y)dydp. \tag{10.16}$$

It follows that the kernel of the operator $\widehat{A}$ is given by formula (10.14).

Proof of (ii). We have

$$K_{\widehat{A}}(x+\tfrac{1}{2}y, x-\tfrac{1}{2}y) = \left(\tfrac{1}{2\pi\hbar}\right)^n \int_{\mathbb{R}^n} e^{\frac{i}{\hbar}p\cdot y}a(x,p)dp; \tag{10.17}$$

fixing x, the Fourier inversion formula yields (10.15). $\qquad\square$

Exercise 206. Show that the function $z \longmapsto e^{-\frac{i}{\hbar}\sigma(z,z_0)}$ is the Weyl symbol of the Heisenberg–Weyl operator $\widehat{T}(z_0)$.

The following consequence of Proposition 205 is obvious:

Corollary 207. *The Weyl symbol of the operator with kernel $\left(\frac{1}{2\pi\hbar}\right)^n (\psi \otimes \overline{\phi})$ is the cross-Wigner transform $W(\psi, \phi)$.*

Proof. It is a straightforward application of formula (10.15). $\qquad\square$

Exercise 208. Find the relation between the kernel $K_{\widehat{A}}$ and the symplectic Fourier transform $a_\sigma = F_\sigma a$ of the symbol of $\widehat{A}$.

There is a simple relation between the L^2 norm of the symbol and that of the kernel. The result is useful when one studies L^2 regularity properties of Weyl operators; we will also use it in Chapter 12 in connection with the calculation of traces of Weyl operators:

Proposition 209. *Assume that $K_{\widehat{A}} \in L^2(\mathbb{R}^n \times \mathbb{R}^n)$. We then have $a \in L^2(\mathbb{R}^n \oplus \mathbb{R}^n)$ and $a_\sigma \in L^2(\mathbb{R}^n \oplus \mathbb{R}^n)$; moreover:*

$$\|a\|_{L^2(\mathbb{R}^{2n})} = \|a_\sigma\|_{L^2(\mathbb{R}^{2n})} = (2\pi\hbar)^{n/2} \|K_{\widehat{A}}\|_{L^2(\mathbb{R}^n \times \mathbb{R}^n)}. \tag{10.18}$$

Proof. The equality $\|a\|_{L^2} = \|a_\sigma\|_{L^2}$ is obvious since the symplectic Fourier transform is unitary. Let us prove that $\|a\|_{L^2} = (2\pi\hbar)^{n/2} \|K_{\widehat{A}}\|_{L^2}$ when $K_{\widehat{A}} \in \mathcal{S}(\mathbb{R}^n \times \mathbb{R}^n)$; the proposition will follow by the density of $\mathcal{S}(\mathbb{R}^n \times \mathbb{R}^n)$ in $L^2(\mathbb{R}^n \times \mathbb{R}^n)$. In view of formula (10.15) the symbol a is, for fixed x, $(2\pi\hbar)^{n/2}$ times the Fourier transform of the function $y \longmapsto K_{\widehat{A}}(x+\frac{1}{2}y, x-\frac{1}{2}y)$ hence, by Plancherel's formula,

$$\int_{\mathbb{R}^n} |a(x,p)|^2 dp = (2\pi\hbar)^n \int_{\mathbb{R}^n} |K_{\widehat{A}}(x + \tfrac{1}{2}y, x - \tfrac{1}{2}y)|^2 dy \tag{10.19}$$

and hence, integrating in x,

$$\int_{\mathbb{R}^{2n}} |a(z)|^2 dz = (2\pi\hbar)^n \int_{\mathbb{R}^n} \left(\int_{\mathbb{R}^n} |K_{\widehat{A}}(x + \tfrac{1}{2}y, x - \tfrac{1}{2}y)|^2 dy \right) dx$$

$$= (2\pi\hbar)^n \iint_{\mathbb{R}^n \times \mathbb{R}^n} |K_{\widehat{A}}(x + \tfrac{1}{2}y, x - \tfrac{1}{2}y)|^2 dx dy$$

where we have applied Fubini's theorem (the integrals are absolutely convergent since $(x,y) \longmapsto K_{\widehat{A}}(x + \frac{1}{2}y, x - \frac{1}{2}y)$ is in $\mathcal{S}(\mathbb{R}^n \times \mathbb{R}^n)$ because $K_{\widehat{A}}$ is). Set now $x' = x + \frac{1}{2}y$ and $y' = x - \frac{1}{2}y$; we have $dx'dy' = dxdy$ hence

$$\int_{\mathbb{R}^{2n}} |a(z)|^2 dz = (2\pi\hbar)^n \iint_{\mathbb{R}^n \times \mathbb{R}^n} |K_{\widehat{A}}(x', y')|^2 dx' dy'$$

which we set out to prove. $\qquad\square$

We have several times insisted on the fact that the "quantization" of classical "observables" should satisfy the Schrödinger prescription: to $a(z) = x_j p_j$ should correspond to the symmetrized operator

$$\widehat{A} = \tfrac{1}{2}(\widehat{X_j P_j} + \widehat{P_j X_j}). \tag{10.20}$$

Let us check that the Weyl correspondence $\widehat{A} \overset{\text{Weyl}}{\longleftrightarrow} a$ satisfies this requirement. We have, taking $a(z) = x_j p_j$ and hence

$$(2\pi\hbar)^n \, \widehat{A}\psi(x) = \tfrac{1}{2} \iint_{\mathbb{R}^{2n}} e^{\frac{i}{\hbar} p \cdot (x-y)} p_j x_j \psi(y) \, dy \, dp$$

$$+ \tfrac{1}{2} \iint_{\mathbb{R}^{2n}} e^{\frac{i}{\hbar} p \cdot (x-y)} p_j y_j \psi(y) \, dy \, dp.$$

Formula (10.20) follows in view of the obvious equalities

$$\left(\tfrac{1}{2\pi\hbar}\right)^n \tfrac{1}{2} x_j \iint_{\mathbb{R}^{2n}} e^{\frac{i}{\hbar} p \cdot (x-y)} p_j \psi(y) \, dy \, dp = \tfrac{1}{2} x_j \widehat{P_j} \psi(x),$$

$$\left(\tfrac{1}{2\pi\hbar}\right)^n \tfrac{1}{2} \iint_{\mathbb{R}^{2n}} e^{\frac{i}{\hbar} p \cdot (x-y)} p_j y_j \psi(y) \, dy \, dp = \tfrac{1}{2} \widehat{P_j}(x_j \psi)(x).$$

Exercise 210. Prove (10.20) using the definition of $\widehat{A}$ in terms of the symbol and the Grossmann–Royer operator.

10.2　Adjoints and products

We are going to see that the symbol of the adjoint of a Weyl operator $\widehat{A}$ can be very easily determined in terms of the symbol of a. This leads to the celebrated property of the Weyl correspondence which says that $\widehat{A} \overset{\text{Weyl}}{\longleftrightarrow} a$ is self-adjoint if and only if a is real. This property is very important in the applications to quantum mechanics, because "quantization" of an observable should precisely lead to a self-adjoint operator.

We also calculate the symbol of the product of two Weyl operators; the result is of particular importance for the study of the applications to deformation quantization ("Bopp calculus").

10.2.1　The adjoint of a Weyl operator

The formal adjoint $\widehat{A}^*$ of an operator $\widehat{A} : \mathcal{S}(\mathbb{R}^n) \longrightarrow \mathcal{S}(\mathbb{R}^n)$ is defined by the formula

$$(\widehat{A}\psi | \phi)_{L^2(\mathbb{R}^n)} = (\psi | \widehat{A}^* \phi)_{L^2(\mathbb{R}^n)}$$

for all $(\psi, \phi) \in \mathcal{S}(\mathbb{R}^n) \times \mathcal{S}(\mathbb{R}^n)$. We are going to determine explicitly the Weyl correspondence $\widehat{A}^* \overset{\text{Weyl}}{\longleftrightarrow} a^*$. For this we need the following elementary lemma:

Lemma 211. *Let b be a function on $\mathbb{R}^n$ such that $(b|W(\psi,\phi))_{L^2(\mathbb{R}^{2n})} = 0$ for all $(\psi,\phi) \in \mathcal{S}(\mathbb{R}^n) \times \mathcal{S}(\mathbb{R}^n)$. Then $b = 0$.*

Proof. In view of formula (10.8) we have, noting that $\overline{W(\psi,\phi)} = W(\phi,\psi)$,

$$(\widehat{B}\phi|\psi))_{L^2} = (b|W(\psi,\phi))_{L^2}$$

where $\widehat{B} \overset{\mathrm{Weyl}}{\longleftrightarrow} b$. If $(b|W(\psi,\phi))_{L^2} = 0$ for all ψ and ϕ we must thus have $\widehat{B}\phi = 0$ for all ϕ hence $\widehat{B} = 0$; but then $b = 0$ since the Weyl correspondence is one-to-one in view of Proposition 197. $\qquad\square$

We have:

Proposition 212. *The adjoint $\widehat{A}^*$ of the Weyl operator $\widehat{A} \overset{\mathrm{Weyl}}{\longleftrightarrow} a$ is the Weyl operator $\widehat{A}^* \overset{\mathrm{Weyl}}{\longleftrightarrow} a^*$ with symbol $a^* = \overline{a}$. In particular, $\widehat{A}$ is formally self-adjoint if and only if a is a real function.*

Proof. Expressing $\widehat{A}$ in terms of the Grossmann–Royer operators we have, in view of formula (10.2),

$$(\widehat{A}\psi|\phi)_{L^2} = \left(\tfrac{1}{\pi\hbar}\right)^n \int_{\mathbb{R}^{2n}} a(z_0) \left(\int_{\mathbb{R}^n} \widehat{T}_{\mathrm{GR}}(z_0)\psi(x)\overline{\phi(x)} \right) dz_0$$

$$= \int_{\mathbb{R}^{2n}} a(z_0)W(\psi,\phi)(z_0)dz_0.$$

Since, by definition, $(\widehat{A}\psi|\phi)_{L^2} = (\psi|\widehat{A}^*\phi)_{L^2}$ we have

$$(\psi|\widehat{A}^*\phi)_{L^2} = \overline{(\widehat{A}^*\phi|\psi)_{L^2}}$$

$$= \int_{\mathbb{R}^{2n}} \overline{a^*}(z_0)W(\psi,\phi)(z_0)dz_0$$

hence

$$(\widehat{A}\psi|\phi)_{L^2} = \int_{\mathbb{R}^{2n}} \overline{a^*}(z_0)W(\psi,\phi)(z_0)dz_0.$$

Applying Lemma 211 to $b = a - \overline{a^*}$ yields $a = \overline{a^*}$ that is $a^* = \overline{a}$. $\qquad\square$

10.2.2 Composition formulas

We now assume that the Weyl operators

$$\widehat{A} = \left(\tfrac{1}{2\pi\hbar}\right)^n \int_{\mathbb{R}^{2n}} a_\sigma(z)\widehat{T}(z)dz,$$

$$\widehat{B} = \left(\tfrac{1}{2\pi\hbar}\right)^n \int_{\mathbb{R}^{2n}} b_\sigma(z)\widehat{T}(z)dz$$

can be composed (this is always the case for instance if a_σ and b_σ are in $\mathcal{S}(\mathbb{R}^n \oplus \mathbb{R}^n)$) and set $\widehat{C} = \widehat{A}\widehat{B}$. Assuming that we can write

$$\widehat{C} = \left(\tfrac{1}{2\pi\hbar}\right)^n \int_{\mathbb{R}^{2n}} c_\sigma(z)\widehat{T}(z)dz$$

we ask: what is c_σ? The answer is giving by the following theorem, which is instrumental in the definition of deformation quantization we will give later in this book:

Theorem 213. *Let* $\widehat{A} \overset{\text{Weyl}}{\longleftrightarrow} a$ *and* $\widehat{B} \overset{\text{Weyl}}{\longleftrightarrow} b$ *be Weyl operators.*

(i) *The product* $\widehat{C} = \widehat{A}\widehat{B}$ *has (when defined) Weyl symbol*

$$c(z) = \left(\tfrac{1}{4\pi\hbar}\right)^{2n} \iint_{\mathbb{R}^{4n}} e^{\frac{i}{2\hbar}\sigma(z',z'')} a(z + \tfrac{1}{2}z')b(z - \tfrac{1}{2}z'')dz'dz''. \tag{10.21}$$

(ii) *The symplectic Fourier transform of* c *is given by*

$$c_\sigma(z) = \left(\tfrac{1}{2\pi\hbar}\right)^n \int_{\mathbb{R}^{2n}} e^{\frac{i}{2\hbar}\sigma(z,z')} a_\sigma(z - z')b_\sigma(z')dz' \tag{10.22}$$

or, equivalently, by

$$c_\sigma(z) = \left(\tfrac{1}{2\pi\hbar}\right)^n \int_{\mathbb{R}^{2n}} e^{-\frac{i}{2\hbar}\sigma(z,z')} a_\sigma(z')b_\sigma(z - z')dz'. \tag{10.23}$$

Proof of (i). Assume that the Weyl symbols a, b of $\widehat{A}$ and $\widehat{B}$ are in $\mathcal{S}(\mathbb{R}^n \oplus \mathbb{R}^n)$. We have

$$K_{\widehat{A}\widehat{B}}(x,y) = \left(\tfrac{1}{2\pi\hbar}\right)^{2n} \iiint_{\mathbb{R}^{3n}} e^{\frac{i}{\hbar}((x-\alpha)\cdot p + (\alpha-y)\cdot p)}$$
$$\times\, a(\tfrac{1}{2}(x+\alpha),\zeta)b(\tfrac{1}{2}(x+y),\xi)d\alpha d\zeta d\xi.$$

In view of formula (10.15) we have

$$c(x,p) = \int e^{-\frac{i}{\hbar}p\cdot u} K_{\widehat{A}\widehat{B}}(x + \tfrac{1}{2}u, x - \tfrac{1}{2}u)du$$

and the symbol of $\widehat{A}\widehat{B}$ is thus

$$c(z) = \left(\tfrac{1}{\pi\hbar}\right)^{2n} \iiiint_{\mathbb{R}^{4n}} e^{\frac{i}{\hbar}Q} a(\tfrac{1}{2}(x+\alpha+\tfrac{1}{2}u),\zeta)$$
$$\times\, b(\tfrac{1}{2}(x+\alpha-\tfrac{1}{2}u),\xi)d\alpha d\zeta du d\xi$$

where the phase Q is given by

$$Q = (x - \alpha + \tfrac{1}{2}u)\cdot\zeta + (\alpha - x + \tfrac{1}{2}u)\cdot\xi - u\cdot p$$
$$= (x - \alpha + \tfrac{1}{2}u)\cdot(\zeta - p) + (\alpha - x + \tfrac{1}{2}u)\cdot(\xi - p).$$

Setting $\zeta' = \zeta - p$, $\xi' = \xi - p$, $\alpha' = \frac{1}{2}(\alpha - x + \frac{1}{2}u)$ and $u' = \frac{1}{2}(\alpha - x - \frac{1}{2}u)$ we have

$$d\alpha d\zeta du d\xi = 2^{2n} d\alpha' \zeta' du' d\xi'$$

and $Q = 2\sigma(u', \xi'; \alpha', \zeta')$, hence

$$c(z) = \left(\tfrac{1}{\pi\hbar}\right)^{2n} \iiiint_{\mathbb{R}^{4n}} e^{\frac{2i}{\hbar}\sigma(u',\xi';\alpha',\zeta')}$$
$$\times a(x + \alpha', p + \zeta')b(x + u', p + \xi')d\alpha' d\zeta' du' d\xi';$$

formula (10.21) follows setting $z' = 2(\alpha', \zeta')$ and $z'' = -2(u', \xi')$.

Proof of (ii). Writing the operators $\widehat{A}$ and $\widehat{B}$ in the usual form

$$\widehat{A} = \left(\tfrac{1}{2\pi\hbar}\right)^{n} \int_{\mathbb{R}^{2n}} a_\sigma(z_0)\widehat{T}(z_0)dz_0,$$
$$\widehat{B} = \left(\tfrac{1}{2\pi\hbar}\right)^{n} \int_{\mathbb{R}^{2n}} b_\sigma(z_1)\widehat{T}(z_1)dz_1$$

we have, using the property

$$\widehat{T}(z_0 + z_1) = e^{-\frac{i}{2\hbar}\sigma(z_0,z_1)}\widehat{T}(z_0)\widehat{T}(z_1)$$

of HW operators,

$$\widehat{T}(z_0)\widehat{B} = \left(\tfrac{1}{2\pi\hbar}\right)^{n} \int_{\mathbb{R}^{2n}} b_\sigma(z_1)\widehat{T}(z_0)\widehat{T}(z_1)dz_1$$
$$= \left(\tfrac{1}{2\pi\hbar}\right)^{n} \int_{\mathbb{R}^{2n}} e^{\frac{i}{2\hbar}\sigma(z_0,z_1)}b_\sigma(z_1)\widehat{T}(z_0 + z_1)dz_1$$

and hence

$$\widehat{A}\widehat{B} = \left(\tfrac{1}{2\pi\hbar}\right)^{2n} \iint_{\mathbb{R}^{4n}} e^{\frac{i}{2\hbar}\sigma(z_0,z_1)}a_\sigma(z_0)b_\sigma(z_1)\widehat{T}(z_0 + z_1)dz_0 dz_1.$$

Setting $z = z_0 + z_1$ and $z' = z_1$ this can be written

$$\widehat{A}\widehat{B} = \left(\tfrac{1}{2\pi\hbar}\right)^{2n} \int_{\mathbb{R}^{2n}} \left(\int_{\mathbb{R}^{2n}} e^{\frac{i}{2\hbar}\sigma(z,z')}a_\sigma(z - z')b_\sigma(z')dz' \right) \widehat{T}(z)dz$$

hence (10.22). Formula (10.23) follows by a trivial change of variables and using the antisymmetry of σ. $\qquad\square$

The composition formulas (10.22), (10.23) are very important in the context of the deformation quantization theory we will study later, and where the function c defined by (10.21) is called the Moyal (or Groenewold–Moyal) starproduct, and is denoted $a \star_\hbar b$.

Exercise 214. Prove formula (10.21) for the Weyl symbol of a product by using the representation (10.2) of $\widehat{A}$ and $\widehat{B}$ in terms of the Grossmann–Royer operators.

10.3   Symplectic covariance of Weyl operators

A striking feature of Weyl calculus is that it is the *only* pseudo-differential operator calculus for which metaplectic covariance holds (see Chapter 10.4 for a discussion of the general notion of pseudo-differential operator). This property is another manifestation of the importance of Weyl operators in quantum mechanics.

10.3.1   Statement and proof of the symplectic covariance property

Recall that the Heisenberg–Weyl and Royer–Grossmann operators $\widehat{T}(z_0)$ and $\widehat{T}_{\mathrm{GR}}(z_0)$ satisfy the following property: for $\widehat{S} \in \mathrm{Mp}(2n, \mathbb{R})$ and $S = \pi^{\mathrm{Mp}}(\widehat{S})$ we have

$$\widehat{S}\widehat{T}(z_0)\widehat{S}^{-1} = \widehat{T}(Sz_0) \ , \ \widehat{S}\widehat{T}_{\mathrm{GR}}(z_0)\widehat{S}^{-1} = \widehat{T}_{\mathrm{GR}}(Sz_0) \tag{10.24}$$

for every $z_0 \in \mathbb{R}^{2n}$.

Theorem 215. *Let $S \in \mathrm{Sp}(2n, \mathbb{R})$ and $\widehat{S} \in \mathrm{Mp}(2n, \mathbb{R})$ be any one of the two metaplectic operators with $S = \pi^{\mathrm{Mp}}(\widehat{S})$. For every Weyl operator $\widehat{A} \overset{\mathrm{Weyl}}{\longleftrightarrow} a$ we have the correspondence*

$$a \circ S \overset{\mathrm{Weyl}}{\longleftrightarrow} \widehat{S}^{-1}\widehat{A}\widehat{S}. \tag{10.25}$$

That is, to the symbol $a_S\,(z) = a(Sz)$ corresponds the Weyl operator $\widehat{S}^{-1}\widehat{A}\widehat{S}$.

Proof. Let us denote by $\widehat{B}$ the Weyl operator with symbol $a \circ S$. In view of formula (10.8) we have

$$\widehat{B}\psi = \int_{\mathbb{R}^{2n}} a_\sigma(Sz)\widehat{T}(z)\psi dz$$

that is, performing the change of variables $Sz \longmapsto z$ and taking into account the fact that $\det S = 1$,

$$\widehat{B}\psi = \int_{\mathbb{R}^{2n}} a_\sigma(z)\widehat{T}(S^{-1}z)\psi dz.$$

By formula (8.11) in Theorem 128 we have $\widehat{S}^{-1}\widehat{T}(z)\widehat{S} = \widehat{T}(S^{-1}z)$ and hence

$$\widehat{B}\psi = \int_{\mathbb{R}^{2n}} a_\sigma(z)\widehat{S}^{-1}\widehat{T}(z)\widehat{S}\psi dz$$

$$= \widehat{S}^{-1} \left(\int_{\mathbb{R}^{2n}} a_\sigma(z)\widehat{T}(z)dz \right) \widehat{S}\psi$$

which is (10.25). $\square$

Problem 216. Give an alternative proof of (10.25) using the formula

$$\widehat{A} = \left(\tfrac{1}{\pi\hbar}\right)^n \int_{\mathbb{R}^{2n}} a(z_0)\widehat{T}_{\mathrm{GR}}(z_0)dz_0$$

expressing the Weyl correspondence in terms of the Grossmann–Royer operators.

As a straightforward consequence of Theorem 128 we obtain the so-called metaplectic covariance formula for the cross-Wigner transform:

Corollary 217. *Let $\psi, \phi \in \mathcal{S}(\mathbb{R}^n)$ and $\widehat{S} \in \mathrm{Mp}(2n, \mathbb{R})$; we denote by S the projection of $\widehat{S}$ on $\mathrm{Sp}(2n, \mathbb{R})$. We have*

$$W(\widehat{S}\psi, \widehat{S}\phi)(z) = W(\psi, \phi)(S^{-1}z) \tag{10.26}$$

and hence in particular

$$W(\widehat{S}\psi)(z) = W\psi(S^{-1}z). \tag{10.27}$$

Proof. In view of formula (10.7) in Proposition 200 we have, since $\widehat{S}$ is unitary,

$$\int_{\mathbb{R}^{2n}} W(\widehat{S}\psi, \widehat{S}\phi)a(z)dz = (\widehat{A}\widehat{S}\psi|\widehat{S}\phi)_{L^2} = (\widehat{S}^{-1}\widehat{A}\widehat{S}\psi|\phi)_{L^2}.$$

In view of (10.25) we have

$$(\widehat{S}^{-1}\widehat{A}\widehat{S}\psi, \phi)_{L^2} = \int_{\mathbb{R}^{2n}} W(\psi, \phi)(z)(a \circ S)(z)dz$$
$$= \int_{\mathbb{R}^{2n}} W(\psi, \phi)(S^{-1}z)a(z)dz$$

which establishes the equality (10.26) since ψ and ϕ are arbitrary; formula (10.27) trivially follows taking $\psi = \phi$. $\qquad\square$

Problem 218. Prove a similar result for the cross-ambiguity function.

10.3.2 Covariance under affine symplectomorphisms

Let us denote by $T(z_0)$ the phase space translation operator $z \longmapsto z + z_0$. It induces a natural action of functions by the formula $T(z_0)a(z) = a(z - z_0)$; this action can be extended into an action on distributions in the obvious way.

Proposition 219. *Let $a \overset{\mathrm{Weyl}}{\longleftrightarrow} \widehat{A}$ with $a \in \mathcal{S}'(\mathbb{R}^n \oplus \mathbb{R}^n)$. Let $T(z_0)a(z) = a(z - z_0)$. We have*

$$T(z_0)a \overset{\mathrm{Weyl}}{\longleftrightarrow} \widehat{T}(z_0)\widehat{A}\widehat{T}(z_0)^{-1} \tag{10.28}$$

where $\widehat{T}(z_0)$ is the Heisenberg–Weyl operator.

Proof. Let $\psi, \phi \in \mathcal{S}(\mathbb{R}^n)$. We first remark that writing (10.8) in terms of distributional brackets, and using the fact that $\overline{W(\psi, \phi)} = W(\phi, \psi)$ we have

$$\langle \widehat{A}\psi, \phi \rangle = \int_{\mathbb{R}^{2n}} a(z)\overline{W(\psi, \phi)}(z)dz = \langle a, W(\phi, \psi) \rangle.$$

We next observe that

$$\langle \widehat{T}(z_0)\widehat{A}\widehat{T}(z_0)^{-1}\psi, \phi \rangle = \langle \widehat{A}\widehat{T}(z_0)^{-1}\psi, \widehat{T}(z_0)^{-1}\phi \rangle$$
$$= \langle a, W(\widehat{T}(z_0)^{-1}\phi, \widehat{T}(z_0)^{-1}\psi) \rangle$$
$$= \langle a, T(-z_0)W(\phi, \psi) \rangle$$

the last equality in view of formula (9.23); since we have

$$\langle a, T(-z_0)W(\phi, \psi) \rangle = \langle T(z_0)a, W(\phi, \psi) \rangle$$

formula (10.28) follows. $\qquad\qquad\qquad\qquad\qquad\qquad\qquad\qquad\qquad\qquad\square$

Exercise 220. Prove directly formula (10.28) using the definition of $\widehat{A}$ in terms of the Heisenberg–Weyl operators. [Hint: use several changes of variables together with the formula $\widehat{T}(z_0)\widehat{T}(z_1) = e^{\frac{i}{\hbar}\sigma(z_0, z_1)}\widehat{T}(z_1)\widehat{T}(z_0)$ (formula (8.8)).]

Recall (Definition 26) that the affine symplectic group $\mathrm{ASp}(2n, \mathbb{R})$ is the semi-direct product $\mathrm{Sp}(2n, \mathbb{R}) \ltimes \mathrm{T}(2n, \mathbb{R})$ of the symplectic group with the translation group $\mathrm{T}(2n, \mathbb{R}) \equiv \mathbb{R}^{2n}$. It is the group generated by symplectic matrices and the translations $T(z_0)$. *Writing* $S_{z_0} = ST(z_0)$ *to each* S_{z_0} one can associate two unitary operators $\pm\widehat{S}_{z_0}$ defined as follows: $\widehat{S}_{z_0} = \widehat{S}\widehat{T}(z_0)$ where $\widehat{S}$ is any one of the two metaplectic operators with projection $S \in \mathrm{Sp}(2n, \mathbb{R})$. The operators $\widehat{S}_{z_0}$ generate a group, the affine (or homogeneous) metaplectic group $\mathrm{AMp}(2n, \mathbb{R})$. Combining Theorem 128 with Proposition 219 we get:

Corollary 221. *The Weyl correspondence is covariant under the action of the affine symplectic group* $\mathrm{ASp}(2n, \mathbb{R})$ *in the sense that if* $\widehat{A} \overset{\mathrm{Weyl}}{\longleftrightarrow} a$ *then*

$$a \circ S_{z_0} \overset{\mathrm{Weyl}}{\longleftrightarrow} \widehat{S}_{z_0}^{-1}\widehat{A}\widehat{S}_{z_0} \qquad\qquad (10.29)$$

for every $S_{z_0} = ST(z_0)$.

It turns out that symplectic covariance is characteristic of the Weyl correspondence: among all other pseudo-differential calculi (as studied in Chapter 14), it is the only one having the properties above. This is a fundamental fact, because it allows the derivation of Schrödinger's equation. We will come back to that property in Chapter 15 when we study the derivation of Schrödinger's equation.

10.4   Weyl operators as pseudo-differential operators

We introduced Weyl operators using concepts from harmonic analysis such as the Heisenberg–Weyl or Grossmann–Royer operators. We now take the point of view of pseudo-differential operators; this approach will be developed in Chapter 14 from a more general point of view using Shubin's theory [147]. The literature on pseudo-differential operators is immense. The main references for what we will be doing here are the books by Wong [163], Nicola and Rodino [131]. The books [40] by Egorov, Komech, and Shubin, contains very valuable material concerning applications of pseudo-differential operators to the theory of partial differential equations. Another good source is Chazarain and Piriou [25] where there is a strong use of microlocal techniques.

We begin with a quick review of the general notion of pseudo-differential operator.

10.4.1   The notion of pseudo-differential operator

We will use the following multi-index notation: for $\alpha = (\alpha_1, \ldots, \alpha_{2n})$ in $\mathbb{N}^{2n}$ we set

$$|\alpha| = \alpha_1 + \cdots + \alpha_{2n} \ , \ \ \partial_z^\alpha = \partial_{z_1}^{\alpha_1} \cdots \partial_{z_{2n}}^{\alpha_{2n}}$$

where $\partial_{z_j}^{\alpha_j} = \partial^{\alpha_j}/\partial x_j^{\alpha_j}$ for $1 \le j \le n$ and $\partial_{z_j}^{\alpha_j} = \partial^{\alpha_j}/\partial \xi_j^{\alpha_j}$ for $n + 1 \le j \le 2n$. We will write $D_x = -i\partial_x$ and $D_x^\alpha = (-i)^{|\alpha|}\partial_x^\alpha$ and use similar notation for D_p and D_p^α.

The starting point of any honest theory of pseudo-differential operators is the search for an extension of partial differential operators. In the early days of the theory one started by remarking that we can write any partial differential operators (with smooth coefficients) on $\mathbb{R}^n$ in the form

$$\widehat{A} = \sum_{|\alpha| \le m} a_\alpha(x) D_x^\alpha \ , \ a_\alpha \in C^\infty(\mathbb{R}^n) \tag{10.30}$$

and define the symbol of $\widehat{A}$ as being the polynomial

$$a(x, \xi) = \sum_{|\alpha| \le m} a_\alpha(x)\xi^\alpha \ , \ \ \xi^\alpha = \xi_1^{\alpha_1} \cdots \xi_n^{\alpha_n}. \tag{10.31}$$

Defining the Fourier transform $\widehat{f}$ of $f \in C_0^\infty(\mathbb{R}^n)$ (the vector space of C^∞ functions with compact support) by

$$\widehat{f}(\xi) = \left(\tfrac{1}{2\pi}\right)^{n/2} \int_{\mathbb{R}^n} e^{-i\xi \cdot x} f(x) dx \tag{10.32}$$

(thus $\widehat{f} = Ff$ with $\hbar = 1$) we can rewrite Definition (10.30) as

$$\widehat{A}f(x) = \int_{\mathbb{R}^n} e^{i\xi \cdot x} a(x, \xi) \widehat{f}(\xi) d\xi. \tag{10.33}$$

Exercise 222. Show that formula (10.33) immediately follows from the Fourier inversion formula

$$f(x) = \left(\tfrac{1}{2\pi}\right)^{n/2} \int_{\mathbb{R}^n} e^{ix \cdot \xi} \widehat{f}(\xi) dx \tag{10.34}$$

and the relation $\xi^\alpha \widehat{f}(\xi) = \widehat{D_x^\alpha f}(\xi)$.

The next step consisted in *defining* pseudo-differential operators as being linear operators of the type (10.33) where a is a (more or less arbitrary) function on $\mathbb{R}^n \oplus \mathbb{R}^n$ called in this context a *symbol* or *amplitude*. Of course, such a definition is vague and rather useless unless one puts some constraints on the classes of symbols that are acceptable. For instance if $a(x, \xi)$ increases "too" fast when $|\xi| \to \infty$ the integral in (10.33) will only be convergent for "very" small classes of functions f. For instance, a reasonable requirement is that one should be able to calculate $\widehat{A}f$ when $f \in C_0^\infty(\mathbb{R}^n)$, and this will be the case if one requires that the symbol not grow faster than a polynomial in ξ. Then, another natural requirement might be that the smoothness of $\widehat{A}f$ corresponds in some way to the smoothness of f, and for this to be true one has to impose conditions on the x-derivatives of the symbol a. The most popular choice in the 1970s was to use the so-called *Hörmander symbol classes* $S_{\rho,\sigma}^m$ (Hörmander [96], Hörmander and Duistermaat [97]):

Definition 223. Let m, ρ, δ be real numbers such that $0 \leq \delta < \rho \leq 1$. The symbol class $S_{\rho,\delta}^m(\mathbb{R}^n \oplus \mathbb{R}^n)$ is the (complex) vector space of all functions $a \in C^\infty(\mathbb{R}^n \oplus \mathbb{R}^n)$ such that for each choice of multi-indices $\alpha, \beta \in \mathbb{N}^n$ and of a compact subset K of $\mathbb{R}^n$ there exists a constant $C_{\alpha,\beta,K} \geq 0$ such that

$$\partial_x^\alpha \partial_\xi^\beta a(x, \xi) \leq C_{\alpha,\beta,K}(1 + |\xi|)^{m-|\beta|} \tag{10.35}$$

for $(x, \xi) \in K \times \mathbb{R}^n$. One writes $S^m(\mathbb{R}^n \oplus \mathbb{R}^n) = S_{1,0}^m(\mathbb{R}^n \oplus \mathbb{R}^n)$. The vector space of pseudo-differential operators

$$\widehat{A}f(x) = \int_{\mathbb{R}^n} e^{i\xi \cdot x} a(x, \xi) \widehat{f}(\xi) d\xi \tag{10.36}$$

with symbol in $S_{\rho,\delta}^m(\mathbb{R}^n \oplus \mathbb{R}^n)$ is denoted $L_{\rho,\delta}^m(\mathbb{R}^n)$. One often calls a the *Kohn–Nirenberg symbol* of $\widehat{A}$.

It is not difficult to show that when $a \in S_{\rho,\delta}^m(\mathbb{R}^n \oplus \mathbb{R}^n)$ the corresponding pseudo-differential operator $\widehat{A}$ maps continuously $C_0^\infty(\mathbb{R}^n)$ into $C^\infty(\mathbb{R}^n)$ (it involves repeated use of Leibniz's rule for the differentiation of a product); moreover any symbol a of the polynomial type (10.31) belongs to the class $S^m(\mathbb{R}^n \oplus \mathbb{R}^n)$, so the definition above achieves the program of generalizing ordinary partial differential operators (at least those with C^∞ coefficients).

So far, so good. The rub comes from the fact that in the elements of the symbol classes $S_{\rho,\delta}^m$ the variables x and ξ are on very different footing, due to the fact that one is more interested in local properties than in global behavior (the cotangent bundle of a manifold is locally identical with $T^*\mathbb{R}^n = \mathbb{R}^n \oplus \mathbb{R}^n$). This dissymmetry conflicts with the phase space approach of both classical and quantum mechanics, where the variables x and ξ play equivalent roles. We will define later in this book a pseudo-differential calculus that avoids this difficulty (the Shubin global calculus, Chapter 14).

Problem 224. Show that every pseudo-differential operator belonging to one of the Hörmander classes $L^m_{\rho,\delta}(\mathbb{R}^n)$ with $0 \leq \delta < \rho \leq 1$ maps (continuously) $\mathcal{S}(\mathbb{R}^n)$ into itself.

The best-known regularity results for these operators are expressed in terms of the usual Sobolev spaces[1] H^s: we have $f \in H^s(\mathbb{R}^n)$ if and only if $\widehat{f}$ is a function satisfying

$$\|f\|^2_s = \int_{\mathbb{R}^n} |\widehat{f}(\xi)|^2 (1 + |\xi|^2)^s d\xi < \infty.$$

It is clear that $H^s(\mathbb{R}^n)$ is a vector space; it is in fact a complex Hilbert space for the sesquilinear form

$$(f, g) \longmapsto (f|g)_s = \int_{\mathbb{R}^n} \widehat{f}(\xi)\overline{\widehat{g}(\xi)}(1 + |\xi|^2)^s d\xi$$

and $f \longmapsto \|f\|_s = (f|f)_s^{1/2}$ is the associated norm. Since pseudo-differential operators are not in general local one cannot in general state regularity results in terms of these spaces, and one considers the following derived spaces:

$$H^s_c(\mathbb{R}^n) = \{f \in H^s(\mathbb{R}^n) : \mathrm{Supp}(f) \text{ is compact}\},$$
$$H^s_{loc}(\mathbb{R}^n) = \{f \in \mathcal{S}'(\mathbb{R}^n) : \varphi f \in H^s_c(\mathbb{R}^n) \text{ for some } \varphi \in C^\infty_0(\mathbb{R}^n)\}.$$

Equipping these spaces with adequate topologies, a classical result is then:

Theorem 225. *Every* $\widehat{A} \in L^m_{1,0}(\mathbb{R}^n)$ *is continuous* $H^s_c(\mathbb{R}^n) \longrightarrow H^{s-m}_{loc}(\mathbb{R}^n)$.

Proof. See any textbook on pseudo-differential operators published before the mid-1980s. $\square$

It should be mentioned that every $\widehat{A} \in L^m_{1,0}(\mathbb{R}^n)$ can be written as $\widehat{A} = \widehat{A}' + R$ where $\widehat{A}' : H^s_c(\mathbb{R}^n) \longrightarrow H^{s-m}_c(\mathbb{R}^n)$ and R is a smoothing operator (i.e., an operator with C^∞ kernel). If an operator maps compactly supported distributions to compactly supported distributions one says it is a proper operator.

10.4.2 The kernel of a Weyl operator revisited

We are going to see that a Weyl operator $\widehat{A} \overset{\text{Weyl}}{\longleftrightarrow} a$ can be represented in the following way:

$$\widehat{A}\psi(x) = \left(\tfrac{1}{2\pi\hbar}\right)^n \iint_{\mathbb{R}^n \times \mathbb{R}^n} e^{\frac{i}{\hbar} p \cdot (x-y)} a(\tfrac{1}{2}(x + y), p)\psi(y)dydp. \tag{10.37}$$

Of course, the right-hand side does not have a well-defined mathematical sense in general; for the "double integral" to be absolutely convergent one has to put

---

[1]Sometimes also called "Bessel potential spaces".

rather stringent conditions on the symbol a and the function ψ. For instance, one can require that $\psi \in \mathcal{S}(\mathbb{R}^n)$ and a decreases sufficiently fast in p. We are going to see that one can, however, give a meaning to this formal expression by using a "mollifier" provided that a belongs to a "good" symbol class. We will refine and improve our results when we discuss the Shubin classes.

Let us show that Weyl operators are just pseudo-differential operators of the type above when a decays rapidly. Recall from Proposition 205 that if $a \in \mathcal{S}(\mathbb{R}^n \oplus \mathbb{R}^n)$ and $\widehat{A} \overset{\text{Weyl}}{\longleftrightarrow} a$ then

$$\widehat{A}\psi(x) = \left(\tfrac{1}{2\pi\hbar}\right)^n \iint_{\mathbb{R}^n \times \mathbb{R}^n} e^{\frac{i}{\hbar}p\cdot(x-y)}a(\tfrac{1}{2}(x+y),p)\psi(y)dydp \tag{10.38}$$

for every $\psi \in \mathcal{S}(\mathbb{R}^n)$. It follows that the kernel of $\widehat{A}$ is given by

$$K_{\widehat{A}}(x,y) = \left(\tfrac{1}{2\pi\hbar}\right)^n \int_{\mathbb{R}^n} e^{\frac{i}{\hbar}p\cdot(x-y)}a(\tfrac{1}{2}(x+y),p)dp. \tag{10.39}$$

Conversely, the expression of the symbol in terms of the kernel is

$$a(x,p) = \int_{\mathbb{R}^n} e^{-\frac{i}{\hbar}p\cdot y}K_{\widehat{A}}(x+\tfrac{1}{2}y,x-\tfrac{1}{2}y)dy. \tag{10.40}$$

Notice that if we interpret formula (10.39) in the distributional sense we recover the fact that the identity operator has symbol $a = 1$. In fact it suffices to observe that in view of the Fourier inversion formula we have

$$\left(\tfrac{1}{2\pi\hbar}\right)^n \iint_{\mathbb{R}^{2n}} e^{\frac{i}{\hbar}p\cdot(x-y)}\psi(y)dydp = \psi(x). \tag{10.41}$$

10.4.3   Justification in the case $a \in S^m$

We are following here closely Chapter 4 in Wong [163]. Let us begin with a regularization result making use of a "mollifier".

Proposition 226. *Let $a \in S^m(\mathbb{R}^n \oplus \mathbb{R}^n) = S^m_{1,0}(\mathbb{R}^n \oplus \mathbb{R}^n)$ and assume that the estimates (10.35) hold uniformly in x, that is*

$$\partial_x^\alpha \partial_\xi^\beta a(x,\xi) \le C_{\alpha,\beta}(1 + |\xi|)^{m-|\beta|}.$$

Let $\theta \in C_0^\infty(\mathbb{R}^n)$ be such that $\theta(0) = 1$. For every $\psi \in \mathcal{S}(\mathbb{R}^n)$ the limit

$$L(x) = \lim_{\varepsilon \to 0+} \iint_{\mathbb{R}^n \times \mathbb{R}^n} e^{\frac{i}{\hbar}p\cdot(x-y)}\theta(\varepsilon p)a(\tfrac{1}{2}(x+y),p)\psi(y)dydp \tag{10.42}$$

exists and is independent of the choice of the function θ.

Proof. Define a function L_ε by the formula

$$L_\varepsilon(x) = \iint_{\mathbb{R}^n \times \mathbb{R}^n} e^{\frac{i}{\hbar} p \cdot (x-y)} \theta(\varepsilon p) a(\tfrac{1}{2}(x+y), p)\psi(y) dy dp.$$

For any integer $N \geq 0$ we have

$$(1 - \Delta_y)^N e^{\frac{i}{\hbar} p \cdot (x-y)} = \hbar^{-N}(1 + |p|^2)^N e^{\frac{i}{\hbar} p \cdot (x-y)}$$

hence, integrating by parts in the y variable,

$$L_\varepsilon(x) = \hbar^N \iint_{\mathbb{R}^n \times \mathbb{R}^n} e^{\frac{i}{\hbar} p \cdot (x-y)} \theta(\varepsilon p)(1 + |p|^2)^{-N}$$
$$\times (1 - \Delta_y)^N \left(a(\tfrac{1}{2}(x+y), p)\psi(y) \right) dy dp.$$

Using Leibniz's formula for the repeated derivatives of a product we have

$$(1 - \Delta_y)^N \left(a(\tfrac{1}{2}(x+y), p)\psi(y) \right) = \sum_{|\alpha| \leq N} \frac{1}{\alpha!} (D_x^\alpha a)(\tfrac{1}{2}(x+y), p) P^\alpha(D)\psi(y) \quad (10.43)$$

where $P_N^\alpha(D)$ is the partial differential operator with symbol $\partial_p^\alpha \left[(1 + |p|^2)^N \right]$. For each fixed $x \in \mathbb{R}^n$ we have

$$\lim_{\varepsilon \to 0+} \left[\theta(\varepsilon p)(1 + |p|^2)^{-N} e^{\frac{i}{\hbar} p \cdot (x-y)} (D_x^\alpha a)(\tfrac{1}{2}(x+y), p) P_N^\alpha(D)\psi(y) \right]$$
$$= (1 + |p|^2)^{-N} e^{\frac{i}{\hbar} p \cdot (x-y)} (D_x^\alpha a)(\tfrac{1}{2}(x+y), p) P_N^\alpha(D)\psi(y).$$

In view of the estimate

$$|(D_x^\alpha a)(\tfrac{1}{2}(x+y), p)| \leq C_\alpha (1 + |p|^2)^m$$

there exists a constant C such that

$$|\theta(\varepsilon p)(1 + |p|^2)^{-N} e^{\frac{i}{\hbar} p \cdot (x-y)} (D_x^\alpha a)(\tfrac{1}{2}(x+y), p) P_N^\alpha(D)\psi(y)|$$
$$\leq C(1 + |p|^2)^{-N}(1 + |p|^2)^m |P_N^\alpha(D)\psi(y)|$$

for y and p in $\mathbb{R}^n$. Since the function

$$(y, p) \longmapsto (1 + |p|^2)^{-N}(1 + |p|^2)^m |P_N^\alpha(D)\psi(y)|$$

is in $L^1(\mathbb{R}^n)$ as soon as $N > (m + n)/2$ it follows from (10.43) and Lebesgue's dominated convergence theorem that $\lim_{\varepsilon \to 0+} L_\varepsilon(x)$ exists and is independent of the choice of θ. $\qquad \square$

Exercise 227. Show, using the formulas in the proof, that the convergence in formula (10.42) is uniform with respect to the variable x.

The limit (10.42) is called an "oscillatory integral", and one often uses the notation

$$L(x) = \left(\tfrac{1}{2\pi\hbar}\right)^n \widetilde{\iint}_{\mathbb{R}^n \times \mathbb{R}^n} e^{\frac{i}{\hbar} p \cdot (x-y)} a(\tfrac{1}{2}(x+y), p)\psi(y)dydp.$$

Provided that some care is taken one can work with oscillatory integrals very much like with ordinary integrals in their evaluation (see Chazarain and Piriou [25] for more on this topic and various extensions).

10.5 Regularity results for Weyl operators

10.5.1 Some general results

Let us begin with a modest goal and see what happens if we assume that a_σ is absolutely integrable. Recall that if A is an operator from a Banach space $\mathcal{B}$ to a Banach space $\mathcal{B}'$ then the operator norm $\|A\|$ is defined by

$$\|A\| = \inf\left\{M : \|Af\|_{\mathcal{B}'} \le M\|f\|_{\mathcal{B}},\ f \in \mathcal{B}\right\}.$$

We have the following precise result:

Proposition 228. *Assume that $a_\sigma \in L^1(\mathbb{R}^n \oplus \mathbb{R}^n)$ [equivalently $a \in FL^1(\mathbb{R}^n \oplus \mathbb{R}^n)$] and $\psi \in \mathcal{S}(\mathbb{R}^n)$. Then:*

(i) *The operator $\widehat{A} \overset{\text{Weyl}}{\longleftrightarrow} a$ is continuous on $\mathcal{S}(\mathbb{R}^n \oplus \mathbb{R}^n)$ for the induced L^2-norm; in fact*

$$\|\widehat{A}\psi\|_{L^2(\mathbb{R}^n)} \le \left(\tfrac{1}{2\pi\hbar}\right)^{2n} \|a_\sigma\|_{L^1(\mathbb{R}^n)}\|\psi\|_{L^2(\mathbb{R}^n)} \tag{10.44}$$

hence the operator norm of $\widehat{A}$ on $L^2(\mathbb{R}^n)$ satisfies

$$\|\widehat{A}\| \le \left(\tfrac{1}{2\pi\hbar}\right)^{2n} \|a_\sigma\|_{L^1(\mathbb{R}^n)}. \tag{10.45}$$

(ii) *$\widehat{A}$ extends into a bounded operator $L^2(\mathbb{R}^n) \longrightarrow L^2(\mathbb{R}^n)$, which will also be denoted $\widehat{A}$.*

Proof. Statement (ii) immediately follows from (i), using the density of $\mathcal{S}(\mathbb{R}^n)$ in $L^2(\mathbb{R}^n)$. Formula (10.45) is equivalent to formula (10.44). The kernel of $\widehat{A}$ is given by

$$K_{\widehat{A}}(x,y) = \left(\tfrac{1}{2\pi\hbar}\right)^{n/2} F_2(\tfrac{1}{2}(x+y), y-x)$$

where F_2 is the Fourier transform in the p variables. By the Fourier inversion formula we have

$$F_2 a(\tfrac{1}{2}(x+y), y-x) = \left(\tfrac{1}{2\pi\hbar}\right)^{n/2} \int_{\mathbb{R}^n} e^{\frac{i}{2\hbar}(x+y)\cdot\xi} F(\xi, y-x)d\xi$$

and hence

$$K_{\widehat{A}}(x,y) = \left(\tfrac{1}{2\pi\hbar}\right)^n \int_{\mathbb{R}^n} e^{\frac{i}{2\hbar}(x+y)\cdot\xi} Fa(\xi, y-x)d\xi.$$

It follows that

$$\int |K_{\widehat{A}}(x,y)|dx \leq \left(\tfrac{1}{2\pi\hbar}\right)^n \iint_{\mathbb{R}^n\times\mathbb{R}^n} |Fa(\xi, y-x)|d\xi dx,$$

$$\int |K_{\widehat{A}}(x,y)|dy \leq \left(\tfrac{1}{2\pi\hbar}\right)^n \iint_{\mathbb{R}^n\times\mathbb{R}^n} |Fa(\xi, y-x)|d\xi dy.$$

Setting $\eta = y - x$ we have

$$\iint_{\mathbb{R}^n\times\mathbb{R}^n} |Fa(\xi, y-x)|d\xi dx = \iint_{\mathbb{R}^n\times\mathbb{R}^n} |Fa(\xi,\eta)|d\xi d\eta$$

hence the two inequalities above can be rewritten, in view of the first equality (8.27), in the form

$$\int_{\mathbb{R}^n} |K_{\widehat{A}}(x,y)|dx \leq \left(\tfrac{1}{2\pi\hbar}\right)^n \|a_\sigma\|_{L^1},$$

$$\int_{\mathbb{R}^n} |K_{\widehat{A}}(x,y)|dy \leq \left(\tfrac{1}{2\pi\hbar}\right)^n \|a_\sigma\|_{L^1}.$$

The rest of the proof of the inequality (10.44) goes as follows: setting $C = (2\pi\hbar)^{-n}\|a_\sigma\|_{L^1}$ we have, using Cauchy–Schwarz's inequality,

$$|\widehat{A}\psi(x)|^2 \leq \int_{\mathbb{R}^n} |K_{\widehat{A}}(x,y)|d^n y \int_{\mathbb{R}^n} |K_{\widehat{A}}(x,y)|\,|\psi(y)|^2 dy$$

$$\leq C^2 \int_{\mathbb{R}^n} |K_{\widehat{A}}(x,y)|\,|\psi(y)|^2 dy$$

and hence

$$\int_{\mathbb{R}^n} |\widehat{A}\psi(x)|^2 dx \leq C^2 \int_{\mathbb{R}^n} \left(\int_{\mathbb{R}^n} |K_{\widehat{A}}(x,y)|dx\right) |\psi(y)|^2 dy,$$

that is

$$\int_{\mathbb{R}^n} |\widehat{A}\psi(x)|^2 dx \leq C^2 \int_{\mathbb{R}^n} |\psi(y)|^2 dy$$

which is precisely the estimate (10.44). $\square$

The study of regularity properties of Weyl operators on $L^2(\mathbb{R}^n)$ has become something of an industry since the early 1990s. One of the most known (and certainly most useful!) results nevertheless goes back to an older paper of Cordes [29]:

Proposition 229. *Assume that the symbol a satisfies the conditions $\partial_x^\alpha \partial_p^\beta a \in L^\infty(\mathbb{R}^n \oplus \mathbb{R}^n)$ for all multi-indices α, β such that $|\alpha|$, $|\beta| \leq [n/2] + 1$. Then the operator $\widehat{A} \overset{\text{Weyl}}{\longleftrightarrow} a$ is bounded on the space $L^2(\mathbb{R}^n)$.*

The conditions on the multi-indices α, β in the result above are rather sharp; see Boulkhemair [21] for a discussion of these conditions and references to previous work (in addition there is a detailed discussion of L^2-boundedness for operators with symbols in the Hörmander classes $S_{\rho,\delta}^m$).

10.5.2 Symbols in L^q spaces

In his book [163] Wong has analyzed in detail regularity results for Weyl operators with symbols in the L^q spaces, and has given conditions for these operators to be continuous, compact, and Hilbert–Schmidt. We refer to Wong's work for details, and limit ourselves here to a few statements.

We begin by proving a boundedness result when the symbol a belongs to the space $L^2(\mathbb{R}^n \oplus \mathbb{R}^n)$ of square integrable functions. It is a nice application of the Moyal identity

$$(W(\psi,\phi)|W(\psi',\phi'))_{L^2(\mathbb{R}^{2n})} = \left(\tfrac{1}{2\pi\hbar}\right)^n (\psi|\psi')_{L^2(\mathbb{R}^n)}\overline{(\phi|\phi')_{L^2(\mathbb{R}^n)}}. \qquad (10.46)$$

Proposition 230. *Let* $\widehat{A} \overset{\text{Weyl}}{\longleftrightarrow} a$ *with* $a \in L^2(\mathbb{R}^n \oplus \mathbb{R}^n)$. *Then the Weyl operator* $\widehat{A}$ *is bounded on* $L^2(\mathbb{R}^n)$ *and we have the estimate*

$$\|\widehat{A}\psi\|_{L^2(\mathbb{R}^n)} \leq \left(\tfrac{1}{2\pi\hbar}\right)^n \|a\|_{L^2(\mathbb{R}^{2n})}\|\psi\|_{L^2(\mathbb{R}^n)} \qquad (10.47)$$

for all functions $\psi \in L^2(\mathbb{R}^n)$.

Proof. It is no restriction to assume that $\widehat{A}\psi \neq 0$. Assume first that $a \in \mathcal{S}(\mathbb{R}^n \oplus \mathbb{R}^n)$. Then, for every $\psi,\phi \in \mathcal{S}(\mathbb{R}^n)$ we have, in view of (10.8),

$$(\widehat{A}\psi|\phi)_{L^2} = \int_{\mathbb{R}^{2n}} a(z)W(\psi,\phi)(z)dz$$

and hence, using successively the Cauchy–Schwarz inequality and the Moyal identity (10.46),

$$|(\widehat{A}\psi|\phi)_{L^2}| \leq \|a\|_{L^2}\|W(\psi,\phi)\|_{L^2}$$
$$\leq \left(\tfrac{1}{2\pi\hbar}\right)^n \|a\|_{L^2}\|\psi\|_{L^2}\|\phi\|_{L^2}.$$

Since $\widehat{A}\psi \in \mathcal{S}(\mathbb{R}^n)$ we may choose $\phi = \widehat{A}\psi$ and the inequality above then becomes

$$\|\widehat{A}\psi\|_{L^2}^2 \leq \left(\tfrac{1}{2\pi\hbar}\right)^n \|a\|_{L^2}\|\psi\|_{L^2}\|\widehat{A}\psi\|_{L^2}.$$

Dividing both sides by $\|\widehat{A}\psi\|_{L^2}$ yields the estimate (10.47) in the considered case. The general case follows in view of the density of $\mathcal{S}(\mathbb{R}^n)$ in $L^2(\mathbb{R}^n)$ and that of $\mathcal{S}(\mathbb{R}^n \oplus \mathbb{R}^n)$ in $L^2(\mathbb{R}^n \oplus \mathbb{R}^n)$. $\qquad \square$

Recall that $L^q(\mathbb{R}^n \oplus \mathbb{R}^n)$ $(1 \leq q < \infty)$ is the Banach space of all measurable functions $a : \mathbb{R}^n \oplus \mathbb{R}^n \longrightarrow \mathbb{C}$ such that

$$\|a\|_q = \left(\int_{\mathbb{R}^{2n}} |a(z)|^q dz \right)^{1/q} < \infty$$

(these spaces will be studied in some detail in Chapter 17).

The result above extends to the case of L^q spaces with $1 \leq q < 2$; we will not prove this here and refer to Wong [163], Theorem 1.1:

Proposition 231. *Let* $\widehat{A} \overset{\text{Weyl}}{\longleftrightarrow} a$ *with* $a \in L^q(\mathbb{R}^n \oplus \mathbb{R}^n)$, $1 \leq q < 2$. *The Weyl operator* $\widehat{A}$ *is bounded on* $L^2(\mathbb{R}^n)$ *and there exists a constant* C_q *only depending on* q *such that*
$$\|\widehat{A}\psi\|_{L^2(\mathbb{R}^n)} \leq C_q \|a\|_{L^q(\mathbb{R}^{2n})} \|\psi\|_{L^2(\mathbb{R}^n)}$$
for all $\psi \in L^2(\mathbb{R}^n)$.

When $2 < q < \infty$ one can no longer expect L^2 boundedness; in fact one can show that for each q such that $2 < q < \infty$ there exists a symbol $a \in L^q(\mathbb{R}^n \oplus \mathbb{R}^n)$ such that $\widehat{A} \overset{\text{Weyl}}{\longleftrightarrow} a$ is not a bounded operator on $L^2(\mathbb{R}^n)$ (for a proof of this result we refer to Wong [163], Theorem 13.1).

Chapter 11

Coherent States and Anti-Wick Quantization

The theory of coherent states plays an important role in various aspects of representation theory, and of, course, in quantum mechanics from which it originates. Historically, the notion of coherent state goes back to Schrödinger's 1926 work [143] on non-dispersing wavepackets for a harmonic oscillator. In 1932 von Neumann [130] considered sets of coherent states associated with a division of phase space into quantum cells. The modern theory was initiated by Glauber's 1963 work [62] in quantum optics (Glauber was awarded the 2005 Nobel Prize in Physics for his contributions; we mention that there was a controversy about priorities involving the physicist Sudarshan, also famous for his work on coherent states). The theory of coherent states has since then been applied to a variety of problems in mathematics and mathematical physics, and has been extended in various directions, for instance within the framework of anti-Wick (also called Toeplitz, or Berezin) quantization (Berezin [13]) which we will study later in this chapter.

11.1 Coherent states

Much of the material of this section is inspired by Littlejohn's seminal paper [117]. We begin with an easy physical motivation of coherent states as "minimum uncertainty wavepackets".

11.1.1 A physical motivation

Consider the Hamiltonian function

$$H(z) = \frac{1}{2m}p^2 + \frac{1}{2}m\omega^2 x^2$$

of the harmonic oscillator in one dimension; the eigenvalues of the corresponding Weyl operator

$$\widehat{H} = -\frac{\hbar}{2m}\frac{\partial^2}{\partial x^2} + \frac{1}{2}m\omega^2 x^2$$

are the numbers $(N + \frac{1}{2})\hbar\omega$ $(N = 0, 1, 2, \ldots)$ and the associated eigenfunctions are the rescaled Hermite functions

$$\psi_N(x) = \sqrt{\frac{1}{2^N N!}}\left(\frac{m\omega}{\pi\hbar}\right)^{1/4}\exp\left(-\frac{m\omega}{2\hbar}x^2\right)H_N\left(x\sqrt{\frac{m\omega}{\hbar}}\right)$$

where the H_N are the Hermite polynomials

$$H_N(x) = (-1)^N e^{x^2}\frac{d^N}{dx^N}e^{x^2}.$$

Let us focus on the ground state ψ_0; it is given by

$$\psi_0(x) = \left(\frac{m\omega}{\pi\hbar}\right)^{1/4}\exp\left(-\frac{m\omega}{2\hbar}x^2\right). \tag{11.1}$$

This function has the following typical property: it is, to use physicists' terminology, a "minimum extension (or uncertainty) wavepacket"; by that we mean that the statistical variances $(\Delta x)_{\psi_0}$ and $(\Delta p)_{\psi_0}$ are such that $(\Delta p)_{\psi_0}(\Delta x)_{\psi_0} = \frac{1}{2}\hbar$ (one says that the Heisenberg inequality is "saturated by ψ_0").

Exercise 232. Show that $(\Delta x)_{\psi_0} = (\hbar/2m\omega)^{1/2}$ and $(\Delta p)_{\psi_0} = (\hbar m\omega/2)^{1/2}$. [Hint: use for instance formula (11.18) to calculate explicitly $W\psi_0(z)$ and then determine the corresponding covariance matrix.]

For more on the physical properties of the coherent state ψ_{z_0} we refer to Messiah's classical (but still very modern) book [123].

11.1.2   Properties of coherent states

Let us introduce the following notation for $x \in \mathbb{R}^n$; we write

$$\psi_0^\hbar(x) = \left(\tfrac{1}{\pi\hbar}\right)^{n/4}e^{-\frac{1}{2\hbar}|x|^2} \tag{11.2}$$

and

$$\psi_{z_0}^\hbar(x) = \widehat{T}(z_0)\psi_0^\hbar(x) \tag{11.3}$$

where $\widehat{T}(z_0)$ is the Heisenberg–Weyl operator. When $n = 1$ the function $\psi_0^\hbar$ is just the ground state (11.1) of the quantum harmonic oscillator when $m = 1$, $\omega = 1$.

Definition 233. The function $\psi_0^\hbar \in \mathcal{S}(\mathbb{R}^n)$ defined by (11.2) is called the standard (or fiducial[1]) coherent state. Let $z_0 = (x_0, p_0)$; the function $\psi_{z_0}^\hbar \in \mathcal{S}(\mathbb{R}^n)$ defined by (11.3) is called the coherent state centered at z_0. (In the quantum mechanical literature these states are often denoted $|0\rangle$ and $|z_0\rangle$, respectively.)

---

[1] In the sense of something taken as an origin or zero of reference – not in the legal sense!

In physics one also sometimes calls $\psi_0^\hbar$ the "vacuum state". In signal analysis (particularly wavelet theory) one often uses the terminology "mother wavelet".

We begin by proving that the set of all translations of the standard coherent states span a dense subset of $L^2(\mathbb{R}^n)$. Our proof is elementary, and relies on Plancherel's formula; Gröchenig [82] (Lemma 1.5.3) gives a related proof of this property, which he then uses to derive Plancherel's formula. More precisely:

Proposition 234. *Let $T(x_0)$ be the translation operator $x \longmapsto x + x_0$ on $\mathbb{R}^n$ and define $T(x_0)\psi(x) = \psi(x - x_0)$. The linear spans of sets $\{T(x_0)\psi : x_0 \in \mathbb{R}^n\}$ and $\{\psi_{z_0}^\hbar : z_0 \in \mathbb{R}^n\}$ are dense in the space $L^2(\mathbb{R}^n)$ of square integrable functions.*

Proof. We will only prove the assertion for the set $\{T(x_0)\psi : x_0 \in \mathbb{R}^n\}$; the case of $\{\psi_{z_0}^\hbar : z_0 \in \mathbb{R}^n\}$ is an immediate adaptation. Let $\psi \in L^2(\mathbb{R}^n)$; since $\psi_\hbar$ is even we have

$$\langle \psi, T(x_0)\psi_\hbar \rangle = \int_{\mathbb{R}^n} \psi(x)\psi^\hbar(x - x_0)dx = \psi * \psi^\hbar(x_0).$$

Thus $\langle \psi, T(x_0)\psi_\hbar \rangle = 0$ for all ψ implies that $\psi * \psi^\hbar = 0$, and hence $F\psi F\psi^\hbar = 0$, that is $F\psi = 0$ since $F\psi^\hbar > 0$. In view of Plancherel's formula $\|F\psi\|_{L^2} = \|\psi\|_{L^2}$ and thus $\psi = 0$ almost everywhere. It follows that there are no non-trivial vectors which are orthogonal to the span of the set $\{\psi_{z_0}^\hbar : z_0 \in \mathbb{R}^n\}$ and hence the span of $\{T(x_0)\psi : x_0 \in \mathbb{R}^n\}$ is dense in $L^2(\mathbb{R}^n)$. $\square$

Note that in the proof above the fact that we were using the standard coherent state was not essential: we could actually replace $\psi^\hbar$ with any function $\phi \in L^2(\mathbb{R}^n)$ such that $F\phi$ never vanishes.

It follows from Proposition 234 that the space of functions spanned by the $\{\psi_{z_0}^\hbar : z_0 \in \mathbb{R}^n\}$ is also dense in the space $L^2(\mathbb{R}^n)$, since it contains the span. Let us develop this property:

Proposition 235. *The coherent states $\psi_{z_0}^\hbar$ have the following properties:*

(i) *They satisfy the generalized orthogonality relations*

$$\left(\tfrac{1}{2\pi\hbar}\right)^n \int_{\mathbb{R}^{2n}} \psi_{z_0}^\hbar(x)\overline{\psi_{z_0}^\hbar(y)}dz_0 = \delta(x - y); \tag{11.4}$$

(ii) *For every $\psi \in L^2(\mathbb{R}^n)$ we have*

$$\|\psi\|_{L^2}^2 = \left(\tfrac{1}{2\pi\hbar}\right)^n \int_{\mathbb{R}^{2n}} |(\psi|\psi_{z_0}^\hbar)_{L^2(\mathbb{R}^n)}|^2 dz_0 \tag{11.5}$$

and

$$\psi(x) = \left(\tfrac{1}{2\pi\hbar}\right)^n \int_{\mathbb{R}^{2n}} (\psi|\psi_{z_0}^\hbar)_{L^2(\mathbb{R}^n)}\psi_{z_0}^\hbar(x)dz_0. \tag{11.6}$$

Proof of (i). We have

$$(\psi, \psi_{z_0}^{\hbar})_{L^2} = e^{\frac{i}{2\hbar}p_0 \cdot x_0} \int_{\mathbb{R}^n} e^{-\frac{i}{\hbar}p_0 \cdot x} \psi(x) e^{-\frac{1}{2\hbar}(x-x_0)^2} dx$$

$$= (2\pi\hbar)^{n/2} e^{\frac{i}{2\hbar}p_0 \cdot x_0} F(\psi\psi^{\hbar}(\cdot - x_0))(p_0)$$

hence, by Plancherel's theorem,

$$\int_{\mathbb{R}^{2n}} |(\psi, \psi_{z_0}^{\hbar})_{L^2}|^2 dz_0 = (2\pi\hbar)^n \int_{\mathbb{R}^n} \|\psi\psi^{\hbar}((\cdot) - x_0)\|_{L^2}^2 dx_0$$

$$= (2\pi\hbar)^n \int_{\mathbb{R}^n} |\psi(x)|^2 \left(\int_{\mathbb{R}^n} |\psi^{\hbar}(x - x_0)|^2 dx_0 \right) dx$$

$$= (2\pi\hbar)^n \|\psi\|_{L^2}^2.$$

Proof of (ii). We have

$$\int_{\mathbb{R}^{2n}} \psi_{z_0}^{\hbar}(x)\overline{\psi_{z_0}^{\hbar}(y)} dz_0 = \int_{\mathbb{R}^{2n}} e^{\frac{i}{\hbar}p_0 \cdot (x-y)} \psi^{\hbar}(x - x_0)\psi^{\hbar}(y - x_0) dz_0$$

$$= \int_{\mathbb{R}^n} \left(\int_{\mathbb{R}^n} e^{\frac{i}{\hbar}p_0 \cdot (x-y)} dp_0 \right) \psi^{\hbar}(x - x_0)\psi^{\hbar}(y - x_0) dx_0$$

$$= (2\pi\hbar)^n \int_{\mathbb{R}^n} \delta(x - y)\psi^{\hbar}(x - x_0)\psi^{\hbar}(y - x_0) dx_0$$

$$= (2\pi\hbar)^n \left(\int_{\mathbb{R}^n} |\psi^{\hbar}(x - x_0)|^2 dx_0 \right) \delta(x - y)$$

$$= (2\pi\hbar)^n \delta(x - y)$$

which proves (11.4). Formula (11.6) follows since we have

$$\psi(x) = \int_{\mathbb{R}^n} \delta(x - y)\psi(y) dy$$

$$= \left(\tfrac{1}{2\pi\hbar}\right)^n \iint_{\mathbb{R}^n \times \mathbb{R}^{2n}} \psi_{z_0}^{\hbar}(x)\overline{\psi_{z_0}^{\hbar}(y)}\psi(y) dy dz_0$$

$$= \left(\tfrac{1}{2\pi\hbar}\right)^n \int_{\mathbb{R}^{2n}} \left(\int_{\mathbb{R}^n} \overline{\psi_{z_0}^{\hbar}(y)}\psi(y) dy \right) \psi_{z_0}^{\hbar}(x) dz_0$$

$$= \left(\tfrac{1}{2\pi\hbar}\right)^n \int_{\mathbb{R}^{2n}} (\psi|\psi_{z_0}^{\hbar})_{L^2(\mathbb{R}^n)}\psi_{z_0}^{\hbar}(x) dz_0. \qquad \square$$

Exercise 236. Restate formulas (11.5) and (11.6) in terms of the cross-ambiguity and Wigner distributions.

The result above leads to interesting expressions in terms of coherent states for the kernel and Weyl symbol of an operator:

Proposition 237. *Let $\widehat{A}$ be a continuous linear operator $\mathcal{S}(\mathbb{R}^n) \longrightarrow \mathcal{S}'(\mathbb{R}^n)$ with kernel $K_{\widehat{A}}$; we have*

$$K_{\widehat{A}}(x,y) = \left(\tfrac{1}{2\pi\hbar}\right)^n \int_{\mathbb{R}^{2n}} \widehat{A}\psi_{z_0}^{\hbar}(x)\overline{\psi_{z_0}^{\hbar}(y)}dz_0 \tag{11.7}$$

and the Weyl symbol $a \overset{\text{Weyl}}{\longleftrightarrow} \widehat{A}$ is given by

$$a(z) = \int_{\mathbb{R}^{2n}} W(\widehat{A}\psi_{z_0}^{\hbar}, \psi_{z_0}^{\hbar})(z)dz_0. \tag{11.8}$$

Proof. We have

$$\widehat{A}\psi_{z_0}^{\hbar}(x) = \int_{\mathbb{R}^n} K_{\widehat{A}}(x,x')\psi_{z_0}^{\hbar}(x')dx'$$

hence, using formula (11.4),

$$\int_{\mathbb{R}^{2n}} \widehat{A}\psi_{z_0}^{\hbar}(x)\overline{\psi_{z_0}^{\hbar}(y)}dz_0 = \int_{\mathbb{R}\times\mathbb{R}^{2n}} K_{\widehat{A}}(x,x')\psi_{z_0}^{\hbar}(x')\overline{\psi_{z_0}^{\hbar}(y)}dx'dz_0$$

$$= (2\pi\hbar)^n \int_{\mathbb{R}^n} K_{\widehat{A}}(x,x')\delta(y-x')dx'$$

$$= (2\pi\hbar)^n K_{\widehat{A}}(x,y)$$

that is (11.7). Formula (11.8) for the symbol readily follows: using successively (10.15) in Proposition 205 and (11.7) we have

$$a(x,p) = \int_{\mathbb{R}^n} e^{-\frac{i}{\hbar}p\cdot y} K_{\widehat{A}}(x + \tfrac{1}{2}y, x - \tfrac{1}{2}y)dy$$

$$= \left(\tfrac{1}{2\pi\hbar}\right)^n \int_{\mathbb{R}^n} e^{-\frac{i}{\hbar}p\cdot y} \left(\int_{\mathbb{R}^{2n}} \widehat{A}\psi_{z_0}^{\hbar}(x + \tfrac{1}{2}y)\overline{\psi_{z_0}^{\hbar}(x - \tfrac{1}{2}y)}dz_0 \right) dy$$

$$= \int_{\mathbb{R}^{2n}} W(\widehat{A}\psi_{z_0}^{\hbar}, \psi_{z_0}^{\hbar})(z)dz_0$$

which concludes the proof. $\qquad\square$

In Section 8.4 we discussed the notion of frame. This notion can be generalized to the continuous case, and this will allow us to interpret the results above. Let us first give the following definition:

Definition 238. Let $\mathcal{H}$ be a complex Hilbert space and (M, μ) a measure space with positive measure μ. A continuous (or generalized) frame in $\mathcal{H}$ is a family of vectors $(\psi_z)_{z\in M}$ in $\mathcal{H}$ such that:

(i) For every $\psi \in \mathcal{H}$ the mapping $z \longmapsto (\psi|\psi_z)_{\mathcal{H}}$ is a measurable function on (M, μ);

(ii) There exist constants $a, b > 0$ such that

$$a\|\psi\|_{\mathcal{H}}^2 \leq \int_M |(\psi|\psi_z)_{\mathcal{H}}|^2 d\mu(z) \leq b\|\psi\|_{\mathcal{H}}^2 \tag{11.9}$$

for every vector $\psi \in \mathcal{H}$.

When $a = b$ (resp. $a = b = 1$) the family $(\psi_z)_{z \in M}$ is called a tight (resp. normalized) continuous frame.

In what follows we choose $\mathcal{H} = L^2(\mathbb{R}^n)$ and $M = \mathbb{R}^{2n}$; μ is the usual Lebesgue measure.

Proposition 239. *The family* $(\psi_z^{\hbar})_{z \in \mathbb{R}^{2n}}$ *is a continuous tight Gabor frame in* $L^2(\mathbb{R}^n)$ *with bound* $(2\pi\hbar)^n$.

Proof. We have by definition $\psi_z^{\hbar} = \widehat{T}(z)\psi_0^{\hbar}$, hence

$$(\psi|\psi_z^{\hbar})_{L^2} = (2\pi\hbar)^n A(\psi, \psi_0^{\hbar})(z)$$

thus the mapping $z \longmapsto (\psi|\psi_z^{\hbar})_{L^2}$ is continuous and hence measurable. In view of formula (11.5) in Proposition 235 we have

$$(2\pi\hbar)^n \|\psi\|_{L^2}^2 = \int_{\mathbb{R}^{2n}} |(\psi|\psi_{z_0}^{\hbar})_{L^2(\mathbb{R}^n)}|^2 dz_0$$

hence $(\psi_z^{\hbar})_{z \in \mathbb{R}^{2n}}$ is a tight frame. $\qquad\square$

Much of what we have said above is not specific to the choice $\psi_{z_0}^{\hbar}$ and remains valid in a much more general setting. We refer to Peremolov's paper [132] for a generalization scheme useful for various physical problems that have dynamical symmetries. From a more abstract point of view one can define a very general notion of coherent states in Hilbert spaces as follows (see Kisil [108]):

Definition 240. Let $\mathcal{H}$ be a Hilbert space and G a Lie group with Haar measure μ acting on $\mathcal{H}$. A family $\{\psi_g \in \mathcal{H} : g \in G\}$ is called a system of coherent states if

(i) There is a representation $T : g \longmapsto Tg$ of the group G by unitary operators on $\mathcal{H}$;

(ii) There is a vector $\psi_0 \in \mathcal{H}$ such that for $\psi_g = T_g\psi_0$ and every $\psi \in \mathcal{H}$ we have

$$\|\psi\|_{\mathcal{H}}^2 = \int_G |(\psi|\psi_g)_{\mathcal{H}}|^2 d\mu. \tag{11.10}$$

The coherent states we have been studying above correspond to the choices $\mathcal{H} = L^2(\mathbb{R}^n)$, $G = \mathbb{H}_n$ (the Heisenberg group); the representation T is of course here the mapping $(z_0, t) \longmapsto \widehat{T}(z_0, t)$ where

$$\widehat{T}(z_0, t)\psi(x) = e^{\frac{i}{\hbar}(tp_0 \cdot x - \frac{1}{2}t^2 p_0 \cdot x_0)}\psi(x - tx_0)$$

(see formula (8.4) in Chapter 8).

11.2 Wigner transforms of Gaussians

In Chapter 6 (Proposition 106) we expressed Hardy's multi-dimensional uncertainty principle in terms of the topological notion of symplectic capacity studied in Chapter 5. We are going to see that similar methods allow us to show that the Wigner transform cannot be "too concentrated" in phase space; the result can be expressed in terms of the symplectic capacity of the "Wigner ellipsoid".

11.2.1 Some explicit formulas

Let us begin by giving a formula allowing us to calculate the Fourier transform

$$F\psi(x) = \left(\tfrac{1}{2\pi\hbar}\right)^{n/2} \int_{\mathbb{R}^n} e^{-\frac{i}{\hbar}x\cdot x'}\,\psi(x')dx'$$

of a complex Gaussian function:

Lemma 241. *Let* $\phi_M(x) = e^{-\frac{1}{2\hbar}Mx^2}$ *where* $M = X + iY$ *is a symmetric complex* $n \times n$ *matrix such that* $X = \operatorname{Re} M > 0$. *We have*

$$F\phi_M(x) = (\det M)^{-1/2}\phi_{M^{-1}}(x) \tag{11.11}$$

where $(\det M)^{-1/2}$ *is given by the formula*

$$(\det M)^{-1/2} = \lambda_1^{-1/2} \cdots \lambda_m^{-1/2},$$

the numbers $\lambda_1^{-1/2},\ldots,\lambda_n^{-1/2}$ *being the square roots with positive real parts of the eigenvalues* $\lambda_1^{-1},\ldots,\lambda_m^{-1}$ *of* M^{-1}.

Proof. It is standard, generalizing from the case $n = 1$ and using a simultaneous diagonalization of X and Y. See, e.g., Folland [59], Appendix A. $\qquad\square$

From now on we denote by $\psi_M^\hbar$ the Gaussian function defined by

$$\psi_M^\hbar(x) = \left(\tfrac{1}{\pi\hbar}\right)^{n/4} (\det X)^{1/4}e^{-\frac{1}{2\hbar}Mx^2} \tag{11.12}$$

where M is as above. The coefficient in front of the exponential is chosen so that $\psi_M^\hbar$ is normalized to unity: $\|\psi_M^\hbar\|_{L^2} = 1$. Gaussians of this type are called "squeezed coherent states"; they will be studied in detail in Chapter 11. Note that since $X > 0$ we have $\psi_M^\hbar \in \mathcal{S}(\mathbb{R}^n)$ and hence $W\psi_M^\hbar \in \mathcal{S}(\mathbb{R}^n \oplus \mathbb{R}^n)$. The following result shows that $W\psi_M^\hbar$ is in fact a phase space Gaussian of a very special type:

Proposition 242. *Let* $M = X + iY$ *and* $\psi_M^\hbar$ *be defined as above.*

(i) *The Wigner transform* $W\psi_M^\hbar$ *is the phase space Gaussian*

$$W\psi_M^\hbar(z) = \left(\tfrac{1}{\pi\hbar}\right)^n e^{-\frac{1}{\hbar}Gz^2} \tag{11.13}$$

where G is the symmetric matrix

$$G = \begin{pmatrix} X + YX^{-1}Y & YX^{-1} \\ X^{-1}Y & X^{-1} \end{pmatrix};$$ (11.14)

(ii) *We have $G \in \mathrm{Sp}(2n, \mathbb{R})$; in fact $G = S^T S$ where*

$$S = \begin{pmatrix} X^{1/2} & 0 \\ X^{-1/2}Y & X^{-1/2} \end{pmatrix}$$ (11.15)

is a symplectic matrix.

Proof of (i). Set $C(X) = (\pi\hbar)^{-n/4} (\det X)^{1/4}$. By definition of the Wigner transform we have

$$W\psi_M^\hbar(z) = \left(\tfrac{1}{2\pi\hbar}\right)^n C(X)^2 \int_{\mathbb{R}^n} e^{-\frac{i}{\hbar}p\cdot y} e^{-\frac{1}{2\hbar}F(x,y)} dy$$ (11.16)

where the phase F is defined by

$$\begin{aligned} F(x, y) &= (X + iY)(x + \tfrac{1}{2}y)^2 + (X - iY)(x - \tfrac{1}{2}y)^2 \\ &= 2Xx \cdot x + 2iYx \cdot y + \tfrac{1}{2}Xy \cdot y \end{aligned}$$

and hence

$$W\psi_M^\hbar(z) = \left(\tfrac{1}{2\pi\hbar}\right)^n e^{-\frac{1}{\hbar}Xx^2} C(X)^2 \int_{\mathbb{R}^n} e^{-\frac{i}{\hbar}(p+Yx)\cdot y} e^{-\frac{1}{4\hbar}Xy^2} dy.$$

Using the Fourier transformation formula (11.11) above with x replaced by $p + Yx$ and M by $\tfrac{1}{2}X$ we get

$$\int_{\mathbb{R}^n} e^{-\frac{i}{\hbar}(p+Yx)\cdot y} e^{-\frac{1}{4\hbar}Xy\cdot y} dy = (2\pi\hbar)^{n/2} \left[\det(\tfrac{1}{2}X)\right]^{-1/2}$$
$$\times C(X)^2 \exp\left[-\tfrac{1}{\hbar}X^{-1}(p + Yx) \cdot (p + Yx)\right].$$

On the other hand we have

$$(2\pi\hbar)^{n/2} \left[\det(\tfrac{1}{2}X)\right]^{-1/2} C(X)^2 = \left(\tfrac{1}{\pi\hbar}\right)^n$$

and hence

$$W\psi_M^\hbar(z) = \left(\tfrac{1}{\pi\hbar}\right)^n e^{-\frac{1}{\hbar}Gz^2}$$

where

$$Gz^2 = (X + YX^{-1})x \cdot x + 2X^{-1}Yx \cdot p + X^{-1}p \cdot p;$$

Proof of (ii). The symmetry of G is obvious, and so is the factorization $G = S^T S$. One immediately verifies that $S^T J S = J$ hence $S \in \mathrm{Sp}(2n, \mathbb{R})$ as claimed.　　□

Exercise 243. Verify directly that $W\psi_M^\hbar$ is normalized to unity by making the change of variables $u = Sz$ in the integral in formula (11.16).

In particular, when $\psi_0^\hbar$ is the "coherent state" defined by

$$\psi_0^\hbar(x) = \left(\tfrac{1}{\pi\hbar}\right)^{n/4} e^{-\frac{1}{2\hbar}|x|^2} \tag{11.17}$$

we immediately get from (11.13) and (11.14) the formula

$$W\psi_0^\hbar(z) = \left(\tfrac{1}{\pi\hbar}\right)^n e^{-\frac{1}{\hbar}|z|^2} \tag{11.18}$$

well known from quantum mechanics.

11.2.2 The cross-Wigner transform of a pair of Gaussians

Let us generalize formula (11.13) by calculating the cross-Wigner transform $W(\psi_M^\hbar, \psi_{M'})$ of a pair of Gaussians of the type above; we recall that the Wigner-Moyal transform of $\psi, \phi \in \mathcal{S}(\mathbb{R}^n)$ is defined by

$$W(\psi, \phi)(z) = \left(\tfrac{1}{2\pi\hbar}\right)^n \int_{\mathbb{R}^n} e^{-\frac{i}{\hbar}p\cdot y}\psi(x + \tfrac{1}{2}y)\overline{\phi(x - \tfrac{1}{2}y)}dy. \tag{11.19}$$

Proposition 244. *Let $\psi_M^\hbar$ and $\psi_{M'}^\hbar$ be Gaussian functions of the type (11.12). We have*

$$W(\psi_M^\hbar, \psi_{M'}^\hbar)(z) = \left(\tfrac{1}{\pi\hbar}\right)^n C_{M,M'}e^{-\frac{1}{\hbar}Fz^2} \tag{11.20}$$

where $C_{M,M'}$ is a constant given by

$$C_{M,M'} = (\det XX')^{1/4}\det\left[\tfrac{1}{2}(M + \overline{M'})\right]^{-1/2} \tag{11.21}$$

and F is the symmetric complex matrix given by

$$F = \begin{pmatrix} 2\overline{M'}(M + \overline{M'})^{-1}M & -i(M - \overline{M'})(M + \overline{M'})^{-1} \\ -i(M + \overline{M'})^{-1}(M - \overline{M'}) & 2(M + \overline{M'})^{-1} \end{pmatrix}. \tag{11.22}$$

Proof. We have

$$W(\psi_M^\hbar, \psi_{M'}^\hbar)(z) = C(X, X')\int_{\mathbb{R}^n} e^{-\frac{i}{\hbar}py}e^{-\frac{1}{2\hbar}\Phi(x,y)}dy$$

where the functions C and Φ are given by

$$C(X, X') = 2^{-n}\left(\tfrac{1}{\pi\hbar}\right)^{2n}(\det XX')^{1/4},$$
$$\Phi(x, y) = M(x + \tfrac{1}{2}y)^2 + \overline{M'}(x - \tfrac{1}{2}y)^2.$$

Let us evaluate the integral

$$I(z) = \int_{\mathbb{R}^n} e^{-\frac{i}{\hbar}py}e^{-\frac{1}{2\hbar}\Phi(x,y)}dy.$$

We have

$$\Phi(x,y) = (M + \overline{M'})x^2 + \frac{1}{4}(M + \overline{M'})y^2 + (M - \overline{M'})x \cdot y$$

and hence

$$I(z) = e^{-\frac{1}{2\hbar}(M+\overline{M'})x^2} \int_{\mathbb{R}^n} e^{-\frac{i}{\hbar}[p-\frac{i}{2}(M-\overline{M'})x]\cdot y} e^{-\frac{1}{8\hbar}(M+\overline{M'})y^2}\, dy.$$

Using the Fourier transformation formula (11.11) we get

$$I(z) = (2\pi\hbar)^{n/2} \det\left[\tfrac{1}{4}(M + \overline{M'})\right]^{-1/2}$$
$$\times \exp\left(-\frac{1}{2\hbar}\left[(M + \overline{M'})x^2 + 4(M + \overline{M'})^{-1}\left(p - \tfrac{1}{2}(M - \overline{M'})x\right)^2\right]\right).$$

A straightforward calculation shows that

$$\tfrac{1}{2}(M + \overline{M'})x^2 + 4(M + \overline{M'})^{-1}\left(p - \tfrac{1}{2}(M - \overline{M'})x\right)^2 = Fz \cdot z$$

where F is the matrix

$$\begin{pmatrix} K & -i(M - \overline{M'})(M + \overline{M'})^{-1} \\ -i(M + \overline{M'})^{-1}(M - \overline{M'}) & 2(M + \overline{M'})^{-1} \end{pmatrix} \qquad (11.23)$$

with left upper block

$$K = \tfrac{1}{2}\left[M + \overline{M'} - (M - \overline{M'})(M + \overline{M'})^{-1}(M - \overline{M'})\right].$$

Using the identity

$$M + \overline{M'} - (M - \overline{M'})(M + \overline{M'})^{-1}(M - \overline{M'}) = 4\overline{M'}(M + \overline{M'})^{-1}M \qquad (11.24)$$

the matrix (11.23) is given by (11.22). We thus have, collecting the constants and simplifying the obtained expression,

$$W(\psi_M^\hbar, \psi_{M'}^\hbar)(z) = \left(\tfrac{1}{\pi\hbar}\right)^n (\det XX')^{1/4} \det\left[\tfrac{1}{2}(M + \overline{M'})\right]^{-1/2} e^{-\frac{1}{\hbar}Fz^2}$$

which we set out to prove. $\qquad\qquad\qquad\qquad\qquad\qquad\qquad\qquad\qquad\qquad\square$

Exercise 245. Check the matrix identity (11.24) above and verify that when $M = M'$ the matrix F is identical to the matrix G in formula (11.14) for the Wigner transform of a Gaussian.

11.3 Squeezed coherent states

We have seen that the usual coherent states are minimum extension wavepackets, in the sense that the Heisenberg inequalities become equalities for these states. However, this property is not characteristic of these coherent states.

11.3.1 Definition and characterization

We define the squeezed coherent states in terms of their Wigner transform:

Definition 246. A function $\psi \in \mathcal{S}(\mathbb{R}^n)$ is called a (normalized) squeezed coherent state if its Wigner transform is

$$W\psi(z) = \left(\tfrac{1}{\pi\hbar}\right)^n e^{-\frac{1}{\hbar}G(z-z_0)^2} \tag{11.25}$$

where $G \in \mathrm{Sp}(2n, \mathbb{R})$ is positive definite: $G = G^T > 0$ and $z_0 \in \mathbb{R}^{2n}$.

Recall that

$$W\psi_0^\hbar(z) = \left(\tfrac{1}{\pi\hbar}\right)^n e^{-\frac{1}{\hbar}|z|^2} \tag{11.26}$$

(formula (11.18) hence, with this terminology, $\psi_0^\hbar$ is itself a squeezed coherent state. Let $\psi_M^\hbar$ be given by (11.12), that is

$$\psi_M^\hbar(x) = \left(\tfrac{1}{\pi\hbar}\right)^{n/4} (\det X)^{1/4} e^{-\frac{1}{2\hbar}Mx^2} \tag{11.27}$$

with $M = X + iY$, X and Y symmetric, $X > 0$. Using Proposition 242 and formula (9.24) in Proposition 174, we see that every function of the type $\widehat{T}(z_0)\psi_M^\hbar$ satisfies (11.25) with

$$G = \begin{pmatrix} X + YX^{-1}Y & YX^{-1} \\ X^{-1}Y & X^{-1} \end{pmatrix}; \tag{11.28}$$

moreover $G = S^T S$ with

$$S = \begin{pmatrix} X^{1/2} & 0 \\ X^{-1/2}Y & X^{-1/2} \end{pmatrix}. \tag{11.29}$$

We are going to see that every squeezed coherent state is of the type $\widehat{T}(z_0)\psi_M^\hbar$ and that it can be obtained from the standard coherent state $\psi_0^\hbar$ using the affine metaplectic group $\mathrm{AMp}(2n, \mathbb{R})$.

Proposition 247. *A function ψ is a squeezed coherent state if and only if there exists $\widehat{S} \in \mathrm{Mp}(2n, \mathbb{R})$ and $z_0 \in \mathbb{R}^{2n}$ such that*

$$\psi = e^{i\gamma}\widehat{T}(z_0)\widehat{S}\psi_0^\hbar$$

where $\psi_0^\hbar$ is the standard coherent state and γ is real.

Proof. We first remark that the relation

$$W(e^{i\gamma}\widehat{T}(z_0)\widehat{S}\psi_0^\hbar)(z) = W(\widehat{S}\psi_0^\hbar)(z - z_0)$$

reduces the proof to case $z_0 = 0$. Let us thus show that if

$$W\psi(z) = \left(\tfrac{1}{\pi\hbar}\right)^n e^{-\frac{1}{\hbar}Gz^2}$$

with $G > 0$ and symplectic, then ψ is equal to $\widehat{S}\psi_0^\hbar$ (up to a complex factor with modulus 1) for some metaplectic operator $\widehat{S}$. We may write $G = S^2$ where $S \in \mathrm{Sp}(2n, \mathbb{R})$ is symmetric (Corollary 33). Thus

$$W\psi(S^{-1}z) = \left(\tfrac{1}{\pi\hbar}\right)^n e^{-\frac{1}{\hbar}|z|^2} = W\psi_0^\hbar(z).$$

In view of the metaplectic covariance formula (10.27) for Wigner transforms we have $W\psi(S^{-1}z) = W(\widehat{S}\psi)(z)$ where $\widehat{S} \in \mathrm{Mp}(2n, \mathbb{R})$ is such that $\pi^{\mathrm{Mp}}(\widehat{S}) = S$; it follows that $\psi = e^{i\gamma}$. We will see in a moment how to calculate explicitly $\widehat{S}\psi_0^\hbar$, but let us first note that an immediate consequence of the discussion above is that the metaplectic group acts on squeezed coherent states. Let us introduce some notation:　　　　　　　　　　　　　　　　　　　　　　　□

Notation 248. The set of all squeezed coherent states is denoted by $\Sigma^\hbar(n)$; the subset consisting of all centered squeezed coherent state is denoted by $\Sigma_0^\hbar(n)$.

It turns out that we have a continuous group action

$$\mathrm{Mp}(2n, \mathbb{R}) \times \Sigma_0^\hbar(n) \longrightarrow \Sigma_0^\hbar(n),$$
$$(\widehat{S}, \psi_M^\hbar) \longmapsto \widehat{S}\psi_M^\hbar.$$

In fact, every $\psi_M^\hbar \in \Sigma_0^\hbar(n)$ can be written $\widehat{S}_M\psi_0^\hbar$ for some $\widehat{S}_M \in \mathrm{Mp}(2n, \mathbb{R})$ hence $\widehat{S}\psi_M^\hbar = (\widehat{S}\widehat{S}_M)\psi_0^\hbar$ is also a squeezed coherent state. The action of $\mathrm{Mp}(2n, \mathbb{R})$ on $\Sigma_0^\hbar(n)$ is transitive: for every pair $(\psi_M^\hbar, \psi_{M'}^\hbar) \in \Sigma_0^\hbar(n) \times \Sigma_0^\hbar(n)$ we have $\psi_M^\hbar = S\psi_{M'}^\hbar$ with $\widehat{S} = \widehat{S}_{M'}^{-1}\widehat{S}_M$ if $\psi_M^\hbar = \widehat{S}_M\psi_0^\hbar$ and $\psi_{M'}^\hbar = \widehat{S}_{M'}\psi_0^\hbar$. These elementary remarks lead to an interesting topological identification of $\Sigma_0^\hbar(n)$: we have

$$\Sigma_0^\hbar(n) \equiv \mathrm{Mp}(2n, \mathbb{R})/U(2n, \mathbb{R}). \tag{11.30}$$

This is easily seen as follows: by the theory of homogeneous spaces there is a bijection of $\Sigma_0^\hbar(n)$ on every coset space $\mathrm{Mp}(2n, \mathbb{R})/St(\psi)$ where

$$St(\psi) = \{\widehat{S} \in \mathrm{Mp}(2n, \mathbb{R}) : \widehat{S}\psi = \psi\}$$

is the stabilizer (or isotropy subgroup) of ψ. Let us choose in particular $\psi = \psi_0^\hbar$, the standard coherent state. The stabilizer of $\psi_0^\hbar$ consists of all metaplectic operators $\widehat{S}$ such that $\widehat{S}\psi_0^\hbar = \psi_0^\hbar$, that is $W(\widehat{S}\psi_0^\hbar) = W\psi_0^\hbar$. Since $W\psi_0^\hbar(z) = (\pi\hbar)^{-n} e^{-\frac{1}{\hbar}|z|^2}$ (formula (11.26) and $W(\widehat{S}\psi_0^\hbar)(z) = W(\psi_0^\hbar)(S^{-1}z)$ the condition $W(\widehat{S}\psi_0^\hbar) = W\psi_0^\hbar$ is equivalent to $|S^{-1}z|^2 = |z|^2$ hence S must be a symplectic rotation, i.e.,

$$S \in \mathrm{Sp}(2n, \mathbb{R}) \cap O(2n, \mathbb{R}) = U(2n, \mathbb{R})$$

whence the identification (11.30).

Problem 249. Show that there is a natural action

$$\mathrm{AMp}(2n, \mathbb{R}) \times \Sigma^\hbar(n) \longrightarrow \Sigma^\hbar(n),$$
$$(\widehat{S}\widehat{T}(z), \psi_{z_0, M}^\hbar) \longmapsto \widehat{S}\widehat{T}(z)\psi_{z_0, M}^\hbar$$

and generalize the discussion above to this case.

11.3.2 Explicit action of $\mathrm{Mp}(2n, \mathbb{R})$ on squeezed states

Recall that the affine metaplectic group $\mathrm{AMp}(2n, \mathbb{R})$ consists of all operators $\widehat{T}(z_0)\widehat{S}$ (or $\widehat{S}\widehat{T}(z_0)$) where $\widehat{S} \in \mathrm{Mp}(2n, \mathbb{R})$ and $z_0 \in \mathbb{R}^{2n}$. We will need the following factorization formula for symplectic matrices (it is a refinement of the polar decomposition result in Proposition 34). It is sometimes called a pre-Iwasawa factorization in the theory of Lie groups.

Proposition 250. *Let* $S = \begin{pmatrix} A & B \\ C & D \end{pmatrix}$ *be a symplectic matrix. We have a unique factorization*

$$S = \begin{pmatrix} I & 0 \\ P & I \end{pmatrix} \begin{pmatrix} L & 0 \\ 0 & L^{-1} \end{pmatrix} \begin{pmatrix} U & V \\ -V & U \end{pmatrix} \tag{11.31}$$

where $P = P^T$, $L = L^T$ *(*$\det L \neq 0$*),* X, *and* Y *are given by the formulas*

$$P = (CA^T + DB^T)(AA^T + BB^T)^{-1} \tag{11.32}$$

$$L = (AA^T + BB^T)^{1/2} \tag{11.33}$$

$$U = (AA^T + BB^T)^{-1/2}A \tag{11.34}$$

$$V = (AA^T + BB^T)^{-1/2}B. \tag{11.35}$$

The proof of this proposition is given in detail in de Gosson [67], Chapter 2, §2.2.2. Note that we have

$$R = \begin{pmatrix} U & V \\ -V & U \end{pmatrix} \in U(2n, \mathbb{R})$$

and hence $RR^T = R^T R = I$ because R is a symplectic rotation (see Subsection 2.3.2).

Exercise 251. Show that every symplectic matrix $S = \begin{pmatrix} A & B \\ C & D \end{pmatrix}$ can be written in the form

$$S = \begin{pmatrix} L & 0 \\ Q & L^{-1} \end{pmatrix} \begin{pmatrix} U & V \\ -V & U \end{pmatrix} \tag{11.36}$$

where Q is given by the formula

$$Q = (CA^T + DB^T)(AA^T + BB^T)^{-1/2}. \tag{11.37}$$

Recall from Proposition 242 that the Wigner transform of a Gaussian

$$\psi_M^\hbar(x) = \left(\tfrac{1}{\pi\hbar}\right)^{n/4} (\det X)^{1/4} e^{-\frac{1}{2\hbar}Mx^2}$$

is given by the formula

$$W\psi_M^\hbar(z) = \left(\tfrac{1}{\pi\hbar}\right)^n e^{-\frac{1}{\hbar}Gz^2} \tag{11.38}$$

where $G = S^T S$ is symplectic matrix, the matrix S being given by

$$S = \begin{pmatrix} X^{1/2} & 0 \\ X^{-1/2}Y & X^{-1/2} \end{pmatrix}, \tag{11.39}$$

that is

$$G = \begin{pmatrix} X + YX^{-1}Y & YX^{-1} \\ X^{-1}Y & X^{-1} \end{pmatrix}. \tag{11.40}$$

The following result is an easy consequence of the formulas above:

Proposition 252. *Let $\widehat{S} \in \mathrm{Mp}(2n, \mathbb{R})$ have projection $S = \begin{pmatrix} A & B \\ C & D \end{pmatrix}$ on $\mathrm{Sp}(2n, \mathbb{R})$. We have*

$$\widehat{S}\psi_0^\hbar(x) = e^{i\gamma} \left(\tfrac{1}{\pi\hbar}\right)^{n/4} (\det X)^{1/4} e^{-\frac{1}{2\hbar}Mx^2} \tag{11.41}$$

where the phase γ is a real constant and $M = X + iY$ where X and Y are real symmetric matrices given by

$$X = (AA^T + BB^T)^{-1}, \tag{11.42}$$
$$Y = (CA^T + DB^T)(AA^T + BB^T)^{-1}. \tag{11.43}$$

Proof. We have $W(\widehat{S}\psi_0^\hbar)(z) = W(\psi_0^\hbar)(S^{-1}z)$ and hence, using formula (11.26),

$$W(\widehat{S}\psi_0^\hbar)(z) = \left(\tfrac{1}{\pi\hbar}\right)^n e^{-\frac{1}{\hbar}(S^{-1})^T S^{-1} z^2}. \tag{11.44}$$

Using formula (2.6) for the inverse of a symplectic matrix we have

$$(S^{-1})^T S^{-1} = \begin{pmatrix} CC^T + DD^T & -DB^T - CA^T \\ -BD^T - AC^T & AA^T + BB^T \end{pmatrix}$$

and hence, comparing with (11.40),

$$X + YX^{-1}Y = CC^T + DD^T \ , \quad YX^{-1} = -DB^T - CA^T,$$
$$X^{-1}Y = -BD^T - AC^T \ , \quad X^{-1} = AA^T + BB^T.$$

Solving this system of matrix equations yields the solutions (11.42) and (11.43). That $\widehat{S}\psi_0^\hbar(x)$ is given by (11.41) follows from the fact that the Wigner transform of

$$\psi_M^\hbar(x) = e^{i\gamma} \left(\tfrac{1}{\pi\hbar}\right)^{n/4} (\det X)^{1/4} e^{-\frac{1}{2\hbar}Mx^2}$$

is given by (11.38). $\qquad\qquad\square$

One can use formula (11.44) to describe the action of $\widehat{S} \in \mathrm{Mp}(2n, \mathbb{R})$ on general squeezed coherent states. However the formulas and calculations are rather lengthy, and in addition it is not immediate that the exact phase factor here is

another method, which has a pleasant geometrical flavor. Let us denote by Σ_n the Siegel half-space, that is

$$\Sigma_n = \{Z : Z = Z^T, \operatorname{Im} M > 0\}$$

where Z denotes a complex $n \times n$ matrix. We have the following interesting result which describes the action of fractional linear transforms on the Siegel half-space:

Proposition 253. *Let $S \in \operatorname{Sp}(2n, \mathbb{R})$ be given by (2.3) and $Z \in \Sigma_n$. Then $\det(A + BZ) \neq 0$, $\det(C + DZ) \neq 0$ and*

$$\alpha(S) = (C + DZ)(A + BZ)^{-1} \in \Sigma_n \tag{11.45}$$

(in particular $\alpha(S)$ is symmetric), and

$$\alpha(SS') = \alpha(S)\alpha(S'). \tag{11.46}$$

The action $\operatorname{Sp}(2n, \mathbb{R}) \times \Sigma_n \longrightarrow \Sigma_n$ thus defined is transitive.

(Proof omitted; see Folland [59] or Littlejohn [117].)

Let $\widehat{S} \in \operatorname{Mp}(2n, \mathbb{R})$ have projection $S = \pi^{\operatorname{Mp}}(\widehat{S}) = \begin{pmatrix} A & B \\ C & D \end{pmatrix}$ on $\operatorname{Sp}(2n, \mathbb{R})$; one then proves that

$$\widehat{S}\psi_0^{\hbar}(x) = \left(\frac{1}{\pi\hbar}\right)^{n/4} \frac{i^{m(\widehat{S})}}{\sqrt{\det(A + iB)}} \exp\left[-\frac{1}{2\hbar}\alpha(S)x^2\right]$$

where the branch cut of the square root of $\det(A + iB)$ is taken to lie just under the positive real axis; $m(\widehat{S})$ is the Maslov index of $\widehat{S}$. It follows from this formula that a squeezed state of the type $\widehat{S}\psi_{z_0}^{\hbar}(x)$ is easily calculated: since by definition $\psi_{z_0}^{\hbar} = \widehat{T}(z_0)\psi_0^{\hbar}$, the metaplectic covariance formula $\widehat{S}\widehat{T}(z_0)\widehat{S}^{-1} = \widehat{T}(Sz_0)$ (see (8.11)) immediately yields

$$\widehat{S}\psi_{z_0}^{\hbar}(x) = \widehat{T}(Sz_0)\widehat{S}\psi_0^{\hbar}(x).$$

The results above can be generalized to arbitrary squeezed coherent states:

Let $\psi_{z_0,M}^{\hbar}$ ($M \in \Sigma_n$), be a squeezed coherent state and $\widehat{S} \in \operatorname{Mp}(2n, \mathbb{R})$, $S = \pi^{\operatorname{Mp}}(\widehat{S})$. We have

$$\widehat{S}\psi_{0,M}^{\hbar} = \psi_{\alpha(S)M}^{\hbar} \quad , \quad \widehat{S}\psi_{z_0,M}^{\hbar} = \widehat{T}(Sz_0)\psi_{\alpha(S)M}^{\hbar}.$$

11.4 Anti-Wick quantization

There are several equivalent ways of defining anti-Wick pseudo-differential operators (also called Toeplitz, or Berezin, operators). In this section we review several possibilities.

11.4.1   Definition in terms of coherent states

We recall that $\psi_0^\hbar$ is the standard coherent state defined by

$$\psi_0^\hbar(x) = \left(\tfrac{1}{\pi\hbar}\right)^{n/4} e^{-\frac{1}{2\hbar}|x|^2}. \tag{11.47}$$

We denote by $\widehat{\Pi}_0^\hbar$ the orthogonal projection operator of $L^2(\mathbb{R}^n)$ onto the ray $\{\lambda\psi_0^\hbar : \lambda \in \mathbb{C}\}$, that is

$$\widehat{\Pi}_0^\hbar\psi = (\psi|\psi_0^\hbar)_{L^2(\mathbb{R}^n)}\psi_0^\hbar. \tag{11.48}$$

The Weyl symbol π_0 of $\widehat{\Pi}_0^\hbar$ is $(2\pi\hbar)^n$ times the Wigner transform of $\psi_0^\hbar$; in view of formula (11.18) we have $W\psi_0^\hbar(z) = (\pi\hbar)^{-n}e^{-|z|^2/\hbar}$ and hence $\pi_0(z) = 2^n e^{-|z|^2/\hbar}$. Consider now the operator $\widehat{\Pi}^\hbar(z_0)$ with Weyl symbol the translated Gaussian

$$\pi_{z_0}(z) = T(z_0)\pi_0(z) = 2^n e^{-\frac{1}{\hbar}|z-z_0|^2}; \tag{11.49}$$

in view of Proposition 219 we have

$$\widehat{\Pi}^\hbar(z_0) = \widehat{T}(z_0)\widehat{\Pi}_0^\hbar\widehat{T}(z_0)^{-1}$$

where $\widehat{T}(z_0)$ is the Heisenberg–Weyl operator. In particular we see that $\widehat{\Pi}^\hbar(z_0)$ is the orthogonal projection in $L^2(\mathbb{R}^n)$ onto the ray $\{\lambda\psi_{z_0}^\hbar : \lambda \in \mathbb{C}\}$ where $\psi_{z_0}^\hbar = \widehat{T}(z_0)\psi_0^\hbar$ (see formula (11.50) below).

The following lemma contains a few useful formulas which we will use to study anti-Wick operators;

Lemma 254. *We have, for every $\psi \in L^2(\mathbb{R}^n)$,*

$$\widehat{\Pi}^\hbar(z_0)\psi = (\psi|\psi_{z_0}^\hbar)_{L^2(\mathbb{R}^n)}\psi_{z_0}^\hbar. \tag{11.50}$$

In particular

$$(\widehat{\Pi}^\hbar(z_0)\psi|\phi)_{L^2(\mathbb{R}^n)} = (\psi|\psi_{z_0}^\hbar)_{L^2(\mathbb{R}^n)}(\psi_{z_0}^\hbar|\phi)_{L^2(\mathbb{R}^n)}. \tag{11.51}$$

We moreover have the identities

$$\widehat{\Pi}^\hbar(z_0)\psi = (2\pi\hbar)^n A(\psi,\psi_0^\hbar)(z_0)\psi_{z_0}^\hbar \tag{11.52}$$

and hence

$$(\widehat{\Pi}^\hbar(z_0)\psi|\phi)_{L^2(\mathbb{R}^n)} = (2\pi\hbar)^{2n} A(\psi,\psi_0^\hbar)(z_0)\overline{A(\psi_0^\hbar,\phi)(z_0)}. \tag{11.53}$$

Proof. Let $\psi \in L^2(\mathbb{R}^n)$; we have

$$\begin{aligned}
\widehat{\Pi}^\hbar(z_0)\psi &= \widehat{T}(z_0)\widehat{\Pi}_0^\hbar(\widehat{T}(z_0)^{-1}\psi) \\
&= (\widehat{T}(z_0)^{-1}\psi|\psi_0^\hbar)_{L^2}\widehat{T}(z_0)\psi_0^\hbar \\
&= (\psi|\widehat{T}(z_0)\psi_0^\hbar)_{L^2}\widehat{T}(z_0)\psi_0^\hbar
\end{aligned}$$

(the last equality because $\widehat{T}(z_0)^{-1} = \widehat{T}(z_0)^*$) hence $\widehat{\Pi^\hbar}(z_0)$ is indeed the orthogonal projection on $\{\lambda\psi_{z_0}^\hbar : \lambda \in \mathbb{C}\}$. Formula (11.51) immediately follows from (11.50). Formula (11.50) can be rewritten

$$\widehat{\Pi^\hbar}(z_0)\psi = (\psi|\widehat{T}(z_0)\psi_0^\hbar)_{L^2(\mathbb{R}^n)}\psi_{z_0}^\hbar$$

and hence, by Definition (9.1) of the cross-ambiguity function,

$$\widehat{\Pi^\hbar}(z_0)\psi = (2\pi\hbar)^n A(\psi, \psi_0^\hbar)(z_0)\psi_{z_0}^\hbar$$

which is precisely formula (11.52). Formula (11.53) follows by a similar argument, rewriting formula (11.51) as

$$(\widehat{\Pi^\hbar}(z_0)\psi|\phi)_{L^2(\mathbb{R}^n)} = (\psi|\widehat{T}(z_0)\psi_0^\hbar)_{L^2}(\widehat{T}(z_0)\psi_0^\hbar|\phi)_{L^2}. \qquad \square$$

These considerations lead to the following definition:

Definition 255. Let $a \in \mathcal{S}(\mathbb{R}^n)$; the anti-Wick operator $\widehat{A}_{\mathrm{aW}}$ with symbol a is defined by

$$\widehat{A}_{\mathrm{aW}}\psi = \int_{\mathbb{R}^{2n}} a(z_0)\widehat{\Pi^\hbar}(z_0)\psi dz_0 \qquad (11.54)$$

or, equivalently, by

$$\widehat{A}_{\mathrm{aW}}\psi = (2\pi\hbar)^n \int_{\mathbb{R}^{2n}} a(z_0)A(\psi, \psi_0^\hbar)(z_0)\psi_{z_0}^\hbar dz_0. \qquad (11.55)$$

We will write $\widehat{A}_{\mathrm{aW}} \overset{\mathrm{AW}}{\longleftarrow} a$ or $a \overset{\mathrm{AW}}{\longleftrightarrow} \widehat{A}_{\mathrm{aW}}$.

The equivalence of Definitions (11.54) and (11.55) of $\widehat{A}_{\mathrm{aW}}$ is immediate taking formula (11.52) in Lemma 254 into account.

We note that if $\psi \in \mathcal{S}(\mathbb{R}^n)$ then the function $z_0 \longmapsto \widehat{\Pi^\hbar}(z_0)\psi$ is in $\mathcal{S}(\mathbb{R}^{2n})$, hence one can expect the definition of $\widehat{A}_{\mathrm{aW}}\psi$ for large classes of symbols. We will see later that Shubin classes are excellent choices, but the following remark is already very useful:

Proposition 256. *The anti-Wick operator $\widehat{A}_{\mathrm{aW}}$ is uniquely defined by the formula*

$$(\widehat{A}_{\mathrm{aW}}\psi|\phi)_{L^2(\mathbb{R}^n)} = (2\pi\hbar)^{2n}(aA(\psi, \psi_0^\hbar)|A(\phi, \psi_0^\hbar))_{L^2(\mathbb{R}^{2n})}. \qquad (11.56)$$

Proof. We have (cf. formula (11.53))

$$(\widehat{A}_{\mathrm{aW}}\psi|\phi)_{L^2} = (2\pi\hbar)^n \int_{\mathbb{R}^{2n}} a(z_0)A(\psi, \psi_0^\hbar)(z_0)(\psi_{z_0}^\hbar|\phi)_{L^2} dz_0$$

$$= (2\pi\hbar)^n \int_{\mathbb{R}^{2n}} a(z_0)A(\psi, \psi_0^\hbar)(z_0)\overline{(\phi|\psi_{z_0}^\hbar)_{L^2}} dz_0$$

$$= (2\pi\hbar)^{2n} \int_{\mathbb{R}^{2n}} a(z_0)A(\psi, \psi_0^\hbar)(z_0)\overline{A(\phi, \psi_0^\hbar)_{L^2}} dz_0,$$

hence (11.56). $\qquad \square$

Immediate – and very pleasant – features of anti-Wick operators are the following self-adjointness and positivity properties:

Proposition 257. *Let a be a symbol defining an anti-Wick operator $\widehat{A}_{\mathrm{aW}}$.*

(i) *If a is real then $\widehat{A}_{\mathrm{aW}}$ is self-adjoint.*

(ii) *If in addition $a \geq 0$ then $\widehat{A}_{\mathrm{aW}} \geq 0$, that is $(\widehat{A}_{\mathrm{aW}}\psi|\psi)_{L^2(\mathbb{R}^n)} \geq 0$ for all $\psi \in L^2(\mathbb{R}^n)$.*

Proof of (i). We have

$$(\widehat{A}_{\mathrm{aW}}\psi|\phi)_{L^2} = \int_{\mathbb{R}^{2n}} a(z_0)(\widehat{\Pi^\hbar}(z_0)\psi|\phi)_{L^2}\,dz_0$$

and, by the sesquilinearity of the L^2 inner product,

$$(\psi|\widehat{A}^*_{\mathrm{aW}}\phi)_{L^2} = \int_{\mathbb{R}^{2n}} \overline{a^*(z_0)}\overline{(\widehat{\Pi^\hbar}(z_0)\phi|\psi)_{L^2}}\,dz_0.$$

(a^* is the symbol of $\widehat{A}^*_{\mathrm{aW}}$). In view of formula (11.51), (11.50) we have

$$\overline{(\widehat{\Pi^\hbar}(z_0)\phi|\psi)_{L^2}} = (\widehat{\Pi^\hbar}(z_0)\psi|\phi)_{L^2}$$

hence $(\widehat{A}_{\mathrm{aW}}\psi|\phi)_{L^2} = (\psi|\widehat{A}^*_{\mathrm{aW}}\phi)_{L^2}$ when a is a real function.

Proof of (ii). We have

$$(\widehat{A}_{\mathrm{aW}}\psi|\psi)_{L^2} = \int_{\mathbb{R}^{2n}} a(z_0)(\widehat{\Pi^\hbar}(z_0)\psi|\psi)_{L^2}\,dz_0;$$

in view of (11.50) we have

$$(\widehat{\Pi^\hbar}(z_0)\psi|\psi)_{L^2} = (\psi|\psi)^2_{L^2} \geq 0$$

hence $(\widehat{A}_{\mathrm{aW}}\psi|\psi)_{L^2} \geq 0$ if $a \geq 0$. $\qquad\square$

11.4.2   The Weyl symbol of an anti-Wick operator

Every anti-Wick operator $\mathcal{S}(\mathbb{R}^n) \longrightarrow \mathcal{S}'(\mathbb{R}^n)$ can be viewed as a Weyl operator. We are going to determine the Weyl symbol of $\widehat{A}_{\mathrm{aW}}$, but let us first recall (Proposition 200) that if $\widehat{A} \overset{\mathrm{Weyl}}{\longleftrightarrow} a$ then we have

$$(\widehat{A}\psi|\phi)_{L^2(\mathbb{R}^n)} = \int_{\mathbb{R}^{2n}} a(z)W(\psi,\phi)(z)\,dz \tag{11.57}$$

for all ψ and ϕ in $\mathcal{S}(\mathbb{R}^n)$.

Proposition 258. *The Weyl symbol a^w of the anti-Wick operator $\widehat{A}_{\mathrm{aW}} \overset{\mathrm{AW}}{\longleftrightarrow} a$ is given by the convolution formula*

$$a^w(z) = \int_{\mathbb{R}^{2n}} e^{-\frac{1}{\hbar}|z-z_0|^2} a(z_0) dz_0. \tag{11.58}$$

Proof. We have, by Definition (11.54),

$$(\widehat{A}_{\mathrm{aW}}\psi|\phi)_{L^2} = \int_{\mathbb{R}^{2n}} a(z_0)(\widehat{\Pi^\hbar}(z_0)\psi|\phi)_{L^2} dz_0$$

that is, taking formulas (11.57) and (11.49) into account:

$$\begin{aligned}
(\widehat{A}_{\mathrm{aW}}\psi|\phi)_{L^2} &= 2^n \int_{\mathbb{R}^{2n}} a(z_0) \left(\int_{\mathbb{R}^{2n}} e^{-\frac{1}{\hbar}|z-z_0|^2} W(\psi,\phi)(z) dz \right) dz_0 \\
&= 2^n \int_{\mathbb{R}^{2n}} \left(\int_{\mathbb{R}^{2n}} a(z_0) e^{-\frac{1}{\hbar}|z-z_0|^2} dz_0 \right) W(\psi,\phi)(z) dz
\end{aligned}$$

hence (11.58), using again (11.57). $\qquad\qquad\square$

We notice that in view of formula (11.58) the Weyl symbol of an anti-Wick operator is real analytic: expanding the exponential in a Taylor series we obtain a power series for $a^w(z)$. This fact shows that not every Weyl operator can be written as an anti-Wick operator. In [28] (Theorem 5.1) Cordero and Nicola have shown that one can build an exact Weyl/anti–Wick correspondence for all Weyl operators whose symbol is real analytic on $\mathbb{R}^{2n}$ and satisfies a certain condition which is expressed in terms of a certain modulation space. In the general case one can prove the following result:

Proposition 259. *Let $\widehat{A} \overset{\mathrm{Weyl}}{\longleftrightarrow} a$ be such that the symbol a satisfies the following conditions: $a \in C^\infty(\mathbb{R}^n \oplus \mathbb{R}^n)$ and for every $\alpha \in \mathbb{N}^{2n}$ there exists a constant $C_\alpha \geq 0$ with*

$$|\partial_z^\alpha a(z)| \leq C_\alpha \langle z \rangle^{m-\rho|\alpha|} \quad \text{for } z \in \mathbb{R}^{2n} \tag{11.59}$$

where $\langle z \rangle = (1 + |z|^2)^{1/2}$. Then there exists an anti-Wick operator $\widehat{B}_{\mathrm{aW}} \overset{\mathrm{AW}}{\longleftrightarrow} b$ where $b \in C^\infty(\mathbb{R}^n \oplus \mathbb{R}^n)$ also satisfies the estimates (11.59) and such that the kernel K of $\widehat{A} - \widehat{B}_{\mathrm{aW}}$ is in $\mathcal{S}(\mathbb{R}^n \times \mathbb{R}^n)$.

We omit the proof of this result here, and refer to Shubin [147] (Theorem 24.2). The conditions (11.59) on the symbol characterize the *Shubin class* $\Gamma_\rho^m(\mathbb{R}^n \oplus \mathbb{R}^n)$ which will be studied in Chapter 14).

11.4.3 Some regularity results

We are going to show that anti-Wick operators are continuous in $\mathcal{S}(\mathbb{R}^n)$ (and hence in $\mathcal{S}'(\mathbb{R}^n)$) if one makes a rather mild assumption of polynomial increase on its symbol (cf. Proposition 259).

Proposition 260. *Assume that a satisfies the conditions (11.59) in Proposition 259. Then $\widehat{A}_{\mathrm{aW}} \overset{\mathrm{AW}}{\longleftrightarrow} a$ is a continuous map $\mathcal{S}(\mathbb{R}^n) \longrightarrow \mathcal{S}(\mathbb{R}^n)$ and hence extends into a continuous map $\mathcal{S}'(\mathbb{R}^n) \longrightarrow \mathcal{S}'(\mathbb{R}^n)$.*

Proof. In view of formula (11.58) for the Weyl symbol a^w of $\widehat{A}_{\mathrm{aW}}$ we have

$$a^w(z) = \int_{\mathbb{R}^{2n}} e^{-\frac{1}{\hbar}|z_0|^2} a(z - z_0)\, dz_0$$

and hence, for every multi-index $\alpha \in \mathbb{N}^{2n}$,

$$z^\alpha \partial_z^\alpha a^w(z) = \int_{\mathbb{R}^{2n}} e^{-\frac{1}{\hbar}|z_0|^2} z^\alpha \partial_z^\alpha a(z - z_0)\, dz_0$$

so that

$$|z^\alpha \partial_z^\alpha a^w(z)| \leq C_\alpha^\hbar \int_{\mathbb{R}^{2n}} e^{-\frac{1}{\hbar}|z_0|^2} \langle z - z_0 \rangle^{m+(1-\rho)|\alpha|}\, dz_0 < \infty$$

for some constant $C_\alpha^\hbar > 0$. It follows that $a^w \in \mathcal{S}(\mathbb{R}^n)$ hence the result in view of Proposition 195. $\qquad\square$

Exercise 261. Prove the result above using directly Definition (11.54) of an anti-Wick operator.

The following operator estimate for the anti-Wick correspondence is very interesting:

Proposition 262. *Let $\widehat{A}_{\mathrm{aW}} \overset{\mathrm{AW}}{\longleftrightarrow} a$ and*

$$\|\widehat{A}_{\mathrm{aW}}\| = \sup_{\psi \in \mathcal{S}(\mathbb{R}^n),\, \psi \neq 0} \frac{\|\widehat{A}_{\mathrm{aW}}\psi\|}{\|\psi\|} \tag{11.60}$$

be the operator norm of $\widehat{A}_{\mathrm{aW}}$. We have

$$\|\widehat{A}_{\mathrm{aW}}\| \leq \sup_{z \in \mathbb{R}^{2n}} |a(z)|. \tag{11.61}$$

Proof. We will prove the estimate (11.61) when a is a real-valued symbol; for the general case we refer to Shubin [147] (Problem 24.4, p. 191). Since a is real the operator $\widehat{A}_{\mathrm{aW}}$ is self-adjoint in view of Proposition 257(i). Let $\lambda \geq 0$; the inequality $\|\widehat{A}_{\mathrm{aW}}\| \leq \lambda$ is equivalent when A is self-adjoint to $\lambda I - \widehat{A}_{\mathrm{aW}} \geq 0$ and $\lambda I + \widehat{A}_{\mathrm{aW}} \geq 0$. These relations obviously hold for $\lambda = \sup_{z \in \mathbb{R}^{2n}} |a(z)|$, hence (11.61). $\qquad\square$

Exercise 263. Apply the proof above to $\mathrm{Re}\, a$ and $\mathrm{Im}\, a$ to show that $\|\widehat{A}_{\mathrm{aW}}\| \leq 2 \sup_{z \in \mathbb{R}^{2n}} |a(z)|$.

Corollary 264. *Let the symbol a satisfy the estimates (11.59) with $m = 0$. Thus $|\partial_z^\alpha a(z)| \leq C_\alpha \langle z \rangle^{-\rho|\alpha|}$ for every multi-index α. The anti-Wick operator $\widehat{A}_{\mathrm{aW}} \xleftrightarrow{\mathrm{AW}} a$ is then bounded on $L^2(\mathbb{R}^n)$.*

Proof. The estimates satisfied by the symbol a imply in particular that it is bounded, and hence

$$\sup_{\psi \in \mathcal{S}(\mathbb{R}^n)} \frac{\|\widehat{A}_{\mathrm{aW}}\psi\|}{\|\psi\|} < \infty.$$

By continuity we also have

$$\sup_{\psi \in L^2(\mathbb{R}^n)} \frac{\|\widehat{A}_{\mathrm{aW}}\psi\|}{\|\psi\|} < \infty$$

hence $\widehat{A}_{\mathrm{aW}}$ is bounded on $L^2(\mathbb{R}^n)$. $\square$

One also proves the following regularity results for anti-Wick operators with symbols in L^q spaces:

Proposition 265. *Let $a \in L^q(\mathbb{R}^n \oplus \mathbb{R}^n)$. Then:*

(i) *The operator $\widehat{A}_{\mathrm{aW}} \xleftrightarrow{\mathrm{AW}} a$ is bounded on $L^2(\mathbb{R}^n)$;*

(ii) *If $1 \leq q \leq 2$ The anti-Wick operator $\widehat{A}_{\mathrm{aW}} : L^2(\mathbb{R}^n) \longrightarrow L^2(\mathbb{R}^n)$ is compact.*

Proof. See Boggiatto and Cordero [15] for a detailed argument. $\square$

Chapter 12

Hilbert–Schmidt and Trace Class Operators

In this chapter we pause to make an excursion to the well-established theory of Hilbert–Schmidt and trace class operators. These are venerable topics from functional analysis; besides their intrinsic interest in mathematics, they are of paramount importance for studying the notion of mixed state in quantum mechanics, as we will see in the next chapter.

We give here a rather succinct treatment of the topic. For details and proofs we refer to Reed and Simon [136], Simon [149], Shubin [147] (Appendix 3), Hörmander [102], §19.1; we are following de Gosson [67] with some additions, modifications and improvements.

12.1 Hilbert–Schmidt operators

In what follows $\mathcal{H}$ is a separable Hilbert space with scalar product $(\cdot|\cdot_j)_{\mathcal{H}}$ and associated norm $\|\cdot\|_{\mathcal{H}}$. We denote by $\mathcal{L}(\mathcal{H})$ the Banach algebra of all bounded operators on $\mathcal{H}$.

We recall the elementary equality

$$\sum_j (u|e_j)_{\mathcal{H}}\overline{(v|e_j)_{\mathcal{H}}} = (u|v), \tag{12.1}$$

valid for all $u, v \in \mathcal{H}$ and all orthonormal bases $(e_j)_j$ of $\mathcal{H}$. When $u = v$ it is called the Bessel equality.

12.1.1 Definition and general properties

Hilbert–Schmidt operators are defined by an integrability condition with respect to an orthonormal basis:

Definition 266. An operator $\widehat{A} \in \mathcal{L}(\mathcal{H})$ is called a Hilbert–Schmidt operator (for short: $\widehat{A}$ is Hilbert–Schmidt) if there exists an orthonormal basis $(e_j)_j$ of $\mathcal{H}$ such that

$$\sum_{j,k} |(\widehat{A}e_j|\widehat{A}e_j)_\mathcal{H} = \sum_j \|\widehat{A}e_j\|_\mathcal{H}^2 < \infty. \tag{12.2}$$

We denote by $\mathcal{L}_2(\mathcal{H})$ the set of all Hilbert–Schmidt operators on $\mathcal{H}$.

If the condition (12.2) holds for one orthonormal basis then it holds for all, and the sum $\sum_j \|\widehat{A}e_j\|_\mathcal{H}^2$ does moreover not depend on the choice of basis. Let us prove this essential property. Let in fact $(f_j)_j$ be an arbitrary orthonormal basis, and write $\widehat{A}e_j = \sum_k (\widehat{A}e_j|f_j)_\mathcal{H} f_k$. Then, using (12.1) with $u = v = \widehat{A}e_j$,

$$\sum_j \|\widehat{A}e_j\|_\mathcal{H}^2 = \sum_{j,k} |(\widehat{A}e_j|f_j)_\mathcal{H}|^2 = \sum_{j,k} |(e_j|\widehat{A}^*f_j)_\mathcal{H}|^2$$

that is, again by (12.1), with this time $u = v = \widehat{A}^*f_j$,

$$\sum_j \|\widehat{A}e_j\|_\mathcal{H}^2 = \sum_{j,k} |(\widehat{A}^*f_j|e_j)_\mathcal{H}|^2 = \sum_k \|\widehat{A}^*f_k\|_\mathcal{H}^2 < \infty.$$

Taking $(f_j)_j = (e_j)_j$ we have $\sum_k \|\widehat{A}^*e_k\|_\mathcal{H}^2 < \infty$ hence the adjoint is also Hilbert–Schmidt; we may thus replace $\widehat{A}$ by $\widehat{A}^*$ in the inequality above, which yields $\sum_k \|\widehat{A}f_k\|_\mathcal{H}^2 < \infty$ as claimed. Notice that we have at the same time proved that $\widehat{A} \in \mathcal{L}(\mathcal{H})$ is Hilbert–Schmidt if and only if $\widehat{A}^*$ is.

Proposition 267. *We have:*

(i) *The set $\mathcal{L}_2(\mathcal{H})$ is a vector subspace of $\mathcal{L}(\mathcal{H})$ and the function $\| \cdot \|_{\mathrm{HS}} \geq 0$ defined by the formula*

$$\|\widehat{A}\|_{\mathrm{HS}}^2 = \sum_i \|\widehat{A}e_i\|_\mathcal{H}^2 \tag{12.3}$$

is a norm on that subspace;

(ii) *$\mathcal{L}_2(\mathcal{H})$ is a two-sided *-ideal in $\mathcal{L}(\mathcal{H})$: if $\widehat{A} \in \mathcal{L}_2(\mathcal{H})$ and $\widehat{B} \in \mathcal{L}(\mathcal{H})$ then $\widehat{A}\widehat{B} \in \mathcal{L}_2(\mathcal{H})$ and $\widehat{B}\widehat{A} \in \mathcal{L}_2(\mathcal{H})$ and we have $\widehat{A}^* \in \mathcal{L}_2(\mathcal{H})$.*

Proof of (i). If $\widehat{A}$ and $\widehat{B}$ are Hilbert–Schmidt operators then $\lambda\widehat{A}$ is trivially a Hilbert–Schmidt operator and $\|\lambda\widehat{A}\|_{\mathrm{HS}} = |\lambda|\|\widehat{A}\|_{\mathrm{HS}}$ for every $\lambda \in \mathbb{C}$; on the other hand

$$\sum_j \|(\widehat{A} + \widehat{B})e_j\|_\mathcal{H}^2 \leq \sum_j \|\widehat{A}e_j\|_\mathcal{H}^2 + \sum_j \|\widehat{B}e_j\|_\mathcal{H}^2 < \infty$$

for every orthonormal basis $(e_j)_j$ hence $\widehat{A} + \widehat{B}$ is also a Hilbert–Schmidt operator and we have

$$\|\widehat{A} + \widehat{B}\|_{\mathrm{HS}}^2 \leq \|\widehat{A}\|_{\mathrm{HS}}^2 + \|\widehat{B}\|_{\mathrm{HS}}^2$$

hence also

$$\|\widehat{A} + \widehat{B}\|_{\mathrm{HS}} \leq \|\widehat{A}\|_{\mathrm{HS}} + \|\widehat{B}\|_{\mathrm{HS}}.$$

Finally, $\|\widehat{A}\|_{\mathrm{HS}} = 0$ is equivalent to $\widehat{A}e_j = 0$ for every index j, that is to $\widehat{A} = 0$.

Proof of (ii). Let us show that $\widehat{B}\widehat{A} \in \mathcal{L}_2(\mathcal{H})$. We have, denoting by $\|\widehat{B}\|$ the operator norm of $\widehat{B}$,

$$\|\widehat{B}\widehat{A}\|_{\mathrm{HS}}^2 = \sum_j \|\widehat{B}\widehat{A}e_j\|_{\mathcal{H}}^2 \leq \|\widehat{B}\| \left(\sum_j \|\widehat{A}e_j\|_{\mathcal{H}}^2 \right) < \infty.$$

Applying the same argument to $\widehat{A}\widehat{B} = (\widehat{B}^*\widehat{A}^*)^*$ shows that $\widehat{A}\widehat{B} \in \mathcal{L}_2(\mathcal{H})$ as well. $\qquad\square$

The norm $\|\cdot\|_{\mathrm{HS}}^2$ is called the "Hilbert–Schmidt norm". The space $\mathcal{L}_2(\mathcal{H})$ is complete for that norm, and hence a Banach space (it is actually even a Hilbert space when equipped with a scalar product that we will define later).

We note the following useful inequality: if $\widehat{A} \in \mathcal{L}_2(\mathcal{H})$ then the operator norm of $\widehat{A}$ satisfies

$$\|\widehat{A}\| = \sup_{\|u\|_{\mathcal{H}} \leq 1} \|\widehat{A}u\|_{\mathcal{H}} \leq \|\widehat{A}\|_{\mathrm{HS}}. \tag{12.4}$$

To prove this inequality, it suffices to note that if $(e_j)_j$ is an orthonormal basis and $u = \sum_j (u|e_j)_{\mathcal{H}} e_j$ then $\widehat{A}u = \sum_j (u|e_j)_{\mathcal{H}} \widehat{A}e_j$ and hence

$$\|\widehat{A}u\|_{\mathcal{H}} \leq \sum_j |(u|e_j)_{\mathcal{H}}| \cdot \|\widehat{A}e_j\|_{\mathcal{H}}.$$

Using the Cauchy–Schwarz inequality for sums we thus have

$$\|\widehat{A}u\|_{\mathcal{H}}^2 \leq \sum_j |(u|e_j)_{\mathcal{H}}|^2 \sum_j \|\widehat{A}e_j\|_{\mathcal{H}}^2 = \|u\|_{\mathcal{H}}^2 \|\widehat{A}\|_{\mathrm{HS}}$$

hence (12.4) taking the supremum for $\|u\|_{\mathcal{H}} \leq 1$.

Simple examples of Hilbert–Schmidt operators are provided by the operators of finite rank; in fact:

Problem 268. Let $\widehat{A}$ be a bounded operator on the Hilbert space $\mathcal{H}$. Prove the following properties: (i) If $\widehat{A}$ is of finite rank, then it is a Hilbert–Schmidt operator; (ii) Every Hilbert–Schmidt operator is the limit of a sequence of compact operators in $\mathcal{L}(\mathcal{H})$; (iii) Hilbert–Schmidt operators are compact operators.

12.1.2   Hilbert–Schmidt operators on $L^2(\mathbb{R}^n)$

The following very important result characterizes all Hilbert–Schmidt operators on $L^2(\mathbb{R}^n)$, and is sometimes taken as their definition. Since by Hilbert–Schmidt these operators are compact (Problem 268 above), that characterization also gives, as a by-product, a sufficient (but of course not necessary) condition for an operator on $L^2(\mathbb{R}^n)$ to be compact.

Theorem 269. *An operator $\widehat{A}$ on $L^2(\mathbb{R}^n)$ is a Hilbert–Schmidt operator if and only it has a kernel $K_{\widehat{A}} \in L^2(\mathbb{R}^n \times \mathbb{R}^n)$, and we have the norm equality*

$$\|\widehat{A}\|_{\mathrm{HS}} = \|K_{\widehat{A}}\|_{L^2(\mathbb{R}^n \times \mathbb{R}^n)}. \tag{12.5}$$

Proof. Let $\widehat{A}$ be a Hilbert–Schmidt operator on $L^2(\mathbb{R}^n)$ and choose an orthonormal basis $(e_i)_i$ in $L^2(\mathbb{R}^n)$. The family $(e_i \otimes e_j)_{i,j}$ of tensor products is an orthonormal basis in $L^2(\mathbb{R}^n \times \mathbb{R}^n)$. Let us now define

$$K(x,y) = \sum_{i,j} (\widehat{A}e_i | e_j)_{L^2} e_j(x) \otimes \overline{e_i(y)}.$$

We have

$$\int_{\mathbb{R}^{2n}} |K(x,y)|^2 dx dy \leq \sum_{i,j} |(\widehat{A}e_i | e_j)_{L^2}|^2 \|e_j \otimes \overline{e_i}\|_{L^2}^2$$

$$= \sum_{i,j} |(\widehat{A}e_i | e_j)_{L^2}|^2,$$

hence $K \in L^2(\mathbb{R}^n \times \mathbb{R}^n)$ using the fact that $\widehat{A}$ is Hilbert–Schmidt. Since $\psi = \sum_i (\psi | e_i)_{L^2} e_i$ and $\widehat{A}e_i = \sum_j (\widehat{A}e_j | e_j)_{L^2} e_j$ we have

$$\widehat{A}\psi(x) = \sum_i (\psi | e_i)_{L^2} \widehat{A}e_i(x)$$

$$= \sum_{i,j} (\psi | e_i)_{L^2} (\widehat{A}e_j | e_j)_{L^2} e_j(x);$$

since on the other hand

$$\int_{\mathbb{R}^n} K(x,y)\psi(y) dy = \sum_{i,j} (\widehat{A}e_i | e_j)_{L^2} (\psi | e_i) e_j(x),$$

by definition of K it follows that we have

$$\widehat{A}\psi(x) = \int_{\mathbb{R}^n} K(x,y)\psi(y) dy \tag{12.6}$$

and hence K is the kernel of the operator $\widehat{A}$. The equality (12.5) now follows from the identity (12.1). Assume conversely that the kernel of $\widehat{A} \in \mathcal{L}(\mathcal{H})$ belongs to $L^2(\mathbb{R}^n \times \mathbb{R}^n)$. We can then find numbers c_{ij} such that $\sum_{i,j} |c_{ij}|^2 < \infty$ and

$$K(x,y) = \sum_{i,j} c_{ij} e_j(x) \otimes \overline{e_i(y)}.$$

Define now the operator $\widehat{A}$ by the equality (12.6); we have

$$\widehat{A}\psi(x) = \sum_{i,j} c_{ij} e_j(x) \int_{\mathbb{R}^n} \overline{e_i(y)} \psi(y) dy$$
$$= \sum_{i,j} c_{ij} (\psi|e_i)_{L^2} e_j(x)$$

and hence, since the basis $(e_i)_i$ is orthonormal,

$$\widehat{A}e_k = \sum_{i,j} c_{ij}(e_k|e_i)_{L^2} e_j = \sum_j c_{kj} e_j$$

so that

$$\|\widehat{A}\|_{\mathrm{HS}}^2 = \sum_k \|\widehat{A}e_k\|_{L^2}^2 = \sum_{j,k} |c_{kj}|^2 < \infty$$

and $\widehat{A}$ is thus Hilbert–Schmidt. $\qquad\square$

12.2 Trace class operators

We begin by recalling some results from elementary functional analysis (see for instance Reed and Simon [136], §6.4). Let $\mathcal{H}$ be a complex Hilbert space, and $\widehat{A} \in \mathcal{L}(\mathcal{H})$ be a positive operator: $(\widehat{A}u|u)_{\mathcal{H}} \geq 0$ for all $u \in \mathcal{H}$. We will write $\widehat{A} \geq 0$. A positive operator on a complex Hilbert space is always self-adjoint: $\widehat{A} = \widehat{A}^*$. There exists a unique $\widehat{B} \in \mathcal{L}(\mathcal{H})$ such that $\widehat{B} \geq 0$ and $\widehat{B}^2 = \widehat{A}$ (in particular $\widehat{B}$ is also self-adjoint). We will write $\widehat{B} = \sqrt{\widehat{A}}$ or $\widehat{B} = \widehat{A}^{1/2}$ and call $\widehat{B}$ the square root of $\widehat{A}$.

12.2.1 The trace of a positive operator

Positive trace class operators have two main advantages: they are easy to study, and they can be used to define trace class operators in the general case. In addition, they are the only ones we will really need when we study the density operator in the next chapter.

Proposition 270. *Let $\widehat{A} \in \mathcal{L}(\mathcal{H})$ be such that $\widehat{A} \geq 0$. Assume that there exists an orthonormal basis $(e_j)_j$ of $\mathcal{H}$ such that $\sum_j (\widehat{A} e_j | e_j)_{\mathcal{H}} < \infty$. Then $\sum_j (\widehat{A} f_j | f_j)_{\mathcal{H}} < \infty$ for every orthonormal basis $(f_j)_j$ of $\mathcal{H}$. More precisely, we have the equality*

$$\sum_j (\widehat{A} e_j | e_j)_{\mathcal{H}} = \sum_j (\widehat{A} f_j | f_j)_{\mathcal{H}}$$

valid for all operators $\widehat{A} \geq 0$, whether of trace class or not.

Proof. Let $(f_j)_j$ be an arbitrary orthonormal basis of $\mathcal{H}$ and set

$$T = \sum_j (\widehat{A} f_j | f_j)_{\mathcal{H}} = \sum_j (\widehat{A}^{1/2} f_j | \widehat{A}^{1/2} f_j)_{\mathcal{H}}.$$

We are going to show that $T = \sum_j (\widehat{A} e_j | e_j)_{\mathcal{H}}$; this will prove the proposition. Taking $u = v = \widehat{A}^{1/2} f_j$ in (12.1) we have

$$(\widehat{A}^{1/2} f_j | \widehat{A}^{1/2} f_j)_{\mathcal{H}} = \sum_k |(\widehat{A}^{1/2} f_j | e_k)_{\mathcal{H}}|^2$$

and hence using the fact that $\widehat{A}^{1/2}$ is symmetric:

$$T = \sum_j \left(\sum_k |(\widehat{A}^{1/2} f_j | e_k)_{\mathcal{H}}|^2 \right) = \sum_k \left(\sum_j |(\widehat{A}^{1/2} e_k | f_j)_{\mathcal{H}}|^2 \right)$$

(interchanging summation signs is allowed because all the terms are positive). Using again (12.1) we have

$$\sum_j |(\widehat{A}^{1/2} e_k | f_j)_{\mathcal{H}}|^2 = (\widehat{A}^{1/2} e_k | \widehat{A}^{1/2} e_k)_{\mathcal{H}} = (\widehat{A} e_k | e_k)_{\mathcal{H}}$$

and hence $T = \sum_k (\widehat{A} e_k | e_k)_{\mathcal{H}}$ which we set out to prove. $\square$

This result motivates the following definition:

Definition 271. The trace of a positive operator $\widehat{A} \in \mathcal{L}(\mathcal{H})$ is

$$\operatorname{Tr} \widehat{A} = \sum_j (\widehat{A} e_j | e_j)_{\mathcal{H}} \tag{12.7}$$

where $(e_j)_j$ is an arbitrary orthonormal basis of $\mathcal{H}$. If $\operatorname{Tr} \widehat{A} < \infty$ one says that $\widehat{A}$ is a trace class operator.

Clearly $\operatorname{Tr} \widehat{A} \geq 0$ (because $(\widehat{A} e_j | e_j)_{\mathcal{H}} \geq 0$ for every j since $\widehat{A} \geq 0$).

Every positive trace class operator is the square of a Hilbert–Schmidt operator:

Proposition 272. *Let $\widehat{A} \in \mathcal{L}(\mathcal{H})$, $\widehat{A} \geq 0$, be of trace class. The square root $\widehat{B} = \sqrt{\widehat{A}}$ is a Hilbert–Schmidt operator on $\mathcal{H}$. Hence every positive trace class operator is the square of a Hilbert–Schmidt operator.*

Proof. Since $\widehat{A} = \widehat{B}^2$ is of trace class we have

$$\sum_j (\widehat{B}e_j|\widehat{B}e_j)_{\mathcal{H}} = \sum_j (\widehat{B}^2 e_j|e_j)_{\mathcal{H}} < \infty$$

for every orthonormal basis $(e_j)_j$ of $\mathcal{H}$, hence $\widehat{B}$ is a Hilbert–Schmidt operator. $\square$

An interesting property of positive trace class operators is their invariance under conjugation with unitary operators:

Proposition 273. *Let $\widehat{A}$ be a positive trace class operator on $\mathcal{H}$ and $\widehat{U}$ a unitary operator on $\mathcal{H}$. Then $\widehat{U}^*\widehat{A}\widehat{U}$ is also a positive trace class operator and we have*

$$\mathrm{Tr}(\widehat{U}^*\widehat{A}\widehat{U}) = \mathrm{Tr}(\widehat{A}). \tag{12.8}$$

Proof. It is clear that $\widehat{U}^*\widehat{A}\widehat{U}$ is a positive operator. The operator $\widehat{A}$ is of trace class if and only if $\sum_j (\widehat{A}e_j|e_j)_{\mathcal{H}} < \infty$ for one (and hence every) orthonormal basis $(e_j)_j$ of $\mathcal{H}$. Since $(\widehat{U}^*\widehat{A}\widehat{U}e_j|e_j)_{\mathcal{H}} = (\widehat{A}\widehat{U}e_j|\widehat{U}e_j)_{\mathcal{H}}$ and $(\widehat{U}e_j)_j$ also is an orthonormal basis of $\mathcal{H}$, it follows that $\widehat{U}^*\widehat{A}\widehat{U}$ is of trace class; formula (12.8) follows from the basis independence of formula (12.7): we have

$$\mathrm{Tr}(\widehat{U}^*\widehat{A}\widehat{U}) = \sum_j (\widehat{U}^*\widehat{A}\widehat{U}e_j|e_j)_{\mathcal{H}} = \sum_j (\widehat{A}\widehat{U}e_j|\widehat{U}e_j)_{\mathcal{H}} = \mathrm{Tr}(\widehat{A})$$

because $(\widehat{U}e_j)_j$ is an orthonormal basis since $\widehat{U}$ is unitary. $\square$

Trace class operators are compact: this will be established in Proposition 280; it is already clear that this is the case for positive trace class operators using Proposition 272 above since Hilbert–Schmidt operators are compact (Problem 268) and the product of two compact operators also is compact. We may thus apply the spectral theory of compact operators to them. Recall the following basic result from functional analysis obtained by amalgamating the Riesz–Schauder and Hilbert–Schmidt theorems:

Theorem 274. *Let $\widehat{A}$ be a compact self-adjoint operator on $\mathcal{H}$. Then*

(i) *The spectrum $\sigma(\widehat{A})$ is discrete and has no limit point except perhaps zero; every non-zero element λ_j of $\sigma(\widehat{A}) = \{\lambda_j : j \in \mathbb{J}\}$ is an eigenvalue with finite multiplicity;*

(ii) *Ordering the λ_j, $j \in \mathbb{J}$ so that $|\lambda_j| \geq |\lambda_{j+1}|$ we have $\lim_{j \to \infty} \lambda_j = 0$ if $\mathbb{J}$ is infinite;*

(iii) *There exists a system of orthonormal eigenvectors $(e_j)_{j \in \mathbb{J}}$ such that $\widehat{A}e_j = \lambda_j e_j$ and the system $(e_j)_{j \in \mathbb{J}}$ is an orthonormal basis of the closure $\overline{\mathrm{Im}\,\widehat{A}} \subset \mathcal{H}$:*

(iv) *For every $u \in \mathcal{H}$ we have*

$$u = v + \sum_{j \in \mathbb{J}} (u|e_j)_{\mathcal{H}} \quad , \quad \widehat{A}u = \sum_{j \in \mathbb{J}} (\widehat{A}u|e_j)_{\mathcal{H}} \tag{12.9}$$

with $v \in \ker \widehat{A}$.

For a proof see, e.g., Dieudonné [35], §11.5; beware of the too concise and therefore somewhat misleading statement in Reed and Simon, Theorem VI.16.

Applying this result to positive trace class operators gives a complete characterization of these operators in terms of orthogonal projections. We begin by remarking that Theorem 274 above implies that, if we denote by $\mathcal{H}_j$ the eigenspace corresponding to the eigenvalue $\lambda_j > 0$, then $\dim \mathcal{H}_j < \infty$ and for $j \neq k$ the spaces $\mathcal{H}_j$ and $\mathcal{H}_k$ are orthogonal. The first formula (12.9) implies that $\mathcal{H}$ splits into the Hilbert sum

$$\mathcal{H} = \mathrm{Ker}\, \widehat{A} \oplus (\mathcal{H}_1 \oplus \mathcal{H}_2 \oplus \cdots). \tag{12.10}$$

Proposition 275. *Let $\widehat{A}$ be a positive self-adjoint operator $\widehat{A}$ of trace class on a Hilbert space $\mathcal{H}$; let $\lambda_1 \geq \lambda_2 \cdots$ be the sequence of eigenvalues of $\widehat{A}$ and $\mathcal{H}_1, \mathcal{H}_2, \ldots$ the corresponding eigenspaces.*

(i) *We have the spectral decomposition formula*

$$\widehat{A} = \sum_j \lambda_j P_j \tag{12.11}$$

where P_j is the orthogonal projection $\mathcal{H} \longrightarrow \mathcal{H}_j$;

(ii) *The trace of $\widehat{A}$ is given by the formula*

$$\mathrm{Tr}(\widehat{A}) = \sum_j \lambda_j \dim \mathcal{H}_j; \tag{12.12}$$

(iii) *Conversely, every operator of the type (12.11) with $\lambda_j > 0$ and P_j being an orthogonal projection operator on a finite-dimensional space is of trace class if we have $\sum_j \lambda_j \dim \mathcal{H}_j < \infty$.*

Proof of (i). Choose an orthonormal basis $(e_{ij})_i$ in each eigenspace $\mathcal{H}_j$ and complete the union $\cup_i (e_{ij})_i$ of these bases into a full orthonormal basis of $\mathcal{H}$ by selecting orthonormal vectors $(f_i)_i$ in $\mathrm{Ker}\, \widehat{A}$ such that $(f_i, e_{jk})_{\mathcal{H}} = 0$ for all j, k. Let u be an arbitrary element of $\mathcal{H}$ and write

$$\widehat{A}u = \sum_i (\widehat{A}u|f_i)_{\mathcal{H}} f_i + \sum_{i,j} (\widehat{A}u|e_{ij})_{\mathcal{H}} e_{ij}.$$

Since $\widehat{A}$ is self-adjoint we have $(\widehat{A}u, f_i)_{\mathcal{H}} = (u, \widehat{A}f_i)_{\mathcal{H}} = 0$ and

$$(\widehat{A}u|e_{ij})_{\mathcal{H}} = (u|\widehat{A}e_{ij})_{\mathcal{H}} = \lambda_j (u|e_{ij})_{\mathcal{H}}.$$

It follows that we have

$$\widehat{A}u = \sum_j \lambda_j \left(\sum_i (u|e_{ij})_{\mathcal{H}} e_{ij} \right).$$

The operator P_j defined by

$$P_j u = \sum_i (u|e_{ij})_{\mathcal{H}} e_{ij}$$

is the orthogonal projection on $\mathcal{H}_j$ hence (12.11).

Proof of (ii). By definition of the trace of a positive operator we have

$$\mathrm{Tr}(\widehat{A}) = \sum_i (\widehat{A}f_i|f_i)_{\mathcal{H}} + \sum_{i,j} (\widehat{A}e_{ij}|e_{ij})_{\mathcal{H}};$$

since $(\widehat{A}f_i, f_i)_{\mathcal{H}} = 0$ and $(\widehat{A}e_{ij}, e_{ij})_{\mathcal{H}} = \lambda_j$ for every index i this reduces to

$$\mathrm{Tr}(\widehat{A}) = \sum_{i,j} \lambda_j (e_{ij}|e_{ij})_{\mathcal{H}} = \sum_j \lambda_j \left(\sum_i (e_{ij}|e_{ij})_{\mathcal{H}} \right) \qquad (12.13)$$

hence (12.12) since the sum indexed by i is equal to the dimension of the eigen-space $\mathcal{H}_j$.

Proof of (iii). Any operator $\widehat{A}$ that can be written in the form (12.11) is self-adjoint because orthogonal projections are self-adjoint operators; moreover the operator $\widehat{A}$ is positive and the condition $\sum_j \lambda_j \dim \mathcal{H}_j < \infty$ is precisely equivalent to $\widehat{A}$ being of trace class in view of the second equality (12.13). $\qquad \square$

12.2.2 General trace class operators

In many texts one defines trace class operators on $\mathcal{H}$ as operators $\widehat{A} \in \mathcal{L}(\mathcal{H})$ such that $|\widehat{A}| = (\widehat{A}^*\widehat{A})^{1/2}$ is of trace class in the sense of the last subsection. In other texts they are defined as products of two Hilbert–Schmidt operators. We give a third definition, which is more in the same spirit as our definition of Hilbert–Schmidt operators. All these definitions are equivalent.

Definition 276. An operator $\widehat{A} \in \mathcal{L}(\mathcal{H})$ is said to be of *trace class* if there exist two orthonormal bases $(e_i)_i$ and $(f_j)_j$ of $\mathcal{H}$ such that

$$\sum_{i,j} |(\widehat{A}e_i|f_j)_{\mathcal{H}}| < \infty. \qquad (12.14)$$

The set of all trace class operators on $\mathcal{H}$ is denoted by $\mathcal{L}_2(\mathcal{H})$.

Obviously $\widehat{A}$ is of trace class if and only if its adjoint $\widehat{A}^*$ is of trace class: we have

$$\sum_{i,j} |(\widehat{A}^* e_i | f_j)_{\mathcal{H}}| = \sum_{i,j} |(e_i | \widehat{A} f_j)_{\mathcal{H}}| = \sum_{i,j} |(\widehat{A} f_j | e_i)_{\mathcal{H}}|$$

and $\sum_{i,j} |(\widehat{A}^* e_i | f_j)_{\mathcal{H}}| < \infty$ if and only if $\sum_{i,j} |(\widehat{A} f_j | e_i)_{\mathcal{H}}| < \infty$.

Definition (12.15) coincides with Definition (12.7) when $\widehat{A} \geq 0$ since in this case $\widehat{A} = |\widehat{A}|$.

We are going to prove that if the condition (12.14) characterizing trace class operators holds for one pair of orthonormal basis, then it holds for all. This property will allow us to prove that $\mathcal{L}_2(\mathcal{H})$ is indeed a vector space, and to define the trace of an element of $\mathcal{L}_2(\mathcal{H})$ by the formula

$$\operatorname{Tr} \widehat{A} = \sum_i (\widehat{A} e_i | e_i)_{\mathcal{H}}. \tag{12.15}$$

Proposition 277. *Suppose that $\widehat{A} \in \mathcal{L}_2(\mathcal{H})$. The following properties hold:*

(i) *We have*

$$\sum_{i,j} |(\widehat{A} e_i | f_j)_{\mathcal{H}}| < \infty \tag{12.16}$$

for all orthonormal bases $(e_i)_i$, $(f_j)_j$ of $\mathcal{H}$ with the same index set;

(ii) *If $(e_i)_i$ and $(f_i)_i$ are two orthonormal bases then*

$$\sum_i (\widehat{A} e_i | e_i)_{\mathcal{H}} = \sum_i (\widehat{A} f_i | f_i)_{\mathcal{H}} \tag{12.17}$$

and both series are absolutely convergent.

(iii) *The set $\mathcal{L}_2(\mathcal{H})$ of all trace class operators is a vector space.*

Proof of (i). Writing Fourier expansions

$$e_i' = \sum_j (e_i' | e_j)_{\mathcal{H}} e_j \ , \ f_\ell' = \sum_k (f_\ell' | f_k)_{\mathcal{H}} f_k$$

we have

$$(\widehat{A} e_i' | f_\ell')_{\mathcal{H}} = \sum_{j,k} (e_i' | e_j)_{\mathcal{H}} \overline{(f_\ell' | f_k)_{\mathcal{H}}} (\widehat{A} e_j | f_k)_{\mathcal{H}} \tag{12.18}$$

and hence, by the triangle inequality,

$$\sum_{i,\ell} |(\widehat{A} e_i' | f_\ell')_{\mathcal{H}}| \leq \sum_{j,k} \left(\sum_{i,\ell} |(e_i' | e_j)_{\mathcal{H}}| \, |(f_\ell' | f_k)_{\mathcal{H}}| \right) |(\widehat{A} e_j | f_k)_{\mathcal{H}}|). \tag{12.19}$$

In view of the trivial inequality $ab \leq \frac{1}{2}(a^2 + b^2)$ we have

$$\sum_{i,\ell} |(e_i'|e_j)_{\mathcal{H}}| \, |(f_\ell'|f_k)_{\mathcal{H}}| \leq \tfrac{1}{2} \sum_i |(e_i'|e_j)_{\mathcal{H}}|^2 + \tfrac{1}{2} \sum_\ell |(f_\ell'|f_k)_{\mathcal{H}}|^2,$$

i.e., since $\sum_i |(e_i'|e_j)_{\mathcal{H}}|^2 = \|e_j\|_{\mathcal{H}}^2 = 1$ and $\sum_\ell |(f_\ell'|f_k)_{\mathcal{H}}|^2 = \|f_k\|_{\mathcal{H}}^2 = 1$,

$$\sum_{i,\ell} |(\widehat{A}e_i'|f_\ell')_{\mathcal{H}}| \leq \sum_{j,k} |(\widehat{A}e_j|f_k)_{\mathcal{H}}|) < \infty$$

which proves (12.16).

Proof of (ii). Assume now that $e_i = f_i$ and $e_i' = f_i'$ for all indices i. In view of (12.18) we have

$$(\widehat{A}e_i'|e_i')_{\mathcal{H}} = \sum_{j,k} (e_i'|e_j)_{\mathcal{H}} \overline{(e_i'|e_k)_{\mathcal{H}}} (\widehat{A}e_j|e_k)_{\mathcal{H}}$$

and hence

$$\sum_i (\widehat{A}e_i'|e_i')_{\mathcal{H}} = \sum_{j,k} \left(\sum_i (e_i'|e_j)_{\mathcal{H}} \overline{(e_i'|e_k)_{\mathcal{H}}} \right) (\widehat{A}e_j|e_k)_{\mathcal{H}}.$$

In view of (12.1)

$$\sum_i (e_i'|e_j)_{\mathcal{H}} \overline{(e_i'|e_k)_{\mathcal{H}}} = (e_j|e_k)_{\mathcal{H}} = \delta_{jk}$$

which establishes (12.17); that the series is absolutely convergent follows from (12.16) with the choice $e_i = f_i$ for all indices i.

Proof of (iii). It is clear that $\lambda \widehat{A} \in \mathcal{L}_2(\mathcal{H})$ if $\lambda \in \mathbb{C}$ and $\widehat{A} \in \mathcal{L}_2(\mathcal{H})$. Let $\widehat{A}, \widehat{B} \in \mathcal{L}_2(\mathcal{H})$; then

$$\sum_i ((\widehat{A} + \widehat{B})e_i|e_i)_{\mathcal{H}} = \sum_i (\widehat{A}e_i|e_i)_{\mathcal{H}} + \sum_i (\widehat{B}e_i|e_i)_{\mathcal{H}}$$

and each sum on the right-hand side is absolutely convergent, implying that $\widehat{A} + \widehat{B} \in \mathcal{L}_2(\mathcal{H})$. $\qquad\square$

Exercise 278. Show that

$$\mathrm{Tr}(\widehat{A}^*) = \overline{\mathrm{Tr}(\widehat{A})} \tag{12.20}$$

(hence $\mathrm{Tr}(\widehat{A})$ is real when $\widehat{A}$ is self-adjoint).

Trace-class operators do not only form a vector space, they also form a normed *algebra*:

Proposition 279. *Let $\widehat{A} \in \mathcal{L}_1(\mathcal{H})$ and $\widehat{B} \in \mathcal{L}(\mathcal{H})$.*

  (i) *We have $\widehat{A}\widehat{B} \in \mathcal{L}_1(\mathcal{H})$ and $\widehat{B}\widehat{A} \in \mathcal{L}_1(\mathcal{H})$, hence $\mathcal{L}_1(\mathcal{H})$ is a two-sided ideal in $\mathcal{L}(\mathcal{H})$; we have $\mathrm{Tr}(\widehat{A}\widehat{B}) = \mathrm{Tr}(\widehat{B}\widehat{A})$.*

  (ii) *The formula $\|\widehat{A}\|_{Tr} = (\mathrm{Tr}(\widehat{A}^*\widehat{A}))^{1/2}$ defines a norm on the algebra $\mathcal{L}_2(\mathcal{H})$ of Hilbert–Schmidt operators on $\mathcal{H}$; that norm is associated to the scalar product $(\widehat{A}|\widehat{B})_{Tr} = \mathrm{Tr}(\widehat{A}^*\widehat{B})$.*

Proof of (i). Let $(e_i)_i$ and $(f_i)_i$ be orthonormal bases of $\mathcal{H}$. Writing $(\widehat{A}\widehat{B}e_i|f_i)_{\mathcal{H}} = (\widehat{B}e_i|\widehat{A}^*f_i)_{\mathcal{H}}$ formula (12.1) with $u = \widehat{B}e_i$ and $v = \widehat{A}^*f_i$ yields

$$(\widehat{A}\widehat{B}e_i|f_i)_{\mathcal{H}} = \sum_j (\widehat{B}e_i|e_j)_{\mathcal{H}}\overline{(\widehat{A}^*f_i|e_j)_{\mathcal{H}}} \tag{12.21}$$

and hence, observing that $|(\widehat{B}e_i|e_j)_{\mathcal{H}}| \leq \|\widehat{B}e_i\|_{\mathcal{H}} \leq \|\widehat{B}\|$,

$$|(\widehat{A}\widehat{B}e_i|f_i)_{\mathcal{H}}| \leq \sum_j |(\widehat{B}e_i|e_j)_{\mathcal{H}}| \cdot |(\widehat{A}^*f_i|e_j)_{\mathcal{H}}| \leq \|\widehat{B}\| \sum_j |(\widehat{A}^*f_i|e_j)_{\mathcal{H}}|.$$

It follows that

$$\sum_i |(\widehat{A}\widehat{B}e_i|e_i)_{\mathcal{H}}| \leq \|\widehat{B}\| \sum_{i,j} |(\widehat{A}^*f_i|e_j)_{\mathcal{H}}| < \infty$$

since $\widehat{A}^*$ is of trace class; it follows that $\widehat{A}\widehat{B} \in \mathcal{L}_1(\mathcal{H})$. The property $\widehat{B}\widehat{A} \in \mathcal{L}_1(\mathcal{H})$ follows by writing $\widehat{B}\widehat{A} = (\widehat{A}^*\widehat{B}^*)^*$. Formula (12.21) implies that

$$\sum_i (\widehat{A}\widehat{B}e_i|e_i)_{\mathcal{H}} = \sum_i (\widehat{B}\widehat{A}e_i|e_i)_{\mathcal{H}}, \tag{12.22}$$

that is $\mathrm{Tr}(\widehat{A}\widehat{B}) = \mathrm{Tr}(\widehat{B}\widehat{A})$.

Proof of (ii). Since $\widehat{A}^*\widehat{A}$ is self-adjoint its trace is real so $(\mathrm{Tr}(\widehat{A}^*\widehat{A}))^{1/2}$ is well defined. If $\|\widehat{A}\|_{\mathrm{Tr}} = 0$ then $(\widehat{A}e_i|e_j)_{\mathcal{H}} = 0$ for all i hence $\widehat{A} = 0$. The relation $\|\lambda\widehat{A}\|_{\mathrm{Tr}} = |\lambda| \|\widehat{A}\|_{\mathrm{Tr}}$ being obvious there only remains to show that the triangle inequality holds. If $\widehat{A}, \widehat{B} \in \mathcal{L}_{\mathrm{Tr}}(\mathcal{H})$ then

$$\begin{aligned}
\|\widehat{A} + \widehat{B}\|_{\mathrm{Tr}}^2 &= \mathrm{Tr}((\widehat{A}^* + \widehat{B}^*)(\widehat{A} + \widehat{B})) \\
&= \mathrm{Tr}(\widehat{A}^*\widehat{A}) + \mathrm{Tr}(\widehat{B}^*\widehat{B}) + \mathrm{Tr}(\widehat{A}^*\widehat{B}) + \mathrm{Tr}(\widehat{B}^*\widehat{A}).
\end{aligned}$$

In view of the formula (12.20) we have

$$\mathrm{Tr}(\widehat{A}^*\widehat{B}) + \mathrm{Tr}(\widehat{B}^*\widehat{A}) = 2\,\mathrm{Re}\,\mathrm{Tr}(\widehat{A}^*\widehat{B})$$

and hence

$$\|\widehat{A} + \widehat{B}\|_{\mathrm{Tr}}^2 = \|\widehat{A}\|_{\mathrm{Tr}}^2 + \|\widehat{B}\|_{\mathrm{Tr}}^2 + 2\,\mathrm{Re}\,\mathrm{Tr}(\widehat{A}^*\widehat{B}). \tag{12.23}$$

We have

$$\mathrm{Tr}(\widehat{A}^*\widehat{B}) = \sum_i (\widehat{B}e_i|\widehat{A}e_i)_{\mathcal{H}}$$

hence, noting that $\mathrm{Re}\,\mathrm{Tr}(\widehat{A}^*\widehat{B}) \le |\mathrm{Tr}(\widehat{A}^*\widehat{B})|$ and using the Cauchy-Schwarz inequality,

$$\mathrm{Re}\,\mathrm{Tr}(\widehat{A}^*\widehat{B}) \le \sum_i (\widehat{B}e_i|\widehat{B}e_i)_{\mathcal{H}}^{1/2}(\widehat{A}e_i|\widehat{A}e_i)_{\mathcal{H}}^{1/2},$$

that is

$$\mathrm{Re}\,\mathrm{Tr}(\widehat{A}^*\widehat{B}) \le \sum_i (\widehat{B}^*\widehat{B}e_i|e_i)_{\mathcal{H}}^{1/2}(\widehat{A}^*\widehat{A}e_i|e_i)_{\mathcal{H}}^{1/2} \le \|\widehat{B}\|_{\mathrm{Tr}}\|\widehat{A}\|_{\mathrm{Tr}}$$

which proves the triangle inequality in view of (12.23). $\square$

An essential feature of trace class operators is that they are compact (an operator $\widehat{A}$ on $\mathcal{H}$ is compact if the image of the unit ball in $\mathcal{H}$ by $\widehat{A}$ is relatively compact). The sum and the product of two compact operators is again a compact operator; in fact compact operators form a two-sided ideal in $\mathcal{L}(\mathcal{H})$.

Proposition 280. *A trace class operator $\widehat{A}$ on a Hilbert space $\mathcal{H}$ is a compact operator.*

Proof. Let (u_j) be a sequence in $\mathcal{H}$ such that $\|u_j\|_{\mathcal{H}} \le 1$ for every j. Let us show that $(\widehat{A}u_j)$ contains a convergent subsequence; this will prove our claim. Let (e_i) be an orthonormal basis of $\mathcal{H}$; writing $u_j = \sum_i (u_j|e_i)_{\mathcal{H}}e_i$ we have

$$\|\widehat{A}u_j\|_{\mathcal{H}}^2 = (\widehat{A}^*\widehat{A}u_j|u_j)_{\mathcal{H}}$$
$$= \sum_{i,k}(u_j|e_i)_{\mathcal{H}}\overline{(u_j|e_k)_{\mathcal{H}}}(\widehat{A}^*\widehat{A}e_k|e_i)_{\mathcal{H}}.$$

Using Cauchy–Schwarz's inequality we have

$$|(u_j|e_i)_{\mathcal{H}}| \le \|u_j\|_{\mathcal{H}}\|e_i\|_{\mathcal{H}} \le 1$$

and hence

$$\|\widehat{A}u_j\|_{\mathcal{H}}^2 \le \sum_{i,k}(\widehat{A}^*\widehat{A}e_k|e_i)_{\mathcal{H}} < \infty;$$

since the operator $\widehat{A}^*\widehat{A}$ is of trace class the sequence $(\widehat{A}u_j)$ is contained in the ball $\widehat{B}(R)$ with $R = \sum_{i,k}(\widehat{A}^*\widehat{A}e_k|e_i)_{\mathcal{H}}$ and thus contains a convergent subsequence as claimed. $\square$

We mentioned at the beginning of this subsection that there are several different definitions of the trace class operators in the literature. In fact:

Proposition 281. *Let $\widehat{A} \in \mathcal{L}(\mathcal{H})$. The following statements are equivalent:*

(i) *$\widehat{A}$ is of trace class;*

(ii) *The modulus $|\widehat{A}| = \sqrt{\widehat{A^*\widehat{A}}}$ is of trace class;*

(iii) *$\widehat{A}$ is the product of two Hilbert–Schmidt operators.*

We do not give the proof of this result here; a proof of the equivalence (i)$\Longleftrightarrow$(ii) is given in Hörmander [102], §19.1. That every trace class operator is the product of two Hilbert–Schmidt operators is easy to see using the polar decomposition of $\widehat{A}$: writing $\widehat{A} = \widehat{U}(\widehat{A}^*\widehat{A})^{1/2}$ we have $\widehat{A} = \widehat{B}^*\widehat{C}$ with $\widehat{B}^* = \widehat{U}(\widehat{A}^*\widehat{A})^{1/4}$ and $\widehat{C} = (\widehat{A}^*\widehat{A})^{1/4}$ and one easily checks that $\widehat{B}^*$ and $\widehat{C}$ are Hilbert–Schmidt operators.

Note the following easy consequence of the result above:

Corollary 282. *Every trace class operator on $L^2(\mathbb{R}^n)$ is the product of two operators with L^2 kernels.*

Proof. It suffices to use (iii) in the proposition above together with Theorem 269. $\square$

12.3   The trace of a Weyl operator

In this section we give several formulas in the case where the involved operators are Weyl operators.

12.3.1   Heuristic discussion

Let $\widehat{A}$ be a trace class operator on $L^2(\mathbb{R}^n)$ with kernel K. It is customary (especially in the physical literature) to calculate the trace of $\widehat{A}$ using the formula

$$\mathrm{Tr}(\widehat{A}) = \int_{\mathbb{R}^n} K(x,x)dx \tag{12.24}$$

which is obviously an extension to the infinite-dimensional case of the usual definition of the trace of a matrix as the sum of its diagonal elements. Needless to say, this formula does not follow directly from the definition of a trace class operator! In fact, even when the integral in (12.24) is absolutely convergent, this formula has no reason to be true in general. The right condition in the case $n = 1$ is the following (Simon [149]): assume that the kernel K is of positive type: this means that

$$\sum_{1 \leq j,k \leq N} \lambda_j \overline{\lambda_k} K(x_j, x_k) \geq 0$$

for all integers N, all $x_j \in \mathbb{R}$ and all $\lambda_j \in \mathbb{C}$ (in particular $K \geq 0$). Then formula (12.24) holds.

On the positive side, Simon [149] notes that if a trace class operator $\widehat{A}$ has kernel K satisfying $\int_{\mathbb{R}^n} |K(x,x)| dx < \infty$ then we are "almost sure" that formula (12.24) holds. Of course this vague statement is not, as Dubin et al. [39] note, a charter allowing carefree calculations!

When $\widehat{A}$ is a Weyl operator $\widehat{A} \overset{\text{Weyl}}{\longleftrightarrow} a$, one then infers from (12.24) that the trace is expressed in terms of the Weyl symbol by the formula

$$\mathrm{Tr}(\widehat{A}) = \left(\tfrac{1}{2\pi\hbar}\right)^n \int_{\mathbb{R}^{2n}} a(z) dz \tag{12.25}$$

(which has no reason to be correct in general!). Heuristically one can argue as follows to justify (12.25). In view of formula (10.14) in Proposition 205 the kernel $K_{\widehat{A}}$ of $\widehat{A}$ is given by

$$K_{\widehat{A}}(x,y) = \left(\tfrac{1}{2\pi\hbar}\right)^n \int_{\mathbb{R}^2} e^{\frac{i}{\hbar} p \cdot (x-y)} a(\tfrac{1}{2}(x+y), p) dp$$

so that $K_{\widehat{A}}(x,x)$ is given by (12.24) and formula (12.25) hence follows.

Another often used formula is the following: assuming that $\widehat{B} \overset{\text{Weyl}}{\longleftrightarrow} b$ also is of trace class, then

$$\mathrm{Tr}(\widehat{A}\widehat{B}) = \left(\tfrac{1}{2\pi\hbar}\right)^n \int_{\mathbb{R}^{2n}} a(z) b(z) dz. \tag{12.26}$$

To justify formula (12.26) one argues as follows: we have

$$\mathrm{Tr}(\widehat{A}\widehat{B}) = \left(\tfrac{1}{2\pi\hbar}\right)^n \int_{\mathbb{R}^{2n}} c(z) dz$$

where $c(z)$ is the Weyl symbol of $\widehat{C} = \widehat{A}\widehat{B}$; in view of formula (10.21) in Theorem 213 we have

$$c(z) = \left(\tfrac{1}{4\pi\hbar}\right)^{2n} \iint_{\mathbb{R}^{4n}} e^{\frac{i}{2\hbar}\sigma(z',z'')} a(z + \tfrac{1}{2}z') b(z - \tfrac{1}{2}z'') dz' dz''.$$

Performing the change of variables $u = z + \tfrac{1}{2}z'$, $v = z - \tfrac{1}{2}z''$ we have $dz' dz'' = 4^{2n} du dv$ and hence

$$c(z) = \left(\tfrac{1}{\pi\hbar}\right)^{2n} \iint_{\mathbb{R}^{4n}} e^{\frac{i}{2\hbar}\sigma(u-z,v-z)} a(u) b(v) du dv$$

$$= \left(\tfrac{1}{\pi\hbar}\right)^{2n} \iint_{\mathbb{R}^{4n}} e^{\frac{2i}{\hbar}\sigma(z,u-v)} \left(e^{\frac{2i}{\hbar}\sigma(u,v)} a(u) b(v) \right) du dv.$$

Integrating $c(z)$ yields

$$\int_{\mathbb{R}^{2n}} c(z) dz = \left(\tfrac{1}{\pi\hbar}\right)^{2n} \iint_{\mathbb{R}^{4n}} \left(\int_{\mathbb{R}^{2n}} e^{\frac{2i}{\hbar}\sigma(z,u-v)} dz \right) e^{\frac{2i}{\hbar}\sigma(u,v)} a(u) b(v) du dv.$$

Now, by the Fourier inversion formula,

$$\int_{\mathbb{R}^{2n}} e^{\frac{2i}{\hbar}\sigma(z,u-v)}dz = \int_{\mathbb{R}^{2n}} e^{\frac{2i}{\hbar}Jz\cdot(u-v)}dz = (2\pi\hbar)^{2n}\delta(2u-2v)$$

and hence

$$\int_{\mathbb{R}^{2n}} c(z)dz = 2^{2n}\iint_{\mathbb{R}^{4n}} \delta(2u-2v)e^{\frac{2i}{\hbar}\sigma(u,v)}a(u)b(v)dudv$$

$$= 2^{2n}\iint_{\mathbb{R}^{4n}} \delta(2u-2v)a(u)b(v)dudv$$

$$= \int_{\mathbb{R}^{4n}} a(u)b(u)du;$$

formula (12.26) follows in view of formula (12.25).

Needless to say, the "derivations" above are formal and one should be extremely cautious when using the "formulas" thus obtained. Shubin [147], §27, discusses a step-by-step procedure for checking such identities, but it is not always easy to use.

Exercise 283. Find the shortcomings in the arguments above, and try to correct as many as possible by imposing conditions on the symbols (and kernels). A good idea is to find out what happens when one assumes that $a \in \mathcal{S}(\mathbb{R}^n \oplus \mathbb{R}^n)$.

12.3.2   Some rigorous results

Here are some rigorous results. Also see the paper by Brislawn [22] for a thorough discussion of the kernels of trace class operators and of the traceability of Hilbert–Schmidt operators.

We begin by giving a rigorous justification of formula (12.26) when the operators $\widehat{A}$ and $\widehat{B}$ are Hilbert–Schmidt:

Proposition 284. *Let* $\widehat{A} \overset{\text{Weyl}}{\longleftrightarrow} a$ *and* $\widehat{B} \overset{\text{Weyl}}{\longleftrightarrow} b$ *be Hilbert–Schmidt operators. We then have*

$$\mathrm{Tr}(\widehat{A}\widehat{B}) = \left(\tfrac{1}{2\pi\hbar}\right)^n \int_{\mathbb{R}^{2n}} a(z)b(z)dz. \tag{12.27}$$

Proof. We first observe that in view of Theorem 269 the kernels $K_{\widehat{A}}$ and $K_{\widehat{B}}$ are square integrable; it then follows from Proposition 209 that we have $a \in L^2(\mathbb{R}^n)$ and $b \in L^2(\mathbb{R}^n)$. Let $(\psi_j)_j$ be an orthonormal basis of $L^2(\mathbb{R}^n)$; we have

$$\mathrm{Tr}(\widehat{A}\widehat{B}) = \sum_{j=1}^{\infty}(\widehat{A}\widehat{B}\psi_j|\psi_j)_{L^2} = \sum_{j=1}^{\infty}(\widehat{B}\psi_j|\widehat{A}^*\psi_j)_{L^2}.$$

Expanding $\widehat{B}\psi_j$ and $\widehat{A}^*\psi_j$ in the basis $(\psi_j)_j$ we have $\widehat{B}\psi_j = \sum_{k=1}^{\infty}(\widehat{B}\psi_j|\psi_k)_{L^2}\psi_k$ and $\widehat{A}^*\psi_j = \sum_{\ell=1}^{\infty}(\psi_j|\widehat{A}\psi_\ell)_{L^2}\psi_\ell$, hence

$$(\widehat{B}\psi_j|\widehat{A}^*\psi_j)_{L^2} = \sum_{k=1}^{\infty}(\widehat{B}\psi_j|\psi_k)_{L^2}\overline{(\widehat{A}\psi_j|\psi_k)_{L^2}}.$$

In view of formula (10.8) in Proposition 200 we have

$$(\widehat{A}\psi_j|\psi_k)_{L^2} = \int_{\mathbb{R}^{2n}} a(z)W(\psi_j,\psi_k)(z)dz = (a|W(\psi_k,\psi_j))_{L^2(\mathbb{R}^{2n})},$$
$$(\widehat{B}\psi_j|\psi_k)_{L^2} = \int_{\mathbb{R}^{2n}} b(z)W(\psi_j,\psi_k)(z)dz = (b|W(\psi_k,\psi_j))_{L^2(\mathbb{R}^{2n})},$$

hence the equality above can be written:

$$(\widehat{B}\psi_j|\widehat{A}^*\psi_j)_{L^2} = \sum_{k=1}^{\infty}(a|W(\psi_k,\psi_j))_{L^2(\mathbb{R}^{2n})}(b|W(\psi_k,\psi_j))_{L^2(\mathbb{R}^{2n})}.$$

Recall now (Proposition 188) that if $(\psi_j)_j$ is an orthonormal basis of $L^2(\mathbb{R}^n)$ then the vectors $\Phi_{j,k} = (2\pi\hbar)^{n/2}W(\psi_k,\psi_j)$ form an orthonormal basis of $L^2(\mathbb{R}^n \oplus \mathbb{R}^n)$; thus

$$\begin{aligned}
\mathrm{Tr}(\widehat{A}\widehat{B}) &= \sum_{j=1}^{\infty}(\widehat{B}\psi_j|\widehat{A}^*\psi_j)_{L^2(\mathbb{R}^n)} \\
&= \left(\tfrac{1}{2\pi\hbar}\right)^n \sum_{1\le j,k<\infty}(a|\Phi_{j,k})_{L^2(\mathbb{R}^{2n})}(b|\Phi_{j,k})_{L^2(\mathbb{R}^{2n})} \\
&= \left(\tfrac{1}{2\pi\hbar}\right)^n (a|\bar{b})_{L^2(\mathbb{R}^{2n})}
\end{aligned}$$

in view of the classical identity (12.1); this proves formula (12.27). $\qquad\square$

Exercise 285. Give an alternative proof of Proposition 284 by justifying the heuristic derivation of formula (12.26) given above when $\widehat{A}$ and $\widehat{B}$ are Hilbert–Schmidt operators. [Hint: show that the symbols a and b are square integrable in view of Theorem 269.]

A very useful criterion is the following (cf. Du and Wong [38], Theorem 2.4.):

Proposition 286. *Let* $\widehat{A} \xleftrightarrow{\text{Weyl}} a$ *be a trace class operator. If* $a \in L^1(\mathbb{R}^n)$ *then*

$$\mathrm{Tr}(\widehat{A}) = \left(\tfrac{1}{2\pi\hbar}\right)^n \int_{\mathbb{R}^{2n}} a(z)dz. \tag{12.28}$$

Equivalently

$$\mathrm{Tr}(\widehat{A}) = a_\sigma(0) \tag{12.29}$$

where $a_\sigma = F_\sigma a$ *is the symplectic Fourier transform of the symbol* a.

Proof. We first observe that the equality $\mathrm{Tr}(\widehat{A}) = a_\sigma(0)$ is obvious since the integral of a is exactly $(2\pi\hbar)^n$ times the symplectic Fourier transform evaluated at 0. Writing $\widehat{A} = \widehat{B}\widehat{C}$ where $\widehat{B}$ and $\widehat{C}$ are Hilbert–Schmidt operators we have, using formula (12.27) in Proposition 284,

$$\mathrm{Tr}(\widehat{A}) = \left(\tfrac{1}{2\pi\hbar}\right)^n \int_{\mathbb{R}^{2n}} b(z)c(z)dz.$$

Let us show that

$$a_\sigma(0) = \left(\tfrac{1}{2\pi\hbar}\right)^n \int_{\mathbb{R}^{2n}} b(z)c(z)dz;$$

formula (12.29) will follow in view of the equality (12.29). We have, in view of formula (10.22) in Theorem 213,

$$a_\sigma(z) = \left(\tfrac{1}{2\pi\hbar}\right)^n \int_{\mathbb{R}^{2n}} e^{\frac{i}{2\hbar}\sigma(z,z')} b_\sigma(z - z')c_\sigma(z')dz'$$

and hence

$$a_\sigma(0) = \left(\tfrac{1}{2\pi\hbar}\right)^n \int_{\mathbb{R}^{2n}} (b_\sigma)^\vee(z)c_\sigma(z)dz$$

$$= \left(\tfrac{1}{2\pi\hbar}\right)^n \left((b_\sigma)^\vee | \overline{c_\sigma}\right)_{L^2(\mathbb{R}^{2n})}$$

with $(b_\sigma)^\vee(z) = b_\sigma(-z)$. Noting that $(b_\sigma)^\vee = (b^\vee)_\sigma$ and $\overline{c_\sigma} = (\overline{c^\vee})_\sigma$ we thus have, since the symplectic Fourier transform is unitary,

$$a_\sigma(0) = \left(\tfrac{1}{2\pi\hbar}\right)^n \left((b^\vee)_\sigma | (\overline{c^\vee})_\sigma\right)_{L^2(\mathbb{R}^{2n})}$$

$$= \left(\tfrac{1}{2\pi\hbar}\right)^n \left(b^\vee | \overline{c^\vee}\right)_{L^2(\mathbb{R}^{2n})}$$

$$= \left(\tfrac{1}{2\pi\hbar}\right)^n \int_{\mathbb{R}^{2n}} b(z)c(z)dz,$$

which was to be proven. $\square$

Notice that in the proof above the assumption that $\widehat{A}$ is a trace class operator is essential; in [38] it is moreover remarked that if $a \in L^1(\mathbb{R}^n)$ but $a \notin L^2(\mathbb{R}^n)$ then $\widehat{A}$ is not a trace class operator.

If one imposes more stringent conditions on the Weyl symbol, one can in addition obtain sufficient conditions for the operator to be of trace class:

Proposition 287. *Let* $\widehat{A} \overset{\text{Weyl}}{\longleftrightarrow} a$ *be a Weyl operator and assume that the symbol a satisfies the following conditions: there exist $m \in \mathbb{R}$ and $\rho \in \mathbb{R}$, $0 < \rho \leq 1$, such that for every multi-index $\alpha \in \mathbb{N}^{2n}$ we can find a constant C_α such that*

$$|\partial_z^\alpha a(z)| \leq C_\alpha(1 + |z|)^{m-\rho|\alpha|}. \tag{12.30}$$

If $m < -2n$ then $\widehat{A}$ is of trace class and we have

$$\mathrm{Tr}\,\widehat{A} = \left(\tfrac{1}{2\pi\hbar}\right)^n \int_{\mathbb{R}^{2n}} a(z)dz. \tag{12.31}$$

Proof. For the proof that the conditions on a imply that $\widehat{A}$ is of trace class see Shubin [147], §27. The condition $m < -2n$ implies that $a \in L^1(\mathbb{R}^n)$ hence it suffices to apply Proposition 286. $\qquad\square$

We remark that the symbols $a \in C^\infty(\mathbb{R}^n \oplus \mathbb{R}^n)$ satisfying the conditions (12.30) form a vector space denoted by $\Gamma^m_\rho(\mathbb{R}^n \oplus \mathbb{R}^n)$; it is one of the Shubin classes of symbols we will study in Chapter 14; these classes play an important role in the study of global properties of pseudo-differential operators.

We finally note the following invariance property of the trace of a Weyl operator under conjugation with metaplectic operators:

Corollary 288. *Assume that* $\widehat{A} \stackrel{\mathrm{Weyl}}{\longleftrightarrow} a$ *of trace class on* $L^2(\mathbb{R}^n)$ *and* $\widehat{S} \in \mathrm{Mp}(2n,\mathbb{R})$*; then* $\widehat{S}\widehat{A}\widehat{S}^{-1} \stackrel{\mathrm{Weyl}}{\longleftrightarrow} a \circ S^{-1}$ *($S = \pi^{\mathrm{Mp}}(\widehat{S})$) is also of trace class and has same trace as* $\widehat{A}$*.*

Proof. It immediately follows from Proposition 273 since metaplectic operators are unitary; that we have $\widehat{S}\widehat{A}\widehat{S}^{-1} \stackrel{\mathrm{Weyl}}{\longleftrightarrow} a \circ S^{-1}$ was proven in Theorem 215. $\qquad\square$

Chapter 13

Density Operator and Quantum States

At first sight the notion of density operator (or density matrix, as it is called in physics) should not lead to any particular difficulty: mathematically, a density operator is just a positive trace class operator with trace equal to one. It turns out that, perhaps somewhat unexpectedly, it is the positivity property which is the most delicate to establish. It turns out that the positivity of a self-adjoint trace class operator is very sensitive to the choice of the value of "Planck's constant" $\hbar$: thus a self-adjoint operator with trace 1 might very well be a positive operator for some values of $\hbar$ and non-positive for other values. We study in this chapter a fundamental tool defined and developed by Narcowich [126, 127, 128] and Narcowich and O'Connell [129] based on earlier work of Kastler [105] and Loupias and Miracle-Sole [118, 119], namely the Narcowich–Wigner spectrum of a self-adjoint trace class operator. Roughly speaking, this set consists of the values of the parameter $\hbar$ for which the operator in question is positive, and hence a "density operator" representing a mixed quantum state.

We mention that much of this chapter can be recast in the language of the theory of C^*-algebras; due to lack of space we do not address this fruitful point of view here and refer to the aforementioned papers of Kastler, Loupias, Miracle-Sole and to the references therein. For a brief discussion (at an elementary level) of the usefulness of the language of C^*-algebras in the study of quantum mechanical states we refer to §11.11 in Hannabuss' book [91].

13.1 The density operator

Density operators (also called "density matrices" in physics) are central objects in quantum mechanics, because they are identified with the "mixed states" of a quantum system. They contain, as a particular case, the "pure states" which are usually described by the wave function.

13.1.1 Pure and mixed quantum states

Let us begin by defining rigorously the notion of density operator in terms of trace class operators.

Definition 289. A density operator (or density matrix) on a separable Hilbert space $\mathcal{H}$ is a bounded operator $\widehat{\rho} : \mathcal{H} \longrightarrow \mathcal{H}$ having the following properties:

  (i) $\widehat{\rho}$ is self-adjoint and semi-definite positive: $\widehat{\rho} = \widehat{\rho}^*$, $\widehat{\rho} \geq 0$;
  (ii) $\widehat{\rho}$ is of trace class and $\mathrm{Tr}(\widehat{\rho}) = 1$.

In quantum mechanics the Hilbert space $\mathcal{H}$ is usually realized as a space of square-integrable functions.

Here is a first example of a density operator. Let us assume that we are in presence of a well-defined quantum state, represented by an element $\psi \neq 0$ of $\mathcal{H}$. Such a state is called a *pure state* in quantum mechanics. It is no restriction to assume that ψ is normalized, that is $\|\psi\|_{\mathcal{H}} = 1$, so that the mathematical expectation of $\widehat{A}$ in the state ψ is

$$\langle \widehat{A} \rangle_\psi = (\widehat{A}\psi | \psi)_{\mathcal{H}}. \tag{13.1}$$

Consider now the projection operator

$$\widehat{\rho}_\psi : \mathcal{H} \longrightarrow \{\alpha\psi : \alpha \in \mathbb{C}\} \tag{13.2}$$

of $\mathcal{H}$ on the "ray" generated by ψ. For each $\phi \in \mathcal{H}$ we have

$$\widehat{\rho}_\psi \phi = \alpha\psi \ , \ \alpha = (\phi | \psi)_{\mathcal{H}}. \tag{13.3}$$

We will call $\widehat{\rho}_\psi$ the *pure density operator* associated with ψ; it is a trace class operator with trace equal to 1.

Exercise 290. Check this last statement in detail.

Observe that when $\mathcal{H} = L^2(\mathbb{R}^n)$ formula (13.3) can be written

$$\widehat{\rho}_\psi \phi(x) = \int_{\mathbb{R}^n} \psi(x)\overline{\psi(y)}\phi(y)dy,$$

hence the kernel of $\widehat{\rho}_\psi$ is just the tensor product

$$K_{\widehat{\rho}_\psi} = \psi \otimes \overline{\psi}. \tag{13.4}$$

We are going to see that the pure density operator $\widehat{\rho}_\psi$ is in this case a Weyl operator whose symbol is (up to a factor) just the Wigner transform of ψ (cf. Corollary 207). Let us restate this property in our new language and notation:

Proposition 291. *Let $\widehat{\rho}_\psi$ be the density operator associated to a pure state ψ by (13.3).*

  (i) *The Weyl symbol ρ_ψ of $\widehat{\rho}_\psi$ and the Wigner transform $W\psi$ of ψ are related by the formula*

$$\rho_\psi(z) = (2\pi\hbar)^n W\psi(z). \tag{13.5}$$

  (ii) *Let $\widehat{A} \overset{\mathrm{Weyl}}{\longleftrightarrow} a$. If $\|\psi\|_{L^2(\mathbb{R}^n)} = 1$ then the expectation value $\langle\widehat{A}\rangle_\psi = (\widehat{A}\psi|\psi)_{L^2(\mathbb{R}^n)}$ of $\widehat{A}$ in the state ψ is given by:*

$$\langle\widehat{A}\rangle_\psi = \left(\tfrac{1}{2\pi\hbar}\right)^n \mathrm{Tr}(\widehat{\rho}_\psi \widehat{A}). \tag{13.6}$$

Proof of (i). In view of formula (10.15) in Proposition 205 the Weyl symbol a_ψ of $\widehat{\rho}_\psi$ is given by

$$\rho_\psi(x,p) = \int_{\mathbb{R}^n} e^{-\frac{i}{\hbar}p\cdot y} K_{\widehat{\rho}_\psi}(x + \tfrac{1}{2}y, x - \tfrac{1}{2}y)dy$$
$$= \int_{\mathbb{R}^n} e^{-\frac{i}{\hbar}p\cdot y} \psi(x + \tfrac{1}{2}y)\overline{\psi(x + \tfrac{1}{2}y)}d^n y$$

that is $\rho_\psi(z) = (2\pi\hbar)^n W\psi(z)$ as claimed.

Proof of (ii). In view of formula (10.8) in Proposition 200 we have

$$\langle\widehat{A}\rangle_\psi = \int_{\mathbb{R}^{2n}} a(z)W\psi(z)dz$$

and formula (13.6) follows from (i) using the expression (12.27) in Proposition 284 giving the trace of the composition of two Weyl operators. $\qquad\square$

We have so far been assuming that the quantum system under consideration was in a well-known state characterized by a function ψ. Suppose for instance that we have the choice between a finite or infinite number of states, described by functions $\psi_1, \psi_2, \ldots$, each ψ_j having a probability α_j to be the "true" description. We can then form a weighted "mixture" of the ψ_j by forming the convex sum

$$\psi = \sum_{j=1}^{\infty} \alpha_j \psi_j \quad,\quad \sum_{j=1}^{\infty} \alpha_j = 1 \quad,\quad \alpha_j \geq 0. \tag{13.7}$$

We will say that ψ is a *mixed state*.

Definition 292. The density operator of the mixed state (13.7) is the self-adjoint operator

$$\widehat{\rho} = \sum_{j=1}^{\infty} \alpha_j \widehat{\rho}_{\psi_j} \tag{13.8}$$

where the real numbers α_j satisfy the conditions (13.7) above.

It is clear that $\widehat{\rho}$ is a density operator in the sense of Definition 289: since trace class operators form a vector space, $\widehat{\rho}$ is indeed of trace class and its trace is 1 since $\mathrm{Tr}(\widehat{\rho}_j) = 1$ and the α_j sum to 1. That $\widehat{\rho} = \widehat{\rho}^*$ is obvious, and the positivity of $\widehat{\rho}$ follows from the fact that $\alpha_j \geq 0$ for each j. We will see below (Corollary 294) that any density operator on $L^2(\mathbb{R}^n)$ can actually be written in the form (13.8).

The following result describes *all* density matrices in a Hilbert space $\mathcal{H}$:

Proposition 293. *An operator $\widehat{\rho}$ on a Hilbert space $\mathcal{H}$ is a density operator if and only if there exists a (finite or infinite) sequence (α_j) of positive numbers and finite-dimensional pairwise orthogonal subspaces $\mathcal{H}_j$ of $\mathcal{H}$ such that*

$$\widehat{\rho} = \sum_j \alpha_j \widehat{\rho}_j \quad and \quad \sum_j \alpha_j \dim \mathcal{H}_j = 1 \tag{13.9}$$

where $\widehat{\rho}_j$ is the orthogonal projection $\mathcal{H} \longrightarrow \mathcal{H}_j$.

Proof. The statement is just Proposition 275, since the orthogonal projections $\widehat{\rho}_j$ are rank-one self-adjoint operators. Since the spaces $\mathcal{H}_j$ are pairwise orthogonal we have $\widehat{\rho}_j\widehat{\rho}_k = 0$ if $j \neq k$ and hence $\widehat{\rho}^2 = \sum_j \alpha_j^2 \widehat{\rho}_j$. $\square$

Specializing to the case where $\mathcal{H} = L^2(\mathbb{R}^n)$ we get:

Corollary 294. *An operator $\widehat{\rho} : L^2(\mathbb{R}^n) \longrightarrow L^2(\mathbb{R}^n)$ is a density operator if and only if there exists a family $(\psi_j)_{j\in\mathbb{J}}$ in $L^2(\mathbb{R}^n)$, a sequence $(\lambda_j)_{j\in\mathbb{J}}$ of numbers $\lambda_j \geq 0$ with $\sum_{j\in\mathbb{J}} \lambda_j = 1$ such that the Weyl symbol ρ of $\widehat{\rho}$ is given by*

$$\rho = \sum_{j\in\mathbb{J}} \lambda_j W\psi_j. \tag{13.10}$$

Proof. Assume that the Weyl symbol of $\widehat{\rho}$ is given by (13.10); in view of Proposition 293 and the discussion preceding it we have

$$\widehat{\rho} = \sum_{j\in\mathbb{J}} \alpha_j \widehat{\rho}_j$$

where $\widehat{\rho}_j$ is the orthogonal projection on the ray $\{\alpha\psi_j : \alpha \in \mathbb{C}\}$. It follows that $\widehat{\rho}$ is a density operator. If conversely $\widehat{\rho}$ is a density operator on $L^2(\mathbb{R}^n)$, then there exist pairwise orthogonal finite-dimensional subspaces $\mathcal{H}_1, \mathcal{H}_2,\ldots$ of $L^2(\mathbb{R}^n)$ such that

$$\widehat{\rho} = \sum_j \alpha_j \widehat{\rho}_{\mathcal{H}_j} \quad with \quad \sum_j m_j \alpha_j = 1$$

with $\widehat{\rho}_j$ the orthogonal projection on $\mathcal{H}_j$ and $m_j = \dim \mathcal{H}_j$. Choose now an orthonormal basis $\psi_1,\ldots,\psi_{m_1}$ of $\mathcal{H}_1$, an orthonormal basis $\psi_{m_1+1},\ldots,\psi_{m_1+m_2+1}$ of $\mathcal{H}_2$, and so on. The Weyl symbol of $\widehat{\rho}$ is

$$\rho = \alpha_1 \sum_{j=1}^{m_1} W\psi_j + \alpha_2 \sum_{j=m_1}^{m_1+m_2+1} W\psi_j + \cdots$$

which is (13.10), setting $\lambda_j = m_j\alpha_j$. $\square$

Let us define the notion of *purity* of a quantum state, which is a measure of how much a quantum state differs from a pure state:

Definition 295. Let $\widehat{\rho}$ be a density operator on $\mathcal{H}$; the number $\mu(\widehat{\rho}) = \mathrm{Tr}(\widehat{\rho}^2)$ is called the "purity of the quantum state" that $\widehat{\rho}$ represents.

The following result justifies the definition above:

Proposition 296. *Let $\widehat{\rho}$ be a density operator on a Hilbert space $\mathcal{H}$. We have*

$$0 \le \mathrm{Tr}(\widehat{\rho}^2) \le \mathrm{Tr}(\widehat{\rho}) \le 1 \tag{13.11}$$

and $\mathrm{Tr}(\widehat{\rho}^2) = 1$ *if and only if $\widehat{\rho}$ is a pure-state density operator.*

Proof. The condition $\sum_j \alpha_j \dim \mathcal{H}_j = 1$ implies that we must have $\alpha_j \le 1$ for each j so that

$$\mathrm{Tr}\,\widehat{\rho}^2 = \sum_j \alpha_j^2 \dim \mathcal{H}_j \le \sum_j \alpha_j \dim \mathcal{H}_j = 1.$$

If $\widehat{\rho}$ is a pure-state density operator then it is a projection of rank 1, hence $\widehat{\rho}^2 = \widehat{\rho}$ and $\mathrm{Tr}(\widehat{\rho}^2) = 1$. Suppose conversely that $\mathrm{Tr}(\widehat{\rho}^2) = 1$, that is

$$\sum_j \alpha_j^2 \dim \mathcal{H}_j = \sum_j \alpha_j \dim \mathcal{H}_j = 1.$$

Since $\dim \mathcal{H}_j > 0$ for every j this equality is only possible if the numbers α_j are either 0 or 1; since the case $\alpha_j = 0$ is excluded it follows that the sum $\sum_j \alpha_j \dim \mathcal{H}_j = 1$ reduces to one single term, say $\alpha_{j_0} \dim \mathcal{H}_{j_0} = 1$ so that $\widehat{\rho} = \alpha_{j_0} P_{j_0}$ and $\widehat{\rho}^2 = \alpha_{j_0}^2 \widehat{\rho}_{j_0}$. The equality $\mathrm{Tr}(\widehat{\rho}) = \mathrm{Tr}(\widehat{\rho}^2) = 1$ can hold if and only if $\alpha_{j_0} = 1$ hence $\dim \mathcal{H}_{j_0} = 1$ and $\widehat{\rho}$ is a projection of rank 1, and hence a pure-state density operator. $\qquad\square$

Recalling (formula (13.4)) that the operator kernel of the density operator of a pure state ψ is just the tensor product $\psi \otimes \overline{\psi}$ we have more generally:

Corollary 297. *Let the density operator $\widehat{\rho}$ be given by formula (13.9) and let $(\psi_{jk})_{j,k}$ be a double-indexed family of orthonormal vectors in $L^2(\mathbb{R}^n)$ such that the subfamily $(\psi_{jk})_k$ is a basis of $\mathcal{H}_j$ for each j.*

(i) The kernel $K_{\widehat{\rho}}$ of $\widehat{\rho}$ is given by

$$K_{\widehat{\rho}}(x,y) = \sum_{j,k} \lambda_j \psi_{jk}(x) \otimes \overline{\psi_{jk}}(y); \tag{13.12}$$

(ii) The Weyl symbol of $\widehat{\rho}$ is given by

$$a(z) = \sum_{j,k} \lambda_j W\psi_{jk}(z) \tag{13.13}$$

($W\psi_{jk}$ the Wigner transform of ψ_{jk}).

Proof. (i) We have

$$\widehat{\rho}\psi = \sum_{j} \lambda_j \widehat{\rho}\psi = \sum_{j,k} \lambda_j (\psi|\psi_{jk})_{\mathcal{H}} \psi_{jk}$$

that is, by definition of the scalar product:

$$\widehat{\rho}\psi = \sum_{j} \lambda_j \widehat{\rho}\psi = \sum_{j,k} \lambda_j \int_{\mathbb{R}^n} \psi_{jk}(x)\psi(y)\overline{\psi_{jk}}(y)dy$$

which is (13.12).

(ii) Formula (13.13) for the symbol immediately follows from (13.12) in view of Proposition 293. $\qquad\square$

13.2   The uncertainty principle revisited

Let us see what the strong uncertainty principle in the Robertson–Schrödinger form becomes in the case of mixed states.

13.2.1   The strong uncertainty principle for the density operator

Recall that the Robertson–Schrödinger inequalities (6.10) are

$$(\Delta X_\alpha)^2 (\Delta P_\alpha)^2 \geq \Delta(X_\alpha, P_\alpha)^2 + \tfrac{1}{4}\hbar^2. \tag{13.14}$$

Let $\widehat{A}$ and $\widehat{B}$ be Weyl operators; we assume that the expectation values

$$\langle \widehat{A} \rangle_{\widehat{\rho}} = \mathrm{Tr}(\widehat{\rho}\widehat{A}) \quad , \quad \langle \widehat{A}^2 \rangle_{\widehat{\rho}} = \mathrm{Tr}(\widehat{\rho}\widehat{A}^2) \tag{13.15}$$

(and similar expressions for $\widehat{B}$) exist and are finite. Setting

$$(\Delta\widehat{A})^2_{\widehat{\rho}} = \langle \widehat{A}^2 \rangle_{\widehat{\rho}} - \langle \widehat{A} \rangle^2_{\widehat{\rho}} \quad , \quad (\Delta\widehat{B})^2_{\widehat{\rho}} = \langle \widehat{B}^2 \rangle_{\widehat{\rho}} - \langle \widehat{B} \rangle^2_{\widehat{\rho}},$$

$$\Delta(\widehat{A}, \widehat{B})_{\widehat{\rho}} = \frac{1}{2}\langle \widehat{A}\widehat{B} + \widehat{B}\widehat{A} \rangle_{\widehat{\rho}} - \langle \widehat{A} \rangle_{\widehat{\rho}}\langle \widehat{B} \rangle_{\widehat{\rho}}$$

we have the following result:

Proposition 298. *Let* $\widehat{A} \overset{\text{Weyl}}{\longleftrightarrow} a$ *and* $\widehat{B} \overset{\text{Weyl}}{\longleftrightarrow} b$ *be two essentially self-adjoint Weyl operators on* $L^2(\mathbb{R}^n)$ *for which the expectation values* (13.15) *are defined. We have*

$$|\langle \widehat{A}\widehat{B} \rangle_{\widehat{\rho}}|^2 = \Delta(\widehat{A}, \widehat{B})^2_{\widehat{\rho}} - \tfrac{1}{4}\langle [\widehat{A}, \widehat{B}] \rangle^2_{\widehat{\rho}} \tag{13.16}$$

where $[\widehat{A}, \widehat{B}] = \widehat{A}\widehat{B} - \widehat{B}\widehat{A}$ *and hence*

$$(\Delta\widehat{A})^2_{\widehat{\rho}}(\Delta\widehat{B})^2_{\widehat{\rho}} \geq \Delta(\widehat{A}, \widehat{B})^2_{\widehat{\rho}} - \tfrac{1}{4}\langle [\widehat{A}, \widehat{B}] \rangle^2_{\widehat{\rho}} . \tag{13.17}$$

Proof. Replacing $\widehat{A}$ and $\widehat{B}$ by $\widehat{A} - \langle \widehat{A} \rangle_{\widehat{\rho}}$ and $\widehat{B} - \langle \widehat{B} \rangle_{\widehat{\rho}}$ we may assume that $\langle \widehat{A} \rangle_{\widehat{\rho}} = \langle \widehat{B} \rangle_{\widehat{\rho}} = 0$ so that (13.16) and (13.17) reduce to, respectively,

$$|\langle \widehat{A}\widehat{B} \rangle_{\widehat{\rho}}|^2 = \tfrac{1}{2}\langle \widehat{A}\widehat{B} + \widehat{B}\widehat{A} \rangle_{\widehat{\rho}}^2 - \tfrac{1}{4}\langle [\widehat{A}, \widehat{B}] \rangle_{\widehat{\rho}}^2 \tag{13.18}$$

and

$$\langle \widehat{A}^2 \rangle_{\widehat{\rho}} \langle \widehat{B}^2 \rangle_{\widehat{\rho}} \geq \tfrac{1}{2}\langle \widehat{A}\widehat{B} + \widehat{B}\widehat{A} \rangle_{\widehat{\rho}}^2 - \tfrac{1}{4}\langle [\widehat{A}, \widehat{B}] \rangle_{\widehat{\rho}}^2. \tag{13.19}$$

Writing $\widehat{A}\widehat{B} = \tfrac{1}{2}(\widehat{A}\widehat{B} + \widehat{B}\widehat{A}) + \tfrac{1}{2}(\widehat{A}\widehat{B} - \widehat{B}\widehat{A})$ we have,

$$\langle \widehat{A}\widehat{B} \rangle_{\widehat{\rho}} = \tfrac{1}{2}\langle \widehat{A}\widehat{B} + \widehat{B}\widehat{A} \rangle_{\widehat{\rho}} + \tfrac{1}{2}\langle \widehat{A}\widehat{B} - \widehat{B}\widehat{A} \rangle_{\widehat{\rho}}.$$

Now, $\Delta(\widehat{A}, \widehat{B})_{\widehat{\rho}}$ is a real number, and $\langle [\widehat{A}, \widehat{B}] \rangle_{\widehat{\rho}}$ is pure imaginary (because $[\widehat{A}, \widehat{B}]^* = -[\widehat{A}, \widehat{B}]$ since $\widehat{A}$ and $\widehat{B}$ are essentially self-adjoint), hence formula (13.18). We next observe that

$$\langle \widehat{A}\widehat{B} \rangle_{\widehat{\rho}} = \sum_{j \in \mathcal{J}} \alpha_j (\widehat{A}\widehat{B}\psi_j | \psi_j)_{L^2} = \sum_{j \in \mathcal{J}} \alpha_j (\widehat{B}\psi_j | \widehat{A}\psi_j)_{L^2}; \tag{13.20}$$

applying the Cauchy–Schwarz inequality to each scalar product $(\widehat{B}\psi_j | \widehat{A}\psi_j)_{L^2}$ we get

$$|\langle \widehat{A}\widehat{B} \rangle_{\widehat{\rho}}|^2 \leq \sum_{j \in \mathcal{J}} \alpha_j \|\widehat{B}\psi_j\|_{L^2} \|\widehat{A}\psi_j\|_{L^2}. \tag{13.21}$$

Since $\langle \widehat{A} \rangle_{\widehat{\rho}}^2 = \langle \widehat{B} \rangle_{\widehat{\rho}}^2 = 0$ we have

$$\|\widehat{A}\psi_j\| = \langle \widehat{A}^2 \rangle_{\psi_j}^{1/2} = (\Delta\widehat{A})_{\psi_j}^2 \quad , \quad \|\widehat{B}\psi_j\| = \langle \widehat{B}^2 \rangle_{\psi_j}^{1/2} = (\Delta\widehat{B})_{\psi_j}^2$$

and the inequality (13.21) is thus equivalent to

$$|\langle \widehat{A}\widehat{B} \rangle_{\widehat{\rho}}| \leq \sum_{j \in \mathcal{J}} \alpha_j \langle \widehat{A}^2 \rangle_{\psi_j}^{1/2} \langle \widehat{B}^2 \rangle_{\psi_j}^{1/2}.$$

Writing $\alpha_j = (\sqrt{\alpha_j})^2$ the Cauchy–Schwarz inequality for sums yields

$$|\langle \widehat{A}\widehat{B} \rangle_{\widehat{\rho}}| \leq \left(\sum_{j \in \mathcal{J}} \alpha_j \langle \widehat{A}^2 \rangle_{\psi_j}^{1/2} \right) \left(\sum_{j \in \mathcal{J}} \alpha \langle \widehat{B}^2 \rangle_{\psi_j}^{1/2} \right) = \langle \widehat{A}^2 \rangle_{\widehat{\rho}} \langle \widehat{B}^2 \rangle_{\widehat{\rho}}$$

hence the inequality (13.19) using formula (13.18). $\square$

Choosing for $\widehat{A}$ the operator of multiplication by x_j and $\widehat{B} = -i\hbar\partial/\partial x_j$ one obtains the usual Robertson–Schrödinger inequalities

$$(\Delta X_j)_{\widehat{\rho}}^2 (\Delta P_j)_{\widehat{\rho}}^2 \geq \Delta(X_j, P_j)_{\widehat{\rho}}^2 + \tfrac{1}{4}\hbar^2 \tag{13.22}$$

for $1 \leq j \leq n$.

Corollary 299. *Assume that $[\widehat{A}, \widehat{B}] = i\hbar$; then*

$$(\Delta \widehat{A})^2_{\widehat{\rho}} (\Delta \widehat{B})^2_{\widehat{\rho}} \geq \Delta(\widehat{A}, \widehat{B})^2_{\widehat{\rho}} + \tfrac{1}{4}\hbar^2. \tag{13.23}$$

If $\widehat{\rho}$ represents a pure state $\psi \in D_{\widehat{A}\widehat{B}} \cap D_{\widehat{B}\widehat{A}}$ then we have equality if and only if the vectors $(\widehat{A} - \langle \widehat{A}\rangle)\psi$ and $(\widehat{B} - \langle \widehat{B}\rangle)\psi$ are collinear.

Proof. The inequality (13.23) immediately follows from the inequality (13.17). Assume that we have equality in (13.23). It is sufficient to consider the case $\langle \widehat{A}\rangle_\psi = \langle \widehat{B}\rangle_\psi = 0$. In view of formula $\langle \widehat{A}\widehat{B}\rangle_\psi = \langle \widehat{A}\widehat{B}\psi|\psi\rangle = \langle \widehat{B}\psi|\widehat{A}\psi\rangle$ this means that the Cauchy–Schwarz inequality reduces to an equality, which implies that $\widehat{A}\psi$ and $\widehat{B}\psi$ are colinear. $\qquad\qquad\square$

Assume in particular that $\widehat{A}\psi = x_1 \psi$ and $\widehat{B}\psi = -i\hbar \partial \psi/\partial x_1$. The inequality (13.22) with $j = 1$ becomes an equality if there exists a complex constant λ_1 such that

$$-i\hbar \frac{\partial \psi}{\partial x_1} = \lambda_1 x_1 \psi;$$

it follows that we must have

$$\psi(x) = C(x_2, \ldots, x_n)e^{-\frac{i}{2\hbar}\lambda_1 x_1^2}$$

for some function C of only the variables $x_2, \ldots, x_n$. Thus, if we require all the Robertson–Schrödinger equalities (13.22) to become equalities we must have

$$\psi(x) = C\exp\left(\frac{i}{2\hbar}\sum_{j=1}^{n}\lambda_j x_j^2\right)$$

where C and the λ_j are complex constants; the condition that ψ be square-integrable requires that $\operatorname{Im}\lambda_j > 0$. Choosing in particular $\lambda_1 = \cdots = \lambda_n = i$ and $C = (\pi\hbar)^{-n/4}$ one obtains the standard coherent state

$$\psi_0^{\hbar}(x) = (\pi\hbar)^{-n/4}e^{-\frac{1}{2\hbar}|x|^2}. \tag{13.24}$$

13.2.2 Sub-Gaussian estimates

We begin by discussing the case of a pure state.

We have seen in Chapter 11, Section 11.2, that the Wigner transform of a Gaussian is itself a Gaussian. More precisely, assume that

$$\psi_M^{\hbar}(x) = \left(\tfrac{1}{\pi\hbar}\right)^{n/4}(\det X)^{1/4}e^{-\frac{1}{2\hbar}Mx^2}$$

where $M = X + iY$ with X and Y symmetric and X positive definite. Then

$$W\psi_M^{\hbar}(z) = \left(\tfrac{1}{\pi\hbar}\right)^{n}e^{-\frac{1}{\hbar}Gz^2} \tag{13.25}$$

where G is the real $2n \times 2n$ matrix given by

$$G = \begin{pmatrix} X + YX^{-1}Y & YX^{-1} \\ X^{-1}Y & X^{-1} \end{pmatrix} = S^T S \tag{13.26}$$

where the symplectic matrix S is given by

$$S = \begin{pmatrix} X^{1/2} & 0 \\ X^{-1/2}Y & X^{-1/2} \end{pmatrix} \in \mathrm{Sp}(2n, \mathbb{R}). \tag{13.27}$$

The following simple geometric remark already hints at the fact that symplectic capacities might already be lurking behind these formulas:

Lemma 300. *If a Gaussian function* $\Psi(z) = Ce^{-\frac{1}{\hbar}Mz^2}$ *on* $\mathbb{R}^{2n}$ *is the Wigner transform of a Gaussian* (11.12) *then the phase-space ellipsoid* $\mathcal{W} = \{z : Mz^2 \leq \hbar\}$ *is the image* $S(B(\sqrt{\hbar}))$ *of the ball* $B(\hbar) : |z| \leq \hbar$ *by some* $S \in \mathrm{Sp}(2n, \mathbb{R})$, *and hence the symplectic capacity of* $\mathcal{W}$ *is* $\pi\hbar = \frac{1}{2}h$.

Proof. (Cf. Proposition 242) We have $M = S^T S$ for some $S \in \mathrm{Sp}(2n, \mathbb{R})$ and hence $\mathcal{W} = S(B(\hbar))$. Let c be an arbitrary symplectic capacity; then, using successively the symplectic invariance of c and the normalization condition $c(B(R)) = \pi R^2$, we get

$$c(\mathcal{W}) = c(S(B(\sqrt{\hbar}))) = c((B(\sqrt{\hbar}))) = \pi\hbar. \qquad \square$$

The following result shows that a Wigner transform cannot be dominated by an arbitrarily sharply peaked phase space Gaussian function. This is of course a phase space version of the uncertainty principle obtained by using Hardy's uncertainty principle. We are following the exposition in de Gosson and Luef [74].

Proposition 301. *Let* $\psi \in L^2(\mathbb{R}^n)$, $\psi \neq 0$, *and assume that there exists* $C > 0$ *such that* $W\psi(z) \leq Ce^{-\frac{1}{\hbar}Mz \cdot z}$. *Then* $c(\mathcal{W}_\Sigma) \geq \frac{1}{2}h$ *where* $\mathcal{W}_\Sigma$ *is the Wigner ellipsoid corresponding to the choice* $\Sigma = \frac{\hbar}{2}M^{-1}$ *(equivalently* $c(\mathcal{B}_M) \geq \frac{1}{2}h$ *where* $\mathcal{B}_M$: $Mz^2 \leq \hbar$).

Proof. In view of Williamson's symplectic diagonalization theorem we can find $S \in \mathrm{Sp}(2n, \mathbb{R})$ such that

$$MSz \cdot Sz = \sum_{j=1}^{n} \lambda_j(x_j^2 + p_j^2)$$

where $\lambda_1 \geq \lambda_2 \geq \cdots \geq \lambda_n$ are the moduli of the eigenvalues $\pm i\lambda$, $\lambda > 0$, of JM. It follows that the assumption $W\psi(z) \leq Ce^{-\frac{1}{\hbar}Mz \cdot z}$ can be rewritten as

$$W\psi(S^{-1}z) \leq C \exp\left(-\frac{1}{\hbar}\sum_{j=1}^{n}\lambda_j(x_j^2 + p_j^2)\right). \tag{13.28}$$

In view of the metaplectic covariance formula (9.29) we have $W\psi(S^{-1}z) = W\widehat{S}\psi(z)$ where $\widehat{S} \in \mathrm{Mp}(n,\mathbb{R})$ has projection S on $\mathrm{Sp}(2n,\mathbb{R})$. Since $\widehat{S}\psi \in L^2(\mathbb{R}^n)$ and $c(\mathcal{W}_\Sigma)$ is a symplectic invariant it is no restriction to assume $S = I$, $\widehat{S} = I$. Integrating the inequality

$$W\psi(z) \le C\exp\left(-\frac{1}{\hbar}\sum_{j=1}^{n}\lambda_j(x_j^2 + p_j^2)\right)$$

in x and p, respectively we get, using the marginal properties formulae (9.17) and (9.18),

$$|\psi(x)| \le C_1\exp\left(-\frac{1}{2\hbar}\sum_{j=1}^{n}\lambda_j x_j^2\right), \tag{13.29}$$

$$|F\psi(p)| \le C_1\exp\left(-\frac{1}{2\hbar}\sum_{j=1}^{n}\lambda_j p_j^2\right) \tag{13.30}$$

for some constant $C_1 > 0$. Let us now introduce the following notation. We set $\psi_1(x_1) = \psi(x_1,0,\ldots,0)$ and denote by F_1 the one-dimensional Fourier transform in the x_1 variable. Now, we first note that (13.29) implies that

$$|\psi_1(x_1)| \le C_1\exp\left(-\frac{\lambda_1}{2\hbar}x_1^2\right). \tag{13.31}$$

On the other hand, by definition of the Fourier transform F,

$$\int_{\mathbb{R}^{n-1}} F\psi(p)dp_2\cdots dp_n = \left(\tfrac{1}{2\pi\hbar}\right)^{n/2}\int_{\mathbb{R}^{n-1}}\left(\int_{\mathbb{R}^n}e^{-\frac{i}{\hbar}p\cdot x}\psi(x)dx\right)dp_2\cdots dp_n;$$

taking into account the Fourier inversion formula this formula can be rewritten as

$$\int_{\mathbb{R}^{n-1}} F\psi(p)dp_2\cdots dp_n = (2\pi\hbar)^{(n-1)/2}\,F_1\psi_1(p_1).$$

It follows that

$$|F_1\psi_1(p_1)| \le \left(\tfrac{1}{2\pi\hbar}\right)^{(n-1)/2}C_1\int\exp\left(-\frac{1}{2\hbar}\sum_{j=1}^{n}\lambda_j p_j^2\right)dp_2\cdots dp_n$$

that is

$$|F_1\psi_1(p_1)| \le C_3\exp\left(-\frac{\lambda_1}{2\hbar}p_1^2\right) \tag{13.32}$$

for some constant $C_3 > 0$. Applying Hardy's uncertainty principle we see that the condition $\lambda_1^2 \le 1$ is both necessary and sufficient for these inequalities to hold (remember that we are using the ordering convention $\lambda_1 \ge \lambda_2 \ge \cdots \ge \lambda_n$); this is equivalent to $c(\mathcal{B}_M) \ge \frac{1}{2}h$. $\qquad\square$

Exercise 302. Use Theorem 301 to show that a Wigner transform $W\psi$ can never have compact support. [Hint: show that if $W\psi$ has compact support then it is dominated by arbitrarily sharply peaked Gaussians.]

The results above allow us to prove the more general result:

13.2.3 Positivity issues and the KLM conditions

Let us now shortly address the following important, deep, and difficult question:

When is a real symmetric $2n \times 2n$ matrix Σ the covariance matrix of a mixed quantum state $\widehat{\rho}$?

In the classical case the answer is simple: Σ is the covariance matrix of some probability density if and only if Σ is positive definite. In the quantum case the situation is much more subtle and difficult than it could appear at first sight, because it is plagued by positivity questions. Let in fact $\widehat{\rho}$ be a self-adjoint operator of trace class with trace $\mathrm{Tr}(\widehat{\rho}) = 1$. The operator $\widehat{\rho}$ is thus a candidate for being a density matrix. However, to be eligible, it must in addition be non-negative, that is we must have $(\widehat{\rho}\psi|\psi)_{L^2} \geq 0$ for all $\psi \in L^2(\mathbb{R}^n)$, and it is this property which is difficult to check.

We are going to present a theoretical characterization of the positivity of a density operator, the KLM conditions. The letters KLM are an acronym for Kastler [105] and Loupias and Miracle-Sole [118, 119] who all three have contributed significantly to a better understanding of the positivity issues for trace class operators. Also see the related paper of Emch [43] on "geometric dequantization" where similar issues are discussed.

Let us first introduce a notation: for a function $a \in \mathcal{S}(\mathbb{R}^n \oplus \mathbb{R}^n)$ we set

$$a_\Diamond(z) = \left(\tfrac{1}{2\pi}\right)^n \int_{\mathbb{R}^n} e^{i\sigma(z,z')}a(z')dz'$$

($a_\Diamond$ can be pronounced "a diamond"). This definition of course also makes sense for $a \in \mathcal{S}'(\mathbb{R}^n \oplus \mathbb{R}^n)$ if one interprets the integral as a distributional bracket. In fact, the function $a_\Diamond(z)$ is related to the symplectic Fourier transform

$$a_\sigma(z) = F_\sigma a(z) = \left(\tfrac{1}{2\pi\hbar}\right)^n \int_{\mathbb{R}^n} e^{-\frac{i}{\hbar}\sigma(z,z')}a(z')dz'$$

by the simple formula

$$a_\Diamond(z)(z) = \hbar^n a_\sigma(-\hbar z). \tag{13.33}$$

Definition 303. Let a be a complex function defined on the symplectic space $(\mathbb{R}^n \oplus \mathbb{R}^n, \sigma)$. Let $\eta \in \mathbb{R}$ be a variable parameter and set

$$\Lambda_{jk}(z_j, z_k) = e^{-\frac{i\eta}{2}\sigma(z_j, z_k)}a_\Diamond(z_j - z_k) \tag{13.34}$$

where $(z_1, \ldots, z_N) \in (\mathbb{R}^{2n})^N$. The set $\mathcal{WS}(a)$ of values of η for which every matrix $\Lambda = (\Lambda_{jk}(z_j, z_k, \eta))_{1 \leq j,k \leq N}$ with N and $(z_1, \ldots, z_N)$ arbitrary is positive semi-definite is called the *Narcowich–Wigner spectrum of a*. If $\eta \in \mathcal{WS}(a)$ one says that a is of *η-positive type*.

Explicitly, $\mathcal{WS}(a)$ is thus the set of all real numbers η such that for every integer $N \geq 1$ and every sequence $(z_1, \ldots, z_N)$ we have

$$\sum_{1 \leq j,k \leq N} \lambda_j \overline{\lambda_k} e^{-\frac{i\eta}{2}\sigma(z_j, z_k)} a_\Diamond(z_j - z_k) \geq 0,$$

$\lambda = (\lambda_1, \ldots, \lambda_N) \in \mathbb{C}^N$. It should be emphasized that in the verification of the condition above it is assumed that a does not depend explicitly on the variable η.

Exercise 304. Show that functions of *η-positive type form a cone: if a and b are of η-positive type then so is $\lambda a + \mu b$ for all $\lambda \geq 0$ and $\mu \geq 0$.*

Notice that if we choose $\eta = 0$ we recover the usual definition of a function of positive type: the function a is of positive type if for each integer N the $N \times N$ matrix with entries $a_\Diamond(z_j - z_k)$ is positive semi-definite. In view of a classical theorem of Bochner this is a well-known sufficient and necessary condition for the continuous function $a_\Diamond$ to be the Fourier transform of a positive measure (see Katznelson [107], p. 137 for a proof of Bochner's theorem). Let us introduce the following terminology from statistical mechanics: a classical state is the datum of a probability density on the phase space $\mathbb{R}^n \oplus \mathbb{R}^n$. We have:

Proposition 305. *Assume that a is continuous on $\mathbb{R}^n \oplus \mathbb{R}^n$ and of 0-positive type. Then a is a positive measure and can thus be identified with a classical state.*

Proof. To say that a is 0-positive type means that $a_\Diamond$ is of positive type in the usual sense, hence $a = (a_\Diamond)_\Diamond$ is a positive measure in view of Bochner's theorem. Normalizing this measure yields a probability density, hence a classical state. $\qquad\square$

When $\eta = \hbar$ we can restate the definition above in terms of the symplectic Fourier transform F_σ:

Proposition 306. *The function $a_\Diamond$ is of $\hbar$-positive type if and only if for every integer $N \geq 1$ and every $(z_1, \ldots, z_N) \in (\mathbb{R}^{2n})^N$ the $N \times N$ matrix $\Lambda' = (\Lambda'_{jk}(z_j, z_k))_{1 \leq j,k \leq N}$ where*

$$\Lambda'_{jk}(z_j, z_k) = e^{\frac{i}{2\hbar}\sigma(z_j, z_k)} F_\sigma a(z_j - z_k) \tag{13.35}$$

is positive semi-definite.

Proof. It is immediate in view of formula (13.33) replacing (z_j, z_k) with $\eta^{-1}(z_k, z_j)$ and noting that $\sigma(z_k, z_j) = -\sigma(z_j, z_k)$. $\qquad\square$

The observant reader has certainly noticed that the right-hand side of (13.35) has some kind of remote resemblance with the formula (8.9)

$$\widehat{T}(z_0 + z_1) = e^{-\frac{i}{2\hbar}\sigma(z_0,z_1)}\widehat{T}(z_0)\widehat{T}(z_1)$$

for Heisenberg–Weyl operators, which we can rewrite as

$$\widehat{T}(z_k - z_j) = e^{-\frac{i}{2\hbar}\sigma(z_j,z_k)}\widehat{T}(z_k)\widehat{T}(-z_j). \tag{13.36}$$

That the Heisenberg–Weyl operators indeed are part of the picture is shown by the following fundamental example of a function of $\hbar$-positive type.

Proposition 307. *The Wigner distribution* $W\psi$ *of* $\psi \in L^2(\mathbb{R}^n)$ *is of* $\hbar$-*positive type:* $\hbar \in \mathcal{WS}(W\psi)$.

Proof. In view of Proposition 306 we have to show that for all $(z_1,\ldots,z_N) \in (\mathbb{R}^{2n})^N$ and $(\lambda_1,\ldots,\lambda_N) \in \mathbb{C}^N$ we have

$$I_N = \sum_{1 \le j,k \le N} \lambda_j\overline{\lambda_k}e^{-\frac{i}{2\hbar}\sigma(z_j,z_k)}F_\sigma W\psi(z_j - z_k)) \ge 0 \tag{13.37}$$

for every complex vector $(\lambda_1,\ldots,\lambda_N) \in \mathbb{C}^N$ and every sequence $(z_1,\ldots,z_N) \in (\mathbb{R}^{2n})^N$. Since the Wigner distribution $W\psi$ and the ambiguity function

$$A\psi(z) = \left(\tfrac{1}{2\pi\hbar}\right)^n (\widehat{T}(-z)\psi|\psi)_{L^2}$$

are obtained from each other by the symplectic Fourier transform F_σ (formula (9.26) in Proposition 175) we have

$$I_N = \sum_{1 \le j,k \le N} \lambda_j\overline{\lambda_k}e^{-\frac{i}{2\hbar}\sigma(z_j,z_k)}A\psi(z_j - z_k)).$$

Let us prove that

$$I_N = \left(\tfrac{1}{2\pi\hbar}\right)^n \left\|\sum_{1 \le j \le N} \lambda_j\widehat{T}(-z_j)\psi\right\|_{L^2}^2; \tag{13.38}$$

the inequality (13.37) will follow. Taking into account the fact that $\widehat{T}(-z_k)^* = \widehat{T}(z_k)$ and using formula (13.36) we have

$$\left\|\sum_{1 \le j \le N} \lambda_j\widehat{T}(z_j)\psi\right\|_{L^2}^2 = \sum_{1 \le j,k \le N} \lambda_j\overline{\lambda_k}(\widehat{T}(-z_j)\psi|\widehat{T}(-z_k)\psi)_{L^2}$$

$$= \sum_{1 \le j,k \le N} \lambda_j\overline{\lambda_k}(\widehat{T}(z_k)\widehat{T}(-z_j)\psi|\psi)_{L^2}$$

$$= \sum_{1 \le j,k \le N} \lambda_j\overline{\lambda_k}e^{\frac{i}{2\hbar}\sigma(z_j,z_k)}(\widehat{T}(z_k - z_j)\psi|\psi)_{L^2}$$

$$= (2\pi\hbar)^n \sum_{1 \le j,k \le N} \lambda_j\overline{\lambda_k}e^{\frac{i}{2\hbar}\sigma(z_j,z_k)}A\psi(z_j - z_k)$$

proving the equality (13.38). $\qquad\square$

One shows (but we will not do it here) that

Proposition 308. *If $a : \mathbb{R}^n \oplus \mathbb{R}^n \longrightarrow \mathbb{C}$ is of $\hbar$-positive type then $a \in L^2(\mathbb{R}^n \oplus \mathbb{R}^n)$.*

(See Loupias and Miracle-Sole [118], Theorem 4.)

The interest of the notion of $\hbar$-positivity comes from the following result:

Theorem 309. *Let $\widehat{A} \xrightarrow{\text{Weyl}} a$ be a self-adjoint trace class operator on $L^2(\mathbb{R}^n)$. The operator $\widehat{A}$ is positive semi-definite (written $\widehat{A} \geq 0$) if and only if the symbol a is of $\hbar$-positive type.*

We will need the following result in our discussion:

Lemma 310. *If $a : \mathbb{R}^n \oplus \mathbb{R}^n \longrightarrow \mathbb{C}$ is continuous and twice continuously differentiable near 0 and of $\hbar$-positive type, then we have*

$$-2\hbar a''(0) + iJ \geq 0 \tag{13.39}$$

where $a''(0) = D^2 a(0)$ is the Hessian matrix of a at 0.

Proof. (Cf. Lemma 2.1 in Narcowich [128]). For $(\lambda_1, \ldots, \lambda_m) \in \mathbb{C}^m$ and $\varepsilon \in \mathbb{R}$ let us set

$$R(\varepsilon) = \sum_{j,k=1}^{m} \overline{\lambda_j} \lambda_k e^{-\frac{i\varepsilon^2}{2\hbar}\sigma(z_j, z_k)} a(\varepsilon(z_j - z_k)).$$

If a is of $\hbar$-positive type we have $R(\varepsilon) \geq 0$ for every ε; choose now the λ_j such that $\sum_j \lambda_j = 0$; then $R(0) = 0$ and $R''(0) \geq 0$. An elementary calculation shows that

$$R''(0) = Z^T(-2a''(0) + i\hbar^{-1}J)Z$$

where $Z = \sum_j \lambda_j z_j \in \mathbb{C}^{2n}$. The λ_j and z_j being arbitrary we thus have $-2a''(0) + i\hbar^{-1}J \geq 0$, proving the lemma. $\square$

Proposition 311. *Let $\widehat{\rho}$ be a density operator.*

(i) *The covariance matrix $\Sigma_{\widehat{\rho}}$ satisfies the strong uncertainty principle:*

$$\Sigma_{\widehat{\rho}} + \tfrac{1}{2}i\hbar J \geq 0; \tag{13.40}$$

(ii) *The Robertson–Schrödinger inequalities hold:*

$$(\Delta X_j)^2_{\widehat{\rho}}(\Delta P_j)^2_{\widehat{\rho}} \geq (\mathrm{Cov}(X_j, P_j)_{\widehat{\rho}})^2 + \tfrac{1}{4}\hbar^2, \tag{13.41}$$

$(j = 1, \ldots, n)$ *and* $(\Delta X_j)^2_{\psi}(\Delta P_k)^2_{\psi} \geq 0$ *if* $j \neq k$.

Proof of (i). The matrix $\Sigma_{\widehat{\rho}} + \tfrac{1}{2}i\hbar J$ is Hermitian since $\Sigma_{\widehat{\rho}}$ is symmetric and the transpose of J is $-J$. We next remark that $\Sigma_{\widehat{\rho}} = \Sigma_{\widehat{\rho_0}}$ where ρ_0 is defined by $\rho_0(z) = \rho(z + \langle z \rangle_\rho)$ with $\langle z \rangle_\rho = (\langle x \rangle_\rho, \langle p \rangle_\rho)$: we have $\langle z \rangle_{\rho_0} = 0$ and hence

$$\mathrm{Cov}(X_j, X_k)_{\rho_0} = \int_{\mathbb{R}^{2n}} x_j x_k \rho_0(z)dz = \mathrm{Cov}(X_j, X_k)_\rho;$$

similarly $\mathrm{Cov}(X_j, P_k)_{\rho_0} = \mathrm{Cov}(X_j, P_k)_\rho$ and $\mathrm{Cov}(P_j, P_k)_{\rho_0} = \mathrm{Cov}(P_j, P_k)_\rho$. It is thus sufficient to prove the proposition for the density operator $\widehat{\rho}_0$. Let us calculate the Hessian matrix $\rho_{0,\sigma}''(0)$. A direct calculation shows that we have

$$\hbar^2 \rho_{0,\sigma}''(0) = (2\pi\hbar)^{-n} \begin{pmatrix} -\Sigma_{PP,\rho_0} & \Sigma_{XP,\rho_0} \\ \Sigma_{PX,\rho_0} & -\Sigma_{XX,\rho_0} \end{pmatrix}$$

and hence

$$\hbar^2 \rho_{0,\sigma}''(0) = \left(\tfrac{1}{2\pi\hbar}\right)^n J\Sigma_{\widehat{\rho}}J. \tag{13.42}$$

Since we have $\rho_\sigma = (2\pi\hbar)^{-n} a_\sigma$ the positivity of $\widehat{\rho}$ implies, taking Proposition 311 and the lemma preceding it into account that

$$M = -2\hbar^{-1} J\Sigma_\rho J + iJ \geq 0 \; ;$$

the condition $M \geq 0$ being equivalent to $J^T M J \geq 0$ the inequality (13.40) follows.

Proof of (ii). It follows from property (i) in view of Theorem 98 in Chapter 6. $\quad\square$

Part III

Pseudo-differential Operators and Function Spaces

Chapter 14

Shubin's Global Operator Calculus

In the applications to quantum mechanics that we have in mind in this book a crucial role is played by non-local effects in symplectic space; for this reason the local theory traditionally developed by many authors working in the theory of partial differential equations is of little use. This chapter is a review of pseudo-differential calculus from the point of view developed in Shubin [147]. The specificity of this calculus is that the symbols satisfy global estimates where the x and p variables are placed on equal footing. This is in strong contrast with the usual pseudodifferential calculus often used in the theory of partial differential equations (especially their microlocal study), and which is less adequate for the study of quantum mechanics in its phase space formulation. The Shubin calculus contains the usual Weyl calculus as a particular case; this remark will be important later when we derive the Schrödinger equation. An excellent source which complements this chapter is Nicola and Rodino [131].

We will again use the multi-index notation $\alpha = (\alpha_1, \ldots, \alpha_{2n}) \in \mathbb{N}^n$, $|\alpha| = \alpha_1 + \cdots + \alpha_{2n}$, and $\partial_z^\alpha = \partial_{x_1}^{\alpha_1} \cdots \partial_{x_n}^{\alpha_n} \partial_{y_1}^{\alpha_{n+1}} \cdots \partial_{y_n}^{\alpha_{2n}}$ if $z = (x, y)$.

14.1 The Shubin classes

We begin by introducing some notation and useful formulas.

14.1.1 Generalities

In Section 1.3.3 we briefly discussed quantization rules of the type

$$px \longrightarrow \tau \widehat{xp} + (1 - \tau)\widehat{px}$$

where τ is an arbitrary real constant, and which generalize the Weyl quantization scheme

$$px \longrightarrow \frac{1}{2}(\widehat{xp} + \widehat{px}).$$

The consideration of such rules leads us to study pseudo-differential operators of
the type

$$\widehat{A}f(x) = \left(\tfrac{1}{2\pi\hbar}\right)^{n} \iint_{\mathbb{R}^{2n}} e^{\frac{i}{\hbar}p\cdot(x-y)} a_\tau((1-\tau)x + \tau y, p)f(y)\,dy\,dp \tag{14.1}$$

where the integral should be understood in some "reasonable" sense. For instance,
this expression makes perfect sense if $f \in \mathcal{S}(\mathbb{R}^n)$ and $a_\tau \in \mathcal{S}(\mathbb{R}^n \oplus \mathbb{R}^n)$ because the
integral is then absolutely convergent. For more general symbols a_τ one can give
a meaning to the expression (14.1) by declaring that the operator $\widehat{A}$ is defined by
the distributional kernel

$$K_{\widehat{A}}(x,y) = \left(\tfrac{1}{2\pi\hbar}\right)^{n/2} (F_2^{-1}a_\tau)((1-\tau)x + \tau y, p)$$

where F_2^{-1} is the inverse Fourier transform in the second set of variables. We
notice that setting $\tau = \tfrac{1}{2}$ and $a_{1/2} = a$ formula (14.1) becomes

$$\widehat{A}f(x) = \left(\tfrac{1}{2\pi}\right)^{n} \iint_{\mathbb{R}^{2n}} e^{i\xi\cdot(x-y)} a(\tfrac{1}{2}(x+y), \xi)f(y)\,dy\,d\xi \tag{14.2}$$

which is the expression (10.38) of a Weyl operator in terms of its symbol when we
choose $\hbar = 1$ (more about that below). To make the notation more compact we
will in fact often assume that $\hbar = 1$, so that we will actually deal with operators
written in the form

$$\widehat{A}f(x) = \left(\tfrac{1}{2\pi}\right)^{n} \iint_{\mathbb{R}^{2n}} e^{i\xi\cdot(x-y)} a_\tau((1-\tau)x + \tau y, \xi)f(y)\,dy\,d\xi \tag{14.3}$$

which is standard in the theory of partial differential operators. In harmonic analysis the preferred choice is to take $\hbar = 1/2\pi$ and to write ω instead of p:

$$\widehat{A}f(x) = \iint_{\mathbb{R}^{2n}} e^{2\pi i\omega\cdot(x-y)} a_\tau((1-\tau)x + \tau y, \omega)f(y)\,dy\,d\omega. \tag{14.4}$$

Each choice has its advantages and disadvantages. But keeping these conventions
in mind, it is easy to translate the properties of each formulation into the other.
Choosing $\tau = 0$ in formula (14.3) yields

$$\widehat{A}f(x) = \left(\tfrac{1}{2\pi}\right)^{n} \iint_{\mathbb{R}^{2n}} e^{i\xi\cdot(x-y)} a_\tau(x, \xi)f(y)\,dy\,d\xi$$

that is

$$\widehat{A}f(x) = \left(\tfrac{1}{2\pi}\right)^{n/2} \int_{\mathbb{R}^{n}} e^{ix\cdot\xi} a(x, \xi)Ff(\xi)\,d\xi \tag{14.5}$$

where

$$Ff(\xi) = \left(\tfrac{1}{2\pi}\right)^{n/2} \int_{\mathbb{R}^{n}} e^{-i\xi\cdot y} f(y)\,dy;$$

this is the conventional definition of a pseudo-differential operator found in most texts dealing with partial differential equations; a is then sometimes called the "Kohn–Nirenberg symbol" of the operator $\widehat{A}$.

We mention for further use the following simple conjugation relation the between usual Weyl operators and operators (14.2) with $\hbar$-dependent symbols (cf. the subsection on the dependence on $\hbar$ of π^{Mp} in Chapter 7):

Lemma 312. *The Weyl operator $\widehat{A} \overset{\mathrm{Weyl}}{\longleftrightarrow} a$ given by*

$$\widehat{A}f(x) = \left(\tfrac{1}{2\pi\hbar}\right)^n \iint_{\mathbb{R}^{2n}} e^{\frac{i}{\hbar}p\cdot(x-y)} a(\tfrac{1}{2}(x+y),p) f(y)\,dy\,dp \qquad (14.6)$$

and the operator (14.2) with symbol $a_{(\hbar)}(z) = a(z\sqrt{\hbar})$, that is

$$\widehat{A}_{(\hbar)}f(x) = \left(\tfrac{1}{2\pi}\right)^n \iint_{\mathbb{R}^{2n}} e^{i\xi\cdot(x-y)} a(\tfrac{1}{2}\sqrt{\hbar}(x+y),\sqrt{\hbar}\xi) f(y)\,dy\,d\xi \qquad (14.7)$$

are related by the formula

$$\widehat{A} = M_{\sqrt{\hbar}}^{-1} \widehat{A}_{(\hbar)} M_{\sqrt{\hbar}}$$

where $M_{\sqrt{\hbar}}f(x) = \hbar^{n/4} f(x\sqrt{\hbar})$.

Proof. We have

$$M_{\sqrt{\hbar}}^{-1}\widehat{A}M_{\sqrt{\hbar}}f(x) = \left(\tfrac{1}{2\pi}\right)^n \iint_{\mathbb{R}^{2n}} e^{i\xi\cdot(\frac{1}{\sqrt{\hbar}}x-y)} a(\tfrac{1}{2}(x+\sqrt{\hbar}y),\sqrt{\hbar}\xi) f(\sqrt{\hbar}y)\,dy\,d\xi;$$

setting $y = y'/\sqrt{\hbar}$ and $\xi = p/\sqrt{\hbar}$ we get

$$M_{\sqrt{\hbar}}^{-1}\widehat{A}M_{\sqrt{\hbar}}f(x) = \left(\tfrac{1}{2\pi\hbar}\right)^n \iint_{\mathbb{R}^{2n}} e^{\frac{i}{\hbar}p\cdot(x-y)} a(\tfrac{1}{2}(x+y'),p) f(y')\,dy'\,dp$$

which establishes the lemma. $\qquad\square$

14.1.2 Definitions and preliminary results

It is convenient – and natural, from the point of view of quantum mechanics – to introduce the Shubin symbol classes. We will be following very closely Shubin's exposition ([147], particularly §23).

We will use throughout this chapter the weight function

$$\langle z \rangle = \sqrt{1 + |z|^2}$$

for $z \in \mathbb{R}^{2n}$.

Definition 313. Let $m \in \mathbb{R}$ and $0 < \rho \leq 1$.

(i) The symbol class $\Gamma_\rho^m(\mathbb{R}^n \oplus \mathbb{R}^n)$ consists of all complex functions $a \in C^\infty(\mathbb{R}^n \oplus \mathbb{R}^n)$ such that for every $\alpha \in \mathbb{N}^{2n}$ there exists a constant $C_\alpha \geq 0$ with

$$|\partial_z^\alpha a(z)| \leq C_\alpha \langle z \rangle^{m-\rho|\alpha|} \quad \text{for } z \in \mathbb{R}^{2n}. \tag{14.8}$$

(ii) The symbol class $\Sigma_{\rho,\delta}^{m,\mu}(\mathbb{R}^n \oplus \mathbb{R}^n)$, $\mu > 0$, $\delta \leq 1/2$, consists of all complex functions $a \in C^\infty(\mathbb{R}^n \oplus \mathbb{R}^n)$ depending continuously on $\hbar \in (0, \varepsilon)$ such that

$$|\partial_z^\alpha a(z, \hbar)| \leq C_\alpha \langle z \rangle^{m-\rho|\alpha|} \hbar^{\mu-\delta|\alpha|} \quad \text{for } z \in \mathbb{R}^{2n}. \tag{14.9}$$

We set

$$\Gamma_\rho^{-\infty}(\mathbb{R}^n \oplus \mathbb{R}^n) = \bigcap_{m \in \mathbb{R}} \Gamma_\rho^m(\mathbb{R}^n \oplus \mathbb{R}^n).$$

We will see later on that the choice of τ is actually irrelevant: if $\widehat{A} \in G_\rho^m(\mathbb{R}^n)$ is given by (14.1) for one choice of τ it is true for all choices of the parameter.

Obviously $\Gamma_\rho^m(\mathbb{R}^n \oplus \mathbb{R}^n)$, $\Sigma_{\rho,\delta}^{m,\mu}(\mathbb{R}^n \oplus \mathbb{R}^n)$, and $\Gamma_\rho^{-\infty}(\mathbb{R}^n \oplus \mathbb{R}^n)$ are complex vector spaces for the usual operations of addition and multiplication by complex numbers. Moreover one easily checks that

$$a \in \Gamma_\rho^m(\mathbb{R}^n \oplus \mathbb{R}^n) \text{ and } b \in \Gamma_\rho^{m'}(\mathbb{R}^n \oplus \mathbb{R}^n) \Longrightarrow ab \in \Gamma_\rho^{m+m'}(\mathbb{R}^n \oplus \mathbb{R}^n);$$

$$a \in \Gamma_\rho^m(\mathbb{R}^n \oplus \mathbb{R}^n) \text{ and } \alpha \in \mathbb{N}^{2n} \Longrightarrow \partial_z^\alpha a \in \Gamma_\rho^{m-|\alpha|}(\mathbb{R}^n \oplus \mathbb{R}^n).$$

The first implication is proved by using the generalized Leibniz rule for the derivatives of a product of functions; the second is obvious in view of the definition of $\Gamma_\rho^m(\mathbb{R}^n \oplus \mathbb{R}^n)$. We also have

$$a \in \Sigma_{\rho,\delta}^{m,\mu}(\mathbb{R}^n \oplus \mathbb{R}^n) \text{ and } b \in \Sigma_{\rho',\delta'}^{m',\mu'}(\mathbb{R}^n \oplus \mathbb{R}^n)$$

$$\Longrightarrow ab \in \Sigma_{\rho'',\delta''}^{m+m',\mu+\mu'}(\mathbb{R}^n \oplus \mathbb{R}^n) \, , \, \rho'' = \min(\rho, \rho'), \, \delta'' = \max(\delta, \delta').$$

The simplest and most typical example is the reduced harmonic oscillator Hamiltonian $H(z) = \frac{1}{2}|z|^2$ which obviously belongs to $\Gamma_1^2(\mathbb{R}^n \oplus \mathbb{R}^n)$. In fact, we may as well choose the "complete" Hamiltonian

$$H(z) = \sum_{j=1}^n \frac{1}{2m_j}(p_j^2 + m_j^2 \omega_j^2 x_j^2)$$

as is shown by the following exercise (the property can also be derived from Proposition 316 below):

Exercise 314. Show that any polynomial function in z of degree m is in $\Gamma_1^m(\mathbb{R}^n \oplus \mathbb{R}^n)$.

The following exercise shows that $\mathcal{S}(\mathbb{R}^n \oplus \mathbb{R}^n)$ can be characterized in terms of the Shubin classes $\Gamma_\rho^m(\mathbb{R}^n \oplus \mathbb{R}^n)$:

Exercise 315. Show that $\Gamma_\rho^{-\infty}(\mathbb{R}^n \oplus \mathbb{R}^n) = \cap_{m \in \mathbb{R}} \Gamma_\rho^m(\mathbb{R}^n \oplus \mathbb{R}^n)$ is identical with the Schwartz space $\mathcal{S}(\mathbb{R}^n \oplus \mathbb{R}^n)$ of rapidly decreasing functions on $\mathbb{R}^n \oplus \mathbb{R}^n$.

The following result is an invariance property which shows that the class $\Gamma_\rho^m(\mathbb{R}^n \oplus \mathbb{R}^n)$ is preserved by linear changes of variables:

Proposition 316. *Let $a \in \Gamma_\rho^m(\mathbb{R}^n \oplus \mathbb{R}^n)$ and ϕ a linear automorphism of $\mathbb{R}^n \oplus \mathbb{R}^n$. Then $\phi^* a = a \circ \phi$ is also in $\Gamma_\rho^m(\mathbb{R}^n \oplus \mathbb{R}^n)$.*

Proof. Let us first show that $|\phi^* a(z)| \le C_\phi \langle z \rangle^m$ for some constant $C_\phi \ge 0$. Diagonalizing the symmetric automorphism $\phi^T \circ \phi$ using an orthogonal transformation we have

$$\lambda_{\min}|z|^2 \le |\phi(z)|^2 \le \lambda_{\max}|z|^2$$

where $\lambda_{\min} > 0$ and $\lambda_{\max} > 0$ are the smallest and largest eigenvalues of $\phi^T \circ \phi$. It follows that

$$\langle \phi(z) \rangle^m \le \max(1, \lambda_{\max}) \langle z \rangle^m$$

if $m \ge 0$, and

$$\langle \phi(z) \rangle^m \le \min(1, \lambda_{\min}) \langle z \rangle^m$$

if $m < 0$. We thus have $|\phi^* a(z)| \le C_F \langle z \rangle^m$. A similar argument shows that for every multi-index α we have an estimate of the type

$$|\partial_z^\alpha (\phi^* a)(z)| \le C_{\alpha,\phi} \langle z \rangle^{m-\rho|\alpha|}$$

where $C_{\alpha,\phi}$ is a constant. $\qquad\square$

This property of invariance under linear transformations allows us to toggle between symbols $(x,\xi) \longmapsto a(x,\xi)$ and $(x,\xi) \longmapsto a(x,\hbar\xi) = a(x,p)$ without changing the symbol class.

Definition 317. The set of operators $\widehat{A}$ defined by

$$\widehat{A}f(x) = \left(\tfrac{1}{2\pi}\right)^n \iint_{\mathbb{R}^{2n}} e^{i\xi \cdot (x-y)} a(\tfrac{1}{2}(x+y), \xi) f(y) \, dy \, d\xi$$

with $a \in \Gamma_\rho^m(\mathbb{R}^n \oplus \mathbb{R}^n)$ for some value of $\tau \in \mathbb{R}$ is denoted by $G_\rho^m(\mathbb{R}^n)$.

The operators $\widehat{A} \in G_\rho^m(\mathbb{R}^n)$ have many interesting regularity properties; the following is important (even if not very surprising):

Proposition 318. *Every operator $\widehat{A} \in G_\rho^m(\mathbb{R}^n)$ is a continuous operator $\mathcal{S}(\mathbb{R}^n) \longrightarrow \mathcal{S}(\mathbb{R}^n)$ and can be extended into a continuous operator $\widehat{A} : \mathcal{S}'(\mathbb{R}^n) \longrightarrow \mathcal{S}'(\mathbb{R}^n)$.*

Proof. That we have a continuous extension $\widehat{A} : \mathcal{S}'(\mathbb{R}^n) \longrightarrow \mathcal{S}'(\mathbb{R}^n)$ follows from the first statement by duality provided that we know that the transpose $\widehat{A}^T$ is also in $G_\rho^m(\mathbb{R}^n)$, this fact is true and will established in Proposition 342. The proof of the continuity property $\widehat{A} : \mathcal{S}(\mathbb{R}^n) \longrightarrow \mathcal{S}(\mathbb{R}^n)$ is classical; it is actually the same that one uses to prove the continuity of operators with symbols in the Hörmander classes. Let us sketch the argument: we write $\widehat{A}$ in the Kohn–Nirenberg form (14.5):

$$\widehat{A}f(x) = \left(\tfrac{1}{2\pi}\right)^{n/2} \int_{\mathbb{R}^n} e^{ix\cdot\xi} a(x,\xi) Ff(\xi)\, d\xi$$

and note that for every integer $N \geq 0$ we have

$$(1+|x|^2)^N \widehat{A}f(x) = \left(\tfrac{1}{2\pi}\right)^{n/2} \int_{\mathbb{R}^n} (1-\Delta_\xi)^N e^{ix\cdot\xi} a(x,\xi) Ff(\xi)\, d\xi$$

$$= \left(\tfrac{1}{2\pi}\right)^{n/2} \int_{\mathbb{R}^n} e^{ix\cdot\xi} (1-\Delta_\xi)^N [a(x,\xi) Ff(\xi)]\, d\xi$$

where the second inequality is obtained using integration by parts. Using Leibniz's formula for the derivatives of a product of functions and the properties of the Fourier transform we thus have

$$(1+|x|^2)^N \widehat{A}f(x) = \sum_{|\alpha|+|\beta|\leq 2N} C_{\alpha\beta} \int_{\mathbb{R}^n} e^{ix\cdot\xi} \partial_\xi^\alpha a(x,\xi) F(x^\beta f)(\xi)\, d\xi$$

where the $C_{\alpha\beta}$ are complex constants. Since $x^\beta f \in \mathcal{S}(\mathbb{R}^n)$ and $|\partial_\xi^\alpha a(x,\xi)| \leq C_\alpha \langle z\rangle^{m-\rho|\alpha|}$ it follows that $|(1+|x|^2)^N \widehat{A}f(x)| \leq C_N$ for some constant C_N. Applying the same argument to

$$\partial_x^\alpha \widehat{A}f(x) = \sum_{\beta\leq\alpha} C'_{\alpha\beta} \int_{\mathbb{R}^n} e^{ix\cdot\xi} \partial_x^\beta a(x,\xi) F(x^{\beta-\alpha} f)(\xi)\, d\xi$$

we have in fact $(1+|x|^2)^N \partial_x^\alpha \widehat{A}f \leq C''_{\alpha,N}$ for all $\alpha \in \mathbb{N}^n$ and $N \geq 0$ and hence $\widehat{A}f \in \mathcal{S}(\mathbb{R}^n)$. Using the expressions above we also get estimates of the semi-norms which imply the continuity of $\widehat{A} : \mathcal{S}(\mathbb{R}^n) \longrightarrow \mathcal{S}(\mathbb{R}^n)$. $\qquad\square$

We next consider the case of $\hbar$-dependent symbols; it is the most important from a quantum mechanical perspective:

Definition 319. The set of operators $\widehat{A}$ defined by

$$\widehat{A}f(x) = \left(\tfrac{1}{2\pi\hbar}\right)^n \iint_{\mathbb{R}^{2n}} e^{\frac{i}{\hbar} p\cdot(x-y)} a(\tfrac{1}{2}(x+y), p, \hbar) f(y)\, dy\, dp$$

with $a \in \Sigma_{\rho,\delta}^{m,\mu}(\mathbb{R}^n \oplus \mathbb{R}^n)$, $\mu > 0$, $\delta \leq 1/2$ is denoted by $S_{\rho,\delta}^{m,\mu}(\mathbb{R}^n \oplus \mathbb{R}^n)$.

The following boundedness results are important; they are consequences of the properties of anti-Wick quantization studied in Chapter 11.

Proposition 320.

(i) Let $a \in \Gamma_\rho^0(\mathbb{R}^n \oplus \mathbb{R}^n)$. The Weyl operator $\widehat{A} \xleftarrow{\text{Weyl}} a$ is bounded on $L^2(\mathbb{R}^n)$.

(ii) Every operator $\widehat{A} \in S_{\rho,\delta}^{0,\mu}(\mathbb{R}^n \oplus \mathbb{R}^n)$ with $\mu > 0$, $\delta \leq 1/2$ is uniformly bounded on $L^2(\mathbb{R}^n)$ for $0 < \hbar \leq \varepsilon$.

Proof of (i). In view of Proposition 259 there exists an anti-Wick operator $\widehat{B}_{\mathrm{aW}} \xleftrightarrow{\mathrm{AW}} b$ with $|\partial_z^\alpha b(z)| \leq C_\alpha \langle z \rangle^{-\rho|\alpha|}$ for every $\alpha \in \mathbb{N}^n$ and $\widehat{A} = \widehat{B}_{\mathrm{aW}} + \widehat{R}$ where the kernel K of $\widehat{R}$ is in $\mathcal{S}(\mathbb{R}^n \times \mathbb{R}^n)$. In view of Corollary 264 the operator $\widehat{B}_{\mathrm{aW}}$ is bounded on $L^2(\mathbb{R}^n)$, so it suffices to show that $\widehat{R}$ is also bounded on $L^2(\mathbb{R}^n)$. Applying the Cauchy–Schwarz inequality to the relation

$$\widehat{R}\psi(x) = \int_{\mathbb{R}^n} K(x,y)\psi(y)dy$$

we get

$$\int_{\mathbb{R}^n} |\widehat{R}\psi(x)|^2 dx \leq \int_{\mathbb{R}^n} |K(x,y)|^2 dxdy \int_{\mathbb{R}^n} |\psi(y)|^2 dy$$

hence $\|\widehat{R}\psi\|_{L^2} \leq \|K\|_{L^2}\|\psi\|_{L^2}$ which proves our assertion.

Proof of (ii)*:* Omitted. $\qquad\qquad\square$

Proposition 321. *Every operator $\widehat{A} \in G_\rho^0(\mathbb{R}^n)$ is bounded on $L^2(\mathbb{R}^n)$*

14.1.3 Asymptotic expansions of symbols

Let us now define and briefly study the notion of asymptotic expansion of a symbol $a \in \Gamma_\rho^m(\mathbb{R}^n \oplus \mathbb{R}^n)$.

Definition 322. Let $(a_j)_j$ be a sequence of symbols $a_j \in \Gamma_\rho^{m_j}(\mathbb{R}^n \oplus \mathbb{R}^n)$ such that $\lim_{j\to+\infty} m \to -\infty$. Let $a \in C^\infty(\mathbb{R}^n \oplus \mathbb{R}^n)$. If for every integer $r \geq 2$ we have

$$a - \sum_{j=1}^{r-1} a_j \in \Gamma_\rho^{\overline{m}_r}(\mathbb{R}^n \oplus \mathbb{R}^n) \tag{14.10}$$

where $\overline{m}_r = \max_{j \geq r} m_j$, we will write $a \sim \sum_{j=1}^\infty a_j$ and call this relation an asymptotic expansion of the symbol a.

The interest of the notion of asymptotic expansion comes from the fact that every sequence of symbols $(a_j)_j$ with $a_j \in \Gamma_\rho^{m_j}(\mathbb{R}^n \oplus \mathbb{R}^n)$, the degrees m_j being strictly decreasing and such that $m_j \to -\infty$ determines a symbol in some $\Gamma_\rho^m(\mathbb{R}^n \oplus \mathbb{R}^n)$, that symbol being unique up to an element of $\mathcal{S}(\mathbb{R}^n \oplus \mathbb{R}^n)$:

Proposition 323. *Let $(a_j)_j$ be a sequence of symbols $a_j \in \Gamma_\rho^{m_j}(\mathbb{R}^n \oplus \mathbb{R}^n)$ such that $m_j > m_{j+1}$ and $\lim_{j\to+\infty} m \to -\infty$. Then:*

(i) *There exists a function a, such that $a \sim \sum_{j=1}^\infty a_j$.*

(ii) *If another function a' is such that $a' \sim \sum_{j=1}^\infty a_j$, then $a - a' \in \mathcal{S}(\mathbb{R}^n \oplus \mathbb{R}^n)$.*

For the proof of this result we refer to Shubin [147], p. 176. Note that property
(ii) immediately follows from the fact that we have

$$\bigcap_{m\in\mathbb{R}} \Gamma_\rho^m(\mathbb{R}^n \oplus \mathbb{R}^n) = \mathcal{S}(\mathbb{R}^n \oplus \mathbb{R}^n)$$

(cf. Exercise 315).

We also have:

Proposition 324. *Let* $a_j \in \Gamma_\rho^{m_j}(\mathbb{R}^n \oplus \mathbb{R}^n)$, $j = 1, 2, \ldots$, *where* $m_j \to -\infty$ *when*
$j \to +\infty$. *Let* $a \in C^\infty(\mathbb{R}^n \oplus \mathbb{R}^n)$ *be such that for any multi-index* α *there exist*
real constants μ_α *and* C_α *such that the following estimate holds:*

$$|\partial_z^\alpha a(z)| \leqq C_\alpha \langle z\rangle^{\mu_\alpha} . \tag{14.11}$$

Also assume that l_j *and* C_j *are such that* $l_j \to -\infty$ *as* $j \to +\infty$ *and that we have*

$$\left| a(z) - \sum_{j=1}^{r-1} a_j(z) \right| \leqq C_r \langle z\rangle^{l_r} . \tag{14.12}$$

Then we have the asymptotic expansion $a \sim \sum_{j=1}^{\infty} a_j$.

The interest of this lies in the fact that the conditions imposed on the function
a are rather weak: it is only required that a and its derivatives are polynomially
bounded at infinity.

14.2   More general operators ... which are not more general!

Formula (14.1) suggests that it could perhaps be interesting to replace the symbol
$a_\tau((1 - \tau)x + \tau y, \xi)$ by some more arbitrary function $a(x, y, \xi)$ and to consider
more general "Fourier integral operators" of the type

$$\widehat{A}f(x) = \left(\tfrac{1}{2\pi}\right)^n \iint_{\mathbb{R}^{2n}} e^{i\xi\cdot(x-y)}a(x, y, \xi)f(y)dyd\xi \tag{14.13}$$

or, equivalently,

$$\widehat{A}f(x) = \left(\tfrac{1}{2\pi\hbar}\right)^n \iint_{\mathbb{R}^{2n}} e^{\frac{i}{\hbar}p\cdot(x-y)}a(x, y, p)f(y)dydp.$$

We will see that nothing is actually gained in generality.

14.2.1 More general classes of symbols

The consideration of such more general operators motivates the introduction of the following extension of the class $\Gamma_\rho^m(\mathbb{R}^n \oplus \mathbb{R}^n)$:

Definition 325. The symbol class $\Pi_\rho^m(\mathbb{R}^{3n})$ consists of all complex functions $a \in C^\infty(\mathbb{R}^{3n})$ such that for every $\alpha, \beta, \gamma \in \mathbb{N}^n$ there exists a constant $C_{\alpha\beta\gamma} \geq 0$ such that the following estimate holds for some $m' \in \mathbb{R}$:

$$|\partial_\xi^\alpha \partial_x^\beta \partial_y^\gamma a(x,y,\xi)| \leq C_{\alpha\beta\gamma} \, \langle u \rangle^{m-\rho|\alpha+\beta+\gamma|} \, \langle x - y \rangle^{m'+\rho|\alpha+\beta+\gamma|} \,. \qquad (14.14)$$

We are using here the notation $u = (x, y, \xi)$ and $\langle u \rangle = (1 + |x|^2 + |y|^2 + |\xi|^2)^{1/2}$.

The following result, and its corollary, shows that $\Gamma_\rho^m(\mathbb{R}^n \oplus \mathbb{R}^n)$ can be used to construct symbols in $\Pi_\rho^m(\mathbb{R}^{3n})$:

Proposition 326. *Let f be a linear map $\mathbb{R}^{2n} \to \mathbb{R}^n$ such that the linear map $\phi : \mathbb{R}^{2n} \to \mathbb{R}^{2n}$ defined by $\phi(x,y) = (f(x,y), x - y)$ is an isomorphism. Let $b \in \Gamma_\rho^m(\mathbb{R}^{2n})$. Define a function $a \in C^\infty(\mathbb{R}^{3n})$ by the formula*

$$a(x, y, \xi) = b(f(x,y), \xi). \qquad (14.15)$$

Then $a \in \Pi_\rho^m(\mathbb{R}^{3n})$.

Proof. The functions $|x| + |y|$ and $|f(x,y)| + |x - y|$ give equivalent norms on $\mathbb{R}^{2n}$. Therefore, for the proof of the proposition it remains to use the easily verified inequality

$$\frac{(1 + |f(x,y)| + |\xi|)^s}{(1 + |f(x,y)| + |x - y| + |\xi|)^s} \leqq C(1 + |x - y|)^{|s|} \qquad (14.16)$$

valid for $s \in \mathbb{R}$, from which the estimates follow for $a(x, y, \xi)$ with $m' = |m|$. $\square$

Exercise 327. Prove Peetre's inequality $(1 + |a - b|)^s \leq 2^{|s|}(1 + |a|)^s(1 + |b|)^{|s|}$.

Exercise 328. Use Peetre's inequality to prove the inequality (14.16).

It readily follows from the result above that:

Corollary 329. *Let $b \in \Gamma_\rho^m(\mathbb{R}^n \oplus \mathbb{R}^n)$ and define*

$$a_1(x, y, \xi) = b(x, \xi) \quad and \quad a_2(x, y, \xi) = b(y, \xi).$$

Then a_1 and a_2 both belong to the symbol class $\Gamma_\rho^m(\mathbb{R}^{3n})$.

Proof. We have $a_1(x, y, \xi) = b(f(x,y), \xi)$ with $f(x,y) = x$ and $\phi(x,y) = (x, x - y)$ obviously is an isomorphism, hence $a_1 \in \Gamma_\rho^m(\mathbb{R}^{3n})$ in view of Proposition 326; a similar argument shows that we have $a_2 \in \Gamma_\rho^m(\mathbb{R}^{3n})$ as well. $\square$

Let us now give a meaning to the expression (14.13) defining the pseudo-differential operator A. The expression

$$\widehat{A}f(x) = \left(\tfrac{1}{2\pi}\right)^n \iint_{\mathbb{R}^{2n}} e^{i\xi\cdot(x-y)} a(x,y,\xi) f(y)\,dy\,d\xi \qquad (14.17)$$

perfectly makes sense if $a \in \Pi_\rho^m(\mathbb{R}^{3n})$ is compactly supported and $f \in C_0^\infty(\mathbb{R}^n)$ (the compactly supported infinitely differentiable functions); the discussion below actually works as well modulo a few minor modifications if we assume $f \in \mathcal{S}(\mathbb{R}^n)$. In fact the integration in the right-hand side of (14.17) is performed over a compact set, so that we get an absolutely convergent integral. To deal with a general $a \in \Pi_\rho^m(\mathbb{R}^{3n})$ we note that in view of the obvious identities

$$\langle x - y \rangle^{-M} \langle \partial_\xi \rangle^M e^{i\xi\cdot(x-y)} = e^{i\xi\cdot(x-y)},$$
$$\langle \xi \rangle^{-N} \langle \partial_y \rangle^N e^{i\xi\cdot(x-y)} = e^{i\xi\cdot(x-y)}$$

we can rewrite (14.17) as

$$\widehat{A}f(x) = \left(\tfrac{1}{2\pi}\right)^n \iint_{\mathbb{R}^{2n}} e^{i\xi\cdot(x-y)} \langle x - y \rangle^{-M}$$
$$\times \langle \partial_\xi \rangle^M \langle \partial_y \rangle^N \left[\langle \xi \rangle^{-N} a(x,y,\xi) f(y) \right] dy\,d\xi.$$

For an arbitrary $a \in \Pi_\rho^m(\mathbb{R}^{3n})$ we can take this expression as a *definition* of the operator A as soon as we have $m - N < -n$ and $m' + m - M < -n$. The estimates (14.14) imply that the double integral is convergent, and thus defines a continuous function Af. One readily verifies that if we increase M and N then we obtain integrals which are also convergent after differentiation with respect to the variable x, hence $A : C_0^\infty(\mathbb{R}^n) \longrightarrow \mathcal{S}(\mathbb{R}^n)$ (respectively $A : \mathcal{S}(\mathbb{R}^n) \longrightarrow \mathcal{S}(\mathbb{R}^n)$). We leave the details to the reader.

14.2.2　A reduction result

The following result, which we dignify as a theorem, is due to Shubin ([147] Theorem 23.2). It is very important because it shows that nothing is really gained by the consideration of the more "exotic" operators of the type (14.17). It shows in fact that every such pseudo-differential operator with symbol in $\Pi_\rho^m(\mathbb{R}^{3n})$ can be expressed as an operator belonging to the class $\Pi G_\rho^m(\mathbb{R}^n)$, and this in infinitely many ways. The proof of this property is long and technical (it makes use of several Taylor expansions), but we reproduce it in Section 14.5 for the sake of completeness.

Theorem 330. *Let τ be an arbitrary real number. Every pseudo-differential operator $\widehat{A} \in G_\rho^m(\mathbb{R}^{3n})$, that is*

$$\widehat{A}f(x) = \left(\tfrac{1}{2\pi}\right)^n \iint_{\mathbb{R}^{2n}} e^{i\xi\cdot(x-y)} a(x,y,\xi) f(y)\,dy\,d\xi \qquad (14.18)$$

or, equivalently, setting $p = \hbar\xi$:

$$\widehat{A}f(x) = \left(\tfrac{1}{2\pi\hbar}\right)^n \iint_{\mathbb{R}^{2n}} e^{\frac{i}{\hbar}p\cdot(x-y)} a(x, y, \hbar^{-1}p) f(y) \, dy \, dp \tag{14.19}$$

with $a \in \Pi_\rho^m(\mathbb{R}^{3n})$, can be written in the form

$$\widehat{A}f(x) = \left(\tfrac{1}{2\pi}\right)^n \iint_{\mathbb{R}^{2n}} e^{i\xi\cdot(x-y)} a_\tau((1-\tau)x + \tau y, \xi) f(y) \, dy \, d\xi \tag{14.20}$$

where $a_\tau \in \Gamma_\rho^m(\mathbb{R}^n \oplus \mathbb{R}^n)$; respectively

$$\widehat{A}f(x) = \left(\tfrac{1}{2\pi\hbar}\right)^n \iint_{\mathbb{R}^{2n}} e^{\frac{i}{\hbar}p\cdot(x-y)} a_\tau((1-\tau)x + \tau y, \hbar^{-1}p) f(y) \, dy \, dp. \tag{14.21}$$

We have in addition the following asymptotic expansion:

$$a_\tau(x,\xi) \sim \sum_{\beta,\gamma} \frac{1}{\beta!\gamma!} \tau^{|\beta|}(1-\tau)^{|\gamma|} \partial_\xi^{\beta+\gamma}(-D_x)^\beta D_y^\gamma a(x, y, \xi)\big|_{y=x}. \tag{14.22}$$

The following consequence of Theorem 330 is obvious; it links the theory of Weyl operators to Shubin's theory of pseudo-differential operators:

Corollary 331. *Every pseudo-differential operator $\widehat{A}$ of the type*

$$\widehat{A}f(x) = \left(\tfrac{1}{2\pi}\right)^n \iint_{\mathbb{R}^{2n}} e^{i\xi\cdot(x-y)} a(x, y, \xi) f(y) \, dy \, d\xi$$

with $a \in \Pi_\rho^m(\mathbb{R}^{3n})$ is a Weyl operator $\widehat{A} \xleftrightarrow{\text{Weyl}} a_w$ with symbol $a_w \in \Gamma_\rho^m(\mathbb{R}^{2n})$ given by the expression

$$a_w(x,\xi) \sim \sum_{\beta,\gamma} \frac{1}{\beta!\gamma!} \left(\frac{1}{2}\right)^{|\beta|+|\gamma|} \partial_\xi^{\beta+\gamma}(-D_x)^\beta D_y^\gamma a(x, y, \xi)\big|_{y=x}.$$

Proof. It suffices to take $\tau = \frac{1}{2}$ in formula (14.22). $\qquad\square$

The following result shows that the correspondence between operators and symbols (for a given τ) is one-to-one and onto.

Proposition 332. *The τ-symbol of an operator $\widehat{A} \in G_\rho^m(\mathbb{R}^{3n})$ is uniquely defined. Thus, for each value of the parameter τ the correspondence $\widehat{A} \xleftrightarrow{\tau} a$ is bijective.*

Proof. Let us first verify that if $\widehat{A}$ has a kernel $K_{\widehat{A}} \in S(\mathbb{R}^n \oplus \mathbb{R}^n)$, then it has a τ-symbol $a_\tau \in S(\mathbb{R}^n \oplus \mathbb{R}^n)$ and the correspondence between kernel and symbol is a one-to-one correspondence. When $a \in S(\mathbb{R}^n \oplus \mathbb{R}^n)$ it follows from formula (14.20) that the correspondence between kernel and τ-symbol is given by

$$K_{\widehat{A}}(x,y) = \left(\tfrac{1}{2\pi}\right)^{n/2} F_{\xi \to x-y}^{-1} a_\tau((1-\tau)x + \tau y, \xi), \tag{14.23}$$

$$a_\tau(v,\xi) = (2\pi)^{n/2} F_{w \to \xi} K_{\widehat{A}}(v + \tau w, v - (1-\tau)w), \tag{14.24}$$

where formula (14.24) is obtained from formula (14.23) by a change of coordinates and the Fourier inversion formula. In particular, for any $K_{\widehat{A}} \in S(\mathbb{R}^{2n})$, we can find $a_\tau(v, \xi) \in \mathcal{S}(\mathbb{R}^n \oplus \mathbb{R}^n)$ using formula (14.24). $\qquad \square$

The following consequence is of course an obvious extension of formulas (10.14) and (10.15) in Proposition 205) relating the Weyl symbol and the kernel of an operator:

Corollary 333. *In the $\hbar$-dependent case the symbol a and the kernel $K_{\widehat{A}}$ are related by the formulas*

$$K_{\widehat{A}}(x, y) = \left(\tfrac{1}{2\pi\hbar}\right)^n \int_{\mathbb{R}^n} e^{\frac{i}{\hbar}p\cdot(x-y)} a((1-\tau)x + \tau y, p) dp, \qquad (14.25)$$

$$a_\tau(x, p) = \int_{\mathbb{R}^n} e^{-\frac{i}{\hbar}p\cdot y} K_{\widehat{A}}(x + \tau y, x - (1-\tau)y) dy \qquad (14.26)$$

(the integrals being understood as Fourier transforms).

Proof. These formulas are just a formal restatement of (14.23) and (14.24) in the case of $\hbar$-dependent τ-pseudodifferential operators. $\qquad \square$

We next show the uniqueness of the τ-symbol in the general case. For this we note that (14.23) is always true when $\widehat{A}$ is given via a τ-symbol $a_\tau(v, \xi)$ and if the partial Fourier transform, which appears in this formula, is understood in the distributional sense. It follows that the inversion formula is also true, leading to (14.24) after the linear change of coordinates (14.62). Furthermore, the uniqueness of a τ-symbol is obvious in view of formula (14.24), taking into account the uniqueness of the kernel $K_{\widehat{A}}$.

In particular, the Weyl correspondence $\widehat{A} \overset{\text{Weyl}}{\longleftrightarrow} a$ is also bijective. More generally:

Corollary 334. *Every operator $\widehat{A} \in G_\rho^m(\mathbb{R}^n)$ can always be written in any of the following three forms:*

$$\widehat{A}u(x) = \left(\tfrac{1}{2\pi}\right)^n \iint_{\mathbb{R}^{2n}} e^{i(x-y)\cdot\xi} a_\ell(x, \xi) u(y) dy d\xi, \qquad (14.27)$$

$$\widehat{A}u(x) = \left(\tfrac{1}{2\pi}\right)^n \iint_{\mathbb{R}^{2n}} e^{i(x-y)\cdot\xi} a_r(y, \xi) u(y) dy d\xi, \qquad (14.28)$$

$$\widehat{A}u(x) = \left(\tfrac{1}{2\pi}\right)^n \iint_{\mathbb{R}^{2n}} e^{i(x-y)\cdot\xi} a_w(\tfrac{1}{2}(x + y), \xi) u(y) dy d\xi. \qquad (14.29)$$

In the expressions above the symbols a_ℓ, a_r and a_w belong to $\Gamma_\rho^m(\mathbb{R}^n \oplus \mathbb{R}^n)$, and are uniquely defined by $\widehat{A}$.

The case of $\hbar$-dependent operators is immediately obtained by the usual modifications; for instance formula (14.29) becomes

$$\widehat{A}\psi(x) = \left(\tfrac{1}{2\pi\hbar}\right)^n \iint_{\mathbb{R}^{2n}} e^{\frac{i}{\hbar}(x-y)\cdot p} a_w(\tfrac{1}{2}(x+y),p)\psi(y)dydp \qquad (14.30)$$

where one recognizes the usual Weyl correspondence $\widehat{A} \overset{\text{Weyl}}{\longleftrightarrow} a_w$.

These results motivate the following definition:

Definition 335. The functions a_ℓ, a_r and a_w in the formulae (14.27)–(14.29) and corresponding to the choices $\tau = 0$, $\tau = 1$, $\tau = \tfrac{1}{2}$ are called, respectively, the left, right and Weyl symbols of operator $\widehat{A}$. When we speak about a "Weyl pseudo-differential operator $\widehat{A}$" it will be implicitly understood that its symbol is a_w, and it will be written a when no confusion can arise.

In this book we mainly use the Weyl symbol; as already emphasized, this choice is in a sense "natural" in quantum mechanics (especially in deformation quantization) because Weyl operators enjoy very nice covariance properties with respect to the symplectic and metaplectic groups.

In ordinary pseudo-differential calculus one has continuity results in terms of the usual Sobolev spaces $H^s(\mathbb{R}^n)$. Since the vocation of the operators studied in this chapter is to incorporate global behavior, it is appropriate to introduce the following variant of the usual Sobolev spaces:

Definition 336. For $s \in \mathbb{R}$ the global Sobolev space $Q^s(\mathbb{R}^n)$ consists of all $f \in \mathcal{S}'(\mathbb{R}^n)$ such that $L_s f \in L^2(\mathbb{R}^n)$ where $L_s \in G_1^s(\mathbb{R}^n)$ is the operator defined by

$$L_s f(x) = \left(\tfrac{1}{2\pi}\right)^{n/2} \int_{\mathbb{R}^n} e^{ix\cdot\xi}\langle z\rangle^{s/2} Ff(\xi)d\xi.$$

The norm on $Q^s(\mathbb{R}^n)$ is defined by $\|f\|_{Q^s(\mathbb{R}^n)} = \|L_s f\|_{L^2(\mathbb{R}^n)}$.

It turns out that the study of $Q^s(\mathbb{R}^n)$ is best understood within the framework of the modulation spaces we will study in Chapter 17. Let us just mention at this point that $Q^s(\mathbb{R}^n)$ can be equipped with an inner product making it into a Hilbert space, and that we have the equalities

$$\bigcap_{s\in\mathbb{R}} Q^s(\mathbb{R}^n) = \mathcal{S}(\mathbb{R}^n) \ \text{ and } \ \bigcup_{s\in\mathbb{R}} Q^s(\mathbb{R}^n) = \mathcal{S}'(\mathbb{R}^n)$$

and that the following regularity result holds:

Proposition 337. *Every operator $\widehat{A}\in\Gamma_\rho^m(\mathbb{R}^n)$ is continuous $Q^s(\mathbb{R}^n) \longrightarrow Q^{s-m}(\mathbb{R}^n)$. In particular $\widehat{A} : \mathcal{S}(\mathbb{R}^n) \longrightarrow \mathcal{S}'(\mathbb{R}^n)$.*

14.3 Relations between kernels and symbols

In Chapter 10 we briefly discussed the relationship between Weyl operators, their kernels, and their symbols. We are giving here a more detailed study within the framework of τ-symbols. This will allow us to make explicit the relation between the Wigner and the Rihaczek transformations.

14.3.1 A general result

We want to be able to compare the symbols a_τ and $a_{\tau'}$ of a given operator $\widehat{A}$. For instance, given the Kohn–Nirenberg symbol of $\widehat{A}$, what is the Weyl symbol of that operator? The following result (Shubin [147], p. 183) gives a method for passing from one symbol to the other:

Proposition 338. *The symbols a_τ and $a_{\tau'}$ of the same operator $\widehat{A} \in G_\rho^m(\mathbb{R}^n)$ are related by the formula*

$$a_\tau(x,\xi) = e^{i(\tau-\tau')D_x^\alpha \cdot D_\xi^\alpha} a_{\tau'}(x,\xi) \tag{14.31}$$

where $D_x^\alpha \cdot D_\xi^\alpha = \sum_j D_{x_j}^{\alpha_j} \cdot D_{\xi_j}^{\alpha_j}$.

   (i) *The following asymptotic formula holds:*

$$a_\tau(x,\xi) \sim \sum_\alpha \frac{1}{\alpha!}(\tau'-\tau)^{|\alpha|}\partial_\xi^\alpha D_x^\alpha a_{\tau'}(x,\xi). \tag{14.32}$$

   (ii) *We have $a_\tau - a_{\tau'} \in \Gamma_\rho^{m-2\rho}(\mathbb{R}^n \oplus \mathbb{R}^n)$.*

Proof. The second statement immediately follows from the first. Notice that the right-hand side of formula (14.31) is well defined because the exponential is a Fourier multiplier. We begin by making the following remark. Consider a τ-pseudodifferential operator with symbol $a \in \Pi_\rho^m(\mathbb{R}^{3n})$:

$$\widehat{A}f(x) = \left(\tfrac{1}{2\pi}\right)^n \iint_{\mathbb{R}^{2n}} e^{i\xi \cdot (x-y)} a_\tau((1-\tau)x + \tau y, \xi)f(y)dyd\xi.$$

The expression of the τ-symbol in terms of the τ'-symbol for a different τ, can be easily obtained from Theorem 330 in the form of an asymptotic series. Indeed, if an operator $\widehat{A}$ has the τ'-symbol $a_{\tau'}(x,\xi)$, that symbol may be determined via

$$a(x,y,\xi) = a_{\tau'}((1-\tau')x + \tau'y, \xi).$$

In view of formula (14.22) in Theorem 330 the τ-symbol has the asymptotic expansion

$$a_\tau(x,\xi) \sim \sum_{\beta,\gamma} \frac{(-1)^{|\beta|}}{\beta!\gamma!}\tau^{|\beta|}(1-\tau)^{|\gamma|}(1-\tau_1)^{|\beta|}\tau'^{|\gamma|}\partial_\xi^{\beta+\gamma} D_x^{\beta+\gamma} a_{\tau'}$$

or

$$a_\tau(x,\xi) \sim \sum_\alpha c_\alpha \partial_\xi^\alpha D_x^\alpha a_{\tau'}(x,\xi),$$

where c_α is a real constant given by

$$c_\alpha = \sum_{\beta+\gamma=\alpha} \frac{(-1)^{|\beta|}}{\beta!\gamma!} \left[\tau(1-\tau')\right]^{|\beta|} \left[(1-\tau)\tau'\right]^{|\gamma|} ; \qquad (14.33)$$

in particular,we have $c_0 = 1$. Now, transforming (14.33) using Newton's binomial formula, we obtain after a few calculations

$$c_\alpha = \frac{1}{\alpha!} \left[(1-\tau)\tau' e - \tau(1-\tau')e\right]^\alpha$$

where $e = (1,1,\ldots,1)$, that is $c_\alpha = \frac{1}{\alpha!}(\tau' - \tau)^{|\alpha|}$ which yields formula (14.32). Formula (14.31) follows. $\qquad\square$

One can also show that we have precise asymptotic expansions relating the symbols left, right, and Weyl symbols a_ℓ, a_r and a_w:

$$a_\ell(x,\xi) \sim \sum_\alpha \frac{1}{\alpha!} \partial_\xi^\alpha D_y^\alpha a(x,y,\xi)|_{y=x}, \qquad (14.34)$$

$$a_r(y,\xi) \sim \sum_\alpha \frac{1}{\alpha!} \partial_\xi^\alpha (-D_x)^\alpha a(x,y,\xi)|_{y=x}, \qquad (14.35)$$

$$a_w(x,\xi) \sim \sum_{\beta,\gamma} \frac{1}{\beta!\gamma!} \left(\frac{1}{2}\right)^{|\beta+\gamma|} \partial_\xi^{\beta+\gamma} (-D_x)^\beta D_y^\gamma a(x,y,\xi)|_{y=x} . \qquad (14.36)$$

14.3.2 Application: Wigner and Rihaczek distributions

Recall from Corollary 207 that the Weyl symbol of the projection operator $P_\psi :$ $L^2(\mathbb{R}^n) \longrightarrow \{\lambda\psi : \lambda \in \mathbb{C}\}$ ($\psi \neq 0$) is (up to a constant) the Wigner transform of ψ. In fact, more generally, the Weyl symbol of the operator with kernel $K_{\psi,\phi} = (2\pi\hbar)^{-n/2}\psi\otimes\phi$ is the cross-Wigner transform $W(\psi,\phi)$. Let us generalize this result to the case of τ-pseudodifferential operators. Let us rewrite in a more tractable form formulas(14.23) and (14.24) relating the τ-symbol and kernel of an operator

$$\widehat{A}f(x) = \left(\tfrac{1}{2\pi}\right)^n \iint_{\mathbb{R}^{2n}} e^{i\xi\cdot(x-y)} a_\tau((1-\tau)x + \tau y,\xi)f(y)dyd\xi.$$

Writing the Fourier transforms as integrals these formulas become

$$K_{\widehat{A},\tau}(x,y) = \left(\tfrac{1}{2\pi}\right)^n \int_{\mathbb{R}^n} e^{i\xi\cdot(x-y)} a_\tau((1-\tau)x + \tau y,\xi)d\xi, \qquad (14.37)$$

$$a_\tau(x,\xi) = \int_{\mathbb{R}^n} e^{-i\xi\cdot y} K_{\widehat{A},\tau}(x + \tau y, x - (1-\tau)y)dy. \qquad (14.38)$$

Having quantum-mechanical applications in perspective we restate this result in terms of operators depending on $\hbar$:

Lemma 339. *The distributional kernel*

$$K_{\widehat{A},\tau}(x,y) = \left(\tfrac{1}{2\pi\hbar}\right)^n \int_{\mathbb{R}^n} e^{\frac{i}{\hbar}p\cdot(x-y)} a_\tau((1-\tau)x + \tau y, p)dp \qquad (14.39)$$

of the operator $\widehat{A}$ defined by

$$\widehat{A}\psi(x) = \left(\tfrac{1}{2\pi\hbar}\right)^n \iint_{\mathbb{R}^{2n}} e^{\frac{i}{\hbar}p\cdot(x-y)} a_\tau((1-\tau)x + \tau y, \xi)\psi(y)dydp \qquad (14.40)$$

is related to the τ-symbol of that operator by the formula

$$a_\tau(x,p) = \int_{\mathbb{R}^n} e^{-\frac{i}{\hbar}p\cdot y} K_{\widehat{A},\tau}(x + \tau y, x - (1-\tau)y)dy. \qquad (14.41)$$

Proof. Let us check this formula directly. We have, using (14.39),

$$K_{\widehat{A},\tau}(x + \tau y, x - (1-\tau)y) = \left(\tfrac{1}{2\pi\hbar}\right)^n \int_{\mathbb{R}^n} e^{\frac{i}{\hbar}p\cdot y} a_\tau(x,p)dp;$$

formula (14.41) follows since the left-hand side is $(2\pi\hbar)^{-n/2}$ the inverse Fourier transform of a_τ. $\square$

We have shown in Corollary 207 that the cross-Wigner transform

$$W(\psi,\phi)(z) = \left(\tfrac{1}{2\pi\hbar}\right)^n \int_{\mathbb{R}^n} e^{-\frac{i}{\hbar}p\cdot y} \psi(x + \tfrac{1}{2}y)\overline{\phi(x - \tfrac{1}{2}y)}dy$$

is the Weyl symbol of the operator $\widehat{A}_{\psi\otimes\overline{\phi}}$ with kernel $(2\pi\hbar)^{-n}\,\psi\otimes\overline{\phi}$. Thus:

$$\widehat{A}_{\psi\otimes\overline{\phi}} \overset{\text{Weyl}}{\longleftrightarrow} W(\psi,\phi).$$

In time-frequency analysis and signal theory one often uses the so-called Rihaczek distribution (also called Kirkwood–Rihaczek distribution in the literature). Its definition, in units where $\hbar = 1/2\pi$, is

$$R(f,g)(x,\omega) = e^{-2\pi i\omega\cdot x} f(x)\overline{\widehat{g}(\omega)}; \qquad (14.42)$$

here $\widehat{g}$ is the unitary Fourier transform given by

$$\widehat{g}(\omega) = \int_{\mathbb{R}^n} e^{-2\pi i\omega\cdot x} g(x)dx.$$

In more general units we may redefine the Rihaczek distribution in the obvious way:

$$R(\psi,\phi)(x,p) = e^{-\frac{i}{\hbar}p\cdot x} \psi(x)\overline{F\phi(p)} \qquad (14.43)$$

where $F\psi$ is the $\hbar$-dependent Fourier transform. (Some authors interchange ψ and ϕ which amounts to replacing $R(\psi, \phi)$ by $R(\phi, \psi)$.) That this distribution is related to the Wigner distribution as a particular case of a τ-dependent symbol was noticed a long time ago by Kozek [110].

Proposition 340. *Let $\widehat{A}_{\psi\otimes\overline{\phi}}$ be the operator with kernel $K = \psi \otimes \overline{\phi}$. We have:*

$$(a_{\psi\otimes\overline{\phi}})\ell = (2\pi\hbar)^{n/2} R(\psi, \phi). \tag{14.44}$$

Proof. The left symbol corresponds to the choice $\tau = 0$ in formula (14.41) and hence

$$(a_{\psi\otimes\overline{\phi}})\ell(z) = \int_{\mathbb{R}^n} e^{-\frac{i}{\hbar}p\cdot y}\psi(x)\overline{\phi(x-y)}dy$$

$$= \psi(x)\int_{\mathbb{R}^n} e^{-\frac{i}{\hbar}p\cdot y}\overline{\phi(x-y)}dy$$

$$= (2\pi\hbar)^{n/2}\psi(x)e^{-\frac{i}{\hbar}p\cdot x}\overline{F\phi(p)}$$

as claimed. $\qquad\qquad\square$

Exercise 341. Compute the right symbol $(a_{\psi\otimes\overline{\phi}})_r$ corresponding to the choice $\tau = 1$.

14.4 Adjoints and products

The Shubin classes "behave well" under the operations of transposition (or of taking the adjoint) and composition. This leads to some useful formulas.

14.4.1 The transpose and adjoint operators

Let us now consider the transposed operator $\widehat{A}^T$, defined by the formula

$$\left\langle \widehat{A}u, v \right\rangle = \left\langle u, \widehat{A}^T v \right\rangle \quad for \quad u, v \in \mathcal{S}(\mathbb{R}^n),$$

where $\langle \cdot, \cdot \rangle$ is the distributional pairing.

Proposition 342. *If the operator $\widehat{A}$ has a τ-symbol a_τ, then $\widehat{A}^T$ has the $(1-\tau)$-symbol $a_{1-\tau}^T$, given by the formula*

$$a_{1-\tau}^T(x, \xi) = a_\tau(x, -\xi). \tag{14.45}$$

Proof. It is an immediate consequence of the equalities

$$\left\langle \widehat{A}u, v \right\rangle = \left(\tfrac{1}{2\pi}\right)^n \int_{\mathbb{R}^{3n}} e^{i(x-y)\cdot\xi}a_\tau((1-\tau)x + \tau y, \xi)u(y)v(x)dydxd\xi$$

$$= \left(\tfrac{1}{2\pi}\right)^n \int_{\mathbb{R}^{3n}} e^{i(y-x)\cdot\xi}a_\tau((1-\tau)x + \tau y, -\xi)u(y)v(x)dydxd\xi. \qquad\square$$

One can also show using the methods above that one has the following asymptotic formula for the symbol of the transpose:

Proposition 343. *If $\widehat{A} \in G_\rho^m(\mathbb{R}^n)$ then $\widehat{A}^T \in G_\rho^m(\mathbb{R}^n)$ and the τ-symbol $a_\tau^T(x, \xi)$ of $\widehat{A}^T$ can be expressed in terms of the τ-symbol $a_\tau(x, \xi)$ of $\widehat{A}$ by the asymptotic formula*

$$a_\tau^T(x, \xi) \sim \sum_\alpha \frac{1}{\alpha!}(1 - 2\tau)^{|\alpha|}\partial_\xi^\alpha D_x^\alpha a_\tau(x, -\xi). \tag{14.46}$$

We omit the proof since we will not need this result (see Shubin [147]).

Defining the formal adjoint of $\widehat{A}^*$ of $\widehat{A}$ by

$$(\widehat{A}u|v)_{L^2(\mathbb{R}^n)} = (u|\widehat{A}^*v)_{L^2(\mathbb{R}^n)} \quad \text{for } u, v \in \mathcal{S}(\mathbb{R}^n) \tag{14.47}$$

one proves similarly that:

Proposition 344. *If $\widehat{A} \in G_\rho^m(\mathbb{R}^n)$, then $\widehat{A}^* \in G_\rho^m(\mathbb{R}^n)$ and the τ-symbol $a_\tau^*(x, \xi)$ of $\widehat{A}^*$ is related to the $(1 - \tau)$-symbol $a_{1-\tau}(x, \xi)$ of $\widehat{A}$ by the formula*

$$a_\tau^*(x, \xi) = \overline{a_{1-\tau}(x, \xi)}, \tag{14.48}$$

and it can be expressed in terms of the τ-symbol $a_\tau(x, \xi)$ of $\widehat{A}$ via the asymptotic series

$$a_\tau \sim \sum_\alpha \frac{1}{\alpha!}(1 - 2\tau)^{|\alpha|}\partial_\xi^\alpha D_x^\alpha \overline{a_\tau}. \tag{14.49}$$

As an immediate consequence of this result we recover the following well-known property of Weyl operators:

Corollary 345. *If $\widehat{A} \in G_\rho^m(\mathbb{R}^n)$, then the Weyl symbol $(a_w)^*$ of the formal adjoint $\widehat{A}^*$ is given by*

$$(a_w)^*(x, \xi) = \overline{a_w(x, \xi)}. \tag{14.50}$$

In particular, the condition $\widehat{A} = \widehat{A}^*$ is equivalent to the real-valuedness of the Weyl symbol $a(x, \xi)$.

14.4.2 Composition formulas

Another important result is the following, which gives asymptotic formulas for composing operators in the Shubin classes:

Proposition 346. *Let $\widehat{A} \in G_\rho^{m_1}(\mathbb{R}^n)$ and $\widehat{B} \in G_\rho^{m_2}(\mathbb{R}^n)$. Then $\widehat{A}\widehat{B} \in G_\rho^{m_1+m_2}(\mathbb{R}^n)$ and if a_{τ_1} is the τ_1-symbol of $\widehat{A}$ and b_{τ_2} the τ_2-symbol of $\widehat{B}$, then the τ-symbol c_τ of $\widehat{A}\widehat{B}$ has the asymptotic expansion*

$$c_\tau \sim \sum_{\alpha,\beta,\gamma,\delta} c_{\alpha\beta\gamma\delta}(\partial_\xi^\alpha D_x^\beta a_{\tau_1})(\partial_\xi^\gamma D_x^\delta b_{\tau_2}) \tag{14.51}$$

where the sum runs over sets of multi-indices α, β, γ and δ such that $\alpha + \gamma = \beta + \delta$; the $c_{\alpha\beta\gamma\delta}$ are complex constants depending on τ, τ_1 and τ_2 such that $c_{0000} = 1$. In particular, we have

$$c_\tau - a_{\tau_1} b_{\tau_2} \in \Gamma_\rho^{m_1+m_2-2\rho}(\mathbb{R}^n \oplus \mathbb{R}^n).$$

Proof. Taking Proposition 326 into account, we see that it suffices to consider only one arbitrary triple of the numbers τ, τ_1 and τ_2. Let us take for simplicity $\tau_1 = 0$, $\tau_2 = 1$. The operator $\widehat{B}$ can be written, using the symbol $b_1(y, \xi)$, as

$$\widehat{B}u(x) = \left(\tfrac{1}{2\pi}\right)^n \iint_{\mathbb{R}^{2n}} e^{i(x-y)\cdot\xi} b_1(y, \xi) u(y) dy d\xi$$

or, equivalently,

$$F(\widehat{B}u)(\xi) = \int_{\mathbb{R}^n} e^{-iy\cdot\xi} b_1(y, \xi) u(y) dy \tag{14.52}$$

where $F(\widehat{B}u)$ is the Fourier transform of $\widehat{B}u$. The operator $\widehat{A}$ has the form

$$\widehat{A}v(x) = \left(\tfrac{1}{2\pi}\right)^n \iint_{\mathbb{R}^{2n}} e^{i(x-y)\cdot\xi} a_0(x, \xi) u(y) dy d\xi \tag{14.53}$$

$$= \left(\tfrac{1}{2\pi}\right)^n \int_{\mathbb{R}^n} e^{ix\cdot\xi} a_0(x, \xi) Fu(\xi) d\xi. \tag{14.54}$$

From (14.52) and (14.53) it follows that

$$\widehat{A}\widehat{B}u(x) = \left(\tfrac{1}{2\pi}\right)^n \iint_{\mathbb{R}^{2n}} e^{i(x-y)\cdot\xi} a_0(x, \xi) b_1(y, \xi) u(y) dy d\xi, \tag{14.55}$$

i.e., $\widehat{A}\widehat{B}$ has symbol

$$c(x, y, \xi) = a_0(x, \xi) b_1(y, \xi) \in \Pi_\rho^{m_1+m_2}(\mathbb{R}^{3n}).$$

It follows that we have $\widehat{A}\widehat{B} \in G_\rho^{m_1+m_2}(\mathbb{R}^n)$. In view of Theorem 330 we have the asymptotic expansion

$$c_\tau(x, \xi) \sim \sum_{\beta,\gamma} \frac{(-1)^{|\beta|}}{\beta! \gamma!} \tau^{|\beta|} (1-\tau)^{|\gamma|} \partial_\xi^{\beta+\gamma} \left[(D_x^\beta a_0(x, \xi))(D_x^\gamma b_1(x, \xi))\right]. \tag{14.56}$$

Using Leibniz's differentiation rule for functions of several variables this formula can be rewritten as

$$c_\tau(x, \xi) \sim \sum \frac{(-1)^{|\beta|}(\beta+\gamma)!}{\beta! \gamma! \delta! \varepsilon!} \tau^{|\beta|} (1-\tau)^{|\gamma|} (\partial_\xi^\delta D_x^\beta a_0)(\partial_\xi^\varepsilon D_x^\gamma b_1), \tag{14.57}$$

where the sum is taken over all $\beta, \gamma, \delta, \varepsilon$ such that $\delta + \varepsilon = \beta + \gamma$, which is the same thing as (14.51). $\qquad\square$

Inserting into (14.57) the expressions for a'_0, a''_1 in terms a'_{τ_1}, a''_{τ_2} we obtain formulae for the coefficients $c_{\alpha\beta\gamma\delta}$ in (14.51), which may sometimes be simplified. For instance, one can show that, similarly to Proposition (346) one has

$$a_0(x,\xi) \sim \sum_\alpha \frac{1}{\alpha!} \partial_\xi^\alpha a'_0(x,\xi) D_x^\alpha a''_0(x,\xi). \tag{14.58}$$

In the case of Weyl operators we get the following important explicit result:

Corollary 347. *If* $\widehat{A} \in G_\rho^{m_1}(\mathbb{R}^n)$ *and* $\widehat{B} \in G_\rho^{m_2}(\mathbb{R}^n)$, *then the Weyl symbol* c^w *of* $\widehat{C} = \widehat{A}\widehat{B}$ *is given by*

$$c^w \sim \sum_{\alpha,\beta} \frac{(-1)^{|\beta|}}{\alpha!\beta!} 2^{-|\alpha+\beta|} (\partial_\xi^\alpha D_x^\beta a_w))(\partial_\xi^\beta D_x^\alpha b^w). \tag{14.59}$$

When $\widehat{A}$, $\widehat{B}$, *and* $\widehat{C}$ *are expressed in the* $\hbar$-*dependent form then this formula becomes*

$$c^w \sim \sum_{\alpha,\beta} \frac{(-1)^{|\beta|}}{\alpha!\beta!} \left(\frac{\hbar}{2}\right)^{|\alpha+\beta|} (\partial_\xi^\alpha D_x^\beta a_w))(\partial_\xi^\beta D_x^\alpha b^w). \tag{14.60}$$

Proof. It is immediate, using formula (14.57) above. $\qquad\square$

The property in the following problem is at the origin of many results in the classical theory of pseudodifferential operators:

Problem 348. Show that the left symbol a_ℓ of an operator $\widehat{A} \in G_\rho^m(\mathbb{R}^n)$ can be expressed in terms of $\widehat{A}$ by the formula

$$a_\ell(x,\xi) = e^{-ix\cdot\xi}\widehat{A}(e^{ix\cdot\xi}) \tag{14.61}$$

where $\widehat{A}$ acts on the variable x.

For the applications to quantum mechanics (especially deformation quantization and its variants which will be studied in Chapter 19) the following result for operators with $\hbar$-dependent symbols is essential:

Theorem 349. *Let* $\widehat{A} \in S_{\rho,\delta}^{m,\mu}(\mathbb{R}^n)$ *and* $\widehat{B} \in S_{\rho',\delta'}^{m',\mu'}(\mathbb{R}^n \oplus \mathbb{R}^n)$. *Then* $\widehat{C} = \widehat{A}\widehat{B} \in S_{\rho'',\delta''}^{m+m',\mu+\mu'}(\mathbb{R}^n)$ *with* $\rho'' = \min(\rho,\rho')$, $\delta'' = \max(\delta,\delta')$ *and the symbol* $c \overset{\text{Weyl}}{\longleftrightarrow} \widehat{C}$ *has the expansion*

$$c = \sum_{|\alpha+\beta|<N} \frac{(-1)^{|\beta|}}{\alpha!\beta!} \left(\frac{-i\hbar}{2}\right)^{|\alpha+\beta|} (\partial_\xi^\alpha \partial_x^\beta a)(\partial_x^\alpha \partial_\xi^\beta a) + \hbar^N r_N$$

with the following condition on the term r_N:

$$r_N \in \Sigma_{\rho,\delta}^{m+m'-N(\rho+\rho'),\mu+\mu'-N(\delta+\delta')}(\mathbb{R}^n \oplus \mathbb{R}^n).$$

The proof of this result is long and technical; see [147], Appendix A.2.5, pp. 245–248.

14.5 Proof of Theorem 330

It is of course sufficient to give the proof in the case $\hbar = 1$.

Setting $v = (1 - \tau)x + \tau y$ and $w = x - y$ in $a(x, y, \xi)$, that is, equivalently,

$$x = v + \tau w \quad , \quad y = v - (1 - \tau)\,w \tag{14.62}$$

we have

$$a(x, y, \xi) = a(v + \tau w, v - (1 - \tau)\,w, \xi). \tag{14.63}$$

Expanding the right-hand side of (14.63) in a Taylor series at $w = 0$, we get $a = a_N + r_N$ where

$$a_N(x, y, \xi) = \sum_{|\beta+\gamma| \le N-1} \frac{(-1)^{|\gamma|}}{\beta!\gamma!} \tau^{|\beta|} (1 - \tau)^{|\gamma|} (x - y)^{\beta+\gamma} (\partial_x^\beta \partial_y^\gamma a)(v, v, \xi) \tag{14.64}$$

and the remainder term r_N is given by the formula

$$r_N(x, y, \xi) = \sum_{|\beta+\gamma|=N} c_{\beta\gamma}(x - y)^{\beta+\gamma} I_{\beta\gamma}(x, y, \xi) \quad \text{with} \tag{14.65a}$$

$$I_{\beta\gamma}(x, y, \xi) = \int_0^1 (1 - t)^{N-1} (\partial_x^\beta \partial_y^\gamma a)(v + t\tau w, v - t(1 - \tau)w, \xi)dt \tag{14.65b}$$

where the $c_{\beta\gamma}$ are constants. In (14.64) the expression $(\partial_x^\beta \partial_y^\gamma a)(v, v, \xi)$ signifies that we have replaced x and y with $v = (1 - \tau)\,x + \tau y$ in the expression $\partial_x^\beta \partial_y^\gamma a(x, y, \xi)$. The expression

$$(\partial_x^\beta \partial_y^\gamma a)(v + t\tau w, v - t(1 - \tau)w, \xi)$$

in (14.65) should be understood in a similar way. We next note that the operator with symbol $(x - y)^{\beta+\gamma}(\partial_x^\beta \partial_y^\gamma a)(v, v, \xi)$ is the same as the one with symbol

$$(-D_\xi)^{\beta+\gamma}(\partial_x^\beta \partial_y^\gamma a)(v, v, \xi) = (-1)^{|\beta|+|\gamma|}(\partial_\xi^{\beta+\gamma} D_x^\beta D_y^\gamma a)(v, v, \xi).$$

It follows from (14.64) that $\widehat{A} = \widehat{A}_N + \widehat{R}_N$ where $\widehat{A}_N$ is an operator with τ-symbol

$$a_N(x, \xi) = \sum_{|\beta+\gamma| \le N-1} \frac{1}{\beta!\gamma!} \tau^{|\beta|}(1 - \tau)^{|\gamma|} \partial_\xi^{\beta+\gamma}(-D_x)^\xi D_y^\gamma a(x, y, \xi)|_{y=x}$$

and $\widehat{R}_N$ is an operator with symbol r_N. Note that the operator $\widehat{R}_N$ is a linear combination of a finite number of terms having symbols of the type

$$\int_0^1 (\partial_\xi^{\beta+\gamma} \partial_x^\beta \partial_y^\gamma a)(v + t\tau w, v - t(1 - \tau)w, \xi)(1 - t)^{N-1}dt \tag{14.66}$$

with $|\beta + \gamma| = N$. Let us now show that the r_N symbol belongs to the class $\Pi_\rho^{m-2N\rho}(\mathbb{R}^{3n})$. For this it suffices to show that this is true for the integrand in (14.66), with all estimates uniform in t (note that this is obvious for each fixed $t \neq 0$ and true for $t = 0$ by Proposition 326). Using the trivial relations

$$v = (1 - \tau)(v + t\tau w) + \tau(v - t(1 - \tau)w),$$
$$tw = (v + t\tau w) - (v - t(1 - \tau)w)$$

it is easy to see that there exists a constant $C > 0$ independent of $t \in [0, 1]$ such that

$$C^{-1} \leq \frac{|v + t\tau w| + |v - t(1 - \tau)w|}{|v| + |tw|} \leq C$$

and we thus have the estimate

$$\left|(\partial_\xi^{\beta+\gamma}\partial_x^\beta\partial_y^\beta a)(v + t\tau w, v - t(1 - \tau)w, \xi)\right|$$
$$\leq C(1 + |v| + |tv| + |\xi|)^{m - 2\rho N}(1 + |tw|)^{m' + 2\rho N}.$$

Since for $m' + 2\rho N \geq 0$ we have the inequality

$$(1 + |tw|)^{m'+2\rho N} \leq (1 + |v| + |tv| + |\xi|)^{m'+2\rho N}(1 + |v| + |\xi|)^{-(m'+2\rho N)},$$

it is clear that if, in addition, $m + m' \geq 0$ and $m - 2\rho N \leq 0$, then

$$\left|(\partial_\xi^{\beta+\gamma}\partial_x^\beta\partial_y^\beta a)(v + t\tau w, v - t(1 - \tau)w, \xi)\right|$$
$$\leq C'(1 + |v| + |\xi|)^{-m'-2\rho N}(1 + |v| + |tw| + |\xi|)^{m'+m}$$
$$\leq C'(1 + |v| + |\xi|)^{m - 2\rho N}(1 + |w|)^{m'+m}$$
$$\leq C'(1 + |v| + |w| + |\xi|)^{m - 2\rho N}(1 + |w|)^{m'+2m+2\rho N}$$

where C' is independent of t. One estimates the derivatives in a similar way. Now, let the symbol $b'(x, \xi) \in \Gamma_\varrho^m(\mathbb{R}^n \oplus \mathbb{R}^n)$ be such that

$$b'(x, \xi) \sim \sum_{N=0}^{\infty}(b_N(x, \xi) - b_{N-1}(x, \xi)).$$

Then, if $\widehat{A}'$ has τ-symbol $b'(x, \xi)$ it is clear that the kernel of the operator $\widehat{A} - \widehat{A}'$ is in $\mathcal{S}(\mathbb{R}^{2n})$.

Part IV

Applications

Chapter 15

The Schrödinger Equation

Schrödinger's equation

$$i\hbar\frac{\partial\psi}{\partial t}(x,t) = \widehat{H}\psi(x,t)$$

is considered by physicists as a postulate that cannot be rigorously derived from Hamiltonian mechanics. It is however well known that in the case of linear Hamiltonian flows, the Schrödinger equation is obtained using the metaplectic representation: this will be proven in the first part of this chapter (Section 15.1). In the second part of this chapter (Section 15.2), which is somewhat tentative in the sense that we pay little attention to domain questions, we will show that the Schrödinger equation can be derived using Stone's theorem on strongly continuous one-parameter groups of unitary operators, if one requires in addition that quantum states satisfy a certain covariance property. This derivation is made possible thanks to Theorem 356 (essentially due to Wong [163]) which says that Weyl calculus is the *only* symplectically covariant pseudo-differential theory. We mention that a more physical version of these results is to appear in de Gosson and Hiley [73] (not surprisingly leading to negative emotional reactions from some physicists); for those interested in the physical aspects (including the scientific ontology) of quantum mechanics, we recommend the texts [92, 93, 94, 95] by Hiley and collaborators.

15.1 The case of quadratic Hamiltonians

Here is again a somewhat technical section; it gives explicit formulae for the solutions of the Schrödinger equation associated with a Hamiltonian function which is quadratic. Most physicists know these explicit formulae, but prefer to invoke the "Feynman path integral" for their derivation. However the Feynman integral as used in Physics has no precise mathematical meaning (even if it may be a useful heuristic tool in some circumstances; see Schulman [144] for an exposition of the techniques related to path integrals).

15.1.1 Preliminaries

It turns out that Schrödinger's equation can be solved explicitly when the operator $\widehat{H}$ is the Weyl operator associated with a Hamiltonian function which is a quadratic polynomial in the position and momentum variables:

$$H(z) = \tfrac{1}{2} M z \cdot z$$

where M is some (arbitrary) real $2n \times 2n$ symmetric matrix. Thus, H is a real quadratic form in the x_j, p_j variables. To H we associate a partial differential operator $\widehat{H}$ by applying the ordering rules (1.34) discussed in the first chapter:

$$x \longrightarrow \widehat{x} \ , \quad p \longrightarrow \widehat{p} \ , \quad px \longrightarrow \frac{1}{2}(\widehat{x}\widehat{p} + \widehat{p}\widehat{x}) \tag{15.1}$$

where $\widehat{x}$ is the operator of multiplication by x and $\widehat{p} = -i\hbar\partial/\partial x$ (this prescription is sufficient to determine unambiguously the operator $\widehat{H}$ since H is a polynomial of degree 2). Writing

$$M = \begin{pmatrix} H_{xx} & H_{xp} \\ H_{px} & H_{pp} \end{pmatrix} \ , \quad H_{px} = H_{xp}^T$$

we have the explicit formula

$$\widehat{H} = -\frac{\hbar^2}{2} H_{pp}\partial_x \cdot \partial_x - i\hbar H_{px}x \cdot \partial_x + \frac{1}{2}H_{xx}x \cdot x - \frac{i}{2}\operatorname{Tr}(H_{px}). \tag{15.2}$$

One verifies, using the fact that

Problem 350. Prove formula (15.2) above and show by a direct calculation that $\widehat{H}^* = \widehat{H}$. What is $\widehat{H}$ when $H(x,p) = p \cdot x$?

The associated Schrödinger equation is

$$i\hbar\frac{\partial \psi}{\partial t}(x,t) = \widehat{H}\psi(x,t)$$

where ψ is a function (or distribution) in the x, t variables. We are going to see that this equation can be explicitly solved using the theory of the metaplectic group. The solutions will in fact be expressed, except for exceptional values of time, as quadratic Fourier transforms.

To prove this remarkable fact we will have to find an explicit description of the Lie algebra of the metaplectic group (on the abstract level, this Lie algebra is identical with that of the symplectic group).

15.1.2 Quadratic Hamiltonians

Let us begin by shortly discussing the properties of quadratic Hamiltonians. These intervene in many interesting problems from classical and quantum mechanics (for instance in the study of motion near equilibrium, or for the calculation of the energy spectrum of an electron in a uniform magnetic field).

Let H be a homogeneous polynomial in $z \in \mathbb{R}^{2n}$ and with coefficients depending on $t \in \mathbb{R}$:

$$H(z,t) = \tfrac{1}{2}H''(t)z \cdot z \tag{15.3}$$

($H''(t) = D_z^2 H(z,t)$ is the Hessian matrix of H in the variables $z = (x,p)$); the associated Hamilton equations can be written

$$\dot{z}(t) = JH''(t)(z(t)). \tag{15.4}$$

Recall the following notation: $(S_{t,t'}^H)$ is the time-dependent flow determined by H, that is if $t \longmapsto z(t)$ is the solution of (15.4) with $z(t') = z'$ then

$$z(t) = S_{t,t'}^H(z'). \tag{15.5}$$

We will set $S_{t,0}^H = S_t^H$. Assume that H does not depend on t; then (S_t^H) is the one-parameter subgroup of $\mathrm{Sp}(2n,\mathbb{R})$ given by

$$S_t^H = e^{tJH''}.$$

Conversely, if (S_t) is an arbitrary one-parameter subgroup of $\mathrm{Sp}(2n,\mathbb{R})$ then $S_t = e^{tX}$ for some $X \in \mathfrak{sp}(2n,\mathbb{R})$ and we have $(S_t) = (S_t^H)$ where H is the quadratic Hamiltonian

$$H(z) = -\tfrac{1}{2}JXz \cdot z. \tag{15.6}$$

The following elementary result gives a useful relation between the Poisson bracket of quadratic Hamiltonian functions and commutators in the symplectic Lie algebra:

Lemma 351. *Let H and K be two quadratic Hamiltonians associated by (15.6) to $X, Y \in \mathfrak{sp}(2n,\mathbb{R})$, that is*

$$H(z) = -\tfrac{1}{2}JXz \cdot z \ , \ \ K(z) = -\tfrac{1}{2}JYz \cdot z.$$

The Poisson bracket $\{H,K\}$ is the quadratic Hamiltonian given by

$$\{H,K\}(z) = -\tfrac{1}{2}J[X,Y]z \cdot z$$

where $[X,Y] = XY - YX$.

Proof. We have

$$\begin{aligned}
\{H,K\}(z) &= -\sigma(X_H(z), X_K(z)) \\
&= -\sigma(Xz, Yz) \\
&= -JXz \cdot Yz \\
&= -Y^T JXz \cdot z.
\end{aligned}$$

Now

$$Y^T J X z \cdot z = \tfrac{1}{2}(Y^T J X - X^T J Y) z \cdot z$$

that is, since $Y^T J$ and $X^T J$ are symmetric,

$$Y^T J X z \cdot z = -\tfrac{1}{2} J (YX - XY) z \cdot z,$$

whence

$$\{H, K\}(z) = \tfrac{1}{2} J (YX - XY) z \cdot z$$

which we set out to prove. $\qquad\qquad\qquad\qquad\qquad\qquad\qquad\qquad\square$

The metaplectic group $\mathrm{Mp}(2n, \mathbb{R})$ is a covering group of $\mathrm{Sp}(2n, \mathbb{R})$; it follows from the general theory of Lie groups that the Lie algebra $\mathfrak{mp}(2n, \mathbb{R})$ of $\mathrm{Mp}(2n, \mathbb{R})$ is isomorphic to $\mathfrak{sp}(2n, \mathbb{R})$ (the Lie algebra of $\mathrm{Sp}(2n, \mathbb{R})$). We are going to construct explicitly an isomorphism $F : \mathfrak{sp}(2n, \mathbb{R}) \longrightarrow \mathfrak{mp}(2n, \mathbb{R})$ making the following diagram commutative:

$$\begin{array}{ccc}
\mathfrak{mp}(2n, \mathbb{R}) & \xrightarrow{F^{-1}} & \mathfrak{sp}(2n, \mathbb{R}) \\
\exp \downarrow & & \downarrow \exp \\
\mathrm{Mp}(2n, \mathbb{R}) & \xrightarrow[\pi^{\mathrm{Mp}}]{} & \mathrm{Sp}(2n, \mathbb{R}) \, .
\end{array} \qquad (15.7)$$

Recall that $\mathfrak{sp}(2n, \mathbb{R})$ denotes the Lie algebra of the symplectic group $\mathrm{Sp}(2n, \mathbb{R})$; we will call $\mathfrak{mp}(2n, \mathbb{R})$ the "metaplectic algebra".

The following result generalizes Lemma 351:

Theorem 352.

(i) *The linear mapping F, which to $X \in \mathfrak{sp}(2n, \mathbb{R})$ associates the anti-Hermitian operator $F(X) = -\frac{i}{\hbar} \widehat{H}$ where $\widehat{H} \overset{\mathrm{Weyl}}{\longleftrightarrow} H$ with H given by (15.6), is injective, and we have the equality*

$$[F(X), F(X')] = F([X, X']) \qquad (15.8)$$

for all X, $X' \in \mathfrak{sp}(2n, \mathbb{R})$;

(ii) *The image $F(\mathfrak{sp}(2n, \mathbb{R}))$ of F is the metaplectic algebra $\mathfrak{mp}(2n, \mathbb{R})$.*

Proof. It is clear that the mapping F is linear and injective. Consider the matrices

$$X_{jk} = \begin{pmatrix} \Delta_{jk} & 0 \\ 0 & -\Delta_{jk} \end{pmatrix}, \; Y_{jk} = \frac{1}{2} \begin{pmatrix} 0 & \Delta_{jk} + \Delta_{kj} \\ 0 & 0 \end{pmatrix},$$

$$Z_{jk} = \frac{1}{2} \begin{pmatrix} 0 & 0 \\ \Delta_{jk} + \Delta_{kj} & 0 \end{pmatrix} \quad (1 \leq j \leq k \leq n)$$

with 1 the only non-vanishing entry at the jth row and kth column; these matrices form a basis of $\mathfrak{sp}(2n, \mathbb{R})$ (you were asked to check this in Problem 79).

For notational simplicity we will assume that $n = 1$ and set $X = X_{11}$, $Y = Y_{11}$, $Z = Z_{11}$:

$$X = \begin{pmatrix} 1 & 0 \\ 0 & -1 \end{pmatrix} , \ Y = \begin{pmatrix} 0 & 1 \\ 0 & 0 \end{pmatrix} , \ Z = \begin{pmatrix} 0 & 0 \\ 1 & 0 \end{pmatrix};$$

the case of general n is studied in an exactly similar way. To the matrices X, Y, Z correspond via formula (15.6) the Hamiltonians

$$H_X = px \ , \ H_Y = \tfrac{1}{2}p^2, \ H_Z = -\tfrac{1}{2}x^2.$$

The operators $\widehat{H}_X = F(X)$, $\widehat{H}_Y = F(Y)$, $\widehat{H}_Z = F(Z)$ form a basis of the vector space $F(\mathfrak{sp}(1))$; they are given explicitly by the formulas

$$\widehat{H}_X = -i\hbar x \partial_x - \tfrac{1}{2}i\hbar \ , \ \widehat{H}_Y = -\tfrac{1}{2}\hbar^2 \partial_x^2 \ , \ \widehat{H}_Z = -\tfrac{1}{2}x^2.$$

Let us show that formula (15.8) holds. In view of the linearity of F it is sufficient to check that

$$[\widehat{H}_X, \widehat{H}_X] = \widehat{H}_{[X,Y]},$$
$$[\widehat{H}_X, \widehat{H}_Z] = \widehat{H}_{[X,Z]},$$
$$[\widehat{H}_Y, \widehat{H}_Z] = \widehat{H}_{[Y,Z]}.$$

We have thus proved that F is a Lie algebra isomorphism.

To show that $F(\mathfrak{sp}(2,\mathbb{R})) = \mathfrak{mp}(2,\mathbb{R})$ it is thus sufficient to check that the one-parameter groups

$$t \longmapsto U_t = e^{-\frac{i}{\hbar}\widehat{H}_X t},$$
$$t \longmapsto V_t = e^{-\frac{i}{\hbar}\widehat{H}_Y t},$$
$$t \longmapsto W_t = e^{-\frac{i}{\hbar}\widehat{H}_Z t}$$

are subgroups of Mp(1). Let $\psi_0 \in \mathcal{S}(\mathbb{R})$ and set $\psi(x,t) = U_t \psi_0(x)$. The function ψ is the unique solution of the Cauchy problem

$$i\hbar \frac{\partial \psi}{\partial t} = -(i\hbar x \partial_x + \tfrac{1}{2}i\hbar)\psi \ , \ \psi(\cdot,0) = \psi_0.$$

A straightforward calculation (using for instance the method of characteristics) yields

$$\psi(x,t) = e^{-t/2}\psi_0(e^{-t}x)$$

hence the group (U_t) is given by $U_t = \widehat{M}_{L(t),0}$ where $L(t) = e^{-t}$ and we thus have $U_t \in \mathrm{Mp}(1)$ for all t. Leaving the detailed calculations to the reader one similarly verifies that

$$V_t \psi_0(x) = \left(\tfrac{1}{2\pi i \hbar t}\right)^{1/2} \int e^{\frac{i}{2\hbar t}(x-x')^2} \psi_0(x')dx',$$

$$W_t \psi_0(x) = e^{-\frac{1}{2\hbar}x^2} \psi_0(x)$$

so that V_t is a quadratic Fourier transform corresponding to the generating function $W = (x-x')^2/2t$ and W_t is the operator $\widehat{V}_{-tI}$; in both cases we have operators belonging to Mp(1). $\square$

We leave it to the reader to check that the diagram (15.7) is commutative.

Exercise 353. Show, using the generators of $\mathfrak{mp}(2n,\mathbb{R})$ and $\mathfrak{sp}(2n,\mathbb{R})$, that $\exp \circ F^{-1} = \pi^{\mathrm{Mp}} \circ \exp$ where $\exp$ is a collective notation for the exponentials $\mathfrak{sp}(2n,\mathbb{R}) \longrightarrow \mathrm{Sp}(2n,\mathbb{R})$ and $\mathfrak{mp}(2n,\mathbb{R}) \longrightarrow \mathrm{Mp}(2n,\mathbb{R})$.

15.1.3 Exact solutions of the Schrödinger equation

Let us apply the result above to the Schrödinger equation associated to a quadratic Hamiltonian (15.3). Since $\mathrm{Mp}(2n,\mathbb{R})$ covers $\mathrm{Sp}(2n,\mathbb{R})$ it follows from the unique path lifting theorem from the theory of covering manifolds that we can lift the path $t \longmapsto S_{t,0}^H = S_t^H$ in a unique way into a path $t \longmapsto \widehat{S}_t^H$ in $\mathrm{Mp}(2n,\mathbb{R})$ such that $\widehat{S}_0^H = I$. Let $\psi_0 \in \mathcal{S}(\mathbb{R}_x^n)$ and set

$$\psi(x,t) = \widehat{S}_t\psi_0(x).$$

It turns out that ψ satisfies Schrödinger's equation

$$i\hbar\frac{\partial\psi}{\partial t} = \widehat{H}\psi$$

where $\widehat{H}$ is the Weyl operator $\widehat{H} \xleftrightarrow{\text{Weyl}} H$. The results in the following problem will be used to prove this property:

Problem 354.

(i) Verify that the operator $\widehat{H}$ is given by

$$\widehat{H} = -\frac{\hbar^2}{2}H_{pp}\partial_x \cdot \partial_x - i\hbar H_{px}x \cdot \partial_x + \frac{1}{2}H_{xx}x \cdot x - \frac{i}{2}\mathrm{Tr}(H_{px}) \qquad (15.9)$$

where $\mathrm{Tr}(H_{px})$ is the trace of the matrix H_{px}. (H_{pp}, H_{px}, H_{xx} are the matrices of second derivatives of H in the corresponding variables.)

(ii) Let $\widehat{H} \xleftrightarrow{\text{Weyl}} H$, $\widehat{K} \xleftrightarrow{\text{Weyl}} K$ where H and K are quadratic Hamiltonians (15.3). Show that

$$[\widehat{H}, \widehat{K}] = i\hbar\widehat{\{H,K\}} = -i\hbar\sigma(X_H, X_K)$$

where $\{\cdot,\cdot\}$ is the Poisson bracket.

Let us now show that $\psi = \widehat{S}_t\psi_0$ is a solution of Schrödinger's equation as claimed:

Corollary 355. *Let* $t \longmapsto \widehat{S}_t$ *be the lift to* $\mathrm{Mp}(2n,\mathbb{R})$ *of the flow* $t \longmapsto S_t^H$. *For every* $\psi_0 \in \mathcal{S}(\mathbb{R}_x^n)$ *the function* ψ *defined by* $\psi(x,t) = \widehat{S}_t \psi_0(x)$ *is a solution of the partial differential equation*

$$i\hbar \frac{\partial \psi}{\partial t} = \widehat{H}\psi \ , \ \ \psi(\cdot, 0) = \psi_0$$

where $\widehat{H} \overset{\mathrm{Weyl}}{\longleftrightarrow} H$. *Equivalently, the function* $t \longmapsto \widehat{S}_t$ *solves the abstract differential equation*

$$i\hbar \frac{d}{dt}\widehat{S}_t = \widehat{H}\widehat{S}_t \ , \ \ \widehat{S}_0 = I.$$

Proof. We have

$$i\hbar \frac{\partial \psi}{\partial t} = i\hbar \lim_{\Delta t \to 0} \left[\frac{1}{\Delta t}(\widehat{S}_{\Delta t} - I)\right] \widehat{S}_t \psi_0,$$

hence it suffices to show that

$$\lim_{\Delta t \to 0} \left[\frac{1}{\Delta t}(\widehat{S}_{\Delta t} - I)\right] f = \widehat{H}f$$

for every function $f \in \mathcal{S}(\mathbb{R}^n)$. But this equality is an immediate consequence of Theorem 352. $\qquad\square$

Here is an example. For instance, in the case $n = 1$ the generating function for the harmonic oscillator Hamiltonian

$$H = \frac{p^2}{2m} + \frac{m\omega^2}{2}x^2 \tag{15.10}$$

is given by

$$W(x, x'; t, t') = \frac{m\omega}{2\sin\omega(t - t')}\left[(x^2 + x'^2)\cos\omega(t - t') - 2xx'\right], \tag{15.11}$$

hence the solution of the corresponding Schrödinger's equation

$$i\hbar \frac{\partial \psi}{\partial t} = \left(-\frac{\hbar^2}{2m}\frac{\partial^2}{\partial x^2} + \frac{m\omega^2}{2}x^2\right)\psi$$

is explicitly given for $t \notin \pi\mathbb{Z}$, by

$$\psi(x, t) = \left(\tfrac{1}{2\pi i\hbar}\right)^{1/2} i^{-[\omega(t-t')/\pi]}\sqrt{\left|\frac{m\omega}{\sin\omega t}\right|} \int_{-\infty}^{+\infty} e^{\frac{i}{\hbar}W(x,x;t,t')}\psi_0(x')dx'.$$

Physicists often see this formula as an application of the so-called "Feynman path integral", which is supposed to be calculable explicitly for the harmonic oscillator (and some other Hamiltonians; see Schulman [144]). However, the path integral is only a heuristic device, and "derivations" using this device are mathematically meaningless. (See Kumano-go [112] for a rigorous theory of the path integral using time-slicing methods.)

15.2   The general case

While the fact that Schrödinger's equation can be derived using the metaplectic representation has been known for a long time in mathematics (at least among people working in harmonic analysis and representation theory), the material we present in this section is new.

15.2.1   Symplectic covariance as a characteristic property of Weyl quantization

We have seen – and very much emphasized! – the fact that Weyl calculus is a symplectically covariant theory. In particular, we proved in Chapter 10, Theorem 128, that if $S \in \mathrm{Sp}(2n, \mathbb{R})$ then

$$\left[\widehat{A} \overset{\text{Weyl}}{\longleftrightarrow} a\right] \iff \left[a \circ S \overset{\text{Weyl}}{\longleftrightarrow} \widehat{S}^{-1}\widehat{A}\widehat{S}\right] \tag{15.12}$$

where $\widehat{S} \in \mathrm{Mp}(2n, \mathbb{R})$ is any one of the two metaplectic operators such that $S = \pi^{\mathrm{Mp}}(\widehat{S})$.

We are going to see that this property is characteristic of Weyl pseudo-differential operators. More precisely: we have introduced in Chapter 14 a very general notion of pseudo-differential operators, of the type

$$A\psi(x) = \left(\tfrac{1}{2\pi\hbar}\right)^n \iint_{\mathbb{R}^n \times \mathbb{R}^n} e^{\frac{i}{\hbar} p \cdot (x-y)} a(x, y, p)\psi(y)\,dy\,dp.$$

We then showed that such operators can be written, for every value of the real parameter τ in the form

$$A\psi(x) = \left(\tfrac{1}{2\pi\hbar}\right)^n \iint_{\mathbb{R}^n \times \mathbb{R}^n} e^{\frac{i}{\hbar} p \cdot (x-y)} a_\tau((1 - \tau)x + \tau y, p)\psi(y)\,dy\,dp$$

extending the quantizations $px \longrightarrow \tau \widehat{xp} + (1 - \tau)\widehat{px}$; the symbol a_τ is uniquely determined by the operator A, the particular choice $\tau = \tfrac{1}{2}$ corresponding to the Weyl correspondence. It is legitimate to ask the question whether the symplectic covariance property (15.12) still holds for the general correspondence $A \longleftrightarrow a_\tau$, at least for some privileged values of the parameter τ. The answer is: no. The Weyl correspondence is the *only* quantization which is symplectically covariant in the sense above. That this property really is *characteristic* of Weyl quantization seems to be somewhat ignored both in mathematics and physics. It will be crucial in our derivation of the Schrödinger equation.

We are going to give a precise statement of this fundamental property below; let us first introduce some notation. Viewing the space of tempered distributions $\mathcal{S}'(\mathbb{R}^n)$ as a space of "classical Hamiltonians", we denote by $\mathcal{L}(\mathcal{S}(\mathbb{R}^n), \mathcal{S}'(\mathbb{R}^n))$ the space of continuous operators $\mathcal{S}(\mathbb{R}^n) \longrightarrow \mathcal{S}'(\mathbb{R}^n)$, and we assume that there is a continuous and linear "quantization mapping"

$$\mathcal{Q} : \mathcal{S}'(\mathbb{R}^n \oplus \mathbb{R}^n) \longrightarrow \mathcal{L}(\mathcal{S}(\mathbb{R}^n), \mathcal{S}'(\mathbb{R}^n)) \tag{15.13}$$

associating to a symbol a an operator $A = \mathcal{Q}(a)$.

The following result is hinted at in Stein's book [153] (§7.6, Chapter 12) and proven in detail in Chapter 30 of Wong [163]. It is really a fundamental result, because it shows that the quantization mapping $\mathcal{Q}$ *must be* the Weyl correspondence if one makes three simple assumptions (we will see in a moment that the third is actually superfluous).

Theorem 356. *Assume that the quantization map* (15.13) *has the three following properties:*

(i) *$\mathcal{Q}$ is continuous in the sense that if (a_k) is a sequence in $\mathcal{S}'(\mathbb{R}^n \oplus \mathbb{R}^n)$ then $\lim_{k\to\infty} a_k = a$ in $\mathcal{S}'(\mathbb{R}^n \oplus \mathbb{R}^n)$ implies $\lim_{k\to\infty} \mathcal{Q}(a_k) = \mathcal{Q}(a)$;*
(ii) *For every $S \in \mathrm{Sp}(2n, \mathbb{R})$ we have $\mathcal{Q}(a \circ S^{-1}) = \widehat{S}\mathcal{Q}(a)\widehat{S}^{-1}$ where $\widehat{S} \in \mathrm{Mp}(2n, \mathbb{R})$ is any of the two metaplectic operators with projection S;*
(iii) *If $a(x, p) = a(x)$ with $a \in L^\infty(\mathbb{R}^n)$ then $\mathcal{Q}(a)$ is multiplication by the function a, that is $\mathcal{Q}(a)\psi = a\psi$.*

Then $\mathcal{Q}$ is the Weyl correspondence $a \xleftrightarrow{\text{Weyl}} \widehat{A}$ that is $\mathcal{Q}(a) = \widehat{A}$ for every $a \in \mathcal{S}'(\mathbb{R}^n \oplus \mathbb{R}^n)$.

It turns out that the requirement (iii) can be relaxed. In fact, in view of the Schwartz kernel theorem together with Theorem 330, condition (i) implies that for every $\tau \in \mathbb{R}$ there exists a symbol $a_\tau \in \mathcal{S}'(\mathbb{R}^n \oplus \mathbb{R}^n)$ such that the quantization map $\mathcal{Q}$ satisfies

$$[Qa]\psi(x) = \left(\tfrac{1}{2\pi\hbar}\right)^n \iint_{\mathbb{R}^n \times \mathbb{R}^n} e^{\frac{i}{\hbar}p\cdot(x-y)} a_\tau((1-\tau)x + \tau y, p)\psi(y)dydp.$$

Suppose that a only depends on x; then the formula above implies that

$$[Qa]\psi(x) = \left(\tfrac{1}{2\pi\hbar}\right)^n \int_{\mathbb{R}^n} \left[\int_{\mathbb{R}^n} e^{\frac{i}{\hbar}p\cdot(x-y)}dp\right] a_\tau((1-\tau)x + \tau y)\psi(y)dy$$

$$= \int_{\mathbb{R}^n} \delta(x - y)a_\tau((1-\tau)x + \tau y)\psi(y)dy$$

$$= a_\tau(x)\psi(x)$$

so that we must have $a_\tau = a$, hence condition (iii) is indeed fulfilled. Summarizing:

If the quantization map $\mathcal{Q}$ is continuous and such that $\mathcal{Q}(a \circ S^{-1}) = \widehat{S}\mathcal{Q}(a)\widehat{S}^{-1}$ then it must be the Weyl correspondence.

15.2.2   Quantum evolution groups and Stone's theorem

Let A be a self-adjoint operator on some Hilbert space, say $L^2(\mathbb{R}^n)$. If A is bounded, we can define the exponential of itA by the usual power series

$$e^{itA} = \sum_{k=0}^{\infty} \frac{(it)^k}{k!} A^k$$

(the series is well defined, being normally convergent). When A is unbounded, one can still define the exponential using a functional calculus of operators. The following properties are well known (see for instance Reed and Simon [136], §VIII.4):

Group 1   *For each $t \in \mathbb{R}$ the operator $U_t = e^{itA}$ is unitary and we have $U_t U_{t'} = U_{t+t'}$;*

Group 2   *The one-parameter group $(U_t)_{t\in\mathbb{R}}$ is strongly continuous: for each $\psi \in L^2(\mathbb{R}^n)$ we have $\lim_{t\to t_0} U_t\psi = U_{t_0}\psi$;*

Group 3   *For $\psi \in D_A$ (the domain of A) we have*

$$\lim_{\Delta t \to 0} \frac{U_{\Delta t}\psi - \psi}{\Delta t} = iA$$

and if the limit above exists for some $\psi \in L^2(\mathbb{R}^n)$ then $\psi \in D_A$.

The operator A is called the infinitesimal generator of the unitary one-parameter group $(U_t)_{t\in\mathbb{R}}$. It turns out that every strongly continuous unitary one-parameter group of operators in $L^2(\mathbb{R}^n)$ arises as the exponential of a self-adjoint operator; this is Stone's theorem:

Theorem 357 (Stone). *Let $(U_t)_{t\in\mathbb{R}}$ be a strongly continuous one-parameter unitary group on $L^2(\mathbb{R}^n)$ (or, on an arbitrary Hilbert space). Then the infinitesimal generator*

$$A = -i \lim_{\Delta t \to 0} \frac{U_{\Delta t}\psi - \psi}{\Delta t} \tag{15.14}$$

is self-adjoint; in particular it is closed and densely defined. The domain D_A consists of all $\psi \in L^2(\mathbb{R}^n)$ for which the limit (15.14) exists, and is invariant under the action of each U_t.

For self-contained proofs of Stone's theorem we refer to Abraham et al. [1], pp. 529–536 or to Reed and Simon [136], §VIII.4.

15.2.3   Application to Schrödinger's equation

Let us now apply Stone's theorem to Schrödinger's equation. We will call a non-zero element ψ_0 of $L^2(\mathbb{R}^n)$ a "quantum state". We assume that the time-evolution of quantum states is determined by a strongly continuous one-parameter group of operators $(U_t)_{t\in\mathbb{R}}$ ("evolution group"). That is, the state at time t is $\psi_t = U_t\psi_0$

where the operators U_t satisfy $U_t U_{t'} = U_{t+t'}$ and the map $\psi_0 \longmapsto \psi_t$ is continuous for every ψ_0. Stone's theorem tells us that the operator

$$\widehat{H} = i\hbar \, \frac{d}{dt} U_t \Psi_0 \bigg|_{t=0} = i\hbar \lim_{\Delta t \to 0} \frac{U_{\Delta t} \Psi_0 - \Psi_0}{\Delta t} \qquad (15.15)$$

exists, is self-adjoint, and is densely defined. We thus have

$$i\hbar \frac{dU_t}{dt} = \widehat{H} U_t. \qquad (15.16)$$

This is an abstract and formal version of Schrödinger's equation.

Definition 358. We will say that $(U_t)_{t \in \mathbb{R}}$ is a "quantum evolution group" associated with the Hamilton function H.

Let us now briefly return to the property of the metaplectic representation of $\mathrm{Sp}(2n, \mathbb{R})$ mentioned above. It can be summarized as follows: to every family (S_t) of symplectic matrices depending smoothly on t and such that $\widehat{S}_0 = I$ we can associate, in a unique way, a family $(\widehat{S}_t)$ of unitary operators on $L^2(\mathbb{R}^n)$ belonging to the metaplectic group $\mathrm{Mp}(2n, \mathbb{R})$ such that $\widehat{S}_0$ is the identity; introducing a rescaling constant $\hbar$ having the dimension of action, that family satisfies the Schrödinger equation

$$i\hbar \frac{d}{dt} \widehat{S}_t = \widehat{H} \widehat{S}_t$$

where $\widehat{H}$ is the self-adjoint operator defined by applying the Weyl rule to the (time-dependent) Hamiltonian H of which (S_t) is the flow. Such a Hamiltonian always exists in view of our discussion of Banyaga's theorem in Chapter 4. That is, the correspondence $H \overset{\mathrm{Weyl}}{\longleftrightarrow} \widehat{H}$ is obtained by simple inspection of the flow $(\phi_t^H) = (S_t)$. We thus have

$$\widehat{S}_t = e^{-i\widehat{H}t/\hbar}$$

(generically $\widehat{H}$ is not a bounded operator on $L^2(\mathbb{R}^n)$ so that the exponential has to be defined using some functional calculus; see, e.g., Reed and Simon [136] §VIII.3). We now ask whether this property has an analogue for paths in the group $\mathrm{Ham}(2n, \mathbb{R})$ of Hamiltonian canonical transformations. Let us introduce the following notation:

- $\mathcal{P}\,\mathrm{Ham}(2n, \mathbb{R})$ *is the set of all one-parameter families (ϕ_t) in* $\mathrm{Ham}(2n, \mathbb{R})$ *depending smoothly on t and passing through the identity at time $t = 0$; in view of Banyaga's theorem such a family of canonical transformations is always the flow (ϕ_t^H) of some (usually time-dependent) Hamiltonian H;*
- $\mathcal{P}U(L^2(\mathbb{R}^n))$ *is the set of all strongly continuous one-parameter families (F_t) of unitary operators on $L^2(\mathbb{R}^n)$ depending smoothly on t and such that F_0 is the identity operator, and having the following property: the domain of the infinitesimal generator $\widehat{H}$ of (F_t) contains the Schwartz space $\mathcal{S}(\mathbb{R}^n)$.*

We will use Stone's theorem to prove the hard part of our main result:

Theorem 359. *There exists a bijective correspondence*

$$\mathcal{C} : PU(L^2(\mathbb{R}^n)) \longleftrightarrow P\operatorname{Ham}(2n,\mathbb{R})$$
$$(\phi_t) \longleftrightarrow (U_t) \tag{15.17}$$

whose restriction to families (S_t) of symplectic matrices reduces to the metaplectic representation, and which has the following symplectic covariance property: for every (ϕ_t) in $P\operatorname{Ham}(2n,\mathbb{R})$ and for every $S \in \operatorname{Sp}(2n,\mathbb{R})$ we have

$$\mathcal{C}(S\phi_t S^{-1}) = (\widehat{S}U_t\widehat{S}^{-1}) \tag{15.18}$$

where S is any of the two operators in $\operatorname{Mp}(2n,\mathbb{R})$ such that $\pi^{\operatorname{Mp}}(\widehat{S}) = S$. This correspondence $\mathcal{C}$ is bijective and we have

$$i\hbar\frac{d}{dt}U_t = \widehat{H}U_t \tag{15.19}$$

where $\widehat{H} \overset{\text{Weyl}}{\longleftrightarrow} H$, the Hamiltonian function H being determined by (ϕ_t).

It is perhaps worth observing that it is always preferable to take the family (U_t) as the fundamental object, rather than $\widehat{H}$ (and hence Schrödinger's equation). This was already remarked by Weyl who noticed that (F_t) is everywhere defined and consists of bounded operators, while $\widehat{H}$ is generically unbounded and only densely defined (see the discussion in Mackey [121] for a discussion of related questions).

Let us begin with derivation of Schrödinger's equation in the case where (ϕ_t) is the Hamiltonian flow determined by a time-independent Hamiltonian function $H = H(z)$. Then $(\phi_t) = (\phi_t^H)$ is a one-parameter group of symplectomorphisms: that is $\phi_t^H \phi_{t'}^H = \phi_{t+t'}^H$. We thus want to associate to (ϕ_t^H) a strongly continuous one-parameter group $(F_t) = (F_t^H)$ of unitary operators on $L^2(\mathbb{R}^n)$ satisfying some additional conditions. We proceed as follows: let $\widehat{H}$ be the operator associated to H by the Weyl correspondence: $\widehat{H} \overset{\text{Weyl}}{\longleftrightarrow} H$ and define $\mathcal{C}(\phi_t^H) = (U_t)$ by $U_t = e^{-it\widehat{H}/\hbar}$. The Weyl operator $\widehat{H}$ is self-adjoint and its domain obviously contains $\mathcal{S}(\mathbb{R}^n)$. Let us show that the covariance property (15.18) holds. We have

$$\mathcal{C}(S\phi_t^H S^{-1}) = \mathcal{C}(\phi_t^{H\circ S^{-1}})$$

in view of formula (1.7), that is, by definition of $\mathcal{C}$,

$$\mathcal{C}(S\phi_t^H S^{-1}) = (e^{-\frac{i}{\hbar}t\widehat{H\circ S^{-1}}}).$$

In view of the symplectic covariance property $a \circ S^{-1} \overset{\text{Weyl}}{\longleftrightarrow} S\widehat{A}S^{-1}$ of Weyl operators we have $\widehat{H \circ S^{-1}} = S\widehat{H}S^{-1}$, and hence

$$\mathcal{C}(S\phi_t^H S^{-1}) = (e^{-\frac{i}{\hbar}tS\widehat{H}S^{-1}}) = (Se^{-\frac{i}{\hbar}t\widehat{H}}S^{-1})$$

which is property (15.18).

Let conversely (F_t) be in $\mathcal{P}U(L^2(\mathbb{R}^n))$; we must show that we can find a unique (ϕ_t) in $\mathcal{P}\operatorname{Ham}(2n,\mathbb{R})$ such that $\mathcal{C}(\phi_t) = (U_t)$. By Stone's theorem and our definition of $\mathcal{P}U(L^2(\mathbb{R}^n))$ there exists a unique self-adjoint operator A, densely defined, and whose domain contains $\mathcal{S}(\mathbb{R}^n)$. Thus A is continuous on $\mathcal{S}(\mathbb{R}^n)$ and for each value of the parameter τ there exists an observable a such that $A \longleftrightarrow a_\tau$. Choose $\tau = \frac{1}{2}$; then $A = \widehat{H} \overset{\text{Weyl}}{\longleftrightarrow} H$ for some function $H = H(x,p)$ and we have $\mathcal{C}(\phi_t^H) = (U_t)$.

There remains to show that the correspondence $\mathcal{C}$ restricts to the metaplectic representation for semigroups $(\phi_t) = (S_t)$ in $\operatorname{Sp}(2n,\mathbb{R})$; but this is clear since (S_t) is generated, as a flow, by a quadratic Hamiltonian, and the unitary one-parameter group of operators determined by such a function precisely consists of metaplectic operators.

We now no longer assume that (ϕ_t) and (F_t) are one-parameter groups. Recall that we have reduced the study of a time-dependent Hamiltonian $H = H(z,t)$ by introducing

$$\widetilde{H}(x,p,t,E) = H(x,p,t) - E \tag{15.20}$$

which is a time-independent Hamiltonian on $(\mathbb{R}^{n+1} \oplus \mathbb{R}^{n+1}) \equiv \mathbb{R}^{2n} \times \mathbb{R}_E \times \mathbb{R}_t$ where E is viewed as a conjugate variable to t. The flow $(\widetilde{\phi}_t^H) = (\phi_t^{\widetilde{H}})$ on $\mathbb{R}^{2n+2}$ generated by $\widetilde{H}$ is related to the time-dependent flow $(\phi_{t,t'}^H)$ by the formula

$$\widetilde{\phi}_t^H(z',t',E') = (\phi_{t,t'}^H(z'), t+t', E_{t,t'}) \tag{15.21}$$

where $E_{t,t'} - E'$ is the variation of the energy in the time interval $[t',t]$. The advantage of this reformulation of the dynamics associated with H is that $(\widetilde{\phi}_t^H)$ is a one-parameter group of canonical transformations of $\mathbb{R}^{2n+2}$. In the operator case we can proceed in a quite similar way, noting that the Weyl operator associated with $\widetilde{H}$ is given by

$$\widehat{\widetilde{H}} = \widehat{H} - i\hbar\frac{\partial}{\partial t}.$$

Of course $\widehat{\widetilde{H}}$ is self-adjoint if and only if $\widehat{H}$ is, which is the case since H is real. We will need the following elementary fact, which is a variant of the method of separation of variables:

Lemma 360. *Let E be an arbitrary real number. The function*

$$\Psi(x,t;t') = \psi(x,t)e^{\frac{i}{\hbar}E(t-t')} \tag{15.22}$$

is a solution of the extended Schrödinger equation

$$i\hbar\frac{\partial\Psi}{\partial t'} = \widehat{\widetilde{H}}\Psi \tag{15.23}$$

if and only if $\psi = \psi(x,t)$ is a solution of the usual Schrödinger equation

$$i\hbar\frac{\partial\psi}{\partial t} = \widehat{H}\psi. \tag{15.24}$$

Proof. We first note the obvious identity

$$i\hbar \frac{\partial \Psi}{\partial t'} = E\Psi. \tag{15.25}$$

Writing for short $\widehat{H}(t) = \widehat{H}(x, -i\hbar\nabla_x, t)$ we have, after a few calculations

$$\left(\widehat{H}(t) - i\hbar\frac{\partial}{\partial t}\right)\Psi = \left[\widehat{H}(t)\psi - i\hbar\frac{\partial \psi}{\partial t}\right] e^{\frac{i}{\hbar}E(t-t')} + E\Psi \tag{15.26}$$

hence (15.23) is equivalent to (15.24) in view of (15.25).   □

This result shows the following: choose an initial function $\psi_0 = \psi_0(x)$ at time $t = 0$ and solve the usual Schrödinger equation (15.24), which yields the solution $\psi = \psi(x, t)$. Then $\Psi = \Psi(x, t; t')$ defined by (15.22) is the solution of the extended Schrödinger equation (15.23) with initial datum $\Psi(x, t; t) = \psi(x, t)$ at time $t' = t$. In terms of flows we can rewrite this as

$$\widetilde{F}_{t'-t}(F_t\psi_0) = (F_t\psi_0)e^{\frac{i}{\hbar}E(t-t')}.$$

Chapter 16

The Feichtinger Algebra

We are now going to address a first topic from the theory of modulation spaces, which was initiated by Feichtinger in the early 1980's: the *Feichtinger algebra* $S_0(\mathbb{R}^n)$ (we will study the general notion of modulation space in the next chapter). The elements of $S_0(\mathbb{R}^n)$ are characterized by the property that their Wigner transform is in $L^1(\mathbb{R}^n \oplus \mathbb{R}^n)$, but it is not obvious at all that, with *this* definition, $S_0(\mathbb{R}^n)$ is a vector space, even less an algebra! We will therefore need an alternative, more tractable, equivalent definition.

The Feichtinger algebra $S_0(\mathbb{R}^n)$ contains continuous non-differentiable functions, such as

$$\psi(x) = \left\{ \begin{array}{c} 1 - |x| \;\; if \; |x| \leq 1 \\ 0 \; if \; |x| > 1 \end{array} \right.$$

and it is thus a good substitute for the Schwartz space $\mathcal{S}(\mathbb{R}^n)$ as long as one is not interested in differentiation properties. It is moreover the smallest Banach space containing $\mathcal{S}(\mathbb{R}^n)$ and being invariant under the action of the inhomogeneous metaplectic group. The dual space $S_0'(\mathbb{R}^n)$ of $S_0(\mathbb{R}^n)$ contains many basic distributions such as the Dirac distribution δ or its translates. These properties, together with the fact that Banach spaces are mathematically easier to deal with than Fréchet spaces, makes the Feichtinger algebra into a tool of choice not only for the study of wavepackets, but also of global regularity properties for quantum mechanical operators. Of a particular interest for the study of the continuous spectrum of such operators is the fact that $(S_0(\mathbb{R}^n), L^2(\mathbb{R}^n), S_0'(\mathbb{R}^n))$ is a Gelfand triple.

16.1 Definition and first properties

We begin by introducing some technical tools. Recall from Chapter 9 (formulas (9.32) and (9.33)) that the short-time Fourier transform

$$V_\phi \psi(z) = \int_{\mathbb{R}^n} e^{-2\pi i p \cdot x'} \psi(x') \overline{\phi(x' - x)} dx' \tag{16.1}$$

and is related to the cross-Wigner transform by the formula

$$W(\psi,\phi)(z) = \left(\tfrac{2}{\pi\hbar}\right)^{n/2} e^{\frac{2i}{\hbar}p\cdot x} V_{\phi^\vee_{\sqrt{2\pi\hbar}}} \psi_{\sqrt{2\pi\hbar}}\left(z\sqrt{\tfrac{2}{\pi\hbar}}\right) \tag{16.2}$$

where $\psi_{\sqrt{2\pi\hbar}}(x) = \psi(x\sqrt{2\pi\hbar})$ and $\phi^\vee(x) = \phi(-x)$; equivalently

$$V_\phi\psi(z) = \left(\tfrac{2}{\pi\hbar}\right)^{-n/2} e^{-i\pi p\cdot x} W(\psi_{1/\sqrt{2\pi\hbar}}, \phi^\vee_{1/\sqrt{2\pi\hbar}})\left(z\sqrt{\tfrac{\pi\hbar}{2}}\right). \tag{16.3}$$

16.1.1 Definition of $S_0(\mathbb{R}^n)$

Let us give a first definition of the Feichtinger algebra. In what follows ϕ will be a non-zero element of $\mathcal{S}(\mathbb{R}^n)$; we will call ϕ a "window".

Definition 361. The Feichtinger algebra $S_0(\mathbb{R}^n)$ (often also denoted by $M^1(\mathbb{R}^n)$) consists of all $\psi \in \mathcal{S}'(\mathbb{R}^n)$ such that $V_\phi\psi \in L^1(\mathbb{R}^n \oplus \mathbb{R}^n)$ for every window ϕ. The number

$$\|\psi\|_{\phi,S_0(\mathbb{R}^n)} = \|V_\phi\psi\|_{L^1(\mathbb{R}^{2n})} = \int_{\mathbb{R}^{2n}} |V_\phi\psi(z)| dz \tag{16.4}$$

is called the STFT norm of ψ relative to the window ϕ.

The notation $M^1(\mathbb{R}^n)$ is used in the context of modulation spaces, of which the Feichtinger algebra is a particular case. The more general modulation spaces $M^q(\mathbb{R}^n)$ and their weighted variants $M^q_v(\mathbb{R}^n)$ will be discussed in the next chapter.

The reader is encouraged to verify the following property:

Exercise 362. Show that $\|\cdot\|_{\phi,S_0}$ indeed is a norm on $S_0(\mathbb{R}^n)$ for each $\phi \in \mathcal{S}(\mathbb{R}^n)$, $\phi \neq 0$.

There are several seemingly obscure points in the definition above. First, while it is obvious that $S_0(\mathbb{R}^n)$ is a vector space (because of the linearity of the mapping $\psi \mapsto V_\phi\psi$) it is much less obvious why it should be an algebra. We will see that $S_0(\mathbb{R}^n)$ is actually an algebra for both pointwise multiplication and for the convolution product. Perhaps an apparently more serious shortcoming of the definition above is the fact that the Feichtinger algebra seems to be defined in terms of infinitely many "windows" ϕ. We are going to see that it actually suffices that $V_\phi\psi \in L^1(\mathbb{R}^n \oplus \mathbb{R}^n)$ for one window! Let us first rewrite Definition 361 in terms of the cross-Wigner transform:

Definition 363. The Feichtinger algebra $S_0(\mathbb{R}^n)$ consists of all $\psi \in \mathcal{S}'(\mathbb{R}^n)$ such that $W(\psi,\phi) \in L^1(\mathbb{R}^n \oplus \mathbb{R}^n)$ for every window ϕ. The number

$$\|\psi\|^\hbar_{\phi,S_0(\mathbb{R}^n)} = \|W(\psi,\phi)\|_{L^1(\mathbb{R}^{2n})} = \int_{\mathbb{R}^{2n}} |W(\psi,\phi)(z)| dz \tag{16.5}$$

is called the Wigner norm of ψ relative to the window ϕ.

Let us verify that both Definitions 361 and 363 are equivalent. In view of formula (9.32) relating the cross-Wigner transform and the STFT the condition $W(\psi, \phi) \in L^1(\mathbb{R}^n)$ is equivalent to

$$\int_{\mathbb{R}^{2n}} |V_{\phi^\vee_{\sqrt{2\pi\hbar}}} \psi_{\sqrt{2\pi\hbar}}(z\sqrt{\tfrac{2}{\pi\hbar}})|dz < \infty$$

that is, performing the change of variables $z \longmapsto \sqrt{\pi\hbar/2}z$, to

$$\int_{\mathbb{R}^{2n}} |V_{\phi^\vee_{\sqrt{2\pi\hbar}}} \psi_{\sqrt{2\pi\hbar}}(z)|dz < \infty.$$

In view of the equality (9.31) in Lemma 178 this inequality can be rewritten as

$$\iint_{\mathbb{R}^n \times \mathbb{R}^n} |V_{\phi^\vee} \psi(x\sqrt{2\pi\hbar}, p/\sqrt{2\pi\hbar})|dpdx < \infty$$

that is, setting $x' = x\sqrt{2\pi\hbar}$ and $p' = p/\sqrt{2\pi\hbar}$, $V_{\phi^\vee}\psi \in L^1(\mathbb{R}^n)$ which is equivalent to $\psi \in S_0(\mathbb{R}^n)$, since ϕ and hence $\phi^\vee$ is arbitrary.

16.1.2 First properties of $S_0(\mathbb{R}^n)$

The main result of this section is the following:

Proposition 364. *Let $\psi \in \mathcal{S}'(\mathbb{R}^n)$ and (ϕ, ϕ') be a pair of windows.*

(i) *Let $\gamma \in \mathcal{S}(\mathbb{R}^n)$ be such that $(\gamma|\phi)_{L^2(\mathbb{R}^n)} \neq 0$. We have*

$$\|\psi\|^\hbar_{\phi', S_0(\mathbb{R}^n)} \leq \frac{2^n}{(\gamma|\phi)_{L^2}} \|\psi\|^\hbar_{\phi, S_0(\mathbb{R}^n)} \|\gamma\|^\hbar_{\phi'^\vee, S_0(\mathbb{R}^n)} \tag{16.6}$$

with $\phi'^\vee(x) = \phi'(-x)$;

(ii) *We have $\psi \in S_0(\mathbb{R}^n)$ if and only if there exists one window ϕ such that $W(\psi, \phi) \in L^1(\mathbb{R}^n \oplus \mathbb{R}^n)$, hence we have $W(\psi, \phi) \in L^1(\mathbb{R}^n \oplus \mathbb{R}^n)$ for all windows ϕ;*

(iii) *The Wigner norms $\| \cdot \|^\hbar_{\phi, S_0(\mathbb{R}^n)}$ (resp. the STFT norms $\| \cdot \|_{\phi, S_0(\mathbb{R}^n)}$) are all equivalent when ϕ ranges over $\mathcal{S}(\mathbb{R}^n)$.*

Proof. Property (ii) follows from property (i): suppose there exists a window ϕ such that $\|\psi\|^\hbar_{\phi, S_0} < \infty$. Choose now an arbitrary window ϕ' and $\gamma \in \mathcal{S}(\mathbb{R}^n)$ such that $(\gamma|\phi)_{L^2} \neq 0$. We then have $\|\psi\|^\hbar_{\phi', S_0} < \infty$ in view of the inequality (16.6). Let us prove (i). Recall that we can express ψ in terms of $W(\psi, \phi)$ using the inversion formula (9.51):

$$\psi(x) = \frac{2^n}{(\gamma|\phi)_{L^2}} \int_{\mathbb{R}^{2n}} W(\psi, \phi)(z_0)\widehat{T}_{\mathrm{GR}}(z_0)\gamma(x)dz_0. \tag{16.7}$$

Let $z' \in \mathbb{R}^{2n}$ and apply $\widehat{T}_{\mathrm{GR}}(z')$ to both sides of this equality; in view of the product formula (8.37) we have

$$\widehat{T}_{\mathrm{GR}}(z')\psi(x) = \frac{2^n}{(\gamma|\phi)_{L^2}} \int_{\mathbb{R}^{2n}} W(\psi,\phi)(z) e^{\frac{2i}{\hbar}\sigma(z,z')} \widehat{T}(2z - 2z')\gamma(x)dz.$$

We now observe that by Definition (9.11) of the cross-Wigner transform we have

$$(\widehat{T}_{\mathrm{GR}}(z')\psi|\phi') = (\pi\hbar)^n W(\psi,\phi')(z')$$

and by Definition (9.1) of the cross-ambiguity function

$$\begin{aligned}
(\widehat{T}(2z - 2z')\gamma|\phi')_{L^2} &= (2\pi\hbar)^n A(\gamma,\phi')(2z' - 2z) \\
&= (\pi\hbar)^n W(\gamma,\phi'^{\vee})(z' - z)
\end{aligned}$$

where the second equality follows from formula (9.27) relating cross-ambiguity and cross-Wigner transforms. Formula (16.7) above thus yields

$$W(\psi,\phi')(z') = \frac{2^n}{(\gamma|\phi)_{L^2}} \int_{\mathbb{R}^{2n}} W(\psi,\phi)(z) e^{\frac{2i}{\hbar}\sigma(z,z')n} W(\gamma,\phi'^{\vee})(z' - z)dz$$

and hence

$$|W(\psi,\phi')(z')| \leq \frac{2^n}{|(\gamma|\phi)_{L^2}|} \int_{\mathbb{R}^{2n}} |W(\psi,\phi)(z)|\,|W(\gamma,\phi'^{\vee})(z' - z)|dz$$

that is

$$|W(\psi,\phi')| \leq \frac{2^n}{|(\gamma|\phi)_{L^2}|} |W(\psi,\phi)| * |W(\gamma,\phi'^{\vee})|. \tag{16.8}$$

Integrating both sides of this inequality with respect to z yields (16.6) in view of the classical inequality $\|F * G\|_{L^1} \leq \|F\|_{L^1}\|G\|_{L^1}$ that is valid for any integrable functions F and G. (iii) That the norms $\| \cdot \|_{\phi,S_0}^{\hbar}$ are equivalent is clear from the inequality (16.6); we leave it to the reader to show that the same is true of the norms $\| \cdot \|_{\phi,S_0}$. $\qquad\square$

The following result shows that the "windows" used in the definition of $S_0(\mathbb{R}^n)$ can themselves be chosen in $S_0(\mathbb{R}^n)$.

Proposition 365. *Let both ψ and ϕ be in $L^2(\mathbb{R}^n)$.*

(i) *If $W(\psi,\phi) \in L^1(\mathbb{R}^n \oplus \mathbb{R}^n)$ then both ψ and ϕ are in $S_0(\mathbb{R}^n)$;*

(ii) *We have $\psi \in S_0(\mathbb{R}^n)$ if and only if $W(\psi,\phi) \in L^1(\mathbb{R}^n \oplus \mathbb{R}^n)$ for one (and hence every) $\phi \in S_0(\mathbb{R}^n)$.*

Proof. Property (ii) immediately follows from (i). Let us prove property (i). The condition that $\psi,\phi \in L^2(\mathbb{R}^n)$ implies that $W(\psi,\phi)$ is a square-integrable and continuous function (Proposition 183). Recall that in the course of the proof of Proposition 364 we proved the inequality (16.8).

Choosing $\gamma = \phi'^\vee$ this inequality becomes

$$|W(\psi, \phi')| \leq \frac{2^n}{|(\gamma|\phi)_{L^2}|} |W(\psi, \phi)| * |W\phi'^\vee|$$

hence, integrating both sides,

$$\|\psi\|^\hbar_{\phi', S_0(\mathbb{R}^n)} \leq \frac{2^n}{|(\gamma|\phi)_{L^2}|} \|W(\psi, \phi)\|_{L^1} \|W\phi'^\vee\|_\infty < \infty$$

which shows that $\psi \in S_0(\mathbb{R}^n)$. Swapping ψ and ϕ the inequality above becomes

$$\|\phi\|^\hbar_{\phi', S_0(\mathbb{R}^n)} \leq \frac{2^n}{|(\gamma|\phi)_{L^2}|} \|W(\phi, \psi)\|_{L^1} \|W(\phi'^\vee)\|_\infty < \infty$$

hence we also have $\phi \in S_0(\mathbb{R}^n)$. $\qquad\qquad\square$

It immediately follows that:

Corollary 366. *A function $\psi \in L^2(\mathbb{R}^n)$ belongs to $S_0(\mathbb{R}^n)$ if and only if $W\psi \in L^1(\mathbb{R}^n \oplus \mathbb{R}^n)$.*

Proof. In view of the statement (i) in Proposition 367 the condition $W\psi \in L^1(\mathbb{R}^n \oplus \mathbb{R}^n)$ implies that $\psi \in S_0(\mathbb{R}^n)$. If conversely $\psi \in S_0(\mathbb{R}^n)$ then $W\psi \in L^1(\mathbb{R}^n \oplus \mathbb{R}^n)$ in view of the statement (ii) in the same Proposition. $\qquad\square$

Note that the condition $W\psi \in L^1(\mathbb{R}^n \oplus \mathbb{R}^n)$ could be taken as the definition of $S_0(\mathbb{R}^n)$ but then it would not be clear at all that $S_0(\mathbb{R}^n)$ is a vector space!

The following result shows that $S_0(\mathbb{R}^n)$ is a subspace of several "nice" spaces of functions (in particular $S_0(\mathbb{R}^n)$ consists of continuous integrable functions):

Proposition 367. *We have the inclusions*

$$S_0(\mathbb{R}^n) \subset C^0(\mathbb{R}^n) \cap L^1(\mathbb{R}^n) \cap F(L^1(\mathbb{R}^n)) \tag{16.9}$$

where $F(L^1(\mathbb{R}^n))$ is the image of $L^1(\mathbb{R}^n)$ by the Fourier transform.

Proof. Recall again the inversion formula

$$\psi(x) = \frac{2^n}{(\gamma|\phi)_{L^2}} \int_{\mathbb{R}^{2n}} W(\psi, \phi)(z_0)\widehat{T}_{\mathrm{GR}}(z_0)\gamma(x)dz_0$$

valid for all $\gamma \in S(\mathbb{R}^n)$ such that $(\gamma|\phi)_{L^2} \neq 0$. Putting $\Delta\psi = \psi(x + \Delta x) - \psi(x)$ we have

$$|\Delta\psi| \leq \frac{2^n}{|(\gamma|\phi)_{L^2}|} \int_{\mathbb{R}^{2n}} |W(\psi, \phi)(z_0)| \, |\widehat{T}_{\mathrm{GR}}(z_0)(\gamma(x + \Delta x) - \gamma(x))|dz_0$$

$$\leq \frac{2^n}{|(\gamma|\phi)_{L^2}|} \|W(\psi, \phi)\|_{L^1} \sup_{z_0} |\widehat{T}_{\mathrm{GR}}(z_0)(\gamma(x + \Delta x) - \gamma(x))|$$

$$= \frac{2^n}{|(\gamma|\phi)_{L^2}|} \|W(\psi, \phi)\|_{L^1} \sup_{x_0} |(\gamma(2x_0 - x - \Delta x) - \gamma(2x_0 - x))|$$

where the last equality follows from the definition of $\widehat{T}_{\mathrm{GR}}(z)$, from which readily follows that $\lim_{\Delta x \to 0} \Delta \psi = 0$; hence $S_0(\mathbb{R}^n) \subset C^0(\mathbb{R}^n)$. Let us next show that $S_0(\mathbb{R}^n) \subset L^1(\mathbb{R}^n)$. Let $\psi \in S_0(\mathbb{R}^n)$. Using again the inversion formula we get,

$$
\begin{aligned}
|\psi(x)| &\leq \frac{2^n}{|(\gamma|\phi)_{L^2}|} \int_{\mathbb{R}^{2n}} |W(\psi,\phi)(z_0)|\,|\gamma(2x_0 - x)|dz_0 \\
&\leq \frac{2^n}{|(\gamma|\phi)_{L^2}|} \int_{\mathbb{R}^{2n}} |W(\psi,\phi)(z_0)|\,|\gamma(2x_0 - x)|dz_0
\end{aligned}
$$

and hence, integrating in x,

$$
\|\psi\|_{L^1} \leq \frac{2^n}{(\gamma|\phi)_{L^2}} \|\psi\|^{\hbar}_{\phi,S_0} \|\gamma\|_{L^\infty} < \infty
$$

so that $\psi \in S_0(\mathbb{R}^n)$. To prove that $S_0(\mathbb{R}^n) \subset F(L^1(\mathbb{R}^n))$ it suffices to note that $S_0(\mathbb{R}^n)$ is invariant under Fourier transform: we have $\psi \in S_0(\mathbb{R}^n)$ if and only if $W(\psi, F^{-1}\phi) \in L^1(\mathbb{R}^n \oplus \mathbb{R}^n)$ for every window ϕ since F is an automorphism $\mathcal{S}(\mathbb{R}^n) \longrightarrow \mathcal{S}(\mathbb{R}^n)$. Now,

$$
W(F^{-1}\psi, F^{-1}\phi)(z) = W(\psi,\phi)(Jz) = W(\psi,\phi)(p,-x)
$$

in view of the symplectic covariance property (10.26) of the cross-Wigner transform. It follows from the inclusion $S_0(\mathbb{R}^n) \subset L^1(\mathbb{R}^n)$ that we have $F^{-1}\psi \in L^1(\mathbb{R}^n)$ and hence $\psi \in F(L^1(\mathbb{R}^n))$ as claimed. $\qquad\square$

16.2 Invariance and Banach algebra properties

We are going to see that the Feichtinger algebra is a Banach algebra; in addition we will prove that this algebra is invariant under the action of the metaplectic group, and that it enjoys a characteristic minimality property for the action of the Heisenberg–Weyl operators.

16.2.1 Metaplectic invariance of the Feichtinger algebra

Another very nice property is that $S_0(\mathbb{R}^n)$ is closed under the action of the inhomogeneous metaplectic group:

Proposition 368. *Let* $\psi \in S_0(\mathbb{R}^n)$, $\widehat{S} \in \mathrm{Mp}(2n,\mathbb{R})$, *and* $z_0 \in \mathbb{R}^n$. *We have*

(i) $\widehat{S}\psi \in S_0(\mathbb{R}^n)$;

(ii) $\widehat{T}(z_0)\psi \in S_0(\mathbb{R}^n)$;

(iii) *In particular* $\psi \in S_0(\mathbb{R}^n)$ *if and only if* $F\psi \in S_0(\mathbb{R}^n)$.

Proof. We have $\psi \in S_0(\mathbb{R}^n)$ if and only if $W\psi \in L^1(\mathbb{R}^n)$. In view of the symplectic covariance property $W(\widehat{S}\psi)(z) = W\psi(S^{-1}z)$ (see (10.27)) of the Wigner function we have $W(\widehat{S}\psi)(z) = W\psi(S^{-1}z)$ where $S \in \mathrm{Sp}(2n, R)$ is the projection of $\widehat{S}$. Now,

$$\int_{\mathbb{R}^{2n}} |W\psi(S^{-1}z)|dz = \int_{\mathbb{R}^{2n}} |W(\widehat{S}\psi)(z)|dz$$

(because $\det S = 1$) and hence $W(\widehat{S}\psi) \in L^1(\mathbb{R}^n)$ if and only if $W\psi \in L^1(\mathbb{R}^n)$. On the other hand

$$W(\widehat{T}(z_0)\psi, \phi) = e^{\frac{i}{\hbar}\sigma(z,z_0)}W(\psi,\phi)(z - \tfrac{1}{2}z_0)$$

in view of formula (9.25) hence, for every window $\phi \in \mathcal{S}(\mathbb{R}^n)$,

$$\int_{\mathbb{R}^{2n}} |W(\widehat{T}(z_0)\psi, \phi)(z)|dz = \int_{\mathbb{R}^{2n}} |W\psi(z - \tfrac{1}{2}z_0)|dz$$
$$= \int_{\mathbb{R}^{2n}} |W\psi(z)|dz$$

so that $W(\widehat{T}(z_0)\psi) \in S_0(\mathbb{R}^n)$ if and only if $W\psi \in L^1(\mathbb{R}^n)$. The statement (iii) follows from the fact that the Fourier transform F is related to the generator $\widehat{J}$ of $\mathrm{Mp}(2n, \mathbb{R})$ by the formula $F = i^{n/2}\widehat{J}$. $\qquad\square$

Corollary 369. *A function $\psi \in S_0(\mathbb{R}^n)$ is bounded and we have $\lim_{z \to \infty} \psi = 0$.*

Proof. Since ψ is continuous it suffices to prove that $\lim_{z \to \infty} \psi = 0$. Since $S_0(\mathbb{R}^n)$ is invariant by Fourier transform, we have $F^{-1}\psi \in S_0(\mathbb{R}^n)$; now $S_0(\mathbb{R}^n) \subset L^1(\mathbb{R}^n)$ hence $\psi = F(F^{-1}\psi)$ has limit 0 at infinity in view of Riemann–Lebesgue's lemma. $\qquad\square$

As we have seen the topology defined on $S_0(\mathbb{R}^n)$ using the Wigner norm

$$\|\psi\|^{\hbar}_{\phi, S_0(\mathbb{R}^n)} = \|W(\psi,\phi)\|_{L^1(\mathbb{R}^n)}$$

is independent of the choice of window ϕ. Let us prove that the normed space $S_0(\mathbb{R}^n)$ is complete.

Proposition 370. *The Feichtinger algebra $S_0(\mathbb{R}^n)$ has the following properties:*

(i) *It is a Banach space for the Wigner norm $\| \cdot \|^{\hbar}_{\phi, S_0}$ (resp. the STFT norm, $\| \cdot \|_{\phi, S_0}$).*

(ii) *The Schwartz space $\mathcal{S}(\mathbb{R}^n)$ is dense in $S_0(\mathbb{R}^n)$.*

Proof of (i). We sketch the proof, and refer to [82], Theorem 11.3.5 for the technical details, replacing the STFT by the cross-Wigner transform. Let (ψ_j) be a Cauchy sequence in $S_0(\mathbb{R}^n)$; then $(\Psi_j) = (W(\psi_j, \phi))$ is a Cauchy sequence in $L^1(\mathbb{R}^n \oplus \mathbb{R}^n)$. The space $L^1(\mathbb{R}^n \oplus \mathbb{R}^n)$ being complete, there exists $\Psi \in L^1(\mathbb{R}^n \oplus \mathbb{R}^n)$ such that

$$\lim_{j \to \infty} \|\Psi - W(\psi_j, \phi)\|_{L^1} = 0.$$

Defining ψ by the formula

$$\psi(x) = \frac{2^n}{\|\phi\|_{L^2}} \int_{\mathbb{R}^{2n}} \Psi(z)\widehat{T}_{\mathrm{GR}}(z)\phi(x)dz \tag{16.10}$$

(cf. inversion formula (9.51)) one then shows that $\psi \in S_0(\mathbb{R}^n)$ and

$$\|\psi - \psi_j\|_{\phi,S_0}^{\hbar} = \lim_{j \to \infty} \|W(\psi - \psi_j, \phi)\|_{L^1}$$
$$= \lim_{j \to \infty} \|\Psi - W(\psi_j, \phi)\|_{L^1} = 0$$

hence $S_0(\mathbb{R}^n)$ is complete as claimed.

Proof of (ii). Let us first show that $\mathcal{S}(\mathbb{R}^n) \subset S_0(\mathbb{R}^n)$. Let $\psi \in \mathcal{S}(\mathbb{R}^n)$; for every window ϕ we have $W(\psi, \phi) \in \mathcal{S}(\mathbb{R}^n \oplus \mathbb{R}^n)$ hence for every $N > 0$ there exists $C_N > 0$ such that

$$|W(\psi, \phi)(z)| \le C_N(1 + |z|)^{-N}.$$

It follows, by definition of the norm $\| \cdot \|_{\phi,S_0}^{\hbar}$ that

$$\|\psi\|_{\phi,S_0}^{\hbar} \le C_N \int_{\mathbb{R}^{2n}} (1 + |z|)^{-N}dz$$

and hence $\|\psi\|_{\phi,S_0}^{\hbar} < \infty$ if we choose $N > 2n$. We defer the proof of the density since it will be proven in a more general setting in next chapter (Proposition 399). $\qquad\qquad\square$

16.2.2   The algebra property of $S_0(\mathbb{R}^n)$

We begin by proving the following result which is interesting by itself:

Proposition 371. *Suppose that* $\psi \in L^1(\mathbb{R}^n)$ *and* $\psi' \in S_0(\mathbb{R}^n)$. *Then* $\psi * \psi' \in S_0(\mathbb{R}^n)$ *and we have*

$$\|\psi * \psi'\|_{\phi,S_0(\mathbb{R}^n)}^{\hbar} \le \|\psi\|_{L^1(\mathbb{R}^n)}\|\psi'\|_{\phi,S_0(\mathbb{R}^n)}^{\hbar} \tag{16.11}$$

for every window $\phi \in \mathcal{S}(\mathbb{R}^n)$. *Thus, if* $\psi \in L^1(\mathbb{R}^n)$ *and* $\psi' \in S_0$ *then* $\psi * \psi' \in S_0(\mathbb{R}^n)$:

$$L^1(\mathbb{R}^n) * S_0(\mathbb{R}^n) \subset S_0(\mathbb{R}^n).$$

Proof. We begin by rewriting the cross-Wigner transform using Definition (9.11) that is

$$W(\psi, \phi)(z) = \left(\tfrac{1}{\pi\hbar}\right)^n (\widehat{T}_{\mathrm{GR}}(z)\psi|\phi)_{L^2},$$

where $\widehat{T}_{\mathrm{GR}}(z_0)$ is the Grossmann–Royer operator. This yields the formula

$$W(\psi, \phi)(z_0) = \left(\tfrac{1}{\pi\hbar}\right)^n e^{-\frac{2i}{\hbar}p_0 \cdot x_0} \int_{\mathbb{R}^n} \psi(2x_0 - x)\phi_{p_0}(x)dx$$

with $\phi_{p_0}(x) = e^{\frac{2i}{\hbar}p_0 \cdot x}\phi(x)$, that is

$$W(\psi, \phi)(z_0) = \left(\tfrac{1}{\pi\hbar}\right)^n e^{-\frac{2i}{\hbar}p_0 \cdot x_0}\psi * \phi_{p_0}(2x_0). \tag{16.12}$$

It follows, in particular, that

$$\|\psi\|_{\phi,S_0}^{\hbar} = \left(\tfrac{1}{2\pi\hbar}\right)^n \int_{\mathbb{R}^n} \|\psi * \phi_{p_0}\|_{L^1}dp_0. \tag{16.13}$$

Formula (16.12) now shows that

$$W(\psi * \psi', \phi)(z_0) = \left(\tfrac{1}{\pi\hbar}\right)^n e^{-\frac{2i}{\hbar}p_0 \cdot x_0}\psi * \psi' * \phi_{p_0}(2x_0)$$

and hence, by (16.13),

$$\|\psi * \psi'\|_{\phi,S_0}^{\hbar} = \left(\tfrac{1}{2\pi\hbar}\right)^n \int_{\mathbb{R}^n} \|\psi * (\psi' * \phi_{p_0})\|_{L^1}dp_0.$$

Since $L^1(\mathbb{R}^n)$ is a convolution algebra we have

$$\|\psi * (\psi' * \phi_{p_0})\|_{L^1} \leq \|\psi\|_{L^1}\|\psi' * \phi_{p_0}\|_{L^1};$$

we obtain the inequality

$$\|\psi * \psi'\|_{\phi,S_0}^{\hbar} \leq \left(\tfrac{1}{2\pi\hbar}\right)^n \|\psi\|_{L^1} \int_{\mathbb{R}^n} \|\psi' * \phi_{p_0}\|_{L^1}dp_0$$

that is, using again (16.13),

$$\|\psi * \psi'\|_{\phi,S_0}^{\hbar} \leq \|\psi\|_{L^1}\|\psi\|_{\phi,S_0}^{\hbar}$$

which we set out to prove. $\square$

The following result motivates the denomination "Feichtinger *algebra*":

Corollary 372. *The Banach space $S_0(\mathbb{R}^n)$ is an algebra for both pointwise multiplication and convolution: if ψ and ψ' are in $S_0(\mathbb{R}^n)$ then $\psi\psi' \in S_0(\mathbb{R}^n)$ and $\psi * \psi' \in S_0(\mathbb{R}^n)$.*

Proof. Since $\psi\psi'$ and $\psi * \psi'$ are interchangeable by the Fourier transform F, and $S_0(\mathbb{R}^n)$ is invariant under F in view of Proposition 368(iii), it is sufficient to show that $\psi * \psi' \in S_0(\mathbb{R}^n)$ if $\psi \in S_0(\mathbb{R}^n)$ and $\psi' \in S_0(\mathbb{R}^n)$. But this follows from the inequality (16.11) since $S_0(\mathbb{R}^n) \subset L^1(\mathbb{R}^n)$ in view of Proposition 367. $\square$

16.3　A minimality property for $S_0(\mathbb{R}^n)$

A remarkable property of the Feichtinger algebra is that of being the smallest Banach space invariant under the action of the Heisenberg–Weyl operators. To prove this we will need an alternative characterization of $S_0(\mathbb{R}^n)$.

16.3.1　Heisenberg–Weyl expansions

Let $\phi \in S_0(\mathbb{R}^n)$ be a window, and consider the series

$$\psi = \sum_{k=1}^{\infty} c_k \widehat{T}(z_k)\phi \tag{16.14}$$

where $(c_k)_k$ is a sequence of complex numbers and $(z_k)_k$ is a sequence of points in $\mathbb{R}^{2n}$. We have

$$|\psi(z)| \le \sum_{k=1}^{\infty} |c_k|\,|\phi(z - z_k)|.$$

We claim that the series is convergent if $\sum_{k=1}^{\infty} |c_k| < \infty$. In fact, since $\phi \in S_0(\mathbb{R}^n)$ is bounded in view of Corollary 369 there exists a constant C_ϕ such that $|\phi(z - z_k)| \le M_\phi$ and hence

$$|\psi(z)| \le M_\phi \sum_{k=1}^{\infty} |c_k| < \infty.$$

Definition 373. The representation (16.14) of ψ is called a "Heisenberg–Weyl expansion (or representation)" and $(\widehat{T}(z_k)\phi)_k$ a "Heisenberg–Weyl frame" (in time-frequency analysis one would rather speak of "Gabor expansions" and "Gabor frames").

Let us now denote by $\mathcal{M}(\mathbb{R}^n)$ the set of all functions ψ that have a Heisenberg–Weyl expansion (16.14); clearly $\mathcal{M}(\mathbb{R}^n)$ is a complex vector space under the usual addition and multiplication by scalars. It is even a normed vector space if one defines

$$\|\psi\|_{\mathcal{M}(\mathbb{R}^n)} = \inf\left\{ \sum_{k=1}^{\infty} |c_k| < \infty : \psi = \sum_{k=1}^{\infty} c_k \widehat{T}(z_k)\phi \right\}.$$

It turns out that $\mathcal{M}(\mathbb{R}^n)$ is identical with $S_0(\mathbb{R}^n)$:

Proposition 374. *We have $\mathcal{M}(\mathbb{R}^n) = S_0(\mathbb{R}^n)$ and there exists a constant $C_\phi > 0$ such that*

$$\frac{1}{C_\phi} \|\psi\|^{\hbar}_{\phi, S_0(\mathbb{R}^n)} \le \|\psi\|_{\mathcal{M}(\mathbb{R}^n)} \le C_\phi \|\psi\|^{\hbar}_{\phi, S_0(\mathbb{R}^n)}$$

hence the norms $\|\cdot\|_{\mathcal{M}(\mathbb{R}^n)}$ and $\|\cdot\|^{\hbar}_{\phi, S_0(\mathbb{R}^n)}$ are equivalent.

Proof. We omit the proof. See Gröchenig [82], Theorem 12.1.8, which is itself based on the paper by Bonsall [17]. $\qquad\square$

This result shows that every $\psi \in S_0(\mathbb{R}^n)$ can be represented in a Heisenberg–Weyl frame by an absolutely convergent series (16.14).

16.3.2 The minimality property

We are going to prove that Feichtinger's algebra is the smallest Banach algebra in $\mathcal{S}'(\mathbb{R}^n)$ which is invariant under the Heisenberg–Weyl operators.

Proposition 375. *Let* $(B(\mathbb{R}^n), \|\cdot\|_B)$ *be a Banach algebra of tempered distributions on* $\mathbb{R}^n$. *Suppose that* $B(\mathbb{R}^n)$ *satisfies the two following conditions:*

(i) *there exists* $C > 0$ *such that* $\|\widehat{T}(z)\psi\| \leq C\|\psi\|$ *for all* $z \in \mathbb{R}^{2n}$ *and* $\psi \in B(\mathbb{R}^n)$;

(ii) $S_0(\mathbb{R}^n) \cap B \neq \{0\}$. *Then* $S_0(\mathbb{R}^n)$ *is continuously embedded in* $B(\mathbb{R}^n)$ *and* $S_0(\mathbb{R}^n)$ *is the smallest algebra having this property.*

Proof. We are following Gröchenig [82]. Let $\phi \in S_0(\mathbb{R}^n) \cap B(\mathbb{R}^n)$, $\phi \neq 0$. In view of Proposition 374 the space $S_0(\mathbb{R}^n)$ consists of all Heisenberg–Weyl expansions

$$\psi = \sum_{k=1}^{\infty} c_k \widehat{T}(z_k)\phi \ , \ \sum_{k=1}^{\infty} |c_k| < \infty.$$

Thus,

$$\|\psi\|_B \leq \sum_{k=1}^{\infty} |c_k| \|\widehat{T}(z_k)\phi\|_B \leq C \left(\sum_{k=1}^{\infty} |c_k| \right) \|\phi\|_B < \infty$$

and hence $\psi \in S_0(\mathbb{R}^n)$ so that $S_0(\mathbb{R}^n) \subset B(\mathbb{R}^n)$. Let us show that this embedding is continuous. taking the infimum of the right-hand side of the inequality

$$\|\psi\|_B \leq C \left(\sum_{k=1}^{\infty} |c_k| \right) \|\phi\|_B$$

for all Heisenberg–Weyl expansions of ψ we get

$$\|\psi\|_B \leq C\|\psi\|_{S_0} \|\phi\|_B$$

which proves the statement. $\qquad\square$

16.4 A Banach Gelfand triple

Dirac already emphasized in his fundamental work [37] the relevance of *rigged Hilbert spaces* for quantum mechanics[1]. Later Schwartz provided an instance of

---

[1] In quantum mechanics, to rig a Hilbert space means simply to equip that Hilbert space with distribution theory [122].

rigged Hilbert spaces based on his class of test functions and on tempered distributions. Later Gelfand and Shilov formalized the construction of Schwartz and Dirac and introduced what is nowadays known as *Gelfand triples*. It has been known for several decades that Dirac's bra-ket formalism is mathematically justified not by the Hilbert space alone, but by the use of Gelfand triples; see de la Madrid [122] (among other things we learn from that paper that the terminology "rigged Hilbert space" is a direct translation of the phrase "*osnashchyonnoe Hilbertovo prostranstvo*" from the original Russian). The prototypical example of a Gelfand triple is

$$(\mathcal{S}(\mathbb{R}^n), L^2(\mathbb{R}^n), \mathcal{S}'(\mathbb{R}^n)).$$

In this section we will study the particular very interesting Gelfand triple

$$(S_0(\mathbb{R}^n), L^2(\mathbb{R}^n), S_0'(\mathbb{R}^n))$$

whose properties are very interesting both from a theoretical and practical point of view. For a very nice review of the topic with many historical remarks see the contribution of Feichtinger et al. in [55].

16.4.1 The dual space $S_0'(\mathbb{R}^n)$

Let us denote by $S_0'(\mathbb{R}^n)$ the dual Banach space of $S_0(\mathbb{R}^n)$. It is the space of all bounded linear functionals on $S_0(\mathbb{R}^n)$. Since $S_0(\mathbb{R}^n)$ is the smallest Banach space isometrically invariant under the action of the affine metaplectic group (and hence under the Heisenberg–Weyl operators) its dual is essentially the largest space of distributions with this property.

The following result characterizes $S_0'(\mathbb{R}^n)$:

Proposition 376. *The Banach space $S_0'(\mathbb{R}^n)$ consists of all $\psi \in S'(\mathbb{R}^n)$ such that $W(\psi, \phi) \in L^\infty(\mathbb{R}^n \oplus \mathbb{R}^n)$ for one (and hence all) window $\phi \in S_0(\mathbb{R}^n)$; the duality bracket is given by the pairing*

$$(\psi, \psi') = \int_{\mathbb{R}^{2n}} W(\psi, \phi)(z)\overline{W(\psi', \phi)(z)}dz \tag{16.15}$$

and the formula

$$\|\psi\|^{\hbar}_{\phi, S_0'(\mathbb{R}^n)} = \sup_{z \in \mathbb{R}^{2n}} |W(\psi, \phi)(z)| \tag{16.16}$$

defines a norm on $S_0'(\mathbb{R}^n)$.

Proof. It is based on the fact that $L^\infty(\mathbb{R}^n \oplus \mathbb{R}^n)$ is the dual space of $L^1(\mathbb{R}^n \oplus \mathbb{R}^n)$; see [82], §11.3. $\square$

It readily follows from this characterization that:

Proposition 377. *The Dirac distribution δ is in $S_0'(\mathbb{R}^n)$; more generally $\delta_a \in S_0'(\mathbb{R}^n)$ where $\delta_a(x) = \delta(x - a)$.*

Proof. We have

$$W(\delta, \phi)(z_0) = \left(\tfrac{1}{\pi\hbar}\right)^n \langle \widehat{T}_{\mathrm{GR}}(z_0)\delta, \overline{\phi}\rangle$$

(formula (9.12)) and

$$\widehat{T}_{\mathrm{GR}}(z_0)\delta(x) = e^{\frac{2i}{\hbar}p_0\cdot(x-x_0)}\delta(2x_0 - x) = e^{\frac{2i}{\hbar}p_0\cdot x_0}\delta(2x_0 - x).$$

It follows that

$$W(\delta, \phi)(z_0) = \left(\tfrac{1}{\pi\hbar}\right)^n e^{\frac{2i}{\hbar}p_0\cdot x_0}\overline{\phi}(2x_0)$$

and hence $|W(\delta, \phi)(z_0)| \le \left(\tfrac{1}{\pi\hbar}\right)^n \|\phi\|_\infty$. It follows that $\delta \in S_0'(\mathbb{R}^n)$. That we also have $\delta_a \in S_0'(\mathbb{R}^n)$ is proven by a similar argument (alternatively one can use formula (9.25) in Proposition 174). One can show that, more generally, the "Dirac comb" $\sum_{a\in\mathbb{Z}^n} \delta_a$ belongs to $S_0'(\mathbb{R}^n)$. $\qquad\square$

16.4.2 The Gelfand triple (S_0, L^2, S_0')

Let us begin by defining rigorously the notion of Banach Gelfand triple (also called rigged, or nested, spaces):

Definition 378. A (Banach) *Gelfand triple* $(\mathcal{B}, \mathcal{H}, \mathcal{B}')$ consists of a Banach space $\mathcal{B}$ which is continuously and densely embedded into a Hilbert space $\mathcal{H}$, which in turn is w^*-continuously and densely embedded into the dual Banach space $\mathcal{B}'$.

In this definition one identifies $\mathcal{H}$ with its dual $\mathcal{H}^*$ and the scalar product on $\mathcal{H}$ thus extends in a natural way into a pairing between $\mathcal{B} \subset \mathcal{H}$ and $\mathcal{B}' \supset \mathcal{H}$.

The use of the Gelfand triple $(S_0(\mathbb{R}^n), L^2(\mathbb{R}^n), S_0'(\mathbb{R}^n))$ not only offers a better description of self-adjoint operators but it also allows a simplification of many proofs. Here is a typical situation, that will be slightly extended in Chapter 18, Subsection 19.1.2. Given a Gelfand triple $(\mathcal{B}, \mathcal{H}, \mathcal{B}')$ one proves that every self-adjoint operator $A : \mathcal{B} \longrightarrow \mathcal{B}$ has a complete family of generalized eigenvalues $(\psi_\alpha)_\alpha = \{\psi_\alpha \in \mathcal{B}' : \alpha \in \mathbb{A}\}$ ($\mathbb{A}$ an index set), defined as follows: for every $\alpha \in \mathbb{A}$ there exists $\lambda_\alpha \in \mathbb{C}$ such that

$$(\psi_\alpha, A\phi) = \lambda_\alpha(\psi_\alpha, \phi) \ \text{ for every } \phi \in \mathcal{B}.$$

Completeness of the family $(\psi_\alpha)_\alpha$ means that there exists at least one ψ_α such that $(\psi_\alpha, \phi) \ne 0$ for every $\phi \in \mathcal{B}$. A basic example, in the case $n = 1$, is the operator $\widehat{x}$ of multiplication by x. This operator has no eigenfunctions in $L^2(\mathbb{R})$, but since $\widehat{x}\delta_a(x) = x\delta(x - a) = a\delta_a(x)$ every $a \in \mathbb{R}$ is a generalized eigenvalue (with associated eigenfunction $\delta_a \in S_0'(\mathbb{R})$).

Given a Gelfand triple $(\mathcal{B}, \mathcal{H}, \mathcal{B}')$ every $\phi \in \mathcal{B}$ has an expansion with respect to the generalized eigenvectors ψ_α which generalizes the usual expansion with respect to a basis of eigenvectors. A classical example is the following: consider the Gelfand triple $(S_0(\mathbb{R}^n), L^2(\mathbb{R}^n), S_0'(\mathbb{R}^n))$ and choose $\widehat{A} = -i\hbar\partial_{x_j}$. The generalized eigenvalues of $\widehat{A}$ are the functions $\chi_p(x) = e^{ip\cdot x/\hbar}$ ($p \in \mathbb{R}^n$) and the corresponding expansion can be written as the Fourier inversion formula

$$\psi(x) = \left(\tfrac{1}{2\pi\hbar}\right)^{n/2} \int_{\mathbb{R}^n} e^{\frac{i}{\hbar} p\cdot x} F\psi(p)dp \tag{16.17}$$

(see Feichtinger et al. in [55] for a detailed discussion of the Fourier transform within the context of the Banach Gelfand triple $(S_0(\mathbb{R}^n), L^2(\mathbb{R}^n), S_0'(\mathbb{R}^n))$).

An important feature of Gelfand triples is the existence of a kernel theorem, which is much more useful both for theoretical and practical purposes than the usual kernel theorem of Schwartz. We denote as usual by $\langle\langle \cdot, \cdot \rangle\rangle$ the distributional bracket for distributions on $\mathbb{R}^n \oplus \mathbb{R}^n$.

Theorem 379 (Feichtinger). *The following properties hold:*

(i) *Every linear bounded operator $A : S_0(\mathbb{R}^n) \longrightarrow S_0'(\mathbb{R}^n)$ has a kernel $K_A \in S_0'(\mathbb{R}^n \times \mathbb{R}^n)$, that is $\langle A\psi, \phi \rangle = \langle\langle K_A, \phi \otimes \psi \rangle\rangle$ for ψ and ϕ in $S_0(\mathbb{R}^n)$.*

(ii) *Conversely, every $K \in S_0'(\mathbb{R}^n \times \mathbb{R}^n)$ defines by the formula above a bounded operator $S_0(\mathbb{R}^n) \longrightarrow S_0'(\mathbb{R}^n)$.*

Formally we can thus write

$$A\psi(x) = \int_{\mathbb{R}^n} K(x,y)\psi(y)dy$$

for some $K \in S_0'(\mathbb{R}^n \times \mathbb{R}^n)$ when $A : S_0(\mathbb{R}^n) \longrightarrow S_0'(\mathbb{R}^n)$ is a continuous operator. This result was announced by Feichtinger in [45] and proven in [52]. See [82], §11.4 for a detailed proof, comments, and various extensions. We will see in the next chapter that this result can be generalized to a whole class of modulation spaces.

For instance, in the example above, one can interpret the Fourier inversion formula (16.17) by saying that the kernel of the inverse Fourier transform is $K(x,p) = (2\pi\hbar)^{n/2} e^{ip\cdot x/\hbar}$.

Chapter 17

The Modulation Spaces M_s^q

If the choice of good symbol classes is essential in any pseudo-differential calculus, so is the choice of good functional spaces between which these operators act. These spaces must reflect regularity properties (in the broad sense) of the operators that are used. For instance, in Hörmander's theory of pseudo-differential operators a standard choice are the Sobolev spaces H^s and their variants. In Shubin theory one has the global spaces Q^s, where one can simultaneously control the behavior in x and its dual variable (these are in fact already modulation spaces). In our case, where we are interested in studying quantum mechanics in phase space, it turns out that the best playground consists of a constellation of spaces called the *modulation spaces* $M_s^q(\mathbb{R}^n) = M_{v_s}^q(\mathbb{R}^n)$ where $1 \leq q \leq \infty$ and v_s is a polynomial weight. These are particular cases of the spaces $M_v^{p,q}(\mathbb{R}^n)$ whose definition goes back to Feichtinger's foundational papers [46, 48, 49]; for a recent review see [50]. In addition, if one denotes by $(M_s^q(\mathbb{R}^n))'$ the dual of $M_s^q(\mathbb{R}^n)$ then $(M_s^q(\mathbb{R}^n), L^2(\mathbb{R}^n), (M_s^q(\mathbb{R}^n))')$ is a *Gelfand triple* which is a natural domain for studying the generalized eigenvalues and eigenvectors of quantum mechanical operators (in particular the continuous spectrum).

17.1 The L^q spaces, $1 \leq q < \infty$

In order to define modulation spaces we have to introduce weighted L^q spaces. The number of books devoted to this topic (at various levels of difficulty) is huge; a limpid recent treatment is given in Chapter 6 of Folland's textbook [60].

We begin by reviewing the L^q theory.

17.1.1 Definitions

We will assume that $\mathbb{R}^m$ is equipped with the Lebesgue measure $dx = dx_1 \cdots dx_m$. Let q a real number such that $1 \leq q < \infty$. If $f : \mathbb{R}^m \longrightarrow \mathbb{C}$ is measurable we set

$$\|f\|_q = \left(\int_{\mathbb{R}^m} |f(x)|^q dx \right)^{1/q} ;$$

$\|f\|_q$ is thus a real number, or ∞. We will write $f \sim g$ when $f - g = 0$ almost everywhere for the measure dx. The relation $\sim$ is an equivalence relation for all functions $\mathbb{R}^m \longrightarrow \mathbb{C}$.

Definition 380. The space $L^q(\mathbb{R}^m)$ consists of all $\sim$ equivalence classes of measurable functions $f : \mathbb{R}^m \longrightarrow \mathbb{C}$ such that $\|f\|_q < \infty$, equipped with the norm $f \longmapsto \|f\|_q$.

That $L^q(\mathbb{R}^m)$ indeed is a vector space is seen as follows. First, it is clear that $\lambda f \in L^q(\mathbb{R}^m)$ if $f \in L^q(\mathbb{R}^m)$, for every $\lambda \in \mathbb{C}$. Let now f and g be two elements of $L^q(\mathbb{R}^m)$. Then

$$|f + g|^q \le (2\sup(f,g))^q \le 2^q(|f|^q + |g|^q) \tag{17.1}$$

(prove this!) and hence

$$\|f + g\|_q \le 2(\|f\|_q + \|g\|_q) < \infty. \tag{17.2}$$

Observe that the inequality (17.2) does not prove that $\| \cdot \|_q$ is a norm: while it is clear that $\|\lambda f\|_q = |\lambda|\,\|f\|_q$ and that $\|f\|_q = 0$ implies $f = 0$ (almost everywhere), we still have to prove the Minkowski inequality

$$\|f + g\|_q \le \|f\|_q + \|g\|_q \tag{17.3}$$

and this will require some extra work. The Minkowski inequality is actually a consequence of the famous *Hölder inequality*:

Proposition 381. *Let q and r be real numbers ≥ 1 such that $1/q + 1/r = 1$. We have*

$$\|fg\|_1 \le \|f\|_q\|g\|_r \tag{17.4}$$

for all $f \in L^q(\mathbb{R}^m)$ and $g \in L^r(\mathbb{R}^m)$.

Proof. We are going to use the elementary inequality

$$qrab \le ra^q + qb^r \tag{17.5}$$

(which is a generalization of the trivial inequality $2ab \le a^2 + b^2$), valid for all nonnegative a, b when $1/q + 1/r = 1$. Choosing $a = |f(x)|$, $b = |g(x)|$ this inequality becomes

$$qr|f(x)|\,|g(x)| \le r|f(x)|^q + q|g(x)|^r$$

and hence, integrating,

$$qr\|fg\|_1 \le r\|f\|_q^q + q\|g\|_r^r.$$

Replacing f and g by $f/\|f\|_q$ and $g/\|g\|_r$ we get

$$\frac{qr}{\|f\|_q\|g\|_r}\|fg\|_1 \le r + q = 1$$

which yields Hölder's inequality (17.4). $\qquad\qquad\qquad\square$

Definition 382. We will call the numbers q and r "conjugate exponents" when they satisfy the conditions $1/q + 1/r = 1$ above.

Exercise 383. Prove the inequality (17.5); show that we have equality if and only if $a = b$. [Hint: study the variations of the function $k(x) = rx^q + qb^r - qrxb$.]

Let us now prove the main result of this subsection:

Theorem 384. *Let q be such that $1 \leq q < \infty$. We have:*

(i) *The mapping $f \longmapsto \|f\|_q$ is a norm on $L^q(\mathbb{R}^m)$;*

(ii) *$L^q(\mathbb{R}^m)$ is a Banach space for the topology defined by this norm.*

Proof. (i) There remains to prove the Minkowski inequality (17.3). The result being obvious for $q = 1$ or $\|f + g\|_q = 0$ we assume $q > 1$ and that $\|f + g\|_q > 0$. We next note that since $|f + g| \leq |f| + |g|$ we have

$$|f + g|^q \leq |f|\,|f + g|^{q-1} + |g|\,|f + g|^{q-1}.$$

Applying Hölder's inequality to $|f|\,|f + g|^{q-1}$ and $|g|\,|f + g|^{q-1}$ we get:

$$\|(f + g)^q\|_1 \leq \|f\|_q \|(f + g)^{q-1}\|_r + \|g\|_q \|(f + g)^{q-1}\|_r$$
$$= (\|f\|_q + \|g\|_q)\|(f + g)^{q-1}\|_r.$$

Since $r(q - 1) = q$ we have

$$\|(f + g)^{q-1}\|_r = \left(\int_{\mathbb{R}^m} |f + g|^{r(q-1)} dx\right)^{1/r}$$
$$= \left(\int_{\mathbb{R}^m} |f + g|^q dx\right)^{1/r}$$

and hence

$$\|(f + g)^q\|_1 \leq (\|f\|_q + \|g\|_q)\|f + g\|_q^{q/r}.$$

Now, $\|(f + g)^q\|_1 = \|f + g\|_q^q$ so that we finally have, dividing both sides by $\|f + g\|_q^{q-1}$,

$$\|f + g\|_q \leq (\|f\|_q + \|g\|_q)\|f + g\|_q^{q/r + 1 - q}$$

which is Minkowski's inequality since $q/r + 1 - q = 0$ because q and r are conjugate exponents. (ii) It is sufficient to show that every absolutely convergent series in $L^q(\mathbb{R}^m)$ is convergent. Let (f_j) be a sequence in $L^q(\mathbb{R}^m)$ and assume that $\sum_{j=1}^{\infty} \|f_j\|_q = M$ (M a real number). Setting $F_k = \sum_{j=1}^{k} |f_j|$ and $F = \sum_{j=1}^{\infty} |f_j|$ we have

$$\|F_k\|_q \leq \sum_{j=1}^{k} \|f_j\|_k \leq M$$

for all indices k, hence

$$\int_{\mathbb{R}^m} F^q dx = \lim_{k\to\infty} \int_{\mathbb{R}^m} F_k^q dx \le M^q$$

so that $F \in L^q(\mathbb{R}^m)$. This implies that we must have $F(x) < \infty$ almost everywhere, and the series $(\sum_{j=1}^k f_j)_k$ is thus convergent almost everywhere. Let $f = \sum_{j=1}^k f_j$ be its limit; we have $|f| \le F$ and hence $f \in L^q(\mathbb{R}^m)$. On the other hand $|f - \sum_{j=1}^k f_j|^q \le (2F)^p$ and $F^p \in L^1(\mathbb{R}^m)$ hence, using the dominated convergence theorem,

$$\lim_{k\to\infty} \|f - \sum_{j=1}^k f_j\|_q^q = \lim_{k\to\infty} \int_{\mathbb{R}^m} |f - \sum_{j=1}^k f_j|^q dx = 0.$$

This proves that the absolutely convergent series $\sum_{j=1}^k f_j$ converges in $L^q(\mathbb{R}^m)$, and we are done. $\qquad\square$

Another important property of the L^q spaces is the following:

Theorem 385. *Let q and r be conjugate exponents, $q > 1$.*

(i) *For each $\Phi \in (L^q(\mathbb{R}^m))^*$ (the dual space of $L^q(\mathbb{R}^m)$) there exists $g_\Phi \in L^r(\mathbb{R}^m)$ such that*

$$\Phi[f] = \int_{\mathbb{R}^m} f g_\Phi dx; \qquad\qquad (17.6)$$

two functions g_Φ and h_Φ satisfying this relation are equal almost everywhere;

(ii) *The dual $(L^q(\mathbb{R}^m))^*$ is isometrically isomorphic to $L^r(\mathbb{R}^m)$.*

Proof. (i) The construction of g is rather lengthy and we omit it here (see Folland [60], §6.2). The duality property (ii) immediately follows from (17.6). Consider the mapping $\Phi \longmapsto g_\Phi$ defined by (17.6). In view of Hölder's inequality (Proposition 381) we have $|\Phi[f]| \le \|f\|_q \|g_\Phi\|_r$ showing that Φ is a continuous linear functional on $L^q(\mathbb{R}^m)$ and hence an element of $(L^q(\mathbb{R}^m))^*$. The proof of the converse is left to the reader. $\qquad\square$

17.1.2 Weighted L^q spaces

The notion of weight plays an essential role in the theory of modulation spaces.

We begin with defining the notion of weight; our definition is rather restrictive, but sufficient for our purposes. For a more general definition see [82], §11.1.

Definition 386. A weight on $\mathbb{R}^{2n}$ is a real, non-negative, and locally integrable function $v : \mathbb{R}^{2n} \longrightarrow \mathbb{C}$ which is in addition sub-multiplicative: $v(z+z') \le v(z)v(z')$ for all z, z' in $\mathbb{R}^{2n}$.

Exercise 387. Show that an even weight v always satisfies $v(0) \ge 1$.

The study of weighted L^q spaces is not new, see for instance Benedik and Panzone [12]. In particular these authors prove a more general result than the following:

Proposition 388. *The weighted spaces $L^q_v(\mathbb{R}^n \oplus \mathbb{R}^n)$ have the two following properties:*

(i) *$L^q_v(\mathbb{R}^n \oplus \mathbb{R}^n)$ is a Banach space;*

(ii) *$L^q_v(\mathbb{R}^n \oplus \mathbb{R}^n)$ is invariant under the translations $T(z_0)\Psi(z) = \Psi(z - z_0)$. In fact:*

$$\|T(z_0)\Psi\|_{L^q_v} \leq v(z_0)\|\Psi\|_{L^q_v}; \tag{17.7}$$

Proof of (i). Let (Ψ_j) be a Cauchy sequence in $L^q_v(\mathbb{R}^n \oplus \mathbb{R}^n)$; then $(\Phi_j) = (v\Psi_j)$ is a Cauchy sequence in $L^q(\mathbb{R}^n \oplus \mathbb{R}^n)$: this immediately follows from the equality $\|\Psi_j\|_{L^q_v} = \|v\Psi_j\|_{L^q}$. Let Φ be its limit and set $\Psi = v^{-1}\Phi$. We claim that $\lim_{j\to\infty} \Psi_j = \Psi$ in $L^q_v(\mathbb{R}^n \oplus \mathbb{R}^n)$. In fact,

$$\|\Psi_j - \Psi\|_{L^q_v} = \|\Phi_j - \Phi\|_{L^q}$$

hence $\lim_{j \longrightarrow \infty} \|\Psi_j - \Psi\|_{L^q_v} = 0$.

Proof of (ii). To prove that $L^q_v(\mathbb{R}^n \oplus \mathbb{R}^n)$ is invariant under the translation operator $T(z_0)$ it suffices to use the submultiplicative property of the weight: we have, for $q < \infty$,

$$\begin{aligned}
\|T(z_0)\Psi\|^q_{L^q_v} &= \int_{\mathbb{R}^{2n}} |\Psi(z - z_0)|^q v(z)^q dz \\
&= \int_{\mathbb{R}^{2n}} |\Psi(z)|^q v(z + z_0)^q dz \\
&\leq v(z_0) \int_{\mathbb{R}^{2n}} |\Psi(z)|^q v(z)^q dz. \qquad \square
\end{aligned}$$

For our purposes it will be sufficient to limit ourselves to choosing for v the standard weight function

$$v_s(z) = (1 + |z|^2)^{s/2} = \langle z \rangle^s. \tag{17.8}$$

Notation 389. When the weight v is given by (17.8) we will write $L^q_s(\mathbb{R}^n \oplus \mathbb{R}^n)$ instead of $L^q_v(\mathbb{R}^n \oplus \mathbb{R}^n)$.

Exercise 390. Verify that the standard weight v_s is submultiplicative.

Recall that for $1 \leq q < \infty$ the space $L^q(\mathbb{R}^n \oplus \mathbb{R}^n)$ consists of all complex functions Ψ on $\mathbb{R}^n \oplus \mathbb{R}^n$ such that

$$\|\Psi\|_{L^q} = \left(\int_{\mathbb{R}^{2n}} |\Psi(z)|^q dz \right)^{1/q} < \infty.$$

The mapping $\Psi \longmapsto \|\Psi\|_{L^q}$ is a norm on $L^q(\mathbb{R}^n \oplus \mathbb{R}^n)$ and $L^q(\mathbb{R}^n \oplus \mathbb{R}^n)$ is a Banach space for the topology thus defined.

Definition 391. The weighted space $L_s^q(\mathbb{R}^n \oplus \mathbb{R}^n)$ consists of all Ψ such that $\langle \cdot \rangle^s \Psi \in L^q(\mathbb{R}^n \oplus \mathbb{R}^n)$ equipped with the norm $\|\Psi\|_{L_s^q} = \|\langle \cdot \rangle^s \Psi\|_{L^q}$.

Exercise 392. Show that $\| \cdot \|_{L_s^q}$ indeed is a norm on $L_v^q(\mathbb{R}^n \oplus \mathbb{R}^n)$. [Hint: Use Exercise 387.]

The following simple result is very useful:

Lemma 393. *The weighted spaces $L_s^q(\mathbb{R}^n \oplus \mathbb{R}^n)$ are invariant under linear changes of variables: if $\Psi \in L_s^q(\mathbb{R}^n \oplus \mathbb{R}^n)$ and $F \in GL(2n, \mathbb{R})$ then $\Psi \circ F \in L_s^q(\mathbb{R}^n \oplus \mathbb{R}^n)$.*

Proof. (Cf. the proof of Proposition 316). Diagonalizing $F^T F$ using an orthogonal transformation we have

$$\lambda_{\min} |z|^2 \leq |F(z)|^2 \leq \lambda_{\max} |z|^2$$

where $\lambda_{\min} > 0$ and $\lambda_{\max} > 0$ are the smallest and largest eigenvalues of $F^T F$. It follows that we have

$$\langle F(z) \rangle^s \leq \max(1, \lambda_{\max}) \langle z \rangle^s$$

if $s \geq 0$, and

$$\langle F(z) \rangle^s \leq \min(1, \lambda_{\min}) \langle z \rangle^s$$

if $s < 0$. $\square$

We will also need the following results about convolutions:

Proposition 394. *Let $s \in \mathbb{R}$.*

(i) *We have the estimate*

$$\|\Psi * \Phi\|_{L_s^q(\mathbb{R}^{2n})} \leq C_s \|\Psi\|_{L_s^1(\mathbb{R}^{2n})} \|\Phi\|_{L_s^q(\mathbb{R}^{2n})} \tag{17.9}$$

and hence

$$L_s^1(\mathbb{R}^n \oplus \mathbb{R}^n) * L_s^q(\mathbb{R}^n \oplus \mathbb{R}^n) \subset L_s^q(\mathbb{R}^n \oplus \mathbb{R}^n).$$

(ii) *If $s > 2n$ then*

$$\|\Psi * \Phi\|_{L_s^\infty(\mathbb{R}^{2n})} \leq C_s \|\Psi\|_{L_s^\infty(\mathbb{R}^{2n})} \|\Phi\|_{L_s^\infty(\mathbb{R}^{2n})} \tag{17.10}$$

and hence

$$L_s^\infty(\mathbb{R}^n \oplus \mathbb{R}^n) * L_s^\infty(\mathbb{R}^n \oplus \mathbb{R}^n) \subset L_s^\infty(\mathbb{R}^n \oplus \mathbb{R}^n).$$

Proof. (Cf. [82], Proposition 11.1.3). (i) Let $\Theta \in L_{-s}^{q'}(\mathbb{R}^n \oplus \mathbb{R}^n)$ with $1/q + 1/q' = 1$. We have, using Fubini's theorem and the inequality (17.7),

$$|(\Psi * \Phi | \Theta)_{L^2}| = \left| \iint_{\mathbb{R}^{4n}} \Psi(u)\Phi(z - u)\overline{\Theta(u)} dz du \right|$$

$$\leq \int_{\mathbb{R}^{2n}} |\Psi(u)| \left(\int_{\mathbb{R}^{2n}} |T(u)\Phi(z)| \, |\Theta(z)| dz \right) du$$

$$\leq \int_{\mathbb{R}^{2n}} |\Psi(u)| \|T(u)\Phi\|_{L_s^q} du \cdot \|\Theta\|_{L_{-s}^{q'}}$$

$$\leq C \left(\int_{\mathbb{R}^{2n}} |\Psi(u)| \langle u \rangle^s du \right) \|\Phi\|_{L_s^q} \|\Theta\|_{L_{-s}^{q'}}.$$

By duality we have

$$\|\Psi * \Phi\|_{L_s^q} = \sup \left\{ |(\Psi * \Phi | \Theta)_{L^2}| : \|\Theta\|_{L_{-s}^{q'}} \leq 1 \right\}$$

$$\leq C_s \|\Psi\|_{L_s^1} \|\Phi\|_{L_s^q},$$

hence the estimate (17.9).

(ii) Let $\Psi, \Phi \in L_s^\infty(\mathbb{R}^n \oplus \mathbb{R}^n)$. We have

$$|\Psi(z)| \leq \|\Psi\|_{L_s^\infty(\mathbb{R}^{2n})} \langle z \rangle^{-s},$$

$$|\Phi(z)| \leq \|\Phi\|_{L_s^\infty(\mathbb{R}^{2n})} \langle z \rangle^{-s}$$

and hence

$$|\Psi * \Phi(z)| \leq \|\Psi v_s\|_{L_s^\infty} \| \langle \cdot \rangle^s \Phi\|_{L^\infty} \int_{\mathbb{R}^{2n}} (1 + |u|^2)^{-s/2}(1 + |z - u|^2)^{-s/2} du$$

where we have used the convolution inequality

$$\langle \cdot \rangle^{-s} * \langle \cdot \rangle^{-s} \leq C_s \langle \cdot \rangle^{-s}. \tag{17.11}$$

(See exercise below.) Integrating in z the estimate (17.10) follows since the integral is absolutely convergent for $s > 2n$. $\qquad\square$

Exercise 395. Prove the convolution inequality (17.11); equivalently:

$$\int_{-\infty}^{\infty} (1 + |t|)^{-s}(1 + |x - t|)^{-s} dt \leq C_s(1 + |x|)^{-s}$$

(in case of emergency see [82], (11.5)).

17.2 The modulation spaces M_s^q

The modulation spaces $M_s^q(\mathbb{R}^n)$ we are going to study in this and the following sections are in a sense rather straightforward extensions of the Feichtinger algebra and of its its dual.

17.2.1 Definition of M_s^q

Recall that the weight v_s is defined by

$$v_s(z) = (1 + |z|^2)^{s/2} = \langle z \rangle^s$$

and that the corresponding weighted L^q spaces are denoted by L_s^q.

Definition 396. The modulation space $M_s^q(\mathbb{R}^n)$ consists of all $\psi \in \mathcal{S}'(\mathbb{R}^n)$ such that $V_\phi \psi \in L_s^q(\mathbb{R}^n \oplus \mathbb{R}^n)$ for every window $\phi \in \mathcal{S}'(\mathbb{R}^n)$ (where $V_\phi \psi$ is the STFT transform).

The definition of $M_s^q(\mathbb{R}^n)$ can be restated in terms of the cross-Wigner transform in the same way as was done for the Feichtinger algebra:

Proposition 397. *We have $\psi \in M_s^q(\mathbb{R}^n)$ if and only if $W(\psi, \phi) \in L_s^q(\mathbb{R}^n \oplus \mathbb{R}^n)$ for every $\phi \in \mathcal{S}(\mathbb{R}^n)$.*

Proof. It is, as in the Feichtinger algebra case, again based on formula (16.2) relating the STFT V_ϕ to the cross-Wigner transform $W(\psi, \phi)$. Recall that we denote by ψ_λ the function defined by $\psi_\lambda(x) = \psi(\lambda x)$. We have $\psi \in M_s^q(\mathbb{R}^n)$ if and only if $V_\phi \psi \in L_s^q(\mathbb{R}^n \oplus \mathbb{R}^n)$ for every $\phi \in \mathcal{S}(\mathbb{R}^n)$, that is if and only if $V_{\phi^\vee_{\sqrt{2\pi\hbar}}} \psi \in L_s^q(\mathbb{R}^n \oplus \mathbb{R}^n)$. Since, in addition, $\psi \in M_s^q(\mathbb{R}^n)$ if and only if $\psi_{\sqrt{2\pi\hbar}} \in M_s^q(\mathbb{R}^n)$, we thus have $\psi \in M_s^q(\mathbb{R}^n)$ if and only if

$$V_{\phi^\vee_{\sqrt{2\pi\hbar}}} \psi_{\sqrt{2\pi\hbar}} \in L_s^q(\mathbb{R}^n \oplus \mathbb{R}^n)$$

or, which amounts to the same,

$$\psi \in M_s^q(\mathbb{R}^n) \Longleftrightarrow 2^n e^{\frac{2i}{\hbar} p \cdot x} V_{\phi^\vee_{\sqrt{2\pi\hbar}}} \psi_{\sqrt{2\pi\hbar}} \in L_s^q(\mathbb{R}^n \oplus \mathbb{R}^n). \tag{17.12}$$

Now, a function Ψ is in $L_s^q(\mathbb{R}^n \oplus \mathbb{R}^n)$ if and only if Ψ_λ is, as follows from the inequality

$$\int_{\mathbb{R}^{2n}} |\langle z \rangle^s \Psi(\lambda z)|^q dz \leq \lambda^{-2nq}(1 + \lambda^{-2})^{s/2} \int_{\mathbb{R}^n} |\langle z \rangle^s \Psi(z)|^q dz$$

obtained by performing the change of variable $z \longmapsto \lambda^{-1} z$ and using the trivial estimate

$$(1 + |\lambda^{-1} z|^2)^{s/2} \leq (1 + \lambda^{-2})^{s/2}(1 + |z|^2)^{s/2}$$

valid for all $s \geq 0$. Combining this property (with $\lambda = \sqrt{2/\pi\hbar}$) and the equivalence (17.12) we thus have $\psi \in M_s^q(\mathbb{R}^n)$ if and only if $W_\phi \psi \in L_s^q(\mathbb{R}^n \oplus \mathbb{R}^n)$. $\square$

We can equip the modulation space $M_s^q(\mathbb{R}^n)$ with a family of norms defined by

$$\|\psi\|_{\phi, M_s^q(\mathbb{R}^n)}^\hbar = \|W(\psi, \phi)\|_{L_s^q(\mathbb{R}^{2n})} = \int_{\mathbb{R}^{2n}} |W(\psi, \phi)(z)|^q \langle z \rangle^s \, dz. \tag{17.13}$$

We have defined $M_s^q(\mathbb{R}^n)$ by requiring that $W(\psi, \phi) \in L_s^q(\mathbb{R}^n \oplus \mathbb{R}^n)$ for every window ϕ. Not surprisingly, taking into account Proposition 364:

Proposition 398. *We have $\psi \in M_s^q(\mathbb{R}^n)$ if and only if $W(\psi, \phi) \in L_s^q(\mathbb{R}^n \oplus \mathbb{R}^n)$ for one window $\phi \in \mathcal{S}(\mathbb{R}^n)$. The topology of $M_s^q(\mathbb{R}^n)$ is defined by using a single norm $\| \cdot \|_{\phi, M_s^q}^\hbar$; moreover all the norms obtained by letting ϕ vary are equivalent.*

Proof. It is similar to that of Proposition 364 with a few technical modification; we therefore leave it to the reader. $\qquad\square$

The following result is the analogue of Proposition 370 where it was stated that the Feichtinger algebra $M^1(\mathbb{R}^n)$ is a Banach space containing $\mathcal{S}(\mathbb{R}^n)$ as a dense subspace.

Proposition 399.

(i) *The modulation space $M_s^q(\mathbb{R}^n)$ is a Banach space for the topology defined by the norm $\| \cdot \|_{\phi, M_s^q}^\hbar$. It is the Feichtinger algebra $S_0(\mathbb{R}^n)$ when $q = 1$.*

(ii) *The Schwartz space $\mathcal{S}(\mathbb{R}^n)$ is a dense subspace of each of the modulation spaces $M_s^q(\mathbb{R}^n)$ for $q < \infty$.*

Proof of (i). See Gröchenig's book [82], Theorem 11.3.5 for a detailed proof using the short-time Fourier transform.

Proof of (ii). Let us first show that $\mathcal{S}(\mathbb{R}^n) \subset M_s^q(\mathbb{R}^n)$. Let $\psi \in \mathcal{S}(\mathbb{R}^n)$; for every window ϕ we have $W(\psi, \phi) \in \mathcal{S}(\mathbb{R}^n \oplus \mathbb{R}^n)$ hence for every $N > 0$ there exists $C_N > 0$ such that $|W(\psi, \phi)(z)| \leq C_N \langle z \rangle^{-N}$. It follows, by definition of the norm $\| \cdot \|_{\phi, M_s^q}^\hbar$ that

$$\|\psi\|_{\phi, M_s^q}^\hbar \leq C_N^q \int_{\mathbb{R}^n \oplus \mathbb{R}^n} (1 + |z|^2)^{(s - qN)/2} dz$$

and hence $\|\psi\|_{\phi, M_s^q}^\hbar < \infty$ if we choose $s - qN < -n$, that is $N > (s + n)/q$. Let us finally prove the density statement. Let us choose an exhaustive sequence (K_j) of compact subsets of $\mathbb{R}^n \oplus \mathbb{R}^n$ (i.e., $K_j \subset \mathring{K}_{j+1}$ and $\mathbb{R}^n \oplus \mathbb{R}^n = \cup_j K_j$) and set $\Psi_j = W(\psi, \phi_j)\chi_j$ where χ_j is the characteristic function of K_j. Also set

$$\psi_j = W_\phi^* \Psi_j = \frac{2^n}{(\gamma | \phi)_{L^2}} \int_{\mathbb{R}^n \oplus \mathbb{R}^n} \Psi_j(z) \widehat{T}_{\mathrm{GR}}(z) \gamma \, dz.$$

Since $W(\psi, \phi) \in \mathcal{S}(\mathbb{R}^n \oplus \mathbb{R}^n)$ we have $\psi_j \in \mathcal{S}(\mathbb{R}^n)$ and

$$\begin{aligned}
\|\psi - \psi_j\|_{\phi, M_s^q}^\hbar &= \|W_\phi^*(W_\phi \psi - \Psi_j)\|_{\phi, M_s^q}^\hbar \\
&\leq C \|W_\phi \psi - \Psi_j\|_{L_s^q}.
\end{aligned}$$

We have $\lim_{j \to \infty} \|W_\phi \psi - \Psi_j\|_{\phi, M_s^q}^\hbar = 0$ and hence also $\lim_{j \to \infty} \|\psi - \psi_j\|_{\phi, M_s^q}^\hbar$ proving our claim. $\qquad\square$

17.2.2   Metaplectic and Heisenberg–Weyl invariance properties

The modulation spaces $M_s^q(\mathbb{R}^n)$ have two remarkable invariance properties, extending the similar properties of the Feichtinger algebra:

Proposition 400. *The modulation spaces $M_s^q(\mathbb{R}^n)$ have the following properties:*

(i) *Each space $M_s^q(\mathbb{R}^n)$ is invariant under the action of the Heisenberg–Weyl operators $\widehat{T}(z)$; in fact there exists a constant $C > 0$ such that*

$$\|\widehat{T}(z)\psi\|_{\phi,M_s^q}^{\hbar} \le C \, \langle z \rangle^s \, \|\psi\|_{\phi,M_s^q}^{\hbar}. \tag{17.14}$$

(ii) *The space $M_s^q(\mathbb{R}^n)$ is invariant under the action of the metaplectic group* $\mathrm{Mp}(2n,\mathbb{R})$*: if $\widehat{S} \in \mathrm{Mp}(2n,\mathbb{R})$ then $\widehat{S}\psi \in M_s^q(\mathbb{R}^n)$ if and only if $\psi \in M_s^q(\mathbb{R}^n)$. In particular $M_s^q(\mathbb{R}^n)$ is invariant under the Fourier transform.*

Proof of (i). The cross-Wigner transform satisfies

$$W(\widehat{T}(z_0)\psi, \phi)(z) = e^{-\frac{i}{\hbar}\sigma(z,z_0)} W(\psi,\phi)(z - \tfrac{1}{2}z_0)$$

(property (9.25)) hence it suffices, in view of Proposition 397, to show that $L_s^q(\mathbb{R}^n \oplus \mathbb{R}^n)$ is invariant under the phase space translation $T(z_0) : z \longmapsto z + z_0$. In view of the submultiplicative property of the weight v_s (cf. Exercise 390) we have, for $q < \infty$,

$$\begin{aligned}
\|T(z_0)\Psi\|_{L_v^q}^q &= \int_{\mathbb{R}^n \oplus \mathbb{R}^n} |\Psi(z - z_0)|^q \, \langle z \rangle^{qs} \, dz \\
&= \int_{\mathbb{R}^n \oplus \mathbb{R}^n} |\Psi(z)|^q \, \langle z + z_0 \rangle^{qs} \, dz \\
&\le \langle z \rangle^s \int_{\mathbb{R}^n \oplus \mathbb{R}^n} |\Psi(z)|^q \, \langle z \rangle^{qs} \, dz,
\end{aligned}$$

hence our claim; the estimate (17.14) follows.

Proof of (ii). In view of Proposition 398 we have $\psi \in M_s^q(\mathbb{R}^n)$ if and only if $W(\psi,\phi) \in L_s^q(\mathbb{R}^n \oplus \mathbb{R}^n)$ for one window $\phi \in \mathcal{S}(\mathbb{R}^n)$; if this property holds, then it holds for all windows. In view of the symplectic covariance formula (10.26) for the Wigner transform we have

$$\begin{aligned}
W(\widehat{S}\psi, \phi) &= W(\widehat{S}\psi, \widehat{S}(\widehat{S}^{-1}\phi))(z) \\
&= W(\psi, (\widehat{S}^{-1}\phi))(S^{-1}z),
\end{aligned}$$

hence $W(\widehat{S}\psi, \phi) \in L_s^q(\mathbb{R}^n \oplus \mathbb{R}^n)$ if and only if the function

$$z \longmapsto W(\psi, (\widehat{S}^{-1}\phi))(S^{-1}z)$$

is in $L_s^q(\mathbb{R}^n \oplus \mathbb{R}^n)$. But this condition is equivalent to $W(\psi, (\widehat{S}^{-1}\phi)) \in L_s^q(\mathbb{R}^n \oplus \mathbb{R}^n)$ in view of Lemma 393, hence $\widehat{S}\psi \in M_s^q(\mathbb{R}^n)$. $\qquad\square$

The following consequence of the result above is interesting:

Corollary 401. *The modulation space $M_s^q(\mathbb{R}^n)$ is invariant under the dilations $\psi \longmapsto \psi_\lambda$ where $\psi_\lambda(x) = \psi(\lambda x)$ where $\lambda \neq 0$. More generally, $M_s^q(\mathbb{R}^n)$ is invariant under every change of variables $x \longmapsto Lx$ $(\det L \neq 0)$.*

Proof. The unitary operators M_L with $M_{L,m}\psi(x) = i^m \sqrt{|\det L|}\psi(Lx)$ $(\det L \neq 0,$ $\arg \det L \equiv m\pi \mod 2\pi)$ belong to $\mathrm{Mp}(2n, \mathbb{R})$; the lemma follows since $M_s^q(\mathbb{R}^n)$ is a vector space. $\qquad\square$

The class of modulation spaces $M_s^q(\mathbb{R}^n)$ contains as particular cases many of the classical function spaces. For instance, $M_s^2(\mathbb{R}^n)$ coincides with the Shubin–Sobolev space

$$Q^s(\mathbb{R}^n) = L_s^2(\mathbb{R}^n) \cap H^s(\mathbb{R}^n)$$

(Shubin [147], p. 45). We also have

$$\mathcal{S}(\mathbb{R}^n) = \bigcap_{s \geq 0} M_s^2(\mathbb{R}^n).$$

17.3 The modulation spaces M_s^∞

We now study the case $q = \infty$.

17.3.1 The weighted spaces L_s^∞

The spaces $L_s^\infty(\mathbb{R}^n \oplus \mathbb{R}^n)$ are defined as follows:

Definition 402. Let Ψ be a complex-valued measurable function on $\mathbb{R}^n \oplus \mathbb{R}^n$. We have $\Psi \in L_s^\infty(\mathbb{R}^n \oplus \mathbb{R}^n)$ if there exists a constant $C > 0$ such that $\mathrm{ess\,sup}(|\Psi|v_s) \leq C$ where "ess sup" stands for "essential supremum".

Equivalently: $|\Psi(z)| \leq Cv_{-s}(z)$ for almost every $z \in \mathbb{R}^n \oplus \mathbb{R}^n$.

It is clear that $L_s^\infty(\mathbb{R}^n \oplus \mathbb{R}^n)$ is a vector space: if $\mathrm{ess\,sup}(|\Psi| \langle \cdot \rangle^s) \leq C$ then $\mathrm{ess\,sup}(|\lambda\Psi|v_s) \leq C|\lambda|$ for $\lambda \in \mathbb{C}$ and if $\mathrm{ess\,sup}(|\Psi'| \langle \cdot \rangle^s) \leq C'$ then

$$\mathrm{ess\,sup}(|\Psi + \Psi'| \langle \cdot \rangle^s) \leq C + C'.$$

The norm on $L_s^\infty(\mathbb{R}^n \oplus \mathbb{R}^n)$ is defined as follows: $\|\Psi\|_{L_s^\infty}$ is the infimum of all constants C such that $\mathrm{ess\,sup}(|\Psi|v_s) \leq C$:

$$\|\Psi\|_{L_s^\infty} = \inf\{C : \mathrm{ess\,sup}(|\Psi| \langle \cdot \rangle^s) \leq C\}. \tag{17.15}$$

Exercise 403. Prove in detail that $\Psi \longmapsto \|\Psi\|_{L_s^\infty}$ defines a norm on the vector space $L_s^\infty(\mathbb{R}^n \oplus \mathbb{R}^n)$.

We invite the reader to prove the completeness of $L_s^\infty(\mathbb{R}^n \oplus \mathbb{R}^n)$ (it is standard, and does not require any unexpected trick):

Problem 404. Prove that $L_s^\infty(\mathbb{R}^n \oplus \mathbb{R}^n)$ is a Banach space for the norm $\|\cdot\|_{L_s^\infty}$. (It is an adaptation of the proof of property (i) in Proposition 388.)

The dual of the Banach space $L_s^\infty(\mathbb{R}^n \oplus \mathbb{R}^n)$ is $L_{-s}^1(\mathbb{R}^n \oplus \mathbb{R}^n)$ where the duality bracket is defined by

$$(\Phi, \Psi) = \int_{\mathbb{R}^n \oplus \mathbb{R}^n} \Phi(z)\overline{\Psi(z)}dz$$

for $\Phi \in L_{-s}^1(\mathbb{R}^n \oplus \mathbb{R}^n)$ and $\Psi \in L_s^\infty(\mathbb{R}^n \oplus \mathbb{R}^n)$. In fact,

$$\begin{aligned}
|(\Phi, \Psi)| &\le \int_{\mathbb{R}^n \oplus \mathbb{R}^n} |\Phi(z)|\,|\Psi(z)|dz \\
&= \int_{\mathbb{R}^n \oplus \mathbb{R}^n} |\Phi(z)|\,\langle z\rangle^{-s}\,\Psi(z)|v_s(z)dz \\
&\le \|\Phi\|_{L_{-s}^1} \|\Psi\|_{L_s^\infty}.
\end{aligned}$$

The following property is the extension to the L_s^∞ case of the invariance property of L_v^q under translations proven in Proposition 388:

Proposition 405. *The space $L_s^\infty(\mathbb{R}^n \oplus \mathbb{R}^n)$ is invariant under translations*

$$T(z_0)\Psi(z) = \Psi(z - z_0),$$

and we have:

$$\|T(z_0)\Psi\|_{L_s^\infty} \le \langle z_0\rangle^s \|\Psi\|_{L_s^\infty} \tag{17.16}$$

for every $\Psi \in L_s^\infty(\mathbb{R}^n \oplus \mathbb{R}^n)$.

Proof. It suffices to prove the estimate (17.16). By definition of the norm on $L_s^\infty(\mathbb{R}^n \oplus \mathbb{R}^n)$ this is equivalent to proving that, if $|\Psi(z - z_0)|v_s(z) \le C$ almost everywhere, then $|\Psi(z)|v_s(z) \le Cv_s(z_0)$ a.e. Now, the condition $|\Psi(z - z_0)|v_s(z) \le C$ a.e. is equivalent to $\Psi(z)|v_s(z + z_0) \le C$ a.e. Noting that in view of the submultiplicativity property of the weight v_s we have, writing $v_s(z) = v_s(z + z_0 - z_0)$,

$$|\Psi(z)|\,\langle z\rangle^s \le |\Psi(z)|\,\langle z + z_0\rangle^s\,\langle z_0\rangle^s \le C\,\langle z_0\rangle^s$$

which concludes the proof. $\qquad\square$

17.3.2 The spaces M_s^∞

The definition is a straightforward adaptation of that of M_s^q for $q < \infty$:

Definition 406. The modulation space $M_s^\infty(\mathbb{R}^n)$ consists of all $\psi \in \mathcal{S}'(\mathbb{R}^n)$ such that $V_\phi\psi \in L_s^\infty(\mathbb{R}^n \oplus \mathbb{R}^n)$ for every window $\phi \in \mathcal{S}'(\mathbb{R}^n)$ (where $V_\phi\psi$ is the STFT transform); equivalently $W(\psi, \phi) \in L_s^\infty(\mathbb{R}^n \oplus \mathbb{R}^n)$.

We leave it to the reader to check that the conditions $V_\phi \psi \in L_s^\infty(\mathbb{R}^n \oplus \mathbb{R}^n)$ and $W(\psi, \phi) \in L_s^\infty(\mathbb{R}^n \oplus \mathbb{R}^n)$ (for all ϕ) are equivalent.

We equip $M_s^\infty(\mathbb{R}^n)$ with the norms

$$\|\psi\|_{\phi, M_s^\infty(\mathbb{R}^n)} = \|W(\psi, \phi)\|_{L_s^\infty(\mathbb{R}^n \oplus \mathbb{R}^n)}.$$

Not very surprisingly, we have:

Proposition 407. *We have $\psi \in M_s^\infty(\mathbb{R}^n)$ if and only if $W(\psi, \phi) \in L_s^\infty(\mathbb{R}^n \oplus \mathbb{R}^n)$ for one window $\phi \in \mathcal{S}(\mathbb{R}^n)$. The norms $\psi \longmapsto \|\psi\|_{\phi, M_s^\infty(\mathbb{R}^n)}$ are equivalent norms on $M_s^\infty(\mathbb{R}^n)$.*

Proof. It is similar to the proof of Proposition 397. $\square$

The spaces $M_s^\infty(\mathbb{R}^n)$ can be seen as non-trivial refinements of the Schwartz space $\mathcal{S}(\mathbb{R}^n)$. In fact, we have the following beautiful result:

Proposition 408. *We have the equalities*

$$\mathcal{S}(\mathbb{R}^n) = \bigcap_{s \geq 0} M_s^\infty(\mathbb{R}^n) \quad , \quad \mathcal{S}'(\mathbb{R}^n) = \bigcup_{s \geq 0} M_{-s}^\infty(\mathbb{R}^n). \tag{17.17}$$

Proof. Let us prove the first equality (17.17). Let $\psi \in \mathcal{S}(\mathbb{R}^n)$; in view of the estimate (9.59) in Proposition 192, for every $N \geq 0$ there exists $C_N > 0$ such that $|W(\psi, \phi)(z)| \leq C_N \langle z \rangle^{-N}$. It follows that

$$|W(\psi, \phi)(z)| \langle z \rangle^s \leq C_N \langle z \rangle^{s-N} \leq C_N$$

if we choose $N \geq s$ and hence $\psi \in M_s^\infty(\mathbb{R}^n)$ for every s. Suppose conversely that $\psi \in M_s^\infty(\mathbb{R}^n)$ for every $s \geq 0$; this is equivalent to $|W(\psi, \phi)(z)| \leq C_s \langle z \rangle^{-s}$ for every s and hence $\psi \in \mathcal{S}(\mathbb{R}^n)$ in view of the implication (iv)$\Longrightarrow$(i) in Proposition 192. Let us now prove the second equality (17.17). First of all it is clear that

$$\bigcup_{s \geq 0} M_s^\infty(\mathbb{R}^n) \subset \mathcal{S}'(\mathbb{R}^n)$$

since, by definition, the elements of each space $M_s^\infty(\mathbb{R}^n)$ are tempered distributions. Let $\psi \in \mathcal{S}'(\mathbb{R}^n)$. Then, by Theorem 190(ii) there exist constants $C \geq 0$ and $\mu \geq 0$ such that $|W(\psi, \phi)(z)| \leq C \langle z \rangle^\mu$ and hence $|W(\psi, \phi)| \langle z \rangle^{-\mu}$ is bounded, so that $\psi \in M_{-s}^\mu(\mathbb{R}^n)$ for some $s \geq 0$. $\square$

17.4 The modulation spaces $M_s^{\infty,1}$

Let us now introduce a different class of modulation spaces, whose elements are excellent candidates for being Weyl symbols. This class contains as a particular case the Sjöstrand classes which were defined by other methods in Sjöstrand [150, 151].

17.4.1 Definition and first properties

We will again use the weight function on $\mathbb{R}^{2n}$ defined for $z \in \mathbb{R}^{2n}$ and $\zeta \in \mathbb{R}^{2n}$ by $\langle z \rangle^s = (1 + |z|^2)^{s/2}$. We will assume that $s \geq 0$.

Definition 409. The modulations space $M_s^{\infty,1}(\mathbb{R}^n \oplus \mathbb{R}^n)$ consists of all tempered distributions $\Psi \in \mathcal{S}'(\mathbb{R}^n \oplus \mathbb{R}^n)$ such that

$$\sup_{z \in \mathbb{R}^{2n}} |W(\Psi, \Phi)(z, \zeta)\langle z \rangle^s| \in L^1(\mathbb{R}^n \oplus \mathbb{R}^n) \tag{17.18}$$

for every $\Phi \in \mathcal{S}(\mathbb{R}^n \oplus \mathbb{R}^n)$. When $s = 0$ the space $M_0^{\infty,1}(\mathbb{R}^n \oplus \mathbb{R}^n) = M^{\infty,1}(\mathbb{R}^n \oplus \mathbb{R}^n)$ is called the Sjöstrand class. It thus consists of all $\Psi \in \mathcal{S}'(\mathbb{R}^n \oplus \mathbb{R}^n)$ such that

$$\sup_{z \in \mathbb{R}^{2n}} |W(\Psi, \Phi)(z, \zeta)| \in L^1(\mathbb{R}^n \oplus \mathbb{R}^n)$$

for every $\Phi \in \mathcal{S}(\mathbb{R}^n \oplus \mathbb{R}^n)$.

Exercise 410. Verify that $M_s^{\infty,1}(\mathbb{R}^n \oplus \mathbb{R}^n)$ is a complex vector space for the usual operations.

The spaces $M_s^{\infty,1}(\mathbb{R}^n \oplus \mathbb{R}^n)$ are usually defined in terms of the short-time Fourier transform $V_\Phi \Psi$ instead of $W(\Psi, \Phi)$. That the choice of definition is irrelevant is easy to prove:

Exercise 411. Show that $\Psi \in M_s^{\infty,1}(\mathbb{R}^n \oplus \mathbb{R}^n)$ if and only if

$$\sup_{z \in \mathbb{R}^{2n}} |V_\Phi \Psi(z, \zeta)\langle z \rangle^s| \in L^1(\mathbb{R}^n \oplus \mathbb{R}^n)$$

by adapting the method in the proof of Proposition 397.

The following result is the analogue of Proposition 399; it shows in particular that it suffices to check the bound (17.18) for one function Φ:

Proposition 412. *We have* $\Psi \in M_s^{\infty,1}(\mathbb{R}^n \oplus \mathbb{R}^n)$ *if and only if* (17.18) *holds for one* $\Phi \in \mathcal{S}(\mathbb{R}^n \oplus \mathbb{R}^n)$, *and*

(i) *The equalities*

$$\|\Psi\|_{M_s^{\infty,1}}^{\Phi} = \int_{\mathbb{R}^{2n}} \sup_{z \in \mathbb{R}^{2n}} |W(\Psi, \Phi)(z, \zeta)\langle z \rangle^s| \, d\zeta$$

define a family of equivalent norms on $M_s^{\infty,1}(\mathbb{R}^n \oplus \mathbb{R}^n)$ *when* Φ *describes* $\mathcal{S}(\mathbb{R}^n \oplus \mathbb{R}^n)$;

(ii) *The space* $M_s^{\infty,1}(\mathbb{R}^n \oplus \mathbb{R}^n)$ *is a Banach space for the topology defined by any of the norms* $\|\cdot\|_{M_s^{\infty,1}}^{\Phi}$ *and* $\mathcal{S}(\mathbb{R}^n \oplus \mathbb{R}^n)$ *is a dense subspace of* $M_s^{\infty,1}(\mathbb{R}^n \oplus \mathbb{R}^n)$.

An important property of the modulation spaces $M_s^{\infty,1}(\mathbb{R}^n \oplus \mathbb{R}^n)$ is their invariance under linear changes of variables:

Proposition 413. *Let $A \in GL(2n, \mathbb{R})$ and set $A^*\Psi = \Psi \circ A$. We have $\Psi \in M_s^{\infty,1}(\mathbb{R}^n \oplus \mathbb{R}^n)$ if and only $A^*\Psi \in M_s^{\infty,1}(\mathbb{R}^n \oplus \mathbb{R}^n)$. There exists a constant $C_A > 0$ such that*

$$\|A^*\Psi\|_{M_s^{\infty,1}}^{\Phi} \leq C_A \|\Psi\|_{M_s^{\infty,1}}^{A^*\Phi} \tag{17.19}$$

for every $\Phi \in \mathcal{S}(\mathbb{R}^n \oplus \mathbb{R}^n)$.

Proof. It is of course sufficient to prove the existence of a constant C_A such that (17.19) holds. Let us set $\Psi' = A^*\Psi$ and choose $\Phi \in \mathcal{S}(\mathbb{R}^n \oplus \mathbb{R}^n)$. We have, by definition of the cross-Wigner transform on $\mathbb{R}^n \oplus \mathbb{R}^n$,

$$W(\Psi', \Phi)(z, \zeta) = \left(\tfrac{1}{2\pi\hbar}\right)^{2n} \int_{\mathbb{R}^{2n}} e^{-\frac{i}{\hbar}\zeta \cdot \eta} \Psi(Az + \tfrac{1}{2}A\eta)\overline{\Phi(z - \tfrac{1}{2}\eta)} d\eta,$$

that is, performing the change of variables $\xi = A\eta$,

$$W(\Psi', \Phi)(z, \zeta) = \left(\tfrac{1}{2\pi\hbar}\right)^{2n} |\det A|^{-1}$$
$$\times \int_{\mathbb{R}^{2n}} e^{-\frac{i}{\hbar}\zeta \cdot A^{-1}\xi} \Psi(Az + \tfrac{1}{2}\xi)\overline{\Phi(z - \tfrac{1}{2}A^{-1}\xi)} d\xi$$

and hence

$$W(\Psi', \Phi)(A^{-1}z, A^T\zeta) = \left(\tfrac{1}{2\pi\hbar}\right)^{2n} |\det A|^{-1}$$
$$\times \int_{\mathbb{R}^{2n}} e^{-\frac{i}{\hbar}\zeta \cdot \xi} \Psi(z + \tfrac{1}{2}\xi)\overline{(A^{-1})^*\Phi(z - \tfrac{1}{2}\xi)} d\xi.$$

It follows that

$$W(\Psi', \Phi)(A^{-1}z, A^T\zeta) = |\det A|^{-1} W(\Psi, (A^{-1})^*\Phi)(z, \zeta),$$

that is

$$W(\Psi', \Phi)(z, \zeta) = |\det A|^{-1} W(\Psi, (A^{-1})^*\Phi)(Az, (A^T)^{-1}\zeta); \tag{17.20}$$

taking the suprema of both sides of this equality we get

$$\sup_{z \in \mathbb{R}^{2n}} |W(\Psi', \Phi)(z, \zeta)\langle z \rangle^s| = |\det A|^{-1} \sup_{z \in \mathbb{R}^{2n}} |W(\Psi, (A^{-1})^*\Phi)(z, \zeta)\langle A^{-1}z \rangle^s|$$

and hence

$$\|A^*\Psi\|_{M_s^{\infty,1}}^{\Phi} = |\det A|^{-1} \int_{\mathbb{R}^{2n}} \sup_{z \in \mathbb{R}^{2n}} |W(\Psi, (A^{-1})^*\Phi)(z, \zeta)\langle A^{-1}z \rangle^s| d\zeta.$$

Since we have $\langle A^{-1}z \rangle^s \leq C(A)\langle z \rangle^s$ for some constant $C(A) > 0$ the estimate (17.19) follows. $\square$

Exercise 414. Derive the equality (17.20) using the symplectic covariance property (10.26) of the cross-Wigner transform.

17.4.2 Weyl operators with symbols in $M_s^{\infty,1}$

For this subsection a good source is Gröchenig's paper [84] (also see the older paper [83] by Gröchenig and Heil), and Gröchenig [82].

It is interesting to view the modulation spaces $M_s^{\infty,1}$ as symbol classes: in contrast to the cases traditionally considered in the literature, membership of a symbol a in $M_s^{\infty,1}(\mathbb{R}^n \oplus \mathbb{R}^n)$ does not imply any smoothness of a. It turns out that this point of view allows us to recover many classical and difficult regularity properties (for instance the Calderón–Vaillancourt theorem) in a rather simple way.

Here is a first very interesting result. It says basically that the Weyl operators with symbols in $M_s^{\infty,1}(\mathbb{R}^n \oplus \mathbb{R}^n)$ preserve phase-space concentration.

Proposition 415. *Suppose that* $a \in M_s^{\infty,1}(\mathbb{R}^n \oplus \mathbb{R}^n)$. *Then the operator* $\widehat{A} \overset{\mathrm{Weyl}}{\longleftrightarrow} a$ *is bounded on each of the modulation spaces* $M_s^q(\mathbb{R}^n)$.

Proof. (Cf. ([82], Theorem 14.5.6). $\square$

The interest of $M_s^{\infty,1}(\mathbb{R}^n \oplus \mathbb{R}^n)$ comes from the following property of the Moyal product (Gröchenig [85]), namely that it equips these spaces with a $*$-algebra structure. Recall that the Moyal product $a \star_\hbar b$ is the Weyl symbol of the product $\widehat{A}\widehat{B}$ of the operators $\widehat{A} \overset{\mathrm{Weyl}}{\longleftrightarrow} a$ and $\widehat{B} \overset{\mathrm{Weyl}}{\longleftrightarrow} b$.

Proposition 416. *Let* $a, b \in M_s^{\infty,1}(\mathbb{R}^n \oplus \mathbb{R}^n)$. *Then* $a \star_\hbar b \in M_s^{\infty,1}(\mathbb{R}^n \oplus \mathbb{R}^n)$. *In particular, for every window* Φ *there exists a constant* $C_\Phi > 0$ *such that*

$$\|a \star_\hbar b\|_{M_s^{\infty,1}}^\Phi \leq C_\Phi \|a\|_{M_s^{\infty,1}}^\Phi \|b\|_{M_s^{\infty,1}}^\Phi.$$

Since obviously $\overline{a} \in M_s^{\infty,1}(\mathbb{R}^n \oplus \mathbb{R}^n)$ if and only if $a \in M_s^{\infty,1}(\mathbb{R}^n \oplus \mathbb{R}^n)$, the property above can be restated in the following concise way:

The modulation space $M_s^{\infty,1}(\mathbb{R}^n \oplus \mathbb{R}^n)$ *is a Banach* $*$*-algebra with respect to the Moyal product* $\star_\hbar$ *and the involution* $a \longmapsto \overline{a}$.

In the case of the Sjöstrand class $M^{\infty,1}(\mathbb{R}^n \oplus \mathbb{R}^n)$ one has the following more precise results:

Proposition 417. *We have the following properties:*

(i) *Every Weyl operator* $\widehat{A} \overset{\mathrm{Weyl}}{\longleftrightarrow} a$ *with* $a \in M^{\infty,1}(\mathbb{R}^n \oplus \mathbb{R}^n)$ *is bounded on* $L^2(\mathbb{R}^n)$;

(ii) *If we have* $\widehat{C} = \widehat{A}\widehat{B}$ *with* $a, b \in M^{\infty,1}(\mathbb{R}^n \oplus \mathbb{R}^n)$ *then* $c \in M^{\infty,1}(\mathbb{R}^n \oplus \mathbb{R}^n)$;

(iii) *If* $\widehat{A}$ *with* $a \in M^{\infty,1}(\mathbb{R}^n \oplus \mathbb{R}^n)$ *is invertible with inverse* $\widehat{B} \overset{\mathrm{Weyl}}{\longleftrightarrow} b$ *then* $b \in M^{\infty,1}(\mathbb{R}^n \oplus \mathbb{R}^n)$.

The Sjöstrand class $M^{\infty,1}(\mathbb{R}^n \oplus \mathbb{R}^n)$ contains, in particular, the symbol class $S_{0,0}^0(\mathbb{R}^n \oplus \mathbb{R}^n)$ consisting of all infinitely differentiable complex functions A on $\mathbb{R}^n \oplus \mathbb{R}^n$ such that $\partial_z^\alpha A$ is bounded for all multi-indices $\alpha \in \mathbb{N}^{2n}$. Property (i) thus extends the L^2-boundedness property of operators with symbols in $S_{0,0}^0(\mathbb{R}^n \oplus \mathbb{R}^n)$. Property (iii) is called the *Wiener property* of $M^{\infty,1}(\mathbb{R}^n \oplus \mathbb{R}^n)$.

Chapter 18

Bopp Pseudo-differential Operators

Bopp pseudo-differential operators are the operators formally obtained from a symbol by the quantization rules

$$x \longrightarrow x + \tfrac{1}{2}i\hbar\partial_p \quad , \quad p \longrightarrow p - \tfrac{1}{2}i\hbar\partial_x \tag{18.1}$$

instead of the usual correspondence $x \longrightarrow x$, $p \longrightarrow -i\hbar\partial_x$. The terminology comes from the fact that the operators $x + \tfrac{1}{2}i\hbar\partial_p$ and $p - \tfrac{1}{2}i\hbar\partial_x$ are called "Bopp shifts" in the physics literature. These operators act, not on functions defined on $\mathbb{R}^n$ as ordinary Weyl operators do, but on functions (or distributions) defined on the phase space $\mathbb{R}^n \oplus \mathbb{R}^n$. The definition of Bopp pseudo-differential operators is sensitive to the choice of symplectic structure on $\mathbb{R}^n \oplus \mathbb{R}^n$; this property will be exploited in the next chapter in the context of non-commutative quantum mechanics where one is led to use other symplectic forms than the standard one.

We will call this quantization procedure "Bopp quantization" in honor of the physicist Fritz Bopp, who was the first to consider (in 1956) operators of this type in his paper [18] where he discussed some statistical implications of quantization. We should also give credit to the mathematical physicist Ryogo Kubo [111] who noticed in 1964 the relationship between operators of this type and Weyl calculus. This possibility has been sporadically discussed in the physics literature (see for instance Brodimas et al. [23] and Balazs and Pauli [4]), but their papers seem to have been unfortunately more or less unnoticed. We will see in the next chapter that the theory of Bopp pseudo-differential operators is a tool of choice for the study of deformation quantization which it reduces to a Weyl calculus of a particular type. We will also see that the study of non-commutative quantum mechanics can also be reduced to Bopp calculus. (Another easy application is the study of generalizations of the magnetic Landau problem.)

The study of phase space pseudo-differential operators (from a slightly different point of view) was initiated in de Gosson [67, 68, 69, 70, 71], and applied to deformation quantization via the theory of modulation spaces in de Gosson and Luef [76, 79].

18.1   Introduction and motivation

To understand what Bopp pseudo-differential calculus is about, let us consider the following simple example. In deformation quantization one studies "star products" of functions defined on phase space. The most commonly used, at least in applications, is the Moyal product $\star_\hbar$, which is one of the cornerstones of deformation quantization, which we will discuss in more detail in the next chapter. By definition $c = a \star_\hbar b$ where

$$c(z) = \left(\tfrac{1}{4\pi\hbar}\right)^{2n} \iint_{\mathbb{R}^{4n}} e^{\frac{i}{2\hbar}\sigma(z',z'')} a(z + \tfrac{1}{2}z')b(z - \tfrac{1}{2}z'')dz'dz''.$$

This is immediately recognized as being the formula giving the symplectic Fourier transform of the symbol of the product $\widehat{C} = \widehat{A}\widehat{B}$ of two Weyl operators (cf. Theorem 213). Equivalently, the symplectic Fourier transform of c is given by the formula

$$c_\sigma(z) = \left(\tfrac{1}{2\pi\hbar}\right)^{n} \int_{\mathbb{R}^{2n}} e^{\frac{i}{2\hbar}\sigma(z,z')} a_\sigma(z - z')b_\sigma(z')dz'.$$

After a few calculations one sees that, in particular,

$$x \star_\hbar a = \left(x + \tfrac{1}{2}i\hbar\partial_p\right) a \quad , \quad p \star_\hbar a = \left(p - \tfrac{1}{2}i\hbar\partial_x\right) a \qquad (18.2)$$

where ∂_p and ∂_x are the gradients in p and x, respectively.

Exercise 418. Prove these formulas. More generally, what is $x^\alpha \star_\hbar a$ (resp. $p^\alpha \star_\hbar a$) when $\alpha \in \mathbb{N}^n$?

The formulas (18.2) suggest that, more generally, the Moyal product $a \star_\hbar b$ of two functions a, b could be rewritten in the form

$$a \star_\hbar b = \widetilde{A}b \qquad (18.3)$$

where $\widetilde{A} = A(\widetilde{x}, \widetilde{p})$ is a pseudo-differential operator formally obtained by the "Bopp quantization rules" $x \longrightarrow \widetilde{x}$ and $p \longrightarrow \widetilde{p}$ where $\widetilde{x}$ and $\widetilde{p}$ are the differential operators

$$\widetilde{x} = x + \tfrac{1}{2}i\hbar\partial_p \quad , \quad \widetilde{p} = p - \tfrac{1}{2}i\hbar\partial_x; \qquad (18.4)$$

writing $\widetilde{z} = (\widetilde{x}, \widetilde{p})$ these relations can be written

$$\widetilde{z} = z + \tfrac{1}{2}i\hbar J\partial_z \qquad (18.5)$$

which has the advantage of making explicit the relation of Bopp quantization with the standard symplectic structure. This also opens the door to more general quantizations associated with non-standard symplectic structures as we will see in Chapter 19. Of course "formula" (18.3) only remains a notation as long as one has not given a working definition of the operator $\widetilde{A} = a(\widetilde{z})$; and it is not at all obvious what this definition should be except when a is a polynomial! Notice that by definition, such operators $\widetilde{A}$ act not on functions (or distributions) defined on $\mathbb{R}^n$ but on functions (or distributions) defined on phase space $\mathbb{R}^n \oplus \mathbb{R}^n$. We can therefore consider the study of such operators as a study of a phase space pseudo-differential calculus.

We will give below a rigorous definition of the Bopp operators $\widetilde{A}$ which can be viewed as a Weyl operator of a very particular type acting on phase space functions. We will see that "Bopp calculus" is intertwined with the usual Weyl calculus by infinitely many partial isometries of $L^2(\mathbb{R}^n)$ onto closed subspaces of $L^2(\mathbb{R}^n)$. This fact has very surprising consequences, and it leads (as a by-product) to simple proofs for many regularity properties for the usual Weyl operators.

18.1.1 Bopp pseudo-differential operators

Recall that the Heisenberg–Weyl operator $\widehat{T}(z_0)$ acts on functions defined on $\mathbb{R}^n$ via the formula
$$\widehat{T}(z_0)\psi(x) = e^{\frac{i}{\hbar}(p_0 \cdot x - \frac{1}{2} p_0 \cdot x_0)}\psi(x - x_0);$$

a natural step is to extend the domain of $\widehat{T}(z_0)$ by letting it act on functions (or distributions) defined on $\mathbb{R}^n \oplus \mathbb{R}^n$ via the formula

$$\widehat{T}(z_0)\Psi(z) = e^{\frac{i}{\hbar}(p_0 \cdot x - \frac{1}{2} p_0 \cdot x_0)}\Psi(z - z_0).$$

This approach was actually initiated in our monograph [67] in connection with the study of the phase space Schrödinger equation (also see de Gosson [69, 68, 70]). For our present purpose, which is the definition of Bopp pseudo-differential operators, we prefer to use a variant of this redefinition of the Heisenberg–Weyl operator:

Definition 419. For $z_0 \in \mathbb{R}^{2n}$ the operator $\widetilde{T}(z_0)$ is defined, for $\Psi \in \mathcal{S}'(\mathbb{R}^n \oplus \mathbb{R}^n)$, by
$$\widetilde{T}(z_0)\Psi(z) = e^{-\frac{i}{\hbar}\sigma(z,z_0)}\Psi(z - \tfrac{1}{2}z_0). \tag{18.6}$$

This choice (as arbitrary as it can seem at first sight!) is consistent with the quantization rules (18.4). This can be seen as follows. Recall from Chapter 8, formula (8.3) that the introduction of the usual Heisenberg–Weyl operator $\widehat{T}(z_0) = e^{-i\sigma(\widehat{z},z_0)}$ can be motivated by the Weyl quantization of the translation Hamiltonian $H_{z_0}(z) = \sigma(z, z_0)$; the operator with Weyl symbol is $\widehat{H}_{z_0}(z) = \sigma(\widehat{z}, z_0)$ with $\widehat{z} = (x, -i\hbar\partial_x)$ and the solution of the corresponding Schrödinger equation

$$i\hbar\frac{\partial}{\partial t}\psi = \widehat{H}_{z_0}\psi \ , \ \ \psi(x,0) = \psi_0(x)$$

is formally given by $\psi(x,t) = e^{it\sigma(\widehat{z},z_0)/\hbar}\psi_0(x)$; a direct calculation then leads to the explicit formula

$$u(x,t) = e^{\frac{i}{\hbar}t\sigma(\widehat{z},z_0)}\psi_0(x) = e^{\frac{i}{\hbar}(t\xi_0 \cdot x - \frac{1}{2}t^2\xi_0 \cdot x_0)}\psi_0(x - tx_0)$$

and hence $\widehat{T}(z_0)\psi(x,0) = \psi(x,1)$. To define the operators $\widetilde{T}(z_0)$ one proceeds exactly in the same way: replacing the Hamiltonian operator $\widehat{H}_{z_0}(z) = \sigma(\widehat{z},z_0)$ with $\widetilde{H}_{z_0}(z) = \sigma(\widetilde{z},z_0)$ where $\widetilde{z} = z + \frac{1}{2}i\hbar J\partial_z$ we are led to the "phase space Schrödinger equation"

$$i\hbar\frac{\partial}{\partial t}\Psi = \sigma(\widetilde{z},z_0)\Psi \ , \ \ \Psi(z,0) = \Psi_0(z)$$

whose solution is

$$\Psi(z,t) = e^{\frac{i}{\hbar}t\sigma(\widetilde{z},z_0)}\Psi_0(z) = e^{-\frac{i}{\hbar}t\sigma(z,z_0)}\Psi_0(z - \tfrac{1}{2}tz_0).$$

We thus have

$$\Psi(z,t) = \widetilde{T}(z_0)\Psi(z) = e^{\frac{i}{\hbar}\sigma(\widetilde{z},z_0)}\Psi_0(z).$$

We also observe that the operators $\widetilde{T}(z_0)$ also appear (albeit in disguise) in the translation formula (9.25) for Heisenberg–Weyl operators:

$$W(\widehat{T}(z_0)\psi,\widehat{T}(z_1)\phi)(z) = e^{-\frac{i}{\hbar}[\sigma(z,z_0-z_1)+\frac{1}{2}\sigma(z_0,z_1)]}W(\psi,\phi)(z - \langle z\rangle)$$

where $\langle z\rangle = (z_0+z_1)/2$. In fact taking $z_1 = 0$ in the formula above we immediately get (18.6). It turns out that the operators $\widetilde{T}(z_0)$ satisfy commutation relations which are similar to

$$\widehat{T}(z_0)\widehat{T}(z_1) = e^{\frac{i}{\hbar}\sigma(z_0,z_1)}\widehat{T}(z_1)\widehat{T}(z_0),$$

$$\widehat{T}(z_0 + z_1) = e^{-\frac{i}{2\hbar}\sigma(z_0,z_1)}\widehat{T}(z_0)\widehat{T}(z_1),$$

which are satisfied by the Heisenberg–Weyl operators (formulae (8.8) and (8.9)). In fact:

Proposition 420. *We have*

$$\widetilde{T}(z_0 + z_1) = e^{-\frac{i}{2\hbar}\sigma(z_0,z_1)}\widetilde{T}(z_0)\widetilde{T}(z_1), \tag{18.7}$$

$$\widetilde{T}(z_1)\widetilde{T}(z_0) = e^{-\frac{i}{\hbar}\sigma(z_0,z_1)}\widetilde{T}(z_0)\widetilde{T}(z_1) \tag{18.8}$$

for all $z_0, z_1 \in \mathbb{R}^{2n}$.

Proof. Formula (18.8) follows from formula (18.7) noting that we can write

$$\widetilde{T}(z_1)\widetilde{T}(z_0) = e^{\frac{i}{2\hbar}\sigma(z_1,z_0)}\widetilde{T}(z_1 + z_0)$$
$$= e^{-\frac{i}{2\hbar}\sigma(z_0,z_1)}\widetilde{T}(z_0 + z_1)$$
$$= e^{-\frac{i}{\hbar}\sigma(z_0,z_1)}\widetilde{T}(z_0)\widetilde{T}(z_1).$$

The proof of (18.7) is similar and left to the reader. $\square$

These relations suggest that $\widetilde{T}(z_0)$ could correspond to a unitary representation of the Heisenberg group on phase space functions. We will see later on that this is indeed the case.

In analogy with the formula

$$\widehat{A}\psi(x) = \left(\tfrac{1}{2\pi\hbar}\right)^n \int_{\mathbb{R}^{2n}} a_\sigma(z_0)\widehat{T}(z_0)\psi(x)dz_0$$

defining a Weyl operator we introduce:

Definition 421. Let $a \in \mathcal{S}'(\mathbb{R}^n \oplus \mathbb{R}^n)$; the operator $\widetilde{A}$ defined for $\Psi \in \mathcal{S}(\mathbb{R}^n \oplus \mathbb{R}^n)$ by

$$\widetilde{A}\Psi(z) = \left(\tfrac{1}{2\pi\hbar}\right)^n \int_{\mathbb{R}^{2n}} a_\sigma(z_0)\widetilde{T}(z_0)\Psi(z)dz_0$$

is called the Bopp operator with symbol a, and we will write $\widetilde{A} \xleftrightarrow{\text{Bopp}} a$ or $a \xleftrightarrow{\text{Bopp}} \widetilde{A}$ ("Bopp correspondence").

The definition of a Bopp operator can be reformulated in many different ways. For instance, in terms of the distributional brackets $\langle\langle \cdot, \cdot \rangle\rangle$ on $\mathbb{R}^n \oplus \mathbb{R}^n$ we have

$$\widetilde{A}\Psi = \left(\tfrac{1}{2\pi\hbar}\right)^n \langle\langle a_\sigma, \widetilde{T}(\cdot)\Psi \rangle\rangle.$$

This immediately shows that the definition of $\widetilde{A}$ still makes sense when the symbol a is a tempered distribution on $\mathbb{R}^n \oplus \mathbb{R}^n$. Thus, the Bopp correspondence $a \xleftrightarrow{\text{Bopp}} \widetilde{A}$ makes sense for $a \in \mathcal{S}'(\mathbb{R}^n \oplus \mathbb{R}^n)$. We will see below that this is not surprising, because Bopp pseudo-differential operators are just Weyl operators of a special type.

Exercise 422. In Chapter 8 we defined the Grossmann–Royer operators by the formula $\widehat{T}_{\mathrm{GR}}(z_0) = \widehat{T}(z_0)\widehat{T}_{\mathrm{GR}}(0)\widehat{T}(z_0)^{-1}$ where $\widehat{T}_{\mathrm{GR}}(0)\psi(x) = \psi(-x)$. Show that the operator $\widetilde{T}_{\mathrm{GR}}(z_0) : \mathcal{S}(\mathbb{R}^n \oplus \mathbb{R}^n) \longrightarrow \mathcal{S}(\mathbb{R}^n \oplus \mathbb{R}^n)$ defined by $\widetilde{T}_{\mathrm{GR}}(z_0) = \widetilde{T}(z_0)\widetilde{T}_{\mathrm{GR}}(0)\widetilde{T}(z_0)^{-1}$ where $\widetilde{T}_{\mathrm{GR}}(0)\Psi(x) = \Psi(-x)$ is explicitly given by

$$\widetilde{T}_{\mathrm{GR}}(z_0)\Psi(z) = e^{-\frac{2i}{\hbar}\sigma(z,z_0)}\Psi(-z+z_0).$$

18.1.2 Bopp operators viewed as Weyl operators

Let us view the linear operator $\widetilde{A}$ as a Weyl operator $\widetilde{A} : \mathcal{S}(\mathbb{R}^n \oplus \mathbb{R}^n) \longrightarrow \mathcal{S}'(\mathbb{R}^n \oplus \mathbb{R}^n)$. We are going to identify the symbol of $\widetilde{A}$ below; we begin by calculating its distributional kernel:

Lemma 423. *The kernel of the operator $\widetilde{A} \xleftrightarrow{\text{Bopp}} a$ is given by the formula*

$$K_{\widetilde{A}}(z, z') = \left(\tfrac{1}{\pi\hbar}\right)^n a_\sigma[2(z - z')]e^{\frac{2i}{\hbar}\sigma(z,z')}. \tag{18.9}$$

Proof. In view of Definition (18.6) of $\widetilde{T}(z_0)\Psi$ we have, performing the change of variable $z_0 = 2(z - z')$,

$$\widetilde{A}\Psi(z) = \left(\tfrac{1}{2\pi\hbar}\right)^n \int_{\mathbb{R}^{2n}} a_\sigma(z_0) e^{-\frac{i}{\hbar}\sigma(z,z_0)}\Psi(z - \tfrac{1}{2}z_0)dz_0$$

$$= \left(\tfrac{1}{\pi\hbar}\right)^n \int_{\mathbb{R}^{2n}} a_\sigma[2(z - z')]e^{\frac{2i}{\hbar}\sigma(z,z')}\Psi(z')dz',$$

hence the kernel $K_{\widetilde{A}}$ of $\widetilde{A}$ is given by formula (18.9). $\qquad\square$

Theorem 424. *Viewing $\widetilde{A}$ as a Weyl operator $\mathcal{S}(\mathbb{R}^n \oplus \mathbb{R}^n) \longrightarrow \mathcal{S}'(\mathbb{R}^n \oplus \mathbb{R}^n)$, its symbol $\widetilde{a} \overset{\mathrm{Weyl}}{\longleftrightarrow} \widetilde{A}$ is obtained from the symbol $a \overset{\mathrm{Weyl}}{\longleftrightarrow} \widehat{A}$ by*

$$\widetilde{a}(z,\zeta) = a(z - \tfrac{1}{2}J\zeta) \tag{18.10}$$

that is, setting $\zeta = (\zeta_x, \zeta_p)$,

$$\widetilde{a}(z,\zeta) = a(x - \tfrac{1}{2}\zeta_p, p + \tfrac{1}{2}\zeta_x). \tag{18.11}$$

Proof. It is a variant of the proof we have given in de Gosson [71] in a slightly different context. In view of formula 10.15 with n replaced by $2n$ we can determine the Weyl symbol $\widetilde{a}$ of $\widetilde{A}$ by the formula

$$\widetilde{a}(z,\zeta) = \int_{\mathbb{R}^{2n}} e^{-\frac{i}{\hbar}\zeta\cdot\eta} K(z + \tfrac{1}{2}\eta, z - \tfrac{1}{2}\eta)d\eta.$$

We have, using the identity $\sigma(z + \tfrac{1}{2}\eta, z - \tfrac{1}{2}\eta) = -\sigma(z,\eta)$,

$$K(z + \tfrac{1}{2}\eta, z - \tfrac{1}{2}\eta) = \left(\tfrac{1}{\pi\hbar}\right)^n a_\sigma(2\eta)e^{-\frac{2i}{\hbar}\sigma(z,\eta)}$$

and hence

$$\widetilde{a}(z,\zeta) = \left(\tfrac{1}{\pi\hbar}\right)^n \int_{\mathbb{R}^{2n}} e^{-\frac{i}{\hbar}\zeta\cdot\eta}e^{-\frac{2i}{\hbar}\sigma(z,\eta)}a_\sigma(2\eta)d\eta$$

that is, setting $\omega = 2\eta$,

$$\widetilde{a}(z,\zeta) = \left(\tfrac{1}{2\pi\hbar}\right)^n \int_{\mathbb{R}^{2n}} e^{-\frac{i}{2\hbar}\zeta\cdot\omega}e^{-\frac{i}{\hbar}\sigma(z,\omega)}a_\sigma(\omega)d\omega.$$

Now we observe that

$$\tfrac{1}{2}\zeta\cdot\omega + \sigma(z,\omega) = \tfrac{1}{2}\zeta\cdot\omega + Jz\cdot\omega$$
$$= J(z - \tfrac{1}{2}J\zeta)\cdot\omega$$
$$= \sigma(z - \tfrac{1}{2}J\zeta,\omega)$$

so that

$$\widetilde{a}(z,\zeta) = \left(\tfrac{1}{2\pi\hbar}\right)^n \int_{\mathbb{R}^{2n}} e^{-\frac{i}{\hbar}\sigma(z - \frac{1}{2}J\zeta,\omega)}a_\sigma(\omega)d\omega.$$

Recalling that the symplectic Fourier transform is its own inverse we thus have

$$\widetilde{a}(z,\zeta) = a(z - \tfrac{1}{2}J\zeta) \tag{18.12}$$

which we set out to prove. $\qquad\square$

Note that (z,ζ) is the generic point of the $4n$-dimensional phase space $\mathbb{R}^{2n} \oplus \mathbb{R}^{2n}$; the variable $\zeta = (\zeta_x, \zeta_p) \in \mathbb{R}^n \times \mathbb{R}^n$ is viewed as the dual variable of $z = (x,p)$.

The result above justifies the interpretation of $\widetilde{A}$ as the operator obtained from the usual Weyl symbol a by the quantization rule

$$z = (x,p) \longrightarrow (x + \tfrac{1}{2}i\hbar\partial_p, p - \tfrac{1}{2}i\hbar\partial_x) = \widetilde{z} \tag{18.13}$$

and thus legitimates the notation

$$\widetilde{A} = a(x + \tfrac{1}{2}i\hbar\partial_p, p - \tfrac{1}{2}i\hbar\partial_x) = a(\widetilde{z})$$

we introduced above.

18.1.3 Adjoints and a composition formula

The usual rules for calculating the adjoint and composing Weyl operators apply to the case of Bopp operators as well.

Proposition 425.

(i) *The Weyl symbol $\widetilde{c}$ of the product $\widetilde{C} = \widetilde{A}\widetilde{B}$ of two Bopp operators is*

$$\widetilde{c}(z,\zeta) = c\left(z - \tfrac{1}{2}J\zeta\right) \tag{18.14}$$

where c is the usual Weyl symbol of the product $\widehat{A}\widehat{B}$. Hence

$$\widetilde{A}\widetilde{B} = \widetilde{\widehat{A}\widehat{B}}. \tag{18.15}$$

(ii) *The symbol of the adjoint $\widetilde{A}^*$ is the complex conjugate $\overline{\widetilde{a}}$ of the symbol $\widetilde{a}$ of $\widetilde{A}$. Hence $\widetilde{A}^*$ is (essentially) self-adjoint if and only if a is real.*

Proof of (i). In view of the composition formulas in Theorem 213 the Weyl symbol of $\widetilde{A}\widetilde{B}$ is given by

$$\widetilde{c}(z,\zeta) = \left(\tfrac{1}{4\pi\hbar}\right)^{4n} \int_{\mathbb{R}^{4n}} e^{\frac{i}{2}\sigma(z',\zeta';z'',\zeta'')} a\left[\tfrac{1}{2}(z + \tfrac{1}{2}z') - J(\zeta + \tfrac{1}{2}\zeta')\right]$$
$$\times\, b\left[\tfrac{1}{2}(z - \tfrac{1}{2}z'') - J(\zeta - \tfrac{1}{2}\zeta'')\right] dz'dz''d\zeta'd\zeta''$$

where ω is the symplectic form on $\mathbb{R}^{4n}$. Defining new variables $u' = \tfrac{1}{2}z' - J\zeta'$ and $u'' = \tfrac{1}{2}z'' - J\zeta''$, this formula becomes

$$\widetilde{c}(z,\zeta) = \left(\tfrac{1}{4\pi\hbar}\right)^{4n} \int_{\mathbb{R}^{4n}} I(u,u'')a(z + u' - \tfrac{1}{2}J\zeta)b(z - u'' - \tfrac{1}{2}J\zeta)dz'dz''du'du''$$

with

$$I(u, u'') = \int_{\mathbb{R}^n} \exp\left[-\frac{i}{2\hbar}(\sigma(z', z'' - u'') - \sigma(u', z''))\right] dz' dz''.$$

Using the properties of the Fourier transform, $I(u, u'')$ is easily calculated and one finds that it is equal to $(4\pi)^{2n} e^{\frac{i}{2}\sigma(u', u')}$; formula (18.15) follows.

Proof of (ii). Part (ii) of the proposition follows from Proposition 212 about the Weyl symbol of the adjoint of an operator and the fact that $\widetilde{a}$ is the Weyl symbol of $\widetilde{A}$ viewed as an operator $\mathcal{S}(\mathbb{R}^n \oplus \mathbb{R}^n) \longrightarrow \mathcal{S}(\mathbb{R}^n \oplus \mathbb{R}^n)$. $\qquad\square$

18.1.4   Symplectic covariance of Bopp operators

The calculus of Bopp pseudo-differential operators should inherit the symplectic covariance properties of the usual Weyl calculus. This is indeed the case:

Proposition 426. *Let* $a\overset{\text{Bopp}}{\longleftrightarrow}\widetilde{A}$ *and* $S \in \mathrm{Sp}(2n, \mathbb{R})$. *We have*

$$a \circ S^{-1}\overset{\text{Bopp}}{\longleftrightarrow}\widetilde{M_S}\widetilde{A}\widetilde{M_S}^{-1} \tag{18.16}$$

where $\widetilde{M_S}$ *is the unitary operator on* $L^2(\mathbb{R}^n \oplus \mathbb{R}^n)$ *defined by* $\widetilde{M_S}\Psi(z) = \Psi(Sz)$. *Equipping* $\mathrm{Sp}(4n, \mathbb{R})$ *with its standard symplectic structure* $\sigma \oplus \sigma$ *we have* $\widetilde{M_S} \in \mathrm{Mp}(4n, \mathbb{R})$ *where* $\mathrm{Mp}(4n, \mathbb{R})$ *is the corresponding metaplectic group.*

Proof. That $\widetilde{M_S} \in \mathrm{Mp}(4n, \mathbb{R})$ is clear, since we have

$$\widetilde{M_S}\Psi(z) = \widehat{M_{S,0}}\Psi(z) = \sqrt{\det S}\,\Psi(Sz)$$

(cf. the notation (7.9) in Chapter 7). With the notation of Chapter 3 (formula (3.16)) set

$$M_S = \begin{pmatrix} S^{-1} & 0 \\ 0 & S^T \end{pmatrix};$$

then M_S is the projection on $\mathrm{Sp}(4n, \mathbb{R})$ of the metaplectic operator $\widetilde{M_S} = \widehat{M_{S,0}}$. To prove the covariance formula (18.16) we recall that the Weyl symbol of $\widetilde{A}$ is given by $\widetilde{a}(z, \zeta) = a(z - \frac{1}{2}J\zeta)$. Let b be the Weyl symbol of the Bopp operator with Weyl symbol $a \circ S^{-1}$; since $S^{-1}J = JS^T$ we have

$$\widetilde{b}(z, \zeta) = a(S^{-1}(z - \tfrac{1}{2}J\zeta)) = \widetilde{a}(M_S(z, \zeta)).$$

This proves formula (18.16) applying the usual symplectic covariance formula (10.25) for Weyl operators in Theorem 128 in our case. $\qquad\square$

We notice that the symplectic covariance formula (18.16) is very simple compared to the covariance formula $a \circ S^{-1}\overset{\text{Weyl}}{\longleftrightarrow}\widehat{S}\widehat{A}\widehat{S}$ because the operator $\widetilde{M_S}$ is just a symplectic change of variables.

18.2 Intertwiners

As a rule, given a symbol a, the Weyl operator $\widehat{A} \overset{\text{Weyl}}{\longleftrightarrow} a$ is less complicated than the corresponding Bopp operator $\widetilde{A} \overset{\text{Bopp}}{\longleftrightarrow} a$, so one would like to deduce the properties of the second from those of the first. For this we first have to find a procedure allowing us to associate to a function $\psi \in L^2(\mathbb{R}^n)$ a function $\Psi \in L^2(\mathbb{R}^n \oplus \mathbb{R}^n)$; that correspondence should be linear, and intertwine in some way the operators $\widehat{A}$ and $\widetilde{A}$. It turns out that there exist many procedures for transforming a function of, say, x into a function of twice as many variables: the well-known and much used Bargmann transform is an archetypical (and probably the oldest) example of such a procedure (see Problem 429 below). However, the Bargmann transform is not sufficient when one wants to recover all the spectral properties of $\widetilde{A}$ from those of $\widehat{A}$. For example, the eigenvalues of $\widetilde{A}$ are generally infinitely degenerate, so it is illusory to attempt to recover the corresponding eigenvectors from those of $\widehat{A}$ using *one* single transform! This difficulty is of course related to the fact that no isometry from $L^2(\mathbb{R}^n)$ to $L^2(\mathbb{R}^n \oplus \mathbb{R}^n)$ can take a basis of the first space to a basis of the other (intuitively $L^2(\mathbb{R}^n)$ is "much smaller" than $L^2(\mathbb{R}^n \oplus \mathbb{R}^n)$). We will overcome this difficulty by constructing an *infinite family* of partial isometries $W_\phi : L^2(\mathbb{R}^n) \longrightarrow L^2(\mathbb{R}^n \oplus \mathbb{R}^n)$ parametrized by the Schwartz space $\mathcal{S}(\mathbb{R}^n)$; these partial isometries are easily defined in terms of the cross-Wigner transform.

18.2.1 Windowed wavepacket transforms

Here is the definition of the wavepacket transforms W_ϕ:

Definition 427. Let $\phi \in \mathcal{S}(\mathbb{R}^n)$ be such that $\|\phi\|_{L^2(\mathbb{R}^n)} = 1$. The linear mapping $W_\phi : \mathcal{S}(\mathbb{R}^n) \longrightarrow S(\mathbb{R}^n \oplus \mathbb{R}^n)$ defined by

$$W_\phi \psi = (2\pi\hbar)^{n/2} W(\psi, \phi) \tag{18.17}$$

where $W(\psi, \phi)$ is the cross-Wigner distribution, is called the wavepacket transform (for short WPT) with window ϕ.

Equivalently, taking into account Definition 171 of the cross-Wigner transform,

$$W_\phi \psi(z) = \left(\tfrac{2}{\pi\hbar}\right)^{n/2} (\widehat{T}_{\mathrm{GR}}(z)\psi | \phi)_{L^2(\mathbb{R}^n)} \tag{18.18}$$

where $\widehat{T}_{\mathrm{GR}}(z)$ is the Grossmann–Royer transform. In view of the formula (9.14) the WPT is explicitly given by

$$W_\phi \psi(z) = \left(\tfrac{1}{2\pi\hbar}\right)^{n/2} \int_{\mathbb{R}^n} e^{-\frac{i}{\hbar} p \cdot y} \psi(x + \tfrac{1}{2}y)\overline{\phi(x - \tfrac{1}{2}y)}\,dy. \tag{18.19}$$

Since $W(\psi, \phi)$ is defined for $\psi \in \mathcal{S}'(\mathbb{R}^n)$ if $\phi \in \mathcal{S}(\mathbb{R}^n)$ the wavepacket transform extends into a mapping $\mathcal{S}'(\mathbb{R}^n) \longrightarrow \mathcal{S}'(\mathbb{R}^n)$. In fact:

Proposition 428. *For every $\phi \in \mathcal{S}(\mathbb{R}^n)$ the mapping $W_\phi : \mathcal{S}(\mathbb{R}^n) \longrightarrow \mathcal{S}(\mathbb{R}^n \oplus \mathbb{R}^n)$ extends into an automorphism*

$$W_\phi : \mathcal{S}'(\mathbb{R}^n) \longrightarrow \mathcal{S}'(\mathbb{R}^n \oplus \mathbb{R}^n)$$

whose inverse $(W_\phi)^{-1}$ is calculated as follows: if $W_\phi \psi = \Psi$ then

$$\psi(x) = \frac{1}{(\gamma|\phi)_{L^2(\mathbb{R}^n)}} \left(\tfrac{2}{\pi\hbar}\right)^{n/2} \int_{\mathbb{R}^{2n}} \Psi(z_0)\widehat{T}_{\mathrm{GR}}(z_0)\gamma(x)dz \qquad (18.20)$$

for each $\gamma \in \mathcal{S}(\mathbb{R}^n)$ such that $(\gamma|\phi)_{L^2(\mathbb{R}^n)} \neq 0$. In particular, since ϕ is normalized,

$$\psi(x) = \left(\tfrac{2}{\pi\hbar}\right)^{n/2} \int_{\mathbb{R}^{2n}} \Psi(z_0)\widehat{T}_{\mathrm{GR}}(z_0)\phi(x)dz. \qquad (18.21)$$

Proof. The fact that W_ϕ can be extended into an automorphism $\mathcal{S}'(\mathbb{R}^n) \longrightarrow \mathcal{S}'(\mathbb{R}^n \oplus \mathbb{R}^n)$ is a consequence of Proposition 192 and its Corollary 194. The inversion formulas (18.20) and (18.21) immediately follow from formula (9.51) in Proposition 184. $\qquad \square$

We mentioned at the beginning of the chapter that the Bargmann transform is a device that allows one to turn functions on configuration space into functions on phase space. In the following problem you are asked to make explicit the relationship between the Bargmann transform and the wavepacket transform for a special Gaussian window.

Problem 429. Bargmann [6] has introduced an integral transform defined on $L^2(\mathbb{R}^n)$ and whose values are functions on the complex space $\mathbb{C}^n$. This transform is defined by the somewhat cumbersome formula

$$B\psi(z) = 2^{n/4} \int_{\mathbb{R}^n} e^{2\pi u \cdot z - \pi u^2 - \frac{\pi}{2} z^2} \psi(u)du.$$

(i) Show that the Bargmann transform is related to the short-time Fourier transform defined in formula (16.1) by

$$V_{\phi_0}\psi(x, -p) = e^{i\pi x \cdot p} B\psi(z)e^{-\pi|z|^2/2} \qquad (18.22)$$

where $z = x + ip$ and $\phi_0(z) = 2^{n/4}e^{-\pi|x|^2}$. [The proof of formula (18.22) is purely computational; the reader who is in a hurry can find a proof in Gröchenig's book [82], pp. 53–54.]

(ii) Deduce from this a relation between the Bargmann transform and W_{ϕ_0}.

The WPT W_ϕ has several important function-analytical properties: let us begin by showing that it is a partial isometry onto a closed subspace of the Hilbert space $L^2(\mathbb{R}^n \oplus \mathbb{R}^n)$.

Theorem 430.

(i) *The standard WPT W_ϕ is an isometry of $L^2(\mathbb{R}^n)$ onto a closed subspace $\mathcal{H}_\phi$ of $L^2(\mathbb{R}^n \oplus \mathbb{R}^n)$ and the adjoint W_ϕ^* of W_ϕ is given by the formula*

$$W_\phi^* \Psi = \left(\tfrac{2}{\pi\hbar}\right)^{n/2} \int_{\mathbb{R}^{2n}} \Psi(z_0) \widehat{T}_{GR}(z_0)\phi \, dz_0 \qquad (18.23)$$

where $\widehat{T}_{GR}(z_0)$ is the Grossmann–Royer operator;

(ii) *The operator $W_\phi^* W_\phi$ is the identity on $L^2(\mathbb{R}^n)$ and*

$$P_\phi = W_\phi W_\phi^* : L^2(\mathbb{R}^n \oplus \mathbb{R}^n) \longrightarrow L^2(\mathbb{R}^n \oplus \mathbb{R}^n)$$

is the orthogonal projection of $L^2(\mathbb{R}^n \oplus \mathbb{R}^n)$ onto $\mathcal{H}_\phi$.

Proof of (i). Let $\psi \in \mathcal{S}(\mathbb{R}^n)$. In view of Moyal's identity (formula (9.41)) in Theorem 182) we have:

$$(W_\phi\psi|W_\phi\psi')_{L^2} = (\psi|\psi')_{L^2}(\phi|\phi)_{L^2}$$

that is, since ϕ has norm 1,

$$(W_\phi\psi|W_\phi\psi')_{L^2} = (\psi|\psi')_{L^2}$$

hence the operator W_ϕ extends into an isometry of $L^2(\mathbb{R}^n)$ onto a subspace $\mathcal{H}_\phi$ of $L^2(\mathbb{R}^n \oplus \mathbb{R}^n)$; that subspace is closed since it is homeomorphic to $L^2(\mathbb{R}^n)$. Let us prove formula (18.23) for the adjoint of W_ϕ. By definition of the adjoint of an operator we have

$$(W_\phi^* \Psi|\psi)_{L^2} = (\Psi|W_\phi\psi)_{L^2},$$

hence it suffices to show that

$$\left(\tfrac{2}{\pi\hbar}\right)^{n/2} \int_{\mathbb{R}^{2n}} \Psi(z_0)(\widehat{T}_{GR}(z_0)\phi|\psi)_{L^2} \, dz_0 = (\Psi|W_\phi\psi)_{L^2}. \qquad (18.24)$$

Now, using Definitions (9.11) and (18.17) we have

$$(\widehat{T}_{GR}(z_0)\phi|\psi)_{L^2} = (\pi\hbar)^n W(\phi,\psi)(z_0) = \left(\tfrac{\pi\hbar}{2}\right)^{n/2} \overline{W_\phi\psi(z_0)}$$

and hence

$$\left(\tfrac{2}{\pi\hbar}\right)^{n/2} \int_{\mathbb{R}^{2n}} \Psi(z_0)(\widehat{T}_{GR}(z_0)\phi|\psi)_{L^2} \, dz_0 = (\Psi|W_\phi\psi)_{L^2}$$

which was to be proven.

Proof of (ii). To prove that $W_\phi^* W_\phi$ is the identity on $L^2(\mathbb{R}^n)$ we choose $\psi \in L^2(\mathbb{R}^n)$ and observe that for every $\psi' \in L^2(\mathbb{R}^n)$ we have

$$(W_\phi^* W_\phi\psi|\psi')_{L^2} = (W_\phi\psi|W_\phi\psi')_{L^2} = (\psi|\psi')_{L^2};$$

it follows that $W_\phi^* W_\phi\psi = \psi$. We have $P_\phi = P_\phi^*$ and $P_\phi P_\phi^* = P_\phi$ hence P_ϕ is an orthogonal projection. Since $W_\phi^* W_\phi$ is the identity on $L^2(\mathbb{R}^n)$ the range of W_ϕ^* is $L^2(\mathbb{R}^n)$ and that of P_ϕ is therefore precisely $\mathcal{H}_\phi$. $\qquad \square$

18.2.2   The intertwining property

The main interest of our wavepacket transforms comes from the fact that they intertwine Bopp operators with the usual Weyl operators.

Proposition 431.

(i) *The operator W_ϕ intertwines the operators $\widetilde{T}(z_0)$ and $\widehat{T}(z_0)$, in the sense that*

$$W_\phi(\widehat{T}(z_0)\psi) = \widetilde{T}(z_0)W_\phi\psi; \tag{18.25}$$

(ii) *and hence*

$$\widetilde{A}W_\phi = W_\phi\widehat{A} \quad and \quad W_\phi^*\widetilde{A} = \widehat{A}W_\phi^*. \tag{18.26}$$

Proof of (i). Making the change of variable $y = y' + x_0$ in Definition (18.17) of W_ϕ we get

$$W_\phi(\widehat{T}(z_0)\psi, \phi)(z) = e^{-\frac{i}{\hbar}\sigma(z, z_0)}W_\phi\psi(z - \tfrac{1}{2}z_0)$$

which is precisely (18.25). Alternatively, this formula is a particular case of property (9.25) of the cross-Wigner transform.

Proof of (ii). Applying W_ϕ to both sides of the formula

$$\widehat{A}\psi = \left(\tfrac{1}{2\pi\hbar}\right)^n \int_{\mathbb{R}^{2n}} a_\sigma(z_0)\widehat{T}(z_0)\psi\, dz_0$$

defining the Weyl operator $\widehat{A} \overset{\text{Weyl}}{\longleftrightarrow} a$, we get

$$W_\phi\widehat{A}\psi = \left(\tfrac{1}{2\pi\hbar}\right)^n \int_{\mathbb{R}^{2n}} a_\sigma(z_0)W_\phi[\widehat{T}(z_0)\psi]dz_0 = \widetilde{A}W_\phi\psi$$

and hence

$$W_\phi\widehat{A}\psi = \left(\tfrac{1}{2\pi\hbar}\right)^n \int_{\mathbb{R}^{2n}} a_\sigma(z_0)[\widetilde{T}(z_0)W_\phi\psi]dz_0 = \widetilde{A}(W_\phi\psi)$$

which is the first equality (18.26). To prove the second equality it suffices to apply this equality to $W_\phi^*\widetilde{A} = (\widetilde{A}^*W_\phi)^*$. $\qquad\square$

Let us have a look at how the windowed wavepacket transform behaves under the action of symplectic linear automorphisms.

Proposition 432. *Let $S \in \mathrm{Sp}(2n, \mathbb{R})$ and $\psi \in L^2(\mathbb{R}^n)$. We have*

$$W_\phi\psi(S^{-1}z) = W_{\widehat{S}\phi}(S\psi)(z) \tag{18.27}$$

where $\widehat{S}$ is any of the two operators in the metaplectic group $\mathrm{Mp}(2n, \mathbb{R})$ covering S.

Proof. It follows immediately from the symplectic covariance formula

$$W(\psi, \phi) \circ S^{-1} = W(\widehat{S}\psi, \widehat{S}\phi)$$

satisfied by the cross-Wigner distribution (Proposition 217). $\qquad\square$

18.3 Regularity results for Bopp operators

As a rule, to each regularity result for a Weyl operator $\widehat{A} \overset{\text{Weyl}}{\longleftrightarrow} a$ corresponds a regularity result for the Bopp operator $\widetilde{A} \overset{\text{Bopp}}{\longleftrightarrow} a$.

18.3.1 Boundedness results

We begin with a simple continuity statement:

Proposition 433. *Let* $a \in L^2(\mathbb{R}^n \oplus \mathbb{R}^n)$. *The operator* $\widetilde{A} \overset{\text{Bopp}}{\longleftrightarrow} a$ *is continuous on* $L^2(\mathbb{R}^n \oplus \mathbb{R}^n) \longrightarrow L^\infty(\mathbb{R}^n \oplus \mathbb{R}^n)$ *and we have the estimate*

$$\|\widetilde{A}\Psi\|_\infty \le \left(\tfrac{1}{\pi\hbar}\right)^n \|a\|_{L^2(\mathbb{R}^{2n})}\|\Psi\|_{L^2(\mathbb{R}^{2n})} \tag{18.28}$$

for all $\Psi \in L^2(\mathbb{R}^n \oplus \mathbb{R}^n)$.

Proof. It is of course sufficient to prove the inequality (18.28). By definition

$$\widetilde{A}\Psi(z) = \left(\tfrac{1}{2\pi\hbar}\right)^n \int_{\mathbb{R}^{2n}} a_\sigma(z_0)\widetilde{T}(z_0)\Psi(z)dz_0$$

with $\widetilde{T}(z_0)\Psi(z) = e^{-\frac{i}{\hbar}\sigma(z,z_0)}\Psi(z - \tfrac{1}{2}z_0)$. Hence, using Cauchy–Schwarz's inequality:

$$|\widetilde{A}\Psi(z)|^2 \le \left(\tfrac{1}{2\pi\hbar}\right)^{2n} \|a\|_{L^2(\mathbb{R}^{2n})} \int_{\mathbb{R}^{2n}} |\Psi(z - \tfrac{1}{2}z_0)|^2 dz_0.$$

Setting $u = z - \tfrac{1}{2}z_0$ in the integral this inequality becomes

$$|\widetilde{A}\Psi(z)|^2 \le \left(\tfrac{1}{\pi\hbar}\right)^{2n} \|a\|_{L^2(\mathbb{R}^{2n})}\|\Psi\|_{L^2(\mathbb{R}^{2n})}$$

which we set out to prove. $\qquad\square$

Let us now introduce the following notation: for an arbitrary window ϕ set

$$\mathcal{L}_\phi^q(\mathbb{R}^{2n}) = W_\phi(M_s^q(\mathbb{R}^n)) \subset L_s^q(\mathbb{R}^{2n}). \tag{18.29}$$

Clearly $\mathcal{L}_\phi(\mathbb{R}^{2n})$ is a closed linear subspace of $L_s^q(\mathbb{R}^{2n})$ (and hence a Banach space).

Proposition 434. *Let* $\widetilde{A} \overset{\text{Bopp}}{\longleftrightarrow} a$ *be associated to the Weyl operator* $\widehat{A} \overset{\text{Weyl}}{\longleftrightarrow} a$. *If* $a \in M_s^{\infty,1}(\mathbb{R}^{2n})$ *then*

$$\widetilde{A}_\omega : \mathcal{L}_\phi^q(\mathbb{R}^{2n}) \longrightarrow \mathcal{L}_\phi^q(\mathbb{R}^{2n})$$

(continuously) for every window $\phi \in \mathcal{S}(\mathbb{R}^n)$.

Proof. Let $U \in \mathcal{L}_\phi^q(\mathbb{R}^{2n})$; by definition there exists $u \in M_s^q(\mathbb{R}^n)$ such that $U = W_\phi u$. In view of the first intertwining relation (18.26) we have

$$\widetilde{A}W_\phi u = W_\phi \widehat{A}u$$

hence $\widehat{A}u \in M_s^q(\mathbb{R}^n)$ and $\widehat{A}$ is bounded. It follows that $W_\phi\widehat{A}u \in \mathcal{L}_{f,\phi}^q(\mathbb{R}^{2n})$. $\qquad\square$

It is worthwhile (and important, in a quantum mechanical context) to note that the spaces $\mathcal{L}_\phi^q(\mathbb{R}^{2n})$ cannot contain functions which are "too concentrated" around a point; this is reminiscent of the uncertainty principle. In particular the Schwartz space $\mathcal{S}(\mathbb{R}^{2n})$ is not contained in any of the $\mathcal{L}_\phi^q(\mathbb{R}^{2n})$. This is an immediate consequence of the Hardy uncertainty principle for the Wigner transform: assume that $u \in \mathcal{S}(\mathbb{R}^n)$ is such that $Wu \le Ce^{-Mz\cdot z}$ for some $C > 0$ and a real matrix $M = M^T > 0$. Consider now the eigenvalues of JM; these are of the form $\pm i\lambda_j$ with $\lambda_j > 0$. Then we must have $\lambda_j \le 1$ for all $j = 1,\ldots,n$. Equivalently, the symplectic capacity $c(\mathcal{W}_M)$ of the "Wigner ellipsoid" $\mathcal{W}_M : Mz \cdot z \le 1$ satisfies $c(\mathcal{W}) \ge \pi$. This result in fact also holds true for the cross-Wigner transform: if $|W(u,\phi)(z)| \le Ce^{-Mz\cdot z}$ for some $\phi \in \mathcal{S}(\mathbb{R}^n)$ then $c(\mathcal{W}) \ge \pi$. Assume now that $U \in \mathcal{L}_{f,\phi}^q(\mathbb{R}^{2n})$ satisfies the sub-Gaussian estimate $|U(z)| \le Ce^{-Mz\cdot z}$; by definition of $\mathcal{L}_\phi^q(\mathbb{R}^{2n})$ this is equivalent to $|W(u,\phi)(z)| \le Ce^{-Mz\cdot z}$ hence the ellipsoid $\mathcal{W}_M$ must have symplectic capacity at least equal to π.

Recall from Chapter 17 that the modulations space $M_s^{\infty,1}(\mathbb{R}^n \oplus \mathbb{R}^n)$ contains the weighted Sjöstrand classes consisting of all tempered distributions $\Psi \in \mathcal{S}'(\mathbb{R}^n \oplus \mathbb{R}^n)$ such that

$$\sup_{z\in\mathbb{R}^{2n}} |W(\Psi,\Phi)(z,\zeta)\langle z\rangle^s| \in L^1(\mathbb{R}^n \oplus \mathbb{R}^n). \tag{18.30}$$

We have seen that $M_s^{\infty,1}(\mathbb{R}^n \oplus \mathbb{R}^n)$ is a Banach $*$-algebra with respect to the Moyal product $\star_\hbar$ and the involution $a \longmapsto \overline{a}$ and that for every window $\Phi \in \mathcal{S}(\mathbb{R}^n \oplus \mathbb{R}^n)$ there exists a constant $C_\Phi > 0$ such that

$$\|a \star_\hbar b\|_{M_s^{\infty,1}}^\Phi \le C_\Phi \|a\|_{M_s^{\infty,1}}^\Phi \|b\|_{M_s^{\infty,1}}^\Phi$$

for $a,b \in M_s^{\infty,1}(\mathbb{R}^n \oplus \mathbb{R}^n)$.

18.3.2 Global hypoellipticity properties

In Chapter 14 we introduced the global Shubin classes of symbols Γ_ρ^m; we have $a \in \Gamma_\rho^m(\mathbb{R}^n \oplus \mathbb{R}^n)$ if and only if a is a complex function in $C^\infty(\mathbb{R}^n \oplus \mathbb{R}^n)$ such that for every $\alpha \in \mathbb{N}^{2n}$ there exists a constant $C_\alpha \ge 0$ with

$$|\partial_z^\alpha a(z)| \le C_\alpha \langle z\rangle^{m-\rho|\alpha|} \quad \text{for } z \in \mathbb{R}^{2n}. \tag{18.31}$$

In the context of Bopp pseudo-differential operators it is interesting to consider symbols belonging to a subclass of $H\Gamma_\rho^{m_1,m_0}$ of Γ_ρ^m; the consideration of these new symbol spaces will enable us to prove both regularity and spectral results.

The following definition goes back to Shubin [147] (Chapter 4):

Definition 435. Let m_0, m_1, and ρ be real numbers such that $m_0 \le m_1$ and $0 < \rho \le 1$. The symbol class $H\Gamma_\rho^{m_1,m_0}(\mathbb{R}^n \oplus \mathbb{R}^n)$ consists of all functions $a \in C^\infty(\mathbb{R}^n \oplus \mathbb{R}^n)$ such that, for $|z|$ sufficiently large, the following properties hold:

$$C_0|z|^{m_0} \le |a(z)| \le C_1|z|^{m_1} \tag{18.32}$$

for some $C_0, C_1 \geq 0$ and, for every $\alpha \in \mathbb{N}^n$ there exists $C_\alpha \geq 0$ such that

$$|\partial_z^\alpha a(z)| \leq C_\alpha |a(z)| \, |z|^{-\rho|\alpha|}. \tag{18.33}$$

We denote by $HG_\rho^{m_1, m_0}(\mathbb{R}^n)$ the class of operators $\widehat{A} : \mathcal{S}(\mathbb{R}^n) \longrightarrow \mathcal{S}'(\mathbb{R}^n)$ with τ-symbols a_τ belonging to $H\Gamma_\rho^{m_1, m_0}(\mathbb{R}^n \oplus \mathbb{R}^n)$.

Thus, $\widehat{A} \in H\Gamma_\rho^{m_1, m_0}(\mathbb{R}^n \oplus \mathbb{R}^n)$ means that for every $\tau \in \mathbb{R}$ there exists $a_\tau \in H\Gamma_\rho^{m_1, m_0}(\mathbb{R}^n \oplus \mathbb{R}^n)$ such that

$$\widehat{A}u(x) = (2\pi)^{-n} \iint_{\mathbb{R}^{2n}} e^{i(x-y)\cdot\xi} a_\tau((1-\tau)x + \tau y, \xi) u(y) dy d\xi;$$

choosing $\tau = \frac{1}{2}$ this means, in particular, that every Weyl operator $\widehat{A} \xleftrightarrow{\text{Weyl}} a$ with $a \in H\Gamma_\rho^{m_1, m_0}(\mathbb{R}^n \oplus \mathbb{R}^n)$ is in $HG_\rho^{m_1, m_0}(\mathbb{R}^n \oplus \mathbb{R}^n)$. It turns out that the condition $a \in H\Gamma_\rho^{m_1, m_0}(\mathbb{R}^n \oplus \mathbb{R}^n)$ is also sufficient, because if $a_\tau \in H\Gamma_\rho^{m_1, m_0}(\mathbb{R}^n \oplus \mathbb{R}^n)$ is true for some τ then it is true for all τ.

The spaces $H\Gamma_\rho^{m_1, m_0}(\mathbb{R}^n \oplus \mathbb{R}^n)$ and $HG_\rho^{m_1, m_0}(\mathbb{R}^n)$ are subspaces of the Shubin classes studied in Chapter 14:

$$H\Gamma_\rho^{m_1, m_0}(\mathbb{R}^n \oplus \mathbb{R}^n) \subset \Gamma_\rho^{m_1}(\mathbb{R}^n \oplus \mathbb{R}^n),$$
$$HG_\rho^{m_1, m_0}(\mathbb{R}^n) \subset G_\rho^{m_1}(\mathbb{R}^n),$$

as trivially follows from their definition. An immediate consequence of Proposition 318 in Chapter 14 is the following:

Proposition 436. *Every operator $\widehat{A} \in H\Gamma_\rho^{m_1, m_0}(\mathbb{R}^n \oplus \mathbb{R}^n)$ is a continuous operator $\mathcal{S}(\mathbb{R}^n) \longrightarrow \mathcal{S}(\mathbb{R}^n)$.*

The classes $HG_\rho^{m_1, m_0}(\mathbb{R}^n \oplus \mathbb{R}^n)$ have an interesting "global hypoellipticity" property, which motivated Shubin's interest (also see the contribution by Boggiatto et al. [16], p. 70). Recall that an operator $\widehat{A} : \mathcal{S}'(\mathbb{R}^n) \longrightarrow \mathcal{S}'(\mathbb{R}^n)$ is C^∞-hypoelliptic if

$$\psi \in \mathcal{S}'(\mathbb{R}^n) \text{ and } \widehat{A}\psi \in C^\infty(\mathbb{R}^n) \Longrightarrow \psi \in C^\infty(\mathbb{R}^n).$$

Shubin's notion of global hypoellipticity is more useful in applications to quantum mechanics than C^∞ hypoellipticity because it incorporates the decay at infinity of the involved functions or distributions.

Definition 437. We will say that a linear operator $\widehat{A} : \mathcal{S}'(\mathbb{R}^n) \longrightarrow \mathcal{S}'(\mathbb{R}^n)$ is "globally hypoelliptic" if we have

$$\psi \in \mathcal{S}'(\mathbb{R}^n) \text{ and } \widehat{A}\psi \in \mathcal{S}(\mathbb{R}^n) \Longrightarrow \psi \in \mathcal{S}(\mathbb{R}^n). \tag{18.34}$$

Shubin ([147], Chapter IV, §23) has proved the existence of a left parametrix of an operator $\widehat{A} \in HG_\rho^{m_0, m_1}(\mathbb{R}^n \oplus \mathbb{R}^n)$; more precisely he shows that:

Proposition 438. *Let $\widehat{A} \in HG_\rho^{m_0,m_1}(\mathbb{R}^n \oplus \mathbb{R}^n)$;*

(i) *There exists an operator $\widehat{B} \in G\Gamma_\rho^{-m_1,-m_0}(\mathbb{R}^n)$ such that $\widehat{B}\widehat{A} = I + \widehat{R}$ where the kernel of $\widehat{R}$ is in $\mathcal{S}(\mathbb{R}^n \times \mathbb{R}^n)$ and hence $\widehat{R} : \mathcal{S}'(\mathbb{R}^n) \longrightarrow \mathcal{S}(\mathbb{R}^n)$;*

(ii) *Any Weyl operator $\widehat{A}$ belonging to the class $HG_\rho^{m_1,m_0}(\mathbb{R}^n \oplus \mathbb{R}^n)$ is globally hypoelliptic.*

The statement (ii) actually immediately follows from the existence of the parametrix $\widehat{B}$: let $\psi \in \mathcal{S}'(\mathbb{R}^n)$ and assume that $\phi = \widehat{A}\psi \in \mathcal{S}(\mathbb{R}^n)$; then $\psi = \widehat{B}\phi - \widehat{R}\psi$. Now it is clear that $\widehat{B}\phi \in \mathcal{S}(\mathbb{R}^n)$ and we have $\widehat{R}\psi \in \mathcal{S}(\mathbb{R}^n)$ in view of the proposition above.

We cannot however use the result above to prove global hypoellipticity properties for the Bopp operator $\widetilde{A} \overset{\text{Bopp}}{\longleftrightarrow} a$: while it is clear from the way in which Bopp operators compose (Proposition 425 above) that we have $\widetilde{B}\widetilde{A} = I + \widetilde{R}$ where $\widetilde{R}$ is the Bopp operator corresponding to the Weyl operator $\widehat{R}$, it is not true in general that the kernel of $\widetilde{R}$ is in $\mathcal{S}(\mathbb{R}^{2n} \times \mathbb{R}^{2n})$.

Exercise 439. Check this statement in detail using the formulas in Theorem 424.

To be able to prove global hypoellipticity results for Bopp operators we can use the following refinement of Proposition 438, also due to Shubin ([147], Chapter IV, §25):

Proposition 440. *Let $\widehat{A} \in HG_\rho^{m_1,m_0}(\mathbb{R}^n \oplus \mathbb{R}^n)$ be such that $\operatorname{Ker} \widehat{A} = \operatorname{Ker} \widehat{A}^* = \{0\}$. Then there exists $\widehat{B} \in HG_\rho^{-m_1,-m_0}(\mathbb{R}^n \oplus \mathbb{R}^n)$ such that $\widehat{B}\widehat{A} = \widehat{A}\widehat{B} = I$ (i.e., $\widehat{B}$ is a true inverse of $\widehat{A}$).*

An immediate consequence of this result is:

Corollary 441. *The Bopp operator $\widetilde{A}$ associated to $a \in G\Gamma_\rho^{m_1,m_0}(\mathbb{R}^n \oplus \mathbb{R}^n)$ such that $\operatorname{Ker} \widehat{A} = \operatorname{Ker} \widehat{A}^* = \{0\}$ is globally hypoelliptic.*

Proof. In view of Proposition 440 the operator $\widehat{A}$ has an inverse $\widehat{B}$ belonging to $HG_\rho^{-m_1,-m_0}(\mathbb{R}^n \oplus \mathbb{R}^n)$. In view of formula (18.15) in Proposition 425 the Bopp operator $\widetilde{B}$ is then an inverse of $\widetilde{A}$. Assume now that $\widetilde{A}\Psi = \Phi \in \mathcal{S}(\mathbb{R}^n \oplus \mathbb{R}^n)$; then $\Psi = \widetilde{B}\Phi$. The result now follows from the observation that $\widetilde{B}$ maps $\mathcal{S}(\mathbb{R}^n \oplus \mathbb{R}^n)$ into $\mathcal{S}(\mathbb{R}^n \oplus \mathbb{R}^n)$ and $\mathcal{S}'(\mathbb{R}^n \oplus \mathbb{R}^n)$ into $\mathcal{S}'(\mathbb{R}^n \oplus \mathbb{R}^n)$. $\qquad\square$

Chapter 19

Applications of Bopp Quantization

In this chapter we study a few selected applications of the techniques and material introduced in the previous chapters. Needless to say, the list of topics we have chosen is not exhaustive, and only limited due to constraints of place and space; they very much reflect the taste – and knowledge... – of the author. Much is part of ongoing research, and the interested reader is invited to consult the bibliographic hints.

We begin by a rather straightforward application of the techniques of Bopp calculus, and then show how it is a tool of choice for understanding deformation quantization from an operator point of view. We thereafter extend Bopp calculus to a topic of current great interest in mathematical physics, namely "noncommutative quantum mechanics" whose study we reduce to that of Bopp calculus.

19.1   Spectral results for Bopp operators

An essential property of the Bopp pseudo-differential operator $\widetilde{A}$ is that it has the same (generalized) eigenvalues as the corresponding Weyl operator $\widehat{A}$. Moreover, the corresponding eigenfunctions are obtained from those of $\widehat{A}$ using wavepacket transforms; this fact implies that in general the eigenvalues of $\widetilde{A}$ have infinite degeneracy.

19.1.1   A fundamental property of the intertwiners

Another essential property of the WPT is the following, which was announced in Proposition 188:

Theorem 442. *Let $(\phi_j)_j$ and $(\psi_j)_j$ be arbitrary orthonormal bases of $L^2(\mathbb{R}^n)$; the vectors $\Phi_{j,k} = W_{\phi_j}\psi_k$ form an orthonormal basis of $L^2(\mathbb{R}^n \oplus \mathbb{R}^n)$. In particular $(W_{\psi_j}\psi_k)_{j,k}$ is such a basis.*

Proof. Using Moyal's identity (9.41) we have

$$
\begin{aligned}
(\Phi_{j,k}|\Phi_{j',k'})_{L^2} &= (W_{\phi_j}\psi_k|W_{\phi_{j'}}\psi_{k'})_{L^2} \\
&= (2\pi\hbar)^n (W(\psi_k,\phi_j)|W(\psi_{k'},\phi_{j'}))_{L^2} \\
&= (\psi_k|\psi_{k'})_{L^2}\overline{(\phi_j|\phi_{j'})_{L^2}}
\end{aligned}
$$

hence the $\Phi_{j,k}$ form an orthonormal system of vectors in $L^2(\mathbb{R}^n \oplus \mathbb{R}^n)$. It is thus sufficient to show that, if $\Psi \in L^2(\mathbb{R}^n \oplus \mathbb{R}^n)$ is orthogonal to the family $(\Phi_{j,k})_{j,k}$ (and hence to all the spaces $\mathcal{H}_{\phi_j}$), then $\Psi = 0$. Assume that $(\Psi|\Phi_{jk})_{L^2(\mathbb{R}^{2n})} = 0$ for all indices j, k. Since we have

$$
(\Psi|\Phi_{jk})_{L^2} = (\Psi|W_{\phi_j}\psi_k)_{L^2} = (W_{\phi_j}^*\Psi|\psi_k)_{L^2}
$$

it follows that $W_{\phi_j}^*\Psi = 0$ for all j since $(\psi_k)_k$ is a basis. Using the sesquilinearity of W_ϕ in ϕ we have in fact $W_\phi^*\Psi = 0$ for all $\phi \in L^2(\mathbb{R}^n)$ since $(\phi_j)_j$ also is a basis. Let us show that this property implies that we must have $\Psi = 0$. Recall (formula (18.23)) that the adjoint of the wavepacket transform W_ϕ^* is given by

$$
W_\phi^*\Psi = \left(\tfrac{2}{\pi\hbar}\right)^{n/2} \int_{\mathbb{R}^{2n}} \Psi(z_0)\widehat{T}_{\mathrm{GR}}(z_0)\phi\,dz_0
$$

where $\widehat{T}_{\mathrm{GR}}(z_0)$ is the Grossmann–Royer operator. Let now ψ be an arbitrary element of $\mathcal{S}(\mathbb{R}^n)$; we have, using Definition (9.11) of the cross-Wigner transform,

$$
\begin{aligned}
(W_\phi^*\Psi|\psi)_{L^2} &= \left(\tfrac{2}{\pi\hbar}\right)^{n/2} \int_{\mathbb{R}^{2n}} \Psi(z)(\widehat{T}_{\mathrm{GR}}(z)\phi|\psi)_{L^2}\,dz \\
&= (2\pi\hbar)^{n/2} \int_{\mathbb{R}^{2n}} \Psi(z)W(\psi,\phi)(z)\,dz.
\end{aligned}
$$

Let us now view $\Psi \in L^2(\mathbb{R}^n \oplus \mathbb{R}^n)$ as the Weyl symbol[1] of an operator $\widehat{A}_\Psi$. In view of formula (10.8) we have

$$
(2\pi\hbar)^{n/2} \int_{\mathbb{R}^{2n}} \Psi(z)W(\psi,\phi)(z)\,dz = (\widehat{A}_\Psi\psi|\phi)_{L^2}
$$

and the condition $W_\phi^*\Psi = 0$ for all $\phi \in \mathcal{S}(\mathbb{R}^n)$ is thus equivalent to $(\widehat{A}_\Psi\psi|\phi)_{L^2} = 0$ for all $\phi, \psi \in \mathcal{S}(\mathbb{R}^n)$. It follows that $\widehat{A}_\Psi\psi = 0$ for all ψ and hence $\widehat{A}_\Psi = 0$. Since the Weyl correspondence is one-to-one we must have $\Psi = 0$ as claimed. $\qquad\square$

Problem 443. State (and prove) a generalization of Theorem 442 to Weyl–Heisenberg frames.

---

[1] I am grateful to Harald Stockinger (NuHAG) for having pointed out this approach to me.

19.1.2 Generalized eigenvalues and eigenvectors of a Bopp operator

In this subsection we give an application of the machinery developed above to spectral results for Bopp operators. In order to avoid delicate domain questions we will deal here with generalized eigenvalues and eigenvectors; more specific results will be studied in the next subsection when we make some additional assumptions on the symbols. In that context the Banach Gelfand triple $(S_0(\mathbb{R}^n), L^2(\mathbb{R}^n), S_0'(\mathbb{R}^n))$ (where $S_0(\mathbb{R}^n)$ is the Feichtinger algebra) is a useful device.

Let us slightly extend the notion of generalized eigenvalues and eigenfunctions we briefly discussed in Subsection 16.4.2:

Definition 444. Let $\widehat{A}$ be an operator on $L^2(\mathbb{R}^n)$. We assume that $\widehat{A}$ is continuous on $\mathcal{S}(\mathbb{R}^n) \longrightarrow \mathcal{S}(\mathbb{R}^n)$, and hence has a continuous extension $\mathcal{S}'(\mathbb{R}^n) \longrightarrow \mathcal{S}'(\mathbb{R}^n)$. Setting $(\psi|\theta) = \langle \psi, \overline{\theta} \rangle$ for $\psi \in \mathcal{S}'(\mathbb{R}^n)$ and $\theta \in \mathcal{S}(\mathbb{R}^n)$ a distribution $\psi \in \mathcal{S}'(\mathbb{R}^n)$ is called a generalized eigenvector of $\widehat{A}$ corresponding to the (generalized) eigenvalue λ if $\psi \neq 0$ and we have

$$(\psi|\widehat{A}^*\theta) = \lambda(\psi|\theta)$$

for all $\theta \in \mathcal{S}(\mathbb{R}^n)$ [observe that we do not require ψ to be in the domain $D_{\widehat{A}}$ of $\widehat{A}$]. We will similarly write $((\Psi|\Theta)) = \langle\langle \Psi, \overline{\Theta} \rangle\rangle$ where $\langle\langle \cdot, \cdot \rangle\rangle$ is the distributional bracket on $\mathbb{R}^n \oplus \mathbb{R}^n$. The distribution $\Psi \in \mathcal{S}'(\mathbb{R}^n \oplus \mathbb{R}^n)$ is a generalized eigenfunction of $\widetilde{A} : \mathcal{S}(\mathbb{R}^n \oplus \mathbb{R}^n) \longrightarrow \mathcal{S}(\mathbb{R}^n \oplus \mathbb{R}^n)$ if $\Psi \neq 0$ and there exists λ such that

$$((\Psi|\widetilde{A}^*\Theta)) = \lambda((\Psi|\Theta))$$

for all $\Theta \in \mathcal{S}(\mathbb{R}^n \oplus \mathbb{R}^n)$.

With this definition and notation we have the following result:

Theorem 445.

(i) *The generalized eigenvalues of the operators $\widehat{A}$ and $\widetilde{A}$ are the same;*

(ii) *Let ψ be a generalized eigenvector of $\widehat{A}$. Then, for every $\phi \in \mathcal{S}(\mathbb{R}^n)$ the vector $\Psi = W_\phi \psi$ satisfies $((\Psi|\widetilde{A}^*\Theta)) = \lambda((\Psi|\Theta))$ for all for all $\Theta \in \mathcal{S}(\mathbb{R}^n \oplus \mathbb{R}^n)$; in particular if $\Psi \neq 0$ it is a generalized eigenvector of $\widetilde{A}$ corresponding to the same generalized eigenvalue.*

(iii) *Conversely, if Ψ is a generalized eigenvector of $\widetilde{A}$ then $\psi = W_\phi^*\Psi$ is a generalized eigenvector of $\widehat{A}$ corresponding to the same eigenvalue.*

Proof. (i) Let us show that if

$$(\psi|\widehat{A}^*\theta) = \lambda(\psi|\theta) \qquad \text{for every } \theta \in \mathcal{S}(\mathbb{R}^n)$$

then

$$((W_\phi\psi|\widetilde{A}^*\Theta)) = \lambda((W_\phi\psi|\Theta)) \quad \text{for every } \Theta \in \mathcal{S}(\mathbb{R}^n \oplus \mathbb{R}^n).$$

We have

$$((W_\phi\psi|\widetilde{A}^*\Theta)) = ((\psi|W_\phi^*\widetilde{A}^*\Theta)) = ((\psi|\widehat{A}^*W_\phi^*\Theta))$$
$$= \lambda((\psi|W_\phi^*\Theta)) = \lambda((W_\phi\psi|\Theta))$$

hence the claim since $W_\phi^*\Theta \in \mathcal{S}(\mathbb{R}^n)$. Suppose conversely that $((\Psi|\widetilde{A}^*\Theta)) = \lambda((\Psi|\Theta))$ for every $\Theta \in \mathcal{S}(\mathbb{R}^n \oplus \mathbb{R}^n)$. We have to show that $(W_\phi^*\Psi|\widehat{A}^*\theta) = \lambda(W_\phi^*\Psi|\theta)$ for every $\theta \in \mathcal{S}(\mathbb{R}^n)$. Now,

$$(W_\phi^*\Psi|\widehat{A}^*\theta) = (\Psi|W_\phi\widehat{A}^*\theta) = (\Psi|\widetilde{A}^*W_\phi\theta)$$
$$= \lambda((\Psi|W_\phi\theta)) = \lambda((W_\phi^*\Psi|\theta))$$

proving our claim. That every eigenvalue of $\widehat{A}$ also is an eigenvalue of $\widetilde{A}$ is clear: if $\widehat{A}\psi = \lambda\psi$ for some $\psi \neq 0$ then

$$\widetilde{A}(W_\phi\psi) = W_\phi\widehat{A}\psi = \lambda W_\phi\psi$$

and $\Psi = W_\phi\psi \neq 0$; this proves at the same time that $W_\phi\psi$ is an eigenvector of $\widetilde{A}$ because W_ϕ has kernel $\{0\}$. (ii) Assume conversely that $\widetilde{A}\Psi = \lambda\Psi$ for $\Psi \in L^2(\mathbb{R}^n \oplus \mathbb{R}^n)$, $\Psi \neq 0$, and $\lambda \in \mathbb{R}$. For every ϕ we have

$$\widehat{A}W_\phi^*\Psi = W_\phi^*\widetilde{A}\Psi = \lambda W_\phi^*\Psi$$

hence λ is an eigenvalue of $\widehat{A}$ and ψ an eigenvector if $\psi = W_\phi^*\Psi \neq 0$. We have

$$W_\phi\psi = W_\phi W_\phi^*\Psi = P_\phi\Psi$$

where P_ϕ is the orthogonal projection on the range $\mathcal{H}_\phi$ of W_ϕ. Assume that $\psi = 0$; then $P_\phi\Psi = 0$ for every $\phi \in \mathcal{S}(\mathbb{R}^n)$, and hence $\Psi = 0$ in view of Theorem 442. $\square$

Let us specialize the results above to the case where $\widetilde{A}$ is (essentially) self-adjoint:

Corollary 446. *Suppose that $\widehat{A}$ is an essentially self-adjoint operator on $L^2(\mathbb{R}^n)$ and that each of the eigenvalues $\lambda_0, \lambda_1, \ldots, \lambda_j, \ldots$ has multiplicity 1. Let $\psi_0, \psi_1, \ldots, \psi_j, \ldots$ be the corresponding sequence of orthonormal eigenvectors. Let Ψ_j be an eigenvector of $\widetilde{A}$ corresponding to the eigenvalue λ_j. There exists a sequence $(\alpha_{j,k})_k$ of complex numbers such that*

$$\Psi_j = \sum_\ell \alpha_{j,\ell}\Psi_{j,\ell} \quad with \quad \Psi_{j,\ell} = W_{\psi_\ell}\psi_j \in \mathcal{H}_j \cap \mathcal{H}_\ell. \tag{19.1}$$

Proof. We know from Theorem 445 above that $\widehat{A}$ and $\widetilde{A}$ have the same eigenvalues and that $\Psi_{j,k} = W_{\psi_k}\psi_j$ satisfies $\widetilde{A}\Psi_{j,k} = \lambda_j\Psi_{j,k}$. Since $\widehat{A}$ is self-adjoint its eigenvectors ψ_j form an orthonormal basis of $L^2(\mathbb{R}^n)$; it follows from Theorem 442 that the $\Psi_{j,k}$ form an orthonormal basis of $L^2(\mathbb{R}^n \oplus \mathbb{R}^n)$, hence there exist non-zero scalars $\alpha_{j,k,\ell}$ such that $\Psi_j = \sum_{k,\ell} \alpha_{j,k,\ell} \Psi_{k,\ell}$. We have, by linearity and using the fact that $\widetilde{A}\Psi_{k,\ell} = \lambda_k \Psi_{k,\ell}$,

$$\widetilde{A}\Psi_j = \sum_{k,\ell} \alpha_{j,k,\ell} \widetilde{A}\Psi_{k,\ell} = \sum_{k,\ell} \alpha_{j,k,\ell} \lambda_k \Psi_{k,\ell}.$$

On the other hand we also have

$$\widetilde{A}\Psi_j = \lambda_j \Psi_j = \sum_{j,k} \alpha_{j,k,\ell} \lambda_j \Psi_{k,\ell}$$

and this is only possible if $\alpha_{j,k,\ell} = 0$ for $k \neq j$; setting $\alpha_{j,\ell} = \alpha_{j,j,\ell}$ formula (19.1) follows. That $\Psi_{j,\ell} \in \mathcal{H}_j \cap \mathcal{H}_\ell$ is clear using the definition of $\mathcal{H}_\ell$ and the sesquilinearity of the cross-Wigner transform. $\qquad\square$

Besides the fact that they intervene in global hypoellipticity questions, one of the main appeals of Shubin's classes $H\Gamma_\rho^{m_1,m_0}$ and $HG_\rho^{m_1,m_0}$ comes from the following property, which is essential for the proof of the main result (Theorem 448) which we will prove in a moment:

Theorem 447. *Let $\widehat{A} \in HG_\rho^{m_1,m_0}(\mathbb{R}^n \oplus \mathbb{R}^n)$ with $m_0 > 0$. If $\widehat{A}$ is formally self-adjoint, that is if $(\widehat{A}u|v)_{L^2} = (u|\widehat{A}v)_{L^2}$ for all $u,v \in C_0^\infty(\mathbb{R}^n)$, then $\widehat{A}$ is essentially self-adjoint and has discrete spectrum in $L^2(\mathbb{R}^n)$. Moreover there exists an orthonormal basis of eigenfunctions $\phi_j \in \mathcal{S}(\mathbb{R}^n)$ $(j = 1, 2, \ldots)$ with eigenvalues $\lambda_j \in \mathbb{R}$ such that $\lim_{j\to\infty} |\lambda_j| = \infty$.*

We remark that the global hypoellipticity property of the operators $\widehat{A} - \lambda I \in HG_\rho^{m_1,m_0}$ automatically implies that the orthonormal basis of eigenfunctions ϕ_j is in the Schwartz space $\mathcal{S}(\mathbb{R}^n)$.

We now have all the elements we need to prove the main result of this section:

Theorem 448. *Let $\widehat{A} \in HG_\rho^{m_1,m_0}(\mathbb{R}^n \oplus \mathbb{R}^n)$.*

(i) *The operators $\widehat{A}$ and $\widetilde{A}$ have the same eigenvalues; if ψ is an eigenfunction of A corresponding to the eigenvalue λ then $\Psi_\phi = W_\phi \psi$ is an eigenfunction of $\widetilde{A}$ corresponding to λ, for every ϕ, and we have $\Psi_\phi \in \mathcal{S}(\mathbb{R}^n \oplus \mathbb{R}^n)$.*

(ii) *Assume in addition that $m_0 > 0$ and that $\widehat{A}$ is formally self-adjoint. Then $\widetilde{A}$ has discrete spectrum $(\lambda_j)_{j\in\mathbb{N}}$ and $\lim_{j\to\infty} |\lambda_j| = \infty$.*

(iii) *The eigenfunctions of $\widetilde{A}$ are in this case given by $\Phi_{jk} = W_{\phi_j}\phi_k$ where the ϕ_j are the eigenfunctions of A.*

(iv) *We have $\Phi_{jk} \in \mathcal{S}(\mathbb{R}^n \oplus \mathbb{R}^n)$ and the Φ_{jk} form an orthonormal basis of $L^2(\mathbb{R}^n \oplus \mathbb{R}^n)$.*

Proof of (i). That every eigenvalue of $\widehat{A}$ also is an eigenvalue of $\widetilde{A}$ is clear: if $\widehat{A}\psi = \lambda\psi$ for some $\psi \neq 0$ then

$$\widetilde{A}(W_\phi\psi) = W_\phi\widehat{A}\psi = \lambda(W_\phi\psi)$$

and $W_\phi\psi \neq 0$ because W_ϕ is injective; this proves at the same time that $W_\phi\psi$ is an eigenfunction of $\widetilde{A}$. Assume conversely that $\widetilde{A}\Psi = \lambda\Psi$ for $\Psi \neq 0$. For every ϕ we have, using the equality $W_\phi^* \widetilde{A} = \widehat{A}W_\phi^*$,

$$\widehat{A}W_\phi^*\Psi = W_\phi^*\widetilde{A}\Psi = \lambda W_\phi^*\Psi,$$

hence λ is an eigenvalue of $\widehat{A}$ and $W_\phi^*\Psi$ will be an an eigenfunction of $\widehat{A}$ if it is different from zero. Let us prove this is indeed the case. Recall that $W_\phi W_\phi^* = P_\phi$ is the orthogonal projection on $\mathcal{H}_\phi$. Assume that $W_\phi^*\Psi = 0$; then $P_\phi\Psi = 0$ for every $\phi \in \mathcal{S}(\mathbb{R}^n)$, and hence $\Psi = 0$ in view of Theorem 442; but this is not possible since Ψ is an eigenfunction. That we have $\Psi_\phi \in \mathcal{S}(\mathbb{R}^n \oplus \mathbb{R}^n)$ is clear since $W_\phi : \mathcal{S}(\mathbb{R}^n) \longrightarrow \mathcal{S}(\mathbb{R}^n \oplus \mathbb{R}^n)$.

Proof of (ii)–(iv). Properties (ii)–(iv) follow immediately from property (i) using the properties of the Shubin classes $HG_\rho^{m_1,m_0}(\mathbb{R}^n \oplus \mathbb{R}^n)$. $\square$

19.1.3   Application: the Landau problem

The "Landau problem" is a classical topic from mathematical physics; it is basically the study of Hamiltonian operators arising in the study of particles moving under the action of a magnetic field (see for instance Landau and Lifshitz [113]). It has mathematically very interesting ramifications, such as the study of the Hall effect (both classical and quantum); see for instance the overview [11] by Bellissard et al. In our study of the spectral properties of the Landau problem for a uniform magnetic field we are following the approach in our paper de Gosson [71] where we showed the strong relationship between this problem and Bopp calculus.

The derivation of the so-called magnetic operator can be found in a multitude of textbooks; we are following here our exposition in [65]. Consider a hydrogen atom placed in a magnetic field $\vec{B} = (B_x, B_y, B_z)$. Neglecting spin and relativistic effects the Hamiltonian function is

$$H(\vec{r}, \vec{p}) = \frac{1}{2m}\left(\vec{p} - \frac{e}{c}\vec{A}\right)^2 - \frac{e^2}{r}$$

where we are using the notation $\vec{r} = (x, y, z)$, $\vec{p} = (p_x p_y, p_z)$, $r = |\vec{r}|$; the vector potential $\vec{A}$ is a solution of the equation $\vec{B} = \nabla_{\vec{r}} \times \vec{A}$ (it is of course not uniquely determined by this relation). Assuming that r is very large (which is the case for instance when the atom is in a highly excited state) we neglect the term e^2/r and the Hamiltonian function becomes

$$H(\vec{r}, \vec{p}) = \frac{1}{2m}\left(\vec{p} - \frac{e}{c}\vec{A}\right)^2 .$$

Supposing that the magnetic field $\vec{B}$ is constant and uniform, and that its direction is the z-axis, the coordinates of the vector potential satisfy the equations

$$\frac{\partial A_y}{\partial z} - \frac{\partial A_z}{\partial y} = \frac{\partial A_x}{\partial z} - \frac{\partial A_z}{\partial x} = 0 \ , \ \frac{\partial A_y}{\partial x} - \frac{\partial A_x}{\partial y} = B_z.$$

Choosing the solution $A_x = -\frac{1}{2}B_z$, $A_y = \frac{1}{2}B_z y$, $A_z = 0$ (it is called the "symmetric gauge") the Hamiltonian takes the explicit form

$$H(\vec{r}, \vec{p}) = \frac{p^2}{2m} + \frac{e^2 B_z^2}{8mc^2}(x^2 + y^2) - \frac{eB_z}{2mc}(xp_y - yp_x) \tag{19.2}$$

with $p = |\vec{p}|$. Since the problem is essentially planar, we can actually assume $p_z = 0$; setting $\omega_L = eB_z/2mc$ ("Larmor frequency") this can be rewritten in the form

$$H_{\mathrm{mag}}(\vec{r}, \vec{p}) = \frac{1}{2m}(p_x^2 + p_y^2) + \frac{m\omega_L^2}{2}(x^2 + y^2) - \omega_L(xp_y - yp_x). \tag{19.3}$$

The corresponding Weyl operator is then given by

$$\widehat{H}_{\mathrm{mag}} = -\frac{\hbar^2}{2m}\Delta_{x,y} - i\hbar\omega_L\left(y\frac{\partial}{\partial x} - x\frac{\partial}{\partial y}\right) + \frac{m\omega_L^2}{2}(x^2 + y^2) \tag{19.4}$$

where $\Delta_{x,y}$ is the Laplace operator in the variables x and y. Notice that the associated Schrödinger equation can be solved exactly (in principle) using the theory of the metaplectic group since the operator $\widehat{H}$ is the "quantization" of the quadratic Hamiltonian (19.3).

A closer look at formula (19.4) reveals that the magnetic operator $\widehat{H}_{\mathrm{mag}}$ is obtained from the simple Hamiltonian function

$$H = \frac{\omega}{2}(p^2 + x^2) \tag{19.5}$$

on $\mathbb{R}^2$ by setting $\omega_L = \sqrt{\omega/m}$ and using the "quantization rules" $x \longrightarrow \widetilde{X}$ and $p \longrightarrow \widetilde{Y}$ where

$$\widetilde{X} = x + \frac{i\hbar}{\sqrt{m\omega}}\frac{\partial}{\partial y} \ , \ \widetilde{Y} = y - \frac{i\hbar}{\sqrt{m\omega}}\frac{\partial}{\partial x}. \tag{19.6}$$

Of course, these rules are just Bopp quantization when $\hbar$ is replaced with $\hbar/\sqrt{m\omega}$ (and p with y). Thus, the theory developed in Chapter 18 applies mutatis mutandis; for instance the intertwining operator W_ϕ should be replaced with a new operator W_ϕ^{mag} defined as follows: first write the cross-Wigner transform by changing $\hbar$ into $\hbar/\sqrt{m\omega}$; this leads to the formula

$$W^{\mathrm{mag}}(\psi, \phi)(z) = \left(\frac{\sqrt{m\omega}}{2\pi\hbar}\right)^n \int_{\mathbb{R}^n} e^{-\frac{i\sqrt{m\omega}}{\hbar}p\cdot y}\psi(x + \tfrac{1}{2}y)\overline{\phi(x - \tfrac{1}{2}y)}dy. \tag{19.7}$$

Then, one replaces the formula $W_\phi \psi = W(\psi, \phi)$ with

$$W_\phi^{\mathrm{mag}} \psi = \left(\tfrac{2\pi\hbar}{\sqrt{m\omega}} \right)^{n/2} W^{\mathrm{mag}}(\psi, \phi).$$

This leads to the following explicit form for the intertwiner W_ϕ^{mag}:

$$W_\phi^{\mathrm{mag}} \psi(z) = \left(\tfrac{\sqrt{m\omega}}{2\pi\hbar} \right)^{n/2} \int_{\mathbb{R}^n} e^{-\frac{i\sqrt{m\omega}}{\hbar} p \cdot y} \psi(x + \tfrac{1}{2}y) \overline{\phi(x - \tfrac{1}{2}y)} dy. \qquad (19.8)$$

Let us choose units in which $\hbar = 1$. The symbol of the Bopp operator (19.5) is in the Shubin class $H\Gamma_1^{2,2}(\mathbb{R}^2)$, hence Theorem 448 applies. The eigenvalues of $\widehat{H}$ are thus those of the operator

$$\widehat{H} = \frac{\omega}{2}(-\partial_x^2 + x^2). \qquad (19.9)$$

These are well known, they are the numbers $\lambda_j = 2j + 1$ $(j = 0, 1, 2, \dots)$; the corresponding eigenfunctions ϕ_j are conveniently rescaled Hermite functions. Using well-known formulae expressing the cross-Wigner transforms of pairs of Hermite functions in terms of Laguerre polynomial $\mathcal{L}_j^k$ of degree j and order k (see, e.g., [163]) one recovers the usual expressions

$$\Phi_{j+k,k}(z) = (-1)^j \frac{1}{\sqrt{2\pi}} \left(\frac{j!}{(j+k)!} \right)^{\frac{1}{2}} 2^{-\frac{k}{2}} z^k \mathcal{L}_j^k(\tfrac{1}{2}|z|^2) e^{-\frac{|z|^2}{4}}$$

and $\Phi_{j,j+k} = \overline{\Phi_{j+k,k}}$ when $k = 0, 1, 2, \dots$ for the eigenfunctions of $\widetilde{H}$ found in the physics literature (see, e.g., Landau and Lifschitz [113]).

Another very interesting property of the Landau Hamiltonian is the following:

Proposition 449. *The partial differential operator*

$$\widehat{H}_{\mathrm{mag}} = -\frac{\hbar^2}{2m}\Delta_{x,y} - i\hbar\omega_L \left(y\frac{\partial}{\partial x} - x\frac{\partial}{\partial y} \right) + \frac{m\omega_L^2}{2}(x^2 + y^2)$$

is globally hypoelliptic: if $\Psi \in \mathcal{S}'(\mathbb{R}^2)$ is such that $\widehat{H}_{\mathrm{mag}}\Psi \in \mathcal{S}(\mathbb{R}^2)$ then $\Psi \in \mathcal{S}(\mathbb{R}^2)$.

Proof. This immediately follows from Corollary 441 since the harmonic oscillator symbol (19.9) is in $H\Gamma_1^{2,2}(\mathbb{R}^2)$ and the corresponding operator has no eigenvalue equal to zero. $\qquad \square$

19.2   Bopp calculus and deformation quantization

We show here that "deformation quantization", which is an alternative way of doing quantum mechanics, is essentially the same thing as Bopp calculus. For a very nice introduction to the topic (readable, following the authors, by an undergraduate student...) see Hancock et al. [90]. Deformation quantization originates from the pioneering work of Weyl, Wigner, Groenewold, Moyal; it is a very active branch not only of physics, but also of mathematics: for instance Kontsevich's work [109] on the deformation quantization of Poisson algebras was part of the reason he was awarded the Fields medal in 1998. More about the historical origins below.

19.2.1 Deformation quantization: motivation

The basic philosophy of deformation quantization is that quantization can be viewed as a deformation of the structure of the algebra of classical observables, rather than a radical change in the nature of the observables. By "quantization" we mean the procedure familiar from quantum mechanics which consists in associating to an "observable" a self-adjoint operator; from a mathematical viewpoint, this is just the Weyl correspondence (or one of its variants) which allows one to associate to a real symbol an essentially self-adjoint operator.

The genesis of the modern theory of deformation quantization roughly goes as follows (we are taking this historical account from Bordemann [19, 20]. In 1974 Flato, Lichnerowicz, and Sternheimer studied in [56, 57] deformations of the Lie algebra structure defined by the Poisson brackets on the algebra of smooth functions on a symplectic manifold. In 1975, Vey [157] pursued their work in a differential context and constructed a deformation on $\mathbb{R}^n \oplus \mathbb{R}^n$ which turns out to be precisely the Moyal bracket. This opened the path to deformation quantization presented in 1976 by Flato, Lichnerowicz, and Sternheimer in [58]. Two years later, in two brilliant papers [9, 10] Bayen, Flato, Fronsdal, Lichnerowicz, and Sternheimer not only posed the mathematical foundations of deformation theory, but also proposed physical applications. They moreover showed that the Moyal star product could be defined on any symplectic manifold which admits a symplectic connection. Their study of star products on manifolds used Gerstenhaber's [61] theory of deformations of associative algebras, where Hochschild cohomology plays a central role. The work of Bayen et al. drew the attention of both physical and mathematical communities to a well-posed mathematical problem of describing and classifying up to some natural equivalence the formal associative differential deformations of the algebra of smooth functions on a manifold; it can therefore be viewed as the birth certificate of deformation quantization.

The notion of Poisson bracket, briefly reviewed in the introductory chapter where the basics of Hamiltonian mechanics were exposed, can be generalized in the following way. Let us call *Poisson algebra* a real vector space A equipped with a commutative and associative algebra structure $(f, g) \longmapsto fg$ and a Lie algebra structure $(f, g) \longmapsto \{g, h\}$ satisfying the Leibniz law

$$\{fg, h\} = f\{g, h\} + \{f, h\}g.$$

Note that since $\{\cdot, \cdot\}$ is a Lie algebra structure, it satisfies in particular the Jacobi identity

$$\{f, \{g, h\}\} + \{h, \{f, g\}\} + \{g, \{h, f\}\} = 0.$$

Keeping this definition in mind, a Poisson manifold is a manifold M such that the space $A = C^\infty(M)$ is a Poisson algebra for the usual (pointwise) multiplication of functions and some prescribed Lie algebra structure. For instance, if $M = \mathbb{R}^n \oplus \mathbb{R}^n$ one can choose the usual Poisson bracket.

Let us denote by $C^\infty(M)[[\hbar]]$ the ring of all formal series in $\hbar$ with coefficients in $C^\infty(M)$; here $\hbar$ is just viewed as a symbol (the "deformation parameter"). An element $f \in C^\infty(M)[[\hbar]]$ can thus be symbolically written

$$f = \sum_{j=0}^{\infty} \left(\tfrac{i\hbar}{2}\right)^j f_k$$

where the f_k's are in $C^\infty(M)$; we prefer to choose $\tfrac{1}{2}i\hbar$ as deformation parameter rather than $\hbar$.

A starproduct on (or formal deformation of) $C^\infty(M)$ is a map

$$\star : C^\infty(M) \times C^\infty(M) \longrightarrow C^\infty(M)[[\hbar]]$$

associating to each pair satisfying (f,g) of function in $C^\infty(M)$ a formal series

$$f \star g = \sum_{j=0}^{\infty} \left(\tfrac{i\hbar}{2}\right)^j C_j(f,g)$$

the following rules:

- Formal associativity: $(f \star g) \star h = f \star (g \star h)$;
- $C_0(f,g) = fg$ and the C_j's are bidifferential operators;
- $C_1(f,g) = 2\{f,g\}$ where $\{\cdot,\cdot\}$ is the Poisson bracket.

We will not pursue the abstract study of deformation quantization here and refer to the aforementioned paper by Kontsevich [109].

19.2.2   The Moyal product and bracket

Recall (formula (10.21)) that by definition $c = a \star_\hbar b$ is the Weyl symbol of the product $\widehat{C} = \widehat{A}\widehat{B}$ with $\widehat{A} \overset{\text{Weyl}}{\longleftrightarrow} a$ and $\widehat{B} \overset{\text{Weyl}}{\longleftrightarrow} b$:

$$a \star_\hbar b \overset{\text{Weyl}}{\longleftrightarrow} \widehat{A}\widehat{B}$$

(assuming that $\widehat{A}\widehat{B}$ is defined as an operator $\mathcal{S}(\mathbb{R}^n) \longrightarrow \mathcal{S}'(\mathbb{R}^n)$). We thus have the explicit formula

$$a \star_\hbar b(z) = \left(\tfrac{1}{4\pi\hbar}\right)^{2n} \iint_{\mathbb{R}^{4n}} e^{\frac{i}{2\hbar}\sigma(z',z'')} a(z + \tfrac{1}{2}z')b(z - \tfrac{1}{2}z'')dz'dz''. \qquad (19.10)$$

This formula can be rewritten in several equivalent ways; for instance the change of variables $u = z + \frac{1}{2}z'$, $v = z - \frac{1}{2}z''$ leads to the formula

$$a \star_\hbar b(z) = \left(\tfrac{1}{\pi\hbar}\right)^{2n} \iint_{\mathbb{R}^{4n}} e^{\frac{2i}{\hbar}\partial\sigma(u,z,v)} a(u)b(v)\,dudv \qquad (19.11)$$

where $\partial\sigma$ is the antisymmetric cocycle defined by

$$\partial\sigma(u,z,v) = \sigma(u,z) - \sigma(u,v) + \sigma(z,v). \qquad (19.12)$$

Exercise 450. Show that for suitable symbols a and b the following identity holds:

$$\int_{\mathbb{R}^{2n}} a \star_\hbar b(z)\,dz = \int_{\mathbb{R}^{2n}} a(z)b(z)\,dz = \int_{\mathbb{R}^{2n}} b \star_\hbar a(z)\,dz.$$

[Hint: use Proposition 284 in Chapter 12.]

The Moyal product is associative: when both sides are defined we have

$$(a \star_\hbar b) \star_\hbar c = a \star_\hbar (b \star_\hbar c) \qquad (19.13)$$

because composition of operators is associative. It is obviously also bilinear:

$$a \star_\hbar (b + c) = a \star_\hbar b + a \star_\hbar c,$$
$$(b + c) \star_\hbar a = b \star_\hbar a + c \star_\hbar a.$$

Note that in general $a \star_\hbar b \neq b \star_\hbar a$.

Exercise 451. Show that $\overline{a \star_\hbar b} = \overline{b} \star_\hbar \overline{a}$.

The important point is that the Moyal product can be defined in terms of Bopp pseudo-differential operators:

Proposition 452. *Let $\widehat{A} \overset{\text{Weyl}}{\longleftrightarrow} a$ and $\widehat{B} \overset{\text{Weyl}}{\longleftrightarrow} b$. We assume that $b \in \mathcal{S}(\mathbb{R}^n \oplus \mathbb{R}^n)$ (or, more generally, that $\widehat{C} = \widehat{A}\widehat{B}$ exists). We have*

$$a \star_\hbar b = \widetilde{A}b. \qquad (19.14)$$

Proof. We have, by definition of the symplectic Fourier transform and the operators $\widetilde{T}(z_0)$,

$$\widetilde{A}b(z) = \left(\tfrac{1}{2\pi\hbar}\right)^{2n} \int_{\mathbb{R}^{2n}} e^{-\frac{i}{\hbar}\sigma(z,z_0)} \left[\int_{\mathbb{R}^{2n}} e^{-\frac{i}{\hbar}\sigma(z_0,z')} a(z')\,dz' \right]$$
$$\times\, b(z - \tfrac{1}{2}z_0)\,dz_0 dz'.$$

Setting $v = z_0$ and $u = 2(z - z')$ and noting that by the antisymmetry of σ

$$\sigma(z,v) + \sigma(v, z + \tfrac{1}{2}u) = -\tfrac{1}{2}\sigma(u,v)$$

the right-hand side is precisely $a \star_\hbar b$. $\qquad\square$

The result above actually reduces the study of deformation quantization to the study of an algebra of pseudo-differential operators (Bopp calculus); this approach was initiated in de Gosson and Luef [76] using the methods in de Gosson [68, 69, 71].

19.3   Non-commutative quantum mechanics

Admittedly, "non-commutative quantum mechanics" (NCQM) is a misnomer: one of the most salient properties of traditional standard quantum mechanics is precisely that it deals with operators which do not commute! We will however stick to this somewhat unfortunate terminology, which is standard among physicists.

The possibility of non-commuting position operators was already put forward by Heisenberg in 1930, and taken up by many physicists (Peierls, Pauli, Snyder). Even though Werner Heisenberg had already hinted, in the 1930s, at the possibility of non-commuting position operators, one can say that the act of birth of NCQM goes back to 1947 when Snyder [152] considered non-commuting coordinates on space-time in order to discard the ultraviolet divergences in quantum field theory. Snyder showed that his approach was compatible with Lorentz invariance, that is with the theory of special relativity. One of the main incentives for studying NCQM comes from the quest for a theory of quantum gravity, and it is widely expected that such a theory will determine a modification of the structure of space-time of some non-commutative nature [30, 36, 145, 155]; also see the excellent paper by Binz et al. [14].

We are going to show that NCQM is just a variant of Bopp quantization, which allows us to reduce its study to Bopp and Weyl calculus.

19.3.1   Background

The Weyl operators corresponding to the symbols x_j and p_j are the operators $\widehat{X}_j = $ *multiplication by* x_j and $\widehat{P}_j = -i\hbar \partial / \partial x_j$. These operators trivially satisfy the "canonical commutation relations"

$$[\widehat{X}_j, \widehat{X}_k] = [\widehat{P}_j, \widehat{P}_k] = 0 \ , \ [\widehat{X}_j, \widehat{P}_k] = i\hbar \delta_{jk} \tag{19.15}$$

on their common domain. Setting $\widehat{Z}_\alpha = \widehat{X}_\alpha$ if $1 \leq \alpha \leq n$ and $\widehat{Z}_\alpha = \widehat{P}_{\alpha-n}$ if $n+1 \leq \alpha \leq 2n$ these relations can be written more compactly as

$$[\widehat{Z}_\alpha, \widehat{Z}_\beta] = i\hbar j_{\alpha\beta} \ , \ 1 \leq \alpha, \beta \leq 2n \tag{19.16}$$

where the $j_{\alpha\beta}$ are the entries of the standard symplectic matrix $J = \begin{pmatrix} 0 & I \\ -I & 0 \end{pmatrix}$. The passage to non-commutative quantum mechanics consists in replacing the operators $\widehat{Z}_\alpha$ by new operators, which we denote by $\widetilde{Z}_\alpha$ so that instead of (19.16) we have the new commutation relations

$$[\widetilde{Z}_\alpha, \widetilde{Z}_\beta] = i\hbar \omega_{\alpha\beta} \ , \ 1 \leq \alpha, \beta \leq 2n \tag{19.17}$$

where $\Omega = (\omega_{\alpha\beta})_{1 \leq \alpha, \beta \leq 2n}$ is the $2n \times 2n$ antisymmetric matrix defined by

$$\Omega = \begin{pmatrix} \hbar^{-1}\Theta & I \\ -I & \hbar^{-1}N \end{pmatrix} \tag{19.18}$$

where $\Theta = (\theta_{\alpha\beta})_{1 \leq \alpha,\beta \leq n}$ and $N = (\eta_{\alpha\beta})_{1 \leq \alpha,\beta \leq n}$ are antisymmetric matrices. Since

$$\Omega = J + \hbar^{-1} \begin{pmatrix} \Theta & 0 \\ 0 & N \end{pmatrix}$$

the matrix Ω can be viewed as a perturbation of the standard symplectic matrix J, and one can thus expect that it is invertible if Θ and n are small enough. In fact, Bastos, Dias, and Prata (see [8]) prove the following:

Proposition 453. *We have* $\det \Omega > 0$ *if the following conditions hold:*

$$\theta_{\alpha\beta}\eta_{\gamma\delta} < \hbar^2 \quad \text{for} \ \ 1 \leq \alpha < \beta \leq n \ \text{and} \ 1 \leq \gamma < \delta \leq n \ . \tag{19.19}$$

We will from now on assume that these conditions are satisfied.

Writing $\widetilde{Z}_\alpha = \widetilde{X}_\alpha$ if $1 \leq \alpha \leq n$ and $\widetilde{Z}_\alpha = \widetilde{P}_{\alpha-n}$ if $n+1 \leq \alpha \leq 2n$, this amounts to replacing the CCR (19.15) with

$$[\widetilde{X}_\alpha, \widetilde{X}_\beta] = \theta_{\alpha\beta} \ , \ [\widetilde{P}_\alpha, \widetilde{P}_\beta] = \eta_{\alpha\beta} \ \ , \ [\widetilde{X}_\alpha, \widetilde{P}_\beta] = i\hbar\delta_{\alpha\beta}. \tag{19.20}$$

Let us now make the following explicit choices for the operators $\widetilde{X}_\alpha$ and $\widetilde{P}_\alpha$: we set

$$\widetilde{X}_\alpha = x_\alpha + \tfrac{1}{2}i\hbar\partial_{p_\alpha} + \tfrac{1}{2}i \sum\nolimits_\beta \theta_{\alpha\beta}\partial_{x_\beta}, \tag{19.21}$$

$$\widetilde{P}_\alpha = p_\alpha - \tfrac{1}{2}i\hbar\partial_{x_\alpha} + \tfrac{1}{2}i \sum\nolimits_\beta \eta_{\alpha\beta}\partial_{p_\beta}. \tag{19.22}$$

These relations suggest that we represent $\widetilde{Z} = (\widetilde{Z}_1, \ldots, \widetilde{Z}_{2n})$ by the vector operator

$$\widetilde{Z} = z + \tfrac{1}{2}i\hbar\Omega\partial_z \tag{19.23}$$

which acts on functions defined on the phase space $\mathbb{R}^n \oplus \mathbb{R}^n$. In analogy with the theory of Bopp operators this formula suggests that we consider generalized Bopp operators of the type

$$\widetilde{A}_\omega = a(z + \tfrac{1}{2}i\hbar\Omega\partial_z)$$

where ω is the symplectic form defined by

$$\omega(z, z') = z \cdot \Omega^{-1}z' = -\Omega^{-1}z \cdot z'. \tag{19.24}$$

Note that the invertibility of the antisymmetric matrix Ω implies that we must have $\det \Omega > 0$ (this follows readily from the properties of the Pfaffian of an antisymmetric matrix). Since $J^T = -J$, the symplectic form ω reduces to the standard symplectic form σ when $\Omega = J$. Let us denote by Σ a linear automorphism of $\mathbb{R}^n \oplus \mathbb{R}^n$ such that $\sigma = \Sigma^*\omega$; equivalently $\Sigma J \Sigma^T = \Omega$. Thus Σ is a symplectomorphism $(\mathbb{R}^n \oplus \mathbb{R}^n, \sigma) \longrightarrow (\mathbb{R}^n \oplus \mathbb{R}^n, \omega)$. Clearly $(\det \Sigma)^2 = \det \Omega > 0$ so we are free to choose Σ such that $\det \Sigma > 0$. The mapping Σ is sometimes called the

"Seiberg–Witten map" in the physics literature; its existence is of course mathematically a triviality, because all symplectic structures with constant coefficients are isomorphic (see [67], §1.1.2). Writing Σ in block-matrix form $\begin{pmatrix} A & B \\ C & D \end{pmatrix}$, the condition $\Sigma J \Sigma^T = \Omega$ is equivalent to the relations

$$AB^T - BA^T = \hbar^{-1}\Theta \; , \; CD^T - DC^T = \hbar^{-1}N \; , \; AD^T - BC^T = I.$$

Of course, the automorphism Σ is not uniquely defined: if $\Sigma^*\omega = \Sigma'^*\omega$ then $\Sigma^{-1}\Sigma' \in \mathrm{Sp}(2n,\mathbb{R})$.

19.3.2 The operators $\widetilde{T}_\omega(z_0)$ and F_ω

Let ω be an arbitrary symplectic form with constant coefficients on $\mathbb{R}^n \oplus \mathbb{R}^n$ (not necessarily the symplectic form (19.24)). It is thus represented by some invertible antisymmetric matrix Ω with $\det\Omega > 0$:

$$\omega(z,z') = z \cdot \Omega^{-1}z' = (\Omega^T)^{-1}z \cdot z'.$$

To ω we associate the operator $\widetilde{T}_\omega(z_0) : \mathcal{S}(\mathbb{R}^n \oplus \mathbb{R}^n) \longrightarrow \mathcal{S}(\mathbb{R}^n \oplus \mathbb{R}^n)$ defined, for $z_0 \in \mathbb{R}^{2n}$, by

$$\widetilde{T}_\omega(z_0)\Psi(z) = e^{-\frac{i}{\hbar}\omega(z,z_0)}\Psi(z - \tfrac{1}{2}z_0). \tag{19.25}$$

Of course $\widetilde{T}_\omega(z_0)$ extends into an operator $\mathcal{S}'(\mathbb{R}^n \oplus \mathbb{R}^n) \longrightarrow \mathcal{S}'(\mathbb{R}^n \oplus \mathbb{R}^n)$ whose restriction to $L^2(\mathbb{R}^n \oplus \mathbb{R}^n)$ is unitary: we have

$$\|\widetilde{T}_\omega(z_0)\Psi\|_{L^2(\mathbb{R}^{2n})} = \|\Psi\|_{L^2(\mathbb{R}^{2n})}$$

for all $\Psi \in L^2(\mathbb{R}^n \oplus \mathbb{R}^n)$. The operators $\widetilde{T}_\omega(z_0)$ are of course strongly reminiscent of the Heisenberg–Weyl operators; the major difference is that they act on functions or distributions) defined on phase space $\mathbb{R}^n \oplus \mathbb{R}^n$, and not on "configuration space" $\mathbb{R}^n$. It turns out that the operators $\widetilde{T}_\omega(z_0)$ satisfy relations similar to those satisfied by the operators $\widetilde{T}(z_0)$ defined in the previous chapter:

Proposition 454. *We have, for all $z_0, z_1 \in \mathbb{R}^{2n}$,*

$$\widetilde{T}_\omega(z_0 + z_1) = e^{-\frac{i}{2\hbar}\omega(z_0,z_1)}\widetilde{T}_\omega(z_0)\widetilde{T}_\omega(z_1), \tag{19.26}$$

$$\widetilde{T}_\omega(z_0)\widetilde{T}_\omega(z_1) = e^{\frac{i}{\hbar}\omega(z_0,z_1)}\widetilde{T}_\omega(z_1)\widetilde{T}_\omega(z_0). \tag{19.27}$$

Proof. It is straightforward, replacing σ by ω in Proposition 420. $\square$

Definition 455. The ω-symplectic transform F_ω is defined by the formula

$$F_\omega a(z) = \left(\tfrac{1}{2\pi\hbar}\right)^n (\det\Omega)^{-1/2} \int_{\mathbb{R}^{2n}} e^{-\frac{i}{\hbar}\omega(z,z')}a(z')dz' \tag{19.28}$$

when $a \in \mathcal{S}(\mathbb{R}^n \oplus \mathbb{R}^n)$.

Clearly F_ω extends in the same way as the ordinary Fourier transform into an involutive automorphism of $\mathcal{S}'(\mathbb{R}^n \oplus \mathbb{R}^n)$ (also denoted by F_ω) and whose restriction to $L^2(\mathbb{R}^n \oplus \mathbb{R}^n)$ is unitary, and if $\omega = \sigma$ we recover the symplectic Fourier transform previously studied.

Let us express the operator $\widetilde{A}_\omega = a(z + \frac{1}{2}i\hbar\Omega\partial_z)$ in terms of $F_\omega a$ and $\widetilde{T}_\omega(z_0)$.

Proposition 456. *Let $\widetilde{A}_\omega$ be the operator on $\mathbb{R}^n \oplus \mathbb{R}^n$ with Weyl symbol*

$$\widetilde{a}_\Omega(z, \zeta) = a(z - \tfrac{1}{2}\Omega\zeta). \tag{19.29}$$

We have

$$\widetilde{A}_\omega = \left(\tfrac{1}{2\pi\hbar}\right)^n (\det \Omega)^{-1/2} \int_{\mathbb{R}^{2n}} F_\omega a(z)\widetilde{T}_\omega(z)dz. \tag{19.30}$$

Proof. Let us denote by $\widetilde{B}$ the right-hand side of (19.30). We have, setting $u = z - \frac{1}{2}z_0$,

$$\widetilde{B}\Psi(z) = \left(\tfrac{1}{2\pi\hbar}\right)^n (\det \Omega)^{-1/2} \int_{\mathbb{R}^{2n}} F_\omega a(z_0)e^{-\frac{i}{\hbar}\omega(z,z_0)}\Psi(z - \tfrac{1}{2}z_0)dz_0$$

$$= \left(\tfrac{2}{\pi\hbar}\right)^n (\det \Omega)^{-1/2} \int_{\mathbb{R}^{2n}} F_\omega a[2(z - u)]e^{\frac{2i}{\hbar}\omega(z,u)}\Psi(u)du,$$

hence the kernel of $\widetilde{B}$ is given by

$$K(z, u) = \left(\tfrac{2}{\pi\hbar}\right)^n (\det \Omega)^{-1/2} F_\omega a[2(z - u)]e^{\frac{2i}{\hbar}\omega(z,u)}.$$

It follows that the Weyl symbol $\widetilde{b}$ of $\widetilde{B}$ is given by

$$\widetilde{b}(z, \zeta) = \int_{\mathbb{R}^{2n}} e^{-\frac{i}{\hbar}\zeta\cdot\zeta'}K(z + \tfrac{1}{2}\zeta', z - \tfrac{1}{2}\zeta')d\zeta'$$

$$= \left(\tfrac{2}{\pi\hbar}\right)^n (\det \Omega)^{-1/2} \int_{\mathbb{R}^{2n}} e^{-\frac{i}{\hbar}\zeta\cdot\zeta'}F_\omega a(2\zeta')e^{-\frac{2i}{\hbar}\omega(z,\zeta')}d\zeta',$$

that is, using the obvious relation

$$\zeta \cdot \zeta' + 2\omega(z, \zeta') = \omega(2z - \Omega\zeta, \zeta')$$

together with the change of variables $z' = 2\zeta'$,

$$\widetilde{b}(z, \zeta) = \left(\tfrac{2}{\pi\hbar}\right)^n (\det \Omega)^{-1/2} \int_{\mathbb{R}^{2n}} e^{-\frac{i}{\hbar}\omega(2z-\Omega\zeta,\zeta')}F_\omega a(2\zeta')d\zeta'$$

$$= \left(\tfrac{1}{2\pi\hbar}\right)^n (\det \Omega)^{-1/2} \int_{\mathbb{R}^{2n}} e^{-\frac{i}{\hbar}\omega(z-\frac{1}{2}\Omega\zeta,z')}F_\omega a(z')dz',$$

that is, using the fact that $F_\omega F_\omega$ is the identity,

$$\widetilde{b}(z, \zeta) = a(z - \tfrac{1}{2}\Omega\zeta) = \widetilde{a}_\Omega(z, \zeta)$$

which concludes the proof. $\qquad\square$

Definition 457. The operator $\widetilde{A}_\omega = a(z + \frac{1}{2}i\hbar\Omega\partial_z)$ will be called the ω-Bopp pseudo-differential operator with symbol a.

We are going to prove a very useful result which reduces the study of the Bopp operators to the case where ω is the standard symplectic form on $\mathbb{R}^n \oplus \mathbb{R}^n$. This result is closely related to the symplectic covariance of Weyl operators under metaplectic conjugation as we will see below.

For Σ a linear automorphism of $\mathbb{R}^n \oplus \mathbb{R}^n$ we define the operator $M_\Sigma : \mathcal{S}'(\mathbb{R}^n \oplus \mathbb{R}^n) \longrightarrow \mathcal{S}'(\mathbb{R}^n \oplus \mathbb{R}^n)$ by

$$M_\Sigma\Psi(z) = \sqrt{|\det\Sigma|}\,\Psi(\Sigma z). \tag{19.31}$$

Clearly M_Σ is a unitary operator when restricted to $L^2(\mathbb{R}^n \oplus \mathbb{R}^n)$: we have

$$\|M_\Sigma\Psi\|_{L^2(\mathbb{R}^{2n})} = \|\Psi\|_{L^2(\mathbb{R}^{2n})}$$

for all $\Psi \in L^2(\mathbb{R}^{2n})$.

The following result shows how the operator $\widetilde{A}_\omega$ is related to the corresponding Bopp operator $\widetilde{A}$ by an intertwining metaplectic operator:

Proposition 458. *Let Σ be a linear automorphism such that $\sigma = \Sigma^*\omega$ and define the automorphism $M_\Sigma : \mathcal{S}'(\mathbb{R}^n \oplus \mathbb{R}^n) \longrightarrow \mathcal{S}'(\mathbb{R}^n \oplus \mathbb{R}^n)$ by*

$$M_\Sigma\Psi(z) = \sqrt{|\det\Sigma|}\,\Psi(\Sigma z). \tag{19.32}$$

We have $M_\Sigma \in Mp(4n, \sigma \oplus \sigma)$ (hence M_Σ is unitary on $L^2(\mathbb{R}^n \oplus \mathbb{R}^n)$). We have

$$M_\Sigma\widetilde{T}_\omega(z_0) = \widetilde{T}(\Sigma^{-1}z_0)M_\Sigma \ ,$$
$$M_\Sigma F_\omega = F_\sigma M_\Sigma \tag{19.33}$$

and hence

$$M_\Sigma\widetilde{A}_\omega = \widetilde{B}M_\Sigma \tag{19.34}$$

where $\widetilde{B} \overset{\text{Bopp}}{\longleftrightarrow} b$ with $b = \Sigma^*a$.

Proof. That $M_\Sigma \in \mathrm{Mp}(4n, \sigma \oplus \sigma)$ is clear (see [67], Chapter 7): it is one of the two metaplectic operators belonging to the fiber of $\begin{pmatrix} \Sigma^{-1} & 0 \\ 0 & \Sigma \end{pmatrix} \in \mathrm{Sp}(4n, \sigma \oplus \sigma)$. We have

$$
\begin{aligned}
M_\Sigma\widetilde{T}_\omega(z_0)\Psi(z) &= \sqrt{|\det\Sigma|}\,e^{-\frac{i}{\hbar}\omega(\Sigma z, z_0)}\Psi(\Sigma z - \tfrac{1}{2}z_0)) \\
&= \sqrt{|\det\Sigma|}\,e^{-\frac{i}{\hbar}\omega(\Sigma z, \Sigma(\Sigma^{-1}z_0))}\Psi(\Sigma(z - \tfrac{1}{2}\Sigma^{-1}z_0)) \\
&= \sqrt{|\det\Sigma|}\,e^{-\frac{i}{\hbar}\sigma(z, \Sigma^{-1}z_0))}\Psi(\Sigma(z - \tfrac{1}{2}\Sigma^{-1}z_0)) \\
&= \widetilde{T}(\Sigma^{-1}z_0)M_\Sigma\Psi(z).
\end{aligned}
$$

The second formula (19.33) follows by a similar argument. Let us prove formula (19.34); using the identities (19.33) we have

$$\widetilde{A}_\omega \Psi(z) = \left(\tfrac{1}{2\pi\hbar}\right)^n (\det\Omega)^{-1/2} \int_{\mathbb{R}^{2n}} F_\omega a(z_0)\widetilde{T}_\omega(z_0)\Psi(z)dz_0$$

and hence

$$M_\Sigma \widetilde{A}_\omega \Psi = \left(\tfrac{1}{2\pi\hbar}\right)^n (\det\Omega)^{-1/2} \int_{\mathbb{R}^{2n}} F_\omega a(z_0) M_\Sigma \left[\widetilde{T}_\omega(z_0)\Psi\right] dz_0$$

$$= \left(\tfrac{1}{2\pi\hbar}\right)^n (\det\Omega)^{-1/2} \int_{\mathbb{R}^{2n}} F_\omega a(z_0)\widetilde{T}(\Sigma^{-1}z_0) M_\Sigma \Psi dz_0$$

$$= \left(\tfrac{1}{2\pi\hbar}\right)^n (\det\Omega)^{-1/2} |\det\Sigma| \int_{\mathbb{R}^{2n}} F_\omega a(\Sigma z_0)\widetilde{T}(z_0) M_\Sigma \Psi dz_0$$

$$= \left(\tfrac{1}{2\pi\hbar}\right)^n (\det\Omega)^{-1/2} |\det\Sigma|^{1/2} \int_{\mathbb{R}^{2n}} M_\Sigma F_\omega a(z_0)\widetilde{T}(z_0) M_\Sigma \Psi dz_0$$

$$= \left(\tfrac{1}{2\pi\hbar}\right)^n (\det\Omega)^{-1/2} |\det\Sigma|^{1/2} \int_{\mathbb{R}^{2n}} F_\sigma(M_\Sigma a)(z_0)\widetilde{T}(z_0) M_\Sigma \Psi dz_0$$

$$= \left(\tfrac{1}{2\pi\hbar}\right)^n (\det\Omega)^{-1/2} |\det\Sigma| \int_{\mathbb{R}^{2n}} F_\sigma b(z_0)\widetilde{T}(z_0) M_\Sigma \Psi dz_0$$

$$= \widetilde{B} M_\Sigma \Psi$$

(the last equality because $|\det\Omega|^{-1/2}|\det\Sigma| = 1$). $\qquad\square$

That we have $M_{S_\sigma}\Psi(z) = \Psi(S_\sigma z)$ is clear since $\det S_\sigma = 1$.

We note that formula (19.34) can be interpreted in terms of the symplectic covariance property of Weyl calculus. To see this, let us equip the double phase space $\mathbb{R}^{2n}\oplus\mathbb{R}^{2n}$ with the symplectic structure $\sigma^\oplus = \sigma\oplus\sigma$. In view of formula (19.29) with $\Omega = J$ the Weyl symbols of operators $\widetilde{A''}$ and $\widetilde{A'}$ are, respectively

$$\widetilde{a'}(z,\zeta) = a\left(\Sigma(z-\tfrac{1}{2}J\zeta)\right) \quad, \quad \widetilde{a''}(z,\zeta) = a\left(\Sigma'(z-\tfrac{1}{2}J\zeta)\right)$$

and hence, using the identities $\Sigma^{-1}\Sigma' = S_\sigma \in \mathrm{Sp}(\mathbb{R}^n\oplus\mathbb{R}^n,\sigma)$ and $S_\sigma J = J(S_\sigma^T)^{-1}$,

$$\widetilde{a''}(z,\zeta) = a'\left(S_\sigma z - \tfrac{1}{2}J(S_\sigma^T)^{-1}\zeta)\right) = \widetilde{a'}(S_\sigma z,(S_\sigma^T)^{-1}\zeta).$$

Let now M_{S_σ} be the automorphism of $\mathbb{R}^{2n}\oplus\mathbb{R}^{2n}$ defined by

$$M_{S_\sigma}(z,\zeta) = (S_\sigma^{-1}z, S_\sigma^T\zeta);$$

formula(19.34) can thus be rewritten

$$\widetilde{A''} = M_{S_\sigma}\widetilde{A'}M_{S_\sigma}^{-1} \quad\text{with}\quad a'' = a'\circ M_{S_\sigma}^{-1}. \tag{19.35}$$

Recall now that each automorphism Σ of $\mathbb{R}^n\oplus\mathbb{R}^n$ induces an element M_Σ of the symplectic group $\mathrm{Sp}(4n,\mathbb{R})$, defined by $m_\Sigma(z,\zeta) = (\Sigma^{-1}z, \Sigma^T\zeta)$, and that M_Σ is the projection of the metaplectic operator $\widehat{M_\Sigma} \in \mathrm{Mp}(\mathbb{R}^{2n}\oplus\mathbb{R}^{2n},\sigma^\oplus)$ defined by (19.31). Formulas (19.35) and (19.35) thus reflect the symplectic covariance property of Weyl calculus.

Bibliography

[1] R. Abraham, J.E. Marsden, and T. Ratiu, *Manifolds, Tensor Analysis, and Applications*, Applied Mathematical Sciences, **75**, Springer, 1988.

[2] R. Abraham and J.E. Marsden, *Foundations of Mechanics,* The Benjamin/Cummings Publishing Company, 2nd edition, 1978.

[3] V. I. Arnold, *Mathematical Methods of Classical Mechanics*, Graduate Texts in Mathematics, 2nd edition, Springer-Verlag, 1989.

[4] N.L. Balazs and H.C. Pauli, *Quantum-Mechanical Operators and Phase-Space Operators*, Z. Physik A, **282** (1977), 221–225.

[5] A. Banyaga, *Sur la structure du groupe des difféomorphismes qui préservent une forme symplectique*, Comm. Math. Helv., **53** (1978), 174–227.

[6] V. Bargmann, *On a Hilbert space of analytic functions and an associated integral transform*, Comm. Pure Appl. Math., **14** (1961), 187–214.

[7] C. Bastos, O. Bertolami, N.C. Dias, and J.N. Prata, *Weyl–Wigner Formulation of Noncommutative Quantum Mechanics*, J. Math. Phys., **49** (2008), 072101.

[8] C. Bastos, N.C. Dias, and J.N. Prata, *Wigner measures in non-commutative quantum mechanics*, Preprint, arXiv:0907.4438v1 [math-ph] 25 July 2009.

[9] F. Bayen, M. Flato, C. Fronsdal, A. Lichnerowicz, and D. Sternheimer, *Deformation Theory and Quantization. I. Deformation of Symplectic Structures*, Annals of Physics, **110** (1978), 111–151.

[10] F. Bayen, M. Flato, C. Fronsdal, A. Lichnerowicz, and D. Sternheimer, *Deformation Theory and Quantization. II. Physical Applications*, Annals of Physics, **111** (1978), 6–110.

[11] J. Bellissard, A. van Helst, and H. Schulz-Baldes, *The noncommutative geometry of the quantum Hall effect*, J. Math. Phys. **35**(10) (1994), 5373–5451.

[12] A. Benedek and R. Panzone, *The space L^p, with mixed norm*, Duke math J., **28** (1961), 301–324.

[13] F.A. Berezin, *Covariant and contravariant symbols of operators*, Math. USSR Izvestia, **6** (1972), 1117–1151.

[14] E. Binz, R. Honegger, and A. Rieckers, *Field-theoretic Weyl quantization as a strict and continuous deformation quantization*, Ann. Henri Poincaré, **5** (2004) 327–346.

[15] P. Boggiatto and E. Cordero, *Anti-Wick Quantization with Symbols in L^p Spaces*, Proc. Amer. Math. Soc., **130**(9) (2002), 2679–2685.

[16] P. Boggiatto, E. Buzano, and L. Rodino, *Global Hypoellipticity and Spectral Theory*, Akademie Verlag, Math. Research, **92** (1996).

[17] F. Bonsall, *Decompositions of functions as sums of elementary functions*, Quart. J. Math. Oxford, ser. (2), **37** (1986), 129–136.

[18] F. Bopp, *La mécanique quantique est-elle une mécanique statistique particulière?* Ann. Inst. H. Poincaré, **15** (1956), 81–112.

[19] M. Bordemann, *Deformation quantization: a mini-lecture. Geometric and topological methods for quantum field theory*, Contemp. Math., **434**, Amer. Math. Soc., Providence, RI, (2007), 3–38.

[20] M. Bordemann, *Deformation Quantization: a survey. International Conference on Non-Commutative Geometry and Physics*. Journal of Physics: Conference Series **103** (2008).

[21] A. Boulkhemair, *L^2 Estimates for Weyl Quantization*, J. Funct. Anal. **165** (1999), 173–204.

[22] C. Brislawn, *Kernels of trace class operators*, Proc. Amer. Math. Soc., **104**(4) (1988), 1181–1190.

[23] G. Brodimas, A. Jannusis, and N. Patargias, *The Density Matrix in Phase Space*, Lettere al Nuovo Cimento, **17**(4) (1976), 119–122.

[24] M. Burdet, M. Perrin, and M. Perroud, *Generating Functions for the Affine Symplectic Group*, Commun. Math. Phys., **58** (1978), 241–254.

[25] J. Chazarain, and A. Piriou, *Introduction à la théorie des équations aux dérivées partielles linéaires*, Gauthier–Villars, Paris, (1981). English translation: *Introduction to the theory of linear partial differential equations*, Studies in Mathematics and its Applications, **14**, North-Holland, 1982.

[26] O. Christensen, *An Introduction to Frames and Riesz Bases*, Birkhäuser (Boston, Basel, Berlin), Series Applied and Numerical Harmonic Analysis (2003).

[27] E. Cordero and F. Nicola, *Strichartz estimates in Wiener amalgam spaces for the Schrödinger equation*, Math. Nachr., **281**(1) (2008), 25–41.

[28] E. Cordero and L. Rodino, *Short-time Fourier analysis of localization operators*, Frames and operator theory in analysis and signal processing, 47–68, Contemp. Math., **451**, Amer. Math. Soc., Providence, RI, 2008.

[29] H.O. Cordes, *On compactness of commutators of multiplications and convolutions, and boundedness of pseudodifferential symbols*, J. Funct. Anal. **18** (1975), 115–131.

[30] F. Delduc, Q. Duret, F. Gieres, and M. Lefrançois, *Magnetic fields in non-commutative quantum mechanics*, J. Phys: Conf., Ser. **103** (2008), 012020.

[31] N.C. Dias and J.N. Prata, *Generalized Weyl–Wigner Map and Vey Quantum Mechanics*, J. Math. Phys., **42** (2001), 5565–5579.

[32] N.C. Dias and J.N. Prata, *Exact master equation for a noncommutative Brownian particle*, Ann. Phys., **324** (2009), 73–96.

[33] N.C. Dias, M. de Gosson, F. Luef, and J.N. Prata, *Deformation Quantization Theory for Non-Commutative Quantum Mechanics*, Preprint 2009, arXiv:0911.1209v1 [math-ph].

[34] N.C, Dias, M. de Gosson, F. Luef, and J.N. Prata, *A pseudo-differential calculus on non-commutative phase space. Intertwining properties and spectral results*, Preprint 2009.

[35] J.A. Dieudonné, *Foundations of Modern Analysis*, Academic Press (1969).

[36] M.R. Douglas and N.A. Nekrasov, *Noncommutative field theory*, Rev. Mod. Phys., **73** (2001), 977–1029.

[37] P.A.M. Dirac, *The Principles of Quantum Mechanics,* Oxford Science Publications, 4th revised edition, (1999).

[38] J. Du, M.W. Wong, *A trace formula for Weyl transforms*, Approx. Theory. Appl. (N. S.) **16**(1) (2000), 41–45.

[39] D.A. Dubin, M.A. Hennings, and T.B. Smith, *Mathematical Aspects of Weyl Quantization and Phase*, World Scientific, (2000).

[40] Yu.V. Egorov, A.I. Komech, and M.A. Shubin, *Elements of the Modern Theory of Partial Differential Equations*, Springer, (1999).

[41] I. Ekeland and H. Hofer, *Symplectic topology and Hamiltonian dynamics*, Mathematische Zeitschrift, Volume **200**(3) (1989), 355–378.

[42] I. Ekeland and H. Hofer, *Symplectic topology and Hamiltonian dynamics*, Mathematische Zeitschrift, **203** (1990), 553–567.

[43] G.G. Emch, Geometric dequantization and the correspondence problem, International Journal of Theoretical Physics, **22**(5) (1983), 397–420.

[44] B.V. Fedosov, *A simple geometrical construction of deformation quantization*, J. Diff. Geom., **40** (1994) 213–238.

[45] H.G. Feichtinger, *Un espace de Banach de distributions tempérées sur les groupes localement compact abéliens*, C. R. Acad. Sci. Paris., Série A–B **290** (17) (1980) A791–A794.

[46] H.G. Feichtinger, *On a new Segal algebra*, Monatsh. Math., **92** (4), (1981), 269–289.

[47] H.G. Feichtinger, *Banach spaces of distributions of Wiener's type and interpolation, in Functional Analysis and Approximation*, Oberwohlfach, (1980), Internat. Ser. Numer. Math.; **60** (1981) Birkhäuser, Basel 153–165.

[48] H.G. Feichtinger, *Banach spaces of distributions of Wiener type and interpolation*, In P. Butzer, S. Nagy, and E. Görlich, editors, *Proc. Conf. Functional Analysis and Approximation, Oberwolfach August 1980*, number **69** in: Internat. Ser. Numer. Math., pages 153–165. Birkhäuser Boston, Basel, (1981).

[49] H.G. Feichtinger, *Modulation spaces on locally compact Abelian groups*, Technical report, January (1983).

[50] H.G. Feichtinger, *Modulation Spaces: Looking Back and Ahead Sample*, Theory Signal Image Process, **5**(2), (2006), 109–140.

[51] H.G. Feichtinger and K. Gröchenig, *Banach spaces related to integrable group representations and their atomic decompositions*, I. J. Funct. Anal., **86**, (1989); 307–340.

[52] H.G. Feichtinger and K. Gröchenig, *Gabor Frames and Time-Frequency Analysis of Distributions*, J. Functional Analysis, **146**(2), (1997), 464–495.

[53] H.G. Feichtinger and N. Kaiblinger, *Varying the Time-Frequency Lattice of Gabor Frames*, Trans. Amer. Math. Soc. **356**(5), (2003), 2001–2023.

[54] H.G. Feichtinger and F. Luef, *Wiener Amalgam Spaces for the Fundamental Identity of Gabor Analysis*, Collect. Math., **57**, (2006), 233–253.

[55] H.G. Feichtinger, F. Luef, and E. Cordero, *Banach Gelfand Triples for Gabor Analysis. in: Pseudo-Differential Operators*, Quantization and Signals. Lecture Notes in Mathematics, Springer, 1949 (2008).

[56] M. Flato, A. Lichnerowicz, and D. Sternheimer, *Déformations 1-différentiables d'algèbres de Lie attachées à une variété symplectique ou de contact*, C. R. Acad. Sci., Paris, Série A 279, (1974), 877–881.

[57] M. Flato, A. Lichnerowicz, and D. Sternheimer, *Déformations 1-différentiables d'algèbres de Lie attachées à une variété symplectique ou de contact*, Compositio Math., **31**, (1975), 47–82.

[58] M. Flato, A. Lichnerowicz, and D. Sternheimer, *Crochet de Moyal-Vey et quantification*, C. R. Acad. Sci., Paris, I, Math. 283, (1976), 19–24.

[59] G.B. Folland, *Harmonic Analysis in Phase space*, Annals of Mathematics studies, Princeton University Press, Princeton, N.J. (1989)

[60] G.B. Folland, *Real Analysis. Modern Techniques and Their Applications*, 2nd edition, Pure and applied mathematics, Wiley (1999).

[61] M. Gerstenhaber, *On the deformation of rings and algebras*, Ann. Math., **79** (1964), 59–103.

[62] R.J. Glauber, *The Quantum Theory of Optical Coherence*, Phys. Rev., **130**(6) (1963), 2529–2539.

[63] H. Goldstein, *Classical Mechanics*, Addison–Wesley, (1950), 2nd edition, (1980), 3d edition, (2002).

[64] M. de Gosson, *Maslov Classes, Metaplectic Representation and Lagrangian Quantization*, Research Notes in Mathematics, **95**, Wiley–VCH, Berlin, (1997).

[65] M. de Gosson, *The Principles of Newtonian and Quantum Mechanics*, Imperial College Press, London, (2001).

[66] M. de Gosson. *On the notion of phase in mechanics*, J. Phys. A: Math. Gen., **37**(29), (2004), 7297–7314.

[67] M. de Gosson, *Symplectic Geometry and Quantum Mechanics*, Birkhäuser, Basel, series "Operator Theory: Advances and Applications" (subseries: "Advances in Partial Differential Equations"), Vol. **166** (2006).

[68] M. de Gosson, *Extended Weyl Calculus and Application to the Phase-Space Schrödinger Equation*, J. Phys., A **38** (2005), no. 19, L325–L329.

[69] M. de Gosson, *Symplectically covariant Schrödinger equation in phase space*, J. Phys., A **38** (2005) no. 42, 9263–9287.

[70] M. de Gosson, *Weyl Calculus in Phase Space and the Torres-Vega and Frederick Equation*, In Quantum theory: reconsideration of foundations, **3**, 300–304, AIP Conf. Proc., 810, Amer. Inst. Phys., Melville, NY, (2006).

[71] M. de Gosson, *Spectral Properties of a Class of Generalized Landau Operators*, Comm. Partial Differential Operators, **33**(11) (2008), 2096–2104.

[72] M. de Gosson, *The Symplectic Camel and the Uncertainty Principle: The Tip of an Iceberg?*, Found. Phys., **99** (2009),194–214.

[73] M. de Gosson and B. Hiley, Imprints of the Quantum World in Classical Mechanics, Preprint 2010.

[74] M. de Gosson and F. Luef, *Quantum States and Hardy's Formulation of the Uncertainty Principle: a Symplectic Approach*, Lett. Math. Phys., **80** (2007), 69–82.

[75] M. de Gosson and F. Luef, *Principe d'Incertitude et Positivité des Opérateurs à Trace; Applications aux Opérateurs Densité*, Ann. H. Poincaré, **9**(2) (2008), 329–346.

[76] M. de Gosson and F. Luef, *A new approach to the $\star$-eigenvalue equation*, Lett. Math. Phys., **85** (2008), 173–183 [on line 16 Aug. 2008; DOI 10.1007/s11005-008-0261-8].

[77] M. de Gosson and F. Luef, *Symplectic Capacities and the Geometry of Uncertainty: the Irruption of Symplectic Topology in Classical and Quantum Mechanics*, Physics Reports, **484** (2009), 131–179 DOI 10.1016/j.physrep.2009.08.001.

[78] M. de Gosson and F. Luef, On the usefulness of modulation spaces in deformation quantization, J. Phys. A: Math. Theor. **42**(31) (2009) 315205 (17 pp).

[79] M. de Gosson and F. Luef, *Spectral and Regularity properties of a Pseudo-Differential Calculus Related to Landau Quantization*, Journal of Pseudo-Differential Operators and Applications, **1**(1) (2010), 3–34 DOI 10.1007/s11868-010-0001-6.

[80] M.J. Gotay and G.A. Isenberg, *The Symplectization of Science*, Gazette des Mathématiciens, **54** (1992), 59–79.

[81] J.M. Gracia-Bondia, *Generalized Moyal Quantization on Homogeneous Symplectic spaces*, Contemporary Mathematics, Amer. Math. Soc., Providence, RI, **134** (1992) 93–114.

[82] K. Gröchenig, *Foundations of Time-Frequency Analysis*, Birkhäuser, Boston, (2000).

[83] K. Gröchenig and C. Heil, *Modulation Spaces and Pseudodifferential Operators*, Integr. Equ. Oper. Theory, **34** (1999), 439–457.

[84] K. Gröchenig, *Composition and spectral invariance of pseudodifferential operators on modulation spaces*, Journal d'analyse mathématique, **98** (2006), 65–82.

[85] K. Gröchenig, *Time-Frequency Analysis on Sjöstrand's Class Rev*, Mat. Iberoamericana, **22** (2) (2006), 703–724.

[86] H.J. Groenewold, *On the principles of elementary quantum mechanics*, Physics, **12** (1946), 405–460.

[87] M. Gromov, *Pseudoholomorphic curves in symplectic manifolds*, Invent. Math., **82** (1985), 307–347.

[88] A. Grossmann, *Parity operators and quantization of δ-functions*, Commun. Math. Phys., **48** (1976), 191–193.

[89] A. Grossmann, G. Loupias, and E.M. Stein, *An algebra of pseudo-differential operators and quantum mechanics in phase space*, Ann. Inst. Fourier, Grenoble **18**(2), (1968) 343–368.

[90] J. Hancock, M.A. Walston, and B. Wynder, *Quantum mechanics another way*, Eur. J. Phys., **25** (2004), 525–534.

[91] K. Hannabuss, *An Introduction to Quantum Theory*, Oxford Science Publications, Oxford University Press, (1997).

[92] B.J. Hiley, *Non-Commutative Geometry, the Bohm Interpretation and the Mind-Matter Relationship*, In Proc. CASYS'2000, Liège, Belgium, Aug. 7–12 (2000)

[93] B.J. Hiley, *Phase Space Descriptions of Quantum Phenomena*, Proc. Int. Conf. Quantum Theory: Reconsideration of Foundations 2, ed. Khrennikov, A., Växjö University Press, Växjö, Sweden, 2003 (267–286).

[94] B.J. Hiley, *On the Relationship between the Wigner-Moyal and Bohm Approaches to Quantum Mechanics: A step to a more General Theory*, Found. Phys., **40** (2010) 365–367.

[95] B.J. Hiley and R.E. Callaghan, *Delayed-choice experiments and the Bohm approach*, Phys. Scr. **74** (2006), 336–348.

[96] L. Hörmander, Fourier integral operators I, Acta Mathematica, **127**(1) (1971), 79–183.

[97] L. Hörmander, J.J. Duistermaat, Fourier integral operators II, Acta Mathematica, **128**(1) (1972), 183–269.

[98] G.H. Hardy, *A theorem concerning Fourier transforms*, J. London. Math. Soc., **8** (1933), 227–231.

[99] S. Helgason, T*he Radon Transform*, Birkhäuser, Progress in Mathematics, (1999).

[100] A.C. Hirshfeld and P. Henselder, *Deformation Quantization in the Teaching of Quantum Mechanics*, American Journal of Physics, (2002), 70–537.

[101] H. Hofer and E. Zehnder, *Symplectic Invariants and Hamiltonian Dynamics*, Birkhäuser Advanced texts, (Basler Lehrbücher, Birkhäuser Verlag, (1994).

[102] L. Hörmander, *The Analysis of Linear Partial Differential Operators III*, Springer–Verlag, Berlin, (1985).

[103] R.L. Hudson, *When is the Wigner quasi-probability density non-negative?*, Rep. Math. Phys., **6** (1974), 249–252.

[104] A.J.E.M. Janssen, *A note on Hudson's theorem about functions with non-negative Wigner distributions*, Siam. J. Math. Anal., **15**(1) (1984), 170–176.

[105] D. Kastler, *The C^*-Algebras of a Free Boson Field*, Commun. math. Phys., **1** (1965), 14–48.

[106] A. Katok, *Ergodic perturbations of degenerate integrable Hamiltonian systems*, Izv. Akad. Nauk., SSSR Ser. Mat. **37** (1973), 539–576, [Russian]. English translation: Math. USSR Izvestija **7** (1973), 535–571.

[107] Y. Katznelson, *An Introduction to Harmonic Analysis*, Dover, New York, (1976).

[108] V.V. Kisil, *Integral representations and Coherent States*, Bull. Belg. Math. Soc., **2** (1995), 529–540.

[109] M. Kontsevich, *Deformation quantization of Poisson manifolds*, Lett. Math. Phys., **66** (3) (2003), 157–216.

[110] W. Kozek, *On the generalized Weyl correspondence and its application to time-frequency analysis of time-varying systems*, Proc. IEEE-SP Int. Symp. on Time-Frequency and Time-Scale Analysis, Oct. (1992), Victoria, British-Columbia, Canada.

[111] R. Kubo, *Wigner Representation of Quantum Operators and Its Applications to Electrons in a Magnetic Field*, Journal of the Physical Society of Japan, **19**(11) (1964), 2127–2139.

[112] N. Kumano-go, *Feynman path integrals as analysis on path space by time slicing approximation*, Bull. Sci. Math., **128** (3) (2004), 197–251.

[113] L.D. Landau, E.M. Lifshitz, *Quantum Mechanics: Nonrelativistic Theory*, Pergamon Press, (1997).

[114] J. Leray, *Lagrangian Analysis and Quantum Mechanics, a mathematical structure related to asymptotic expansions and the Maslov index*, MIT Press, Cambridge, Mass., (1981); translated from *Analyse Lagrangienne* RCP 25, Strasbourg Collège de France, (1976–1977).

[115] P. Libermann and C.-M. Marle, *Symplectic Geometry and Analytical Mechanics,* D. Reidel Publishing Company, (1987).

[116] E.H. Lieb, *Integral bounds for radar ambiguity functions and Wigner distributions,* J. Math. Phys. **31**(3) (1990), 594–599.

[117] R.G. Littlejohn, *The semiclassical evolution of wave packets,* Physics Reports **138**(4–5) (1986), 193–291.

[118] G. Loupias and S. Miracle-Sole, C^**-Algèbres des systèmes canoniques, I,* Commun. math. Phys. **2** (1966), 31–48.

[119] G. Loupias and S. Miracle-Sole, C^**-Algèbres des systèmes canoniques, II,* Ann. Inst. Henri Poincaré **6**(1) (1967), 39–58.

[120] Yu.I. Lyubarski, *Frames in the Bargmann space of entire functions,* In Entire and subharmonic functions, 167–180, Amer. Math. Soc., Providence RI (1992).

[121] G.W. Mackey. Mathematical Foundations of Quantum Mechanics, Courier Dover Publications, (2004).

[122] R. de la Madrid, *The role of the rigged Hilbert space in quantum mechanics,* Eur. J. Phys., **26** (2005), 287–312.

[123] A. Messiah, *Quantum Mechanics* (two volumes). North–Holland Publ. Co., (1991); translated from the French; original title Mécanique Quantique. Dunod, Paris, (1961).

[124] J. Moser, *On the volume element of a manifold,* Trans. Amer. Math. Soc., **120** (1965), 286–294.

[125] J.E. Moyal, *Quantum mechanics as a statistical theory,* Proc. Camb. Phil. Soc., **45** (1947), 99–124.

[126] F.J. Narcowich, *Conditions for the convolution of two Wigner distributions to be itself a Wigner distribution,* Journal of Mathematical Physics, **29**(9) (1988), 2036–2041.

[127] F.J. Narcowich, *Distributions of $\hbar$-positive type and applications,* Journal of Mathematical Physics, **30**(11), (1989), 2565–2573.

[128] F.J. Narcowich, *Geometry and uncertainty,* Journal of Mathematical Physics, **31**(2) (1990), 354–364.

[129] F.J. Narcowich and R.F. O'Connell, *Necessary and sufficient conditions for a phase-space function to be a Wigner distribution,* Phys. Rev. A **34**(1), (1986), 1–6.

[130] J. von Neumann, *Mathematische Grundlagen der Quantenmechanik,* Springer, Berlin (1932).

[131] F. Nicola, L. Rodino. Global Pseudo-differential Calculus on Euclidean Spaces. Birkhäuser, Series: Pseudo-Differential Operators, 2010.

[132] A.M. Peremolov, *Generalized coherent states and some of their applications,* Sov. Phys. Usp. **20**(9) (1977), 703–720.

[133] L. Polterovich, *The Geometry of the Group of Symplectic Diffeomorphisms*, Lectures in Mathematics, Birkhäuser, (2001).

[134] J. Radon, *Über die Bestimmung von Funktionen durch ihre Integralwerte längs gewisser Mannigfaltigkeiten*, Berichte über die Verhandlungen der Sächsischen Akademie der Wissenschaften **69** (1917) 262–277.

[135] J. Radon and P.C. Parks (translator), *On the determination of functions from their integral values along certain manifolds*, IEEE Transactions on Medical Imaging **5** (1986) 170–176.

[136] M. Reed and B. Simon, *Methods of Modern Mathematical Physics*, Academic Press, New York, (1972).

[137] H.P. Robertson, *The uncertainty principle*, Physical Reviews, **34** (1929), 163–164.

[138] A. Royer, *Wigner functions as the expectation value of a parity operator*, Phys. Rev., *A* **15** (1977), 449–450.

[139] G.A. Sardanashvily, *Hamiltonian time-dependent mechanics*, Journal of Mathematical Physics, **39**(5) (1998), 2714–2729.

[140] W. Schempp, *Harmonic Analysis on the Heisenberg Nilpotent Lie Group*, Pitman Research Notes in Mathematics, **147**, Longman Scientific and Technical, (1986).

[141] W. Schempp, *Magnetic Resonance Imaging*, Mathematical Foundations and Applications, Wiley, (1997).

[142] E. Schrödinger, *Zum Heisenbergschen Unschärfeprinzip*, Berliner Berichte (1930), 296–303 [English translation: A. Angelow, M.C. Batoni, *About Heisenberg Uncertainty Relation*, Bulgarian Journal of Physics 26 5/6 (1999), 193–203, and http://arxiv.org/abs/quant-ph/9903100.

[143] E. Schrödinger, *Der stetige Übergang von der Mikro- zur Makromechanik*, Naturwissenschaften, **14**(28) (1926), 664–666.

[144] L.S. Schulman, *Techniques and Applications of Path Integration*, Dover, (2005).

[145] N. Seiberg and E. Witten, *String theory and noncommutative geometry*, JHEP, **09** (1999), 032.

[146] K. Seip and R. Wallstén, *Density theorems for sampling and interpolation in the Bargmann–Fock space. II*, J. Reine Angew. Math **429** (1992), 107–113.

[147] M.A. Shubin, *Pseudodifferential Operators and Spectral Theory*, Springer-Verlag, (1987) [original Russian edition in Nauka, Moskva (1978)].

[148] K.F. Siburg, *Symplectic capacities in two dimensions*, Manuscripta Mathematica, **78** (1993), 149–163.

[149] B. Simon, *Trace Ideals and their Applications*, Cambridge U. P., Cambridge (1979).

[150] J. Sjöstrand, *An algebra of pseudodifferential operators*, Math. Res. Lett., **1**(2) (1994), 185–192.

[151] J. Sjöstrand, *Wiener type algebras of pseudodifferential operators*, In Séminaire sur les Équations aux Dérivées Partielles, (1994–1995), École Polytech., Palaiseau, Exp. No. IV, 21.

[152] H. Snyder, *Quantized Space-Time*, Phys. Rev., **71** (1947), 38–41.

[153] E.M. Stein, *Harmonic Analysis: Real Variable Methods, Orthogonality, and Oscillatory Integrals*, Princeton University Press, (1993).

[154] J. Struckmeier, *Hamiltonian dynamics on the symplectic extended phase space for autonomous and non-autonomous systems*, J. Phys. A: Math. Gen., **38** (2005), 1257–1278.

[155] R. Szabo, *Quantum Field Theory on Noncommutative Spaces*, Phys. Rep., **378** (2003), 207.

[156] S. Thangavelu, *Lectures on Hermite and Laguerre Expansions*, Princeton University Press, (1993).

[157] J. Vey, *Déformation du crochet de Poisson sur une variété symplectique*, Comment. Math. Helvet., **50** (1975), 421–454.

[158] N. Wallach, *Lie Groups: History, Frontiers and Applications*, **5**. Symplectic Geometry and Fourier Analysis, Math. Sci. Press, Brooklyn, MA, (1977).

[159] D.F. Walnut, *Continuity properties of the Gabor frame operator*, J. Math. Anal. Appl. **165**(2) (1992), 479–504.

[160] H. Weyl, *Quantenmechanik und Gruppentheorie*, Zeitschrift für Physik., **46** (1927).

[161] E. Wigner, *On the quantum correction for thermodynamic equilibrium*, Phys. Rev., **40** (1932), 799–755.

[162] J. Williamson, *On the algebraic problem concerning the normal forms of linear dynamical systems*, Amer. J. of Math., **58** (1936), 141–163.

[163] M.W. Wong, *Weyl Transforms*. Springer, 1998.

[164] M.W. Wong, *Weyl Transforms and a Degenerate Elliptic Partial Differential Equation*, Proc. R. Soc., A **461** (2005), 3863–3870.

[165] M.W. Wong, *Weyl Transforms, the Heat Kernel and Green Function of a Degenerate Elliptic Operator*, Annals of Global Analysis and Geometry, **28** (2005), 271–283.

[166] M.W. Wong, *The Heat Equation for the Hermite Operator on the Heisenberg Group*, Hokkaido Math. J., **34** (2005), 393–404.

[167] H.P. Yuen, *Multimode two-photon coherent states, in International Symposium on Spacetime Symmetries*, edited by Y.S. Kim and W.W. Zachary North-Holland., Amsterdam (1989), 309–313.

Index

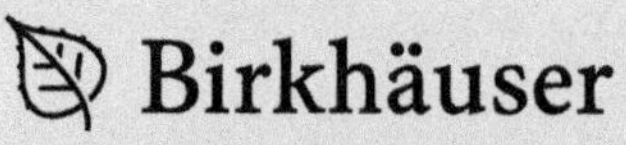 **Birkhäuser** | **www.birkhauser-science.com**

Pseudo-Differential Operators (PDO)
Theory and Applications

This series is devoted to the publication of current research in operator theory, with particular emphasis on applications to classical analysis and the theory of integral equations, as well as to numerical analysis, mathematical physics and mathematical methods in electrical engineering.

Edited by
M. W. Wong, York University, Canada
In cooperation with an international editorial board

■ **PDO 8: Unterberger, A.**, Pseudodifferential Analysis, Automorphic Distributions in the Plane and Modular Forms (2011).
ISBN 978-3-0348-0165-2

Pseudodifferential analysis, introduced in this book in a way adapted to the needs of number theorists, relates automorphic function theory in the hyperbolic half-plane Π to automorphic distribution theory in the plane. Spectral-theoretic questions are discussed in one or the other environment: in the latter one, the problem of decomposing automorphic functions in Π according to the spectral decomposition of the modular Laplacian gives way to the simpler one of decomposing automorphic distributions in $\mathbb{R}^2$ into homogeneous components. The Poincaré summation process, which consists in building automorphic distributions as series of g-transforms, for $g \in SL(2;\mathbb{Z})$, of some initial function, say in $S(\mathbb{R}^2)$, is analyzed in detail. Several new ideas are far from being pushed to the end, and call for many possible generalizations, only hinted at.

■ **PDO 6: Gupur, G.**, Functional Analysis Methods for Reliability Models (2011).
ISBN 978-3-0348-0100-3

The main goal of this book is to introduce readers to functional analysis methods, in particular, time dependent analysis, for reliability models. Understanding the concept of reliability is of key importance – schedule delays, inconvenience, customer dissatisfaction, and loss of prestige and even weakening of national security are common examples of results that are caused by unreliability of systems and individuals.

The book begins with an introduction to C_0-semigroup theory. Then, after a brief history of reliability theory, methods that study the well-posedness, the asymptotic behaviors of solutions and reliability indices for varied reliability models are presented. Finally, further research problems are explored.
Functional Analysis Methods for Reliability Models is an excellent reference for graduate students and researchers in operations research, applied mathematics and systems engineering.

■ **PDO 5: Wong, M.W.**, Discrete Fourier Analysis (2011).
ISBN 978-3-0348-0115-7

This textbook presents basic notions and techniques of Fourier analysis in discrete settings. Written in a concise style, it is interlaced with remarks, discussions and motivations from signal analysis.
The first part is dedicated to topics related to the Fourier transform, including discrete time-frequency analysis and discrete wavelet analysis. Basic knowledge of linear algebra and calculus is the only prerequisite. The second part is built on Hilbert spaces and Fourier series and culminates in a section on pseudo-differential operators, providing a lucid introduction to this advanced topic in analysis. Some measure theory language is used, although most of this part is accessible to students familiar with an undergraduate course in real analysis.
Discrete Fourier Analysis is aimed at advanced undergraduate and graduate students in mathematics and applied mathematics. Enhanced with exercises, it will be an excellent resource for the classroom as well as for self-study.

Made in the USA
Monee, IL
07 July 2026

56564656R00201